Air Pollution and Plant Life

Second edition

Edited by

J.N.B. BELL
Imperial College of Science, Technology and Medicine, University of London

and

M. TRESHOW
University of Utah

JOHN WILEY & SONS, LTD

Baffins Lane, Chichester,
West Sussex, PO19 1UD, England

National 01243 779777
International (+44) 1243 779777
e-mail (for orders and customer service enquiries): cs-books@wiley.co.uk
Visit our Home Page on http://www.wiley.co.uk

Other Wiley Editorial Offices

John Wiley & Sons, Inc., 605 Third Avenue,
New York, NY 10158-0012, USA

Wiley-VCH Verlag GmbH, Pappelallee 3,
D-69469 Weinheim, Germany

John Wiley & Sons Australia, Ltd, 33 Park Road, Milton,
Queensland 4064, Australia

John Wiley & Sons (Asia) Pte Ltd, 2 Clementi Loop #02-01,
Jin Xing Distripark, Singapore 129809

John Wiley & Sons (Canada) Ltd, 22 Worcester Road,
Rexdale, Ontario M9W 1L1, Canada

British Library Cataloguing in Publication Data

A catalogue record for this book is available from the British Library

ISBN 0 471 49090 3 (HB) 0 471 49091 1 (PB)

Typeset in 10/12pt Times by Laserwords Private Limited, Chennai, India
Printed and bound in Great Britain by TJ International, Padstow, Cornwall
This book is printed on acid-free paper responsibly manufactured from sustainable forestry, in which at least two trees are planted for each one used for paper production.

Contents

List of Contributors

Ashenden, Trevor	Centre for Ecology and Hydrology, Bangor Research Unit, University College of North Wales, Deniol Road, Bangor, Gwynedd, LL5 7UP, UK.
Ashmore, Michael	Department of Environmental Science, University of Bradford, West Yorkshire BD7 1DP, UK.
Barnes, Jeremy	Department of Agriculture and Environmental Science, Newcastle University, Ridley Building, Newcastle upon Tyne, NE1 7RU, UK.
Bates, Jeffrey	Department of Biological Sciences, Imperial College of Science, Technology and Medicine, Silwood Park, Ascot, Berkshire, SL5 7PY, UK.
Battarbee, Richard	Geography Department, University College London, Gower Street, London, WC1E 6BT, UK.
Bell, Nigel	Department of Biological Sciences, Imperial College of Science, Technology and Medicine, Silwood Park, Ascot, Berkshire, SL5 7PY, UK.
Bender, Juergen	Institute for Agroecology, Federal Agricultural Research Centre (FAL), Bundesallee 50, D-38116 Braunschweig, Germany.
Bobbink, Roland	Department of Plant Ecology & Evolutionary Biology, P.O. Box 800.84, 3508 TB Utrecht-NL., The Netherlands.
Sabine Braun	Institute for Applied Plant Biology, Sandgrubenstrasse 25, CH4124, Schönenbuch, Switzerland.
Collins, Christopher	Department of Environmental Science and Technology, Imperial College of Science, Technology and Medicine, Silwood Park, Ascot, Berkshire, SL5 7PY, UK.
Colvile, Roy	Department of Environmental Science and Technology, Imperial College of Science, Technology and Medicine, Royal School of Mines Building, Prince Consort Road, London SW7 2BP, UK.
Davison, Alan	Department of Agriculture & Environmental Science, Newcastle University, Ridley Building, Newcastle upon Tyne NE1 7RU, UK.

Farmer, Andrew	Institute for European Environmental Policy, Dean Bradley House, 52 Horseferry Road, London SW18 2AL, UK.
Fangmeier, Andreas	Institut für Landschafts – und Pflanzenökologie, Universität Hohenheim, Stuttgart, Germany.
Fowler, Dave	Centre for Ecology and Hydrology, Bush Estate, Penicuick, Midlothian, EH26 0QB, UK.
Flückiger, Walter	Institute for Applied Plant Biology, Sandgrubenstrasse 25, CH 4124, Schönenbuch, Switzerland.
Harriman, Ronald	Environmental Studies, Freshwater Fisheries Laboratory, Pitlochry, Perthshire, UK.
Hiltbrunner, Erika	Institute for Applied Plant Biology, Sandgrubenstrasse 25, CH 4124, Schönenbuch, Switzerland.
Innes, John	Faculty of Forestry, University of British Columbia, Vancouver, BC, Canada.
Jäger, Hans-Juergen	Institute for Plant Ecology, Heinrich-Buf-Ring 38, University of Giessen, 35392, Giessen, Germany.
Krupa, Sagar	Department of Plant Pathology, 495 Borlough Hall, 1991 Buford Circle, University of Minnesota, St. Paul MN 55108, USA.
Legge, Allan	Biosphere Solutions, 1601-11th Avenue NW, Calgary, Alberta T2N 1H1, Canada.
Long, Stephen	Department of Crop Science & Plant Biology, University of Illinois, 379 ERML, 1201 W. Gregory Drive, Urbana, IL 61801, USA.
Mansfield, Terry	Department of Biological Sciences, Lancaster University, Lancaster LA1 4YW, UK.
Marshall, Fiona	Department of Environmental Science and Technology, Imperial College of Science, Technology and Medicine, Silwood Park, Ascot, Berkshire, SL5 7PY, UK.
McCune, Del	61 Felt Road, Keene, New Hampshire 03431-2105, USA.
Gina Mills	Centre for Ecology and Hydrology, Bangor Research Unit, University College of North Wales, Deiniol Road, Bangor, LL57 2UP, UK.
Monteith, Don	Geography Department, University College London, Gower Street, London, WC1E 6BT
Naidu, Shawna	Department of Crop Science & Plant Biology, University of Illinois, 379 ERML, 1201 W. Gregory Drive, Urbana, IL 61801, USA.
Runeckles, Victor	Faculty of Agricultural Sciences, University of British Columbia, Vancouver, B.C., Canada.

Skelly, John	Department of Plant Pathology, College of Agricultural Sciences, Pennsylvannia State University, 210 Buckhout Laboratory, University Park, PA 16802-4507, USA.
Treshow, Michael	Department of Biology, University of Utah, 201 Biology Building, Salt Lake City, Utah 84112, U.S.A.
Weigel, Hans-Joachim	Institute for Agroecology, Federal Agricultural Research. Centre, Bundesallee 50, D-38116, Braunschweig, Germany
Weinstein, Len	Boyce Thompson Institute, Tower Road, Ithaca, NY14853, USA

Dedication

This book is dedicated to the memory of Christopher Barker, who died under sudden and tragic circumstances during the final stages of the editing procedure. Chris was an exceptional research student of Nigel Bell and his colleague, Sally Power, and had just completed his PhD at the time of his untimely death. His research was concerned with the influence of both nitrogen deposition and management on lowland heaths, and would certainly have been quoted extensively in Chapter 12 had this book been published a year or two later.

Acknowledgement

The collation of the individual chapters and the editing of the final version of this book owe their success very much to the dedication and efficiency of Nigel Bell's secretary, Mrs. Anne Elliott of Imperial College at Silwood Park. Her invaluable assistance and support are acknowledged with gratitude and affection.

1 Introduction

J.N.B. BELL AND M. TRESHOW

It is 18 years since the first edition of this book was produced. Now we have reached the millennium it is timely to produce a new totally revised second edition. Since 1984 massive changes have occurred in the areas of concern in the general field of air pollution impacts on plant life, with a range of new issues appearing on the research agenda. Furthermore, over the same period there has been a major expansion of higher education in the general field of environmental science, at both undergraduate and postgraduate levels. Thus we felt that it was highly opportune to produce a new edition which is aimed primarily at final year undergraduates and Masters students, as well as providing a comprehensive introduction for PhD students and other research workers entering this field for the first time. We were also very keen to produce a publication with a purchase price within the average student's pocket. Over the last 10 years or so a number of books on air pollution effects on vegetation have been published. However, these all tend either to contain a series of chapters on particular specialist topics or are strongly biased towards certain aspects of the field, and thus we believe this second edition fills a niche in the market and hope that it will prove as popular as its predecessor.

At the time of the publication of the first edition, the great bulk of research was concerned with impacts of SO_2, NO_2 and O_3 on crop productivity in the developed world, particularly North America. Thus the National Crop Loss Assessment Network (NCLAN) programme was in full swing, utilising open-top chambers and dose/response and economic models to attempt an assessment of monetary losses arising from air pollutant induced reductions of major crops across the USA as a whole. At the same time much of the research into effects on crops in Europe was piecemeal with a lack of international, and in most cases, national integration. It was not until a year or two later that the European equivalent of NCLAN emerged, in the form of the European Open-Top Chambers Network (EOTCN). In fact the latter coincided with a general fall in interest in most European countries as well as in North America on the subject of air pollution impacts on agricultural crops: in the USA the NCLAN programme had cost a lot of money but produced some definitive results, while agricultural surpluses in western Europe doubtless mitigated against much concern for this problem.

As interest in crops declined, it was more than counteracted by a massive increase into research on the effects of air pollutants on forest trees. This had been the subject of some of the earliest research on air pollution impacts on vegetation in Germany in the nineteenth century (see Chapter 2) and continued over the first three-quarters of the

Air Pollution and Plant Life, second edition. Edited by J.N.B. Bell and M. Treshow. ISBN 0 471 49090 3 (HB), 0 471 49091 1 (PB).

twentieth century in both Europe and North America. However, much of this research was largely devoted to problems around point sources or very close to industrial areas, with the exception of the growing concerns from the 1960s onwards of widespread oxidant damage to forests in the USA, particularly in California. The appearance of a mysterious apparent decline in forest health in Germany, followed by elsewhere in Europe, and also in North America resulted in a massive switch towards research into impacts of air pollution on trees. Large research programmes were inaugurated in both continents, accompanied by a concomitant decline in crop research. This issue was driven by powerful political forces, not the least in Europe and the issue remains far from resolved and it has been argued strongly that a non-existent problem was being addressed (see Chapter 15). An examination of our first edition indicates a low profile for this issue: there was one chapter on 'long range transport of air pollutants and acid precipitation', otherwise no further indication of the subject. Thus in this new edition we have attempted to address this important subject via Chapter 16, covering acidification effects on aquatic vegetation and Chapter 15 on 'Forest Decline'. In the most recent years attention has moved on to some extent from trees to other forms of natural vegetation. This is primarily concerned with ozone in view of its widespread rural distribution, this being covered in Chapter 6, and with excess nitrogen deposition (Chapter 12).

As an overview of the book, we will now put the other chapters in context, particularly with respect to developments since 1984. Chapter 2 provides a historical overview of the development of our understanding of air pollution impacts both pre-1984 and up to the present day. Since the first edition the vast bulk of research efforts into air pollution have not been concerned with impacts on vegetation or, at least until recently, on human health. Indeed they have been concerned with the identification, elucidation and quantification of emissions sources and the subsequent dispersion and atmospheric transformation of the pollutants concerned. These subjects are covered in Chapter 3, where the enormous strides in understanding air pollutant pathways and transformations that have taken place in the last 17 years are covered in some depth. These have been driven primarily by concerns over long-distance transboundary air pollution and the requirement to understand who produces what, where it goes to and in what form it reaches sensitive receptors. Equally important is the subject of Chapter 4, which received scant recognition in the first edition. The link between atmospheric burdens and deposition to surfaces, particularly uptake into vegetation, has been subject to major investigation over these years, with new understanding developing on issues such as co-deposition of NH_3 and acid gases, and the importance of determining fluxes rather than atmospheric concentrations, when developing dose/response models for effects on plants.

Unlike the first edition we have followed these initial chapters with a consideration of impacts of individual major pollutant categories. The realisation since 1984 that oxidants are the major pollutants of concern not only in North America, but also in Europe, as well as increasingly in developing countries (Chapter 21) is reflected in the number of pages devoted to this topic in Chapters 5 and 6, covering physiological/biochemical and whole plant/community effects, respectively. These chapters are followed by one on nitrogen oxides (Chapter 7), where much of the concern is involved with mixtures with other pollutants (Chapter 14), and on sulphur dioxide (Chapter 8), which many view as a past or minimal problem in the developed world, but is becoming a very serious issue in many developing countries (Chapter 21). Chapter 9 covers fluorides, a topic of considerable concern around specific industries and again, particularly in the developing world, where much of the problem lies in poisoning of herbivorous animals. Since 1984 there has been a spectacular upsurge in research into environmental aspects of volatile organic compounds (VOCs), yet little on their direct effects on vegetation, which may be greater than realised, this subject being covered in Chapter 10. Another little understood area is covered in Chapter 11 – the effects of particulates. This is a topic which has turned full wheel in terms of human health, with a return to the concerns of the 1950s over coal smoke being replaced by those with vehicle emissions and other forms of PM_{10} ($<10\ \mu$), but advances in terms of vegetation impacts are minimal. Chapter 12 addresses an issue which has grown out of all proportion since 1984: this is the deposition of N in both oxidised and reduced forms, with indications of widespread impacts on (semi-)natural ecosystems, amounting to a problem of eutrophication in addition to acidification, which was the matter of concern in the past. Chapter 13 covers impacts of wet deposited acidity: at the time of the first edition this had been the subject of intensive investigation in terms of acid rain *(sensu stricto)* effects on agricultural crops; subsequently interest switched to a possible role in forest decline, with identification of the potential importance of the hitherto neglected occult deposition in the form of fog, mist and cloud.

Chapter 14 follows on directly from Vic Runeckles' 1984 chapter on effects of pollutant combinations – a subject still relatively unexplored, but of immense importance if the real effects of air pollutants in the field are to be elucidated and quantified. The next two chapters have already been discussed with respect to their importance for long-distance transport of pollutants. Probably the most severe and widespread effects of air pollution on plants, in the developed world at least, occur on bryophytes and lichens. Yet, this topic has generally received scant attention, except in their use as bioindicators of air quality. Chapter 17 covers this issue, with some emphasis on the pitfalls of biomonitoring. The interpretation of field and quasi-field (e.g. open-top chamber) studies has always been fraught with difficulty as a result of the interactions that can occur with environmental conditions prevailing at the time. This important subject is covered in Chapter 18. Since the first edition, an equally or more important topic has emerged in that air pollution can predispose plants to damage by other, more familiar, environmental stresses and, indeed it has been argued that such impacts could be more important than the direct impacts of air pollutants on plant performance. The major advances in recent years in these subjects are discussed in Chapters 19 and 20, for abiotic and biotic stresses, respectively. Perhaps of greatest concern for the topic of this book is the growing evidence that the most serious impacts may be taking place in the developing world, notably on agricultural productivity. This subject was effectively ignored in the first edition, but Chapter 21 in this book addresses the subject, reviewing the evidence that air pollution presents a threat to food security in developing countries, where agricultural surpluses are most definitely not a problem!

Ultimately, the vast bulk of research into air pollutant impacts on vegetation is aimed at generating data which can be employed in the formulation of pollution control policy, whether at national or international levels. This important issue is covered in Chapter 22, which includes a description of a range of air quality standards and guidelines, and the way in which vegetation studies have been used to develop these. The most intransigent of all air pollution issues are those that are truly global, notably in the form of increasing concentrations of greenhouse gases, and perhaps now to a lesser extent stratospheric O_3 depletion. When planning this book we deliberated for some time whether or not to include a substantial element on the effects of rising CO_2 levels and increased ultra-violet radiation on vegetation. In the event we decided that these subjects were so vast that they warranted a separate book and that their inclusion would seriously dilute the main theme, which is the same as that of the first edition. However, we felt that this topic deserved inclusion in the form of the interactions of CO_2 and ultra-violet radiation with the pollutants covered in this book, and this is covered in Chapter 23. Finally, in Chapter 24 we have included a concluding chapter which draws on the individual contributions, pointing towards future research priorities and directions of this massive environmental issue which is certainly not going to be resolved in the near future.

We hope that this updated edition of Mike Treshow's 1984 edition will prove useful to students, teachers and research workers for a good many years. All the contributors are top international scientists, carefully chosen for their expertise in the subject area covered by their chapters.

2 Historical perspectives

M. TRESHOW AND J.N.B. BELL

The quiet, muggy fall air clung over the Los Angeles Basin in California that historic day in 1945. The County Agent was puzzled. Plant pathologists from the University of California Citrus Experiment Station in Riverside were called out. Never before had they seen the silvery bronze glaze that appeared on the leaves of vegetable crops throughout the Basin. The leafy crops, especially near the oil refineries in the Dominguez area were ruined. The glazing, silvering and bronzing, sometimes coupled with necrosis of the lower leaf surfaces, appeared on such species as endive, lettuce, spinach and romaine (Middleton *et al.*, 1950).

A decade later, on the eastern seaboard of the United States, another mysterious syndrome appeared. First observed on the tobacco leaves grown for cigar-wrappers, the characteristic, straw-coloured flecks and lesions rendered the leaves totally unsuited for cigars (Wanta *et al.*, 1961). The disease characteristically appeared following periods of high humidity and temperature inversions common in the summer and fall months. Similar symptoms subsequently appeared on numerous other crops both in the east and in California (Richards *et al.*, 1958). Pathologists suspected that weather was somehow involved, but the specific cause remained a mystery. The disease was called weather fleck.

Because of the diversity of crops that were affected, and the absence of known biotic or viral pathogens, thoughts turned to air pollutants as a possible cause. But if so, which ones?

The earliest concerns

The first reports of air pollution problems appear to have been made by writers in ancient Rome, who were aware of adverse effects on human health. Perhaps the best documented examples of concern about air pollution occurred in Medieval London, where importation of coal by sea from collieries of north east England and its combustion in various industrial processes led to legislation aimed at prohibiting the latter as early as the thirteenth century. Needless to say the legislation proved completely ineffective and by the seventeenth century ever increasing coal combustion in London led to a serious and growing problem (Brimblecombe, 1987). It is at this time that the first records appear to have been made of damage to vegetation. In 1661, the English diarist, John Evelyn, published his famous treatise, *Fumifugium: Or the Inconvenience of the*

Air Pollution and Plant Life, second edition. Edited by J.N.B. Bell and M. Treshow. ISBN 0 471 49090 3 (HB), 0 471 49091 1 (PB).

Aer and Smoake of London Dissipated, in which he described the contemporary air pollution problems in the English capital, making recommendations for their amelioration. *Fumifugium* contains graphic descriptions of effects on vegetation, such as '...Our Anemonies and many other choycest Flowers, will by no Industry be made to blow in London, or the Precincts of it, unless they be raised on a Hot-bed and governed with extraordinary Artifice to accelerate their springing; imparting a bitter and ungrateful Tast to those few wretched Fruits, which never arriving to their desired maturity, seem, like the Apples of Sodome, to fall even to dust, when they are but touched.' Fascinatingly, Evelyn also describes what can be described as the first experiment, albeit inadvertant, on air pollution impacts on plants, when coal smoke was eliminated in London one summer as a result of the English Civil War stopping the coastal trade concerned, with him noting how the trees produced unprecedented quantities of high quality fruit. Clearly air quality deteriorated even further over the next 100 years, as a preface to a second edition of *Fumifugium*, written in 1772 noted 'It would now puzzle the most skilful gardener to keep fruit trees alive in these places: the complaint at this time would be, not that the trees were without fruit, but that they would not bear even leaves'.

After almost another century had passed, the early chemical industry caused such serious air pollution damage to vegetation in parts of northern England that the world's first air pollution control inspectorate was established in 1863. This was not concerned with coal smoke and its associated phytotoxic gases, but with hydrochloric acid gas emanating from alkali works. Hence the establishment of the 'Alkali Inspectorate', with its first inspector being Robert Angus Smith, who is remembered for his measurements of rain chemistry in industrial areas of north west England, subsequently published in 1872 in his book, *Air and Rain: The Beginnings of a Chemical Climatology*, in which he coined the term 'acid rain' which was to become such a controversial environmental issue after the lapse of another century.

Sulphur dioxide

When forests and other vegetation began to die out in the vicinity of smelters in Central Europe and Britain in the late 1800s, biologists looked toward emissions from the smelters that were often central to the area impacted. Copious volumes of sulphurous compounds in ores high in sulphur, together with arsenic and various heavy metals were all released in the dense smoke produced in the smelting process. Numerous studies were conducted to determine just which of these compounds was most offensive to the forests. As trees declined and died out in areas surrounding the smelters, it was immediately apparent to such early forest pathologists as R. Hartig that something in the smoke was responsible (Hartig, 1894). Various components were considered. Most popular was sulphuric acid, but the odiferous hydrogen sulphide was also suspected. The injurious components of the smoke were in doubt until 1866 when Morren, and shortly afterwards, Von Schroeder (1883) in Germany, proved that sulphurous acid was an important toxic agent responsible for much of the damage. Arsenic, lead, zinc, copper and other metallic emissions were of concern but less harmful directly. It was only after much deliberation and study that sulphur dioxide was established as the specific pathogen (Sorauer, 1886).

Meanwhile, in the Swansea Valley in Wales, some 600 furnaces were smelting ores along the banks of the River Tawe. The toxic smoke seemed to be the natural companion of prosperity, so was often tolerated much to the frustration of foresters and the farmers whose crops were damaged (Sorauer, 1914).

When rich copper deposits were discovered in the United States, the ores were first sent to Swansea for refining. But late in the nineteenth century, smelters were established in Tennessee and Montana. Since the early smelters in the USA as well as Europe consisted of little more than piles of lumber or coal alternating with the ore and set on fire to 'roast', it should have been no surprise that the low hanging smoke would impact any vegetation encompassed. The devastated forests that persist and even today slowly continue to recover, were largely a product of this era.

The obvious damage stimulated considerable research: first to ascertain the cause of the damage, and second to assess the economic aspects. In other words, to determine the amount of monetary losses that should be assessed not only on forests but often to the agricultural crops grown on farms near the smelters. In one situation near Selby, California, a Smelter Commission was established; their published report was one of the earlier instances in which scientific expertise assisted in assessing the losses. The study also prompted recovery of sulphur and sulphuric acid by-products.

Some of the early research initiated to learn more about the effects of sulphur dioxide on plants incorporated the use of glass-sided cabinets. In response to smoke damage near Anaconda, Montana, Haywood (1910) utilised such chambers in early fumigation studies.

Other studies were stimulated by injury around a smelter at Trail, British Columbia (Committee on Trail Smelter Smoke, 1939). This was the first instance in which an international commission was formed to study and determine the cause and scope of the problem, and foreshadowed the problems of acid rain in both Europe and North America in the latter part of the twentieth century, resulting from long distance trans-boundary transport of pollutants.

Much of the early research concerning the effects of sulphur dioxide on crops was conducted by Dr Moyer Thomas and colleagues at the American Smelting and Refining Company in Salt Lake City, Utah. Research here reached a new level of sophistication in the 1930s with the development of the Thomas Autometer that measured changes in conductivity of the sulphuric acid solution produced by sulphuric dioxide, or more accurately, the sulphurous anhydride absorbed in acidified water (Thomas and Hill, 1935).

When Dr. Morris Katz summarised the total SO_2 emissions on a worldwide basis in 1961 (Katz, 1961), smelting of copper, lead and zinc ores accounted for roughly 65% of the SO_2 emissions, and oil refining and power plants the remainder. Continually improving recovery methods, their economic feasibility, litigation, and public pressure have rendered the role of smelters as major contributors of SO_2 to be negligible today, at least in industrialised countries.

The sulphur dioxide story extends well beyond smelters and forests, for there were many other sources and targets. Power generating plants, some burning thousands of tons of high sulphur coal each day, have been another source of sulphur dioxide. Again, damage to vegetation provided an impetus for technology whose application has minimised this threat over the years to where emissions of SO_2 are no longer a problem at modern facilities.

The problems described in *Fumifugium* continued to be a major problem in British cities over more than the first half of the twentieth century. A particularly valuable contribution to understanding the impacts of domestic and industrial coal smoke pollution was published by Cohen and Ruston (1925). This describes a major research exercise in the northern English city of Leeds just before World War I, involving air pollution monitoring and experiments with plants. The latter included transect studies along a gradient of smoke and SO_2 pollution, in which various plant species were exposed in standard cultures at stations from outside the city to the most polluted industrial area. The results were dramatic, with falls in dry weight of up to 70% from the former to the latter. Such problems for urban gardening and park management remained until SO_2 and smoke levels fell dramatically from the 1960s onwards, partly as a result of the effects of the UK Clean Air Act of 1956, itself being a response to the outcry following the deaths of over 4000 people in a traditional London 'pea-souper smog' in 1952.

The classic symptoms of injury have been described and reviewed over the decades and are well known to consist largely of yellowing of the leaves followed by the browning or necrosis of sensitive plants in more severe instances (O'Gara, 1922; Treshow and Anderson 1989). Over the years of progressive controls though, such symptoms have become less pervasive.

The question long prevailed as to what might be the effect of SO_2 on agricultural crops at concentrations too low to cause actual visible injury to the leaves (Stoklasa, 1923). Pioneering work in the 1930s and reviewed in 1956 by Moyer Thomas showed that photosynthesis could be inhibited in the absence of leaf markings. However, the results of fumigation studies in both the USA and Canada appeared to indicate that yield of crops was not reduced in the absence of visible injury and this view held sway until the early 1970s when Bell and Clough (1973) showed reductions in yield of perennial ryegrass (*Lolium perenne*) when fumigated with SO_2 at levels much lower than those previously believed harmful, but without any clear visible symptoms. Subsequently, it became apparent that much of the early fumigation chamber work was flawed in that the air flows employed were much slower than normally prevailing in the field, resulting in a much lower dose of SO_2 to the leaf surface than would occur under realistic conditions.

Atmospheric fluorides

It was the1940s. The world was at war and aircraft production was at its zenith. Near an aluminium reduction plant in California, the margins of leaves on apricot and fig tress developed brown scorching, or necrosis. Similar symptoms were soon reported on Italian prune trees and gladioli near another aluminium plant in western Washington.

As early as 1943, pine forests surrounding still another aluminium plant in Washington showed extensive browning of the needles and considerable mortality (Adams *et al.*, 1952). The disease was ubiquitously referred to simply as Pine Blight. Following considerable conjecture and research, but rather limited literature review that could have answered many questions biologists concluded that an air pollutant, specifically atmospheric fluoride, was involved.

Much earlier literature from Germany had reported that similar leaf necrosis in the areas near superphosphate and glass factories appeared in the 1890s (Mayrhofer, 1893). Reports had also shown that gases from open roasting at the aluminium works at Bodesburg damaged pine trees in the surrounding forests in the Rhone Valley in Switzerland (Hasselhoff and Lindau, 1903).

Expansion of the aluminium industry during World War II, and of mining and processing of phosphate deposits in Tennessee and Florida, generated a tremendous increase in fluoride emissions in the USA, together with attendant impacts on crops and in certain cases the livestock, that fed on forage high in fluorides.

The widespread occurrence of such effects provided the impetus not only for extensive research programmes to understand better plant and animal responses to fluorides, but the development of technology to curb the emissions. By the 1980s, the latter had rendered fluoride pollution in industrialised countries largely to historic interest.

The knowledge and research experience gained in the study of fluoride effects in the 1950s provided a strong foundation for studying the next type of air pollution that soon engulfed the industrial countries. Furthermore, fluoride research provided the impetus for establishing a number of laboratories that became central to further research in the field when the focus of study shifted.

Photochemical pollution

When thoughts in Los Angeles and the eastern USA turned to atmospheric issues, it was quite natural to consider SO_2 first, clearly the most studied and best understood air pollutant. In California, especially, there was a number of sources of this well-studied pollutant, largely refineries, smelters, and the combustion products of coal and petroleum (Katz, 1961).

With the appearance of symptoms of unknown origins, the suspicion of SO_2 injury provided impetus for study and legislation directed toward its control from industries small and large. While this programme was successful in reducing the amount of SO_2, it did nothing to reduce the glazing or bronzing on such crops as endive, lettuce or spinach. Nor did it improve the visibility or lessen the sore throats, running noses, smarting eyes or headaches that often accompanied the hazy, irritating atmosphere that was becoming so pervasive in the Los Angeles area. 'Smog' was as bad as ever.

Of course, there were other speculations as to the cause of these symptoms including one, nitrogen oxides drifting in from the ocean, that came curiously close to hitting the mark. Oxidants associated with automobile exhaust also were given serious consideration. Early exposure studies tested a miscellany of chemicals including pentane, hexane, heptane, formic acid, acrolein, pyridine, aldehydes, acetic acids and various organic chlorides. None, however, provided any consistent patterns that simulated the effects of the natural smog (Haagen-Smit and Fox, 1956).

This period of the late 1940s was critical not only in the appearance of a new type of air pollution, the Los Angeles smog, but in the identification of the chemical composition of the atmosphere of other large cities and recording many types of pollutants.

The development of analytical instrumentation such as the infra-red spectrophotometer was critical to opening an era of measuring photochemical air pollution and understanding its nature. Emissions from auto exhausts, as well as pollution from the oil refineries were given serious consideration, and together with oxidised hydrocarbons, became the primary suspects in the formation of smog.

At the time though, the exact chemicals that caused the plant injury remained unknown. They were clearly associated with periods of high smog intensity, but what was the phytotoxic component of this smog? Concentrations of sulphur dioxide, carbon monoxide, oxidant, aldehydes, hydrocarbons, nitrogen oxides and ozone were all elevated on smoggy days.

In 1949, fumigation studies were begun at the California Institute of Technology Earhart Plant Research Laboratory in which plants were exposed to a number of these chemicals. But none produced the characteristic symptoms (Haagen-Smit *et al.*, 1952). The research though led to learning that when gasoline vapours were combined with ozone, the fumigations produced injury symptoms rather similar to those found in the field (Middleton *et al.*, 1955, 1956). When ozone was introduced into the exhaust stream of a small engine, an artificial 'smog' was produced. When released into a small chamber containing the plants to be studied, characteristic symptoms of smog injury were produced on such sensitive plants as spinach, romaine, endive, lettuce and mustard.

When Ed Stephens and his colleagues at the Citrus Experiment Station (Stephens *et al.*, 1956) analysed the 'smog' products in a new 250–500 m infra-red cell, they found that the ozone was formed as nitrogen dioxide disappeared. Another, unidentifiable compound arose that was considered to be particularly important. They called it compound X. A few years later they found it consisted of a group of peroxyacyl radicals that in turn could add nitric oxides forming peroxyacyl nitrates, or PAN. This series of compounds was subsequently found to be the principal constituent causing the glazing type of plant damage.

In the course of the studies, the small engine that was used to provide the exhaust gases occasionally ran out of gasoline, although on occasion the ozone generator continued to function. Thus the plants were inadvertently being exposed solely to ozone. When this happened, necrotic flecks sometimes appeared on the leaves of pinto bean plants.

Researchers recognised the similarity of these symptoms to those of weather fleck on tobacco in the east. Weather fleck had become an especially serious problem of the cigar-wrapper tobacco crops in the Connecticut Valley although the disorder soon extended from North Carolina to Ontario. Other crops were affected to a lesser degree with symptoms consisting of leaf yellowing, defoliation and loss of productivity. Controlled fumigation studies established that ozone could produce all of these symptoms (Heggestad and Middleton, 1959).

After identifying the etiology of weather fleck and similar disorders, there still remained serious questions as to the sources of ozone and how it was formed, as well as the concentrations that were harmful. Also, years earlier, the door had been opened to the traditional question: what, if any, were the adverse effects before the appearance of any symptoms visible to the naked eye. Such effects have traditionally been referred to as chronic, hidden or subliminal injury.

Another quandary confronting the researcher was how ozone concentrations could be measured so that the potential plant injury could be assessed and concentrations compared to the extent of plant injury. It was known that rubber was sensitive to cracking from oxidants such as ozone. Researchers in California had a rubber formulated that was especially sensitive to such cracking. Sheets of this rubber were cut into narrow strips that were bent over in two to provide a stretched area that was most subject to cracking (Bradley and Haagen-Smit, 1950).

The depths of cracks formed, typically over a 24-hour period, were measured microscopically. This provided an early field test for relative ozone 'concentrations' before the development of the Mast recorder in the 1960s that was based on the oxidation of potassium iodide, and other instruments developed subsequently in the 1970s that were based on chemiluminescent techniques that were still more reliable. In the 1980s ultraviolet spectroanalysis largely replaced former methods. During the 1960s, ozone injury became widely recognised throughout the USA on numerous crops and forests. Agricultural areas of Canada were also impacted. By 1970, ozone concentrations were found to be elevated in Europe and injury reported on indicator plants and forest species (Bell and Cox, 1975; Guderian, 1977).

Subsequently O_3 was demonstrated to be the most important phytotoxic pollutant in Europe as well as in North America, with studies using filtered chambers or chemical protectants demonstrating growth reductions and/or visible injury in many locations (Jäger *et al.*, 1993; Benton *et al.*, 1995). Over the same period it also became apparent that there were major oxidant problems in Japan, but it took much longer to recognise that this presented a significant environmental issue in the developing world. This is unfortunate, because growing emissions of precursors and meteorological conditions favourable to its formation are both prevalent in many developing countries. Yet, there is little O_3 measurement carried out in such places, and even this is normally confined to the cities. However, evidence is now accruing, either on a predicted basis or by experimental demonstration that there may be major O_3 effects on crop growth in China (Chameides *et al.*, 1999), Egypt, India and Pakistan (Ashmore and Marshall, 1998).

Air quality standards

An air quality standard, that could be regulated by law and hopefully achieved, could only be set once some basic parameters were established.

Concentrations of ozone or other pollutants at which injury first appeared, would have to be determined, and reliable analytical methods would be required to measure the pollutants.

Of course, even if such concentrations and thresholds of effects could be determined, it would still be quite another matter to regulate emissions to meet them.

Such was the major objection heard to establishing air quality criteria when such a regulation was first proposed and discussed in the 1960s (Atkisson and Gaines, 1970).

The initial regulations were developed in California largely in response to concerns over human health, but similar air quality programmes were soon adopted nationally. Over the course of considerable debate, such criteria were set by the US Congress with the passage of the Air Quality Act of 1967. The original standard of 0.12 pphm (120 ppb) for ozone for one hour was based partly on the best available data, coupled with political concerns and input, was based more on rough conjecture than scientific fact.

By 1976, criteria documents designated in the Clean Air Act of 1970, had been developed for the six 'criteria' pollutants (photochemical oxidants, carbon monoxide, particulate matter, sulphur oxides, nitrogen oxides, and non-methane hydrocarbons). A seventh 'criteria' pollutant, lead, was added in 1977.

In Europe the first international air quality standards were introduced by the European Commission in 1980 for SO_2 and suspended particulates, but these were aimed at protecting human health. An important development occurred a few years later when the World Health Organisation recognised ecological damage as being relevant to human health and introduced air quality guidelines for Europe which included the former as well as the latter, and these have recently been revised (WHO, 2000). With growing concern over transboundary air pollution in Europe, the United Nations Economic Commission for Europe (UNECE) has subsequently developed 'Critical Levels' for protection of different categories of vegetation, the pollutants concerned being SO_2, NO_x, NH_3, and O_3 (Sanders *et al.*, 1995). The validity of basing a standard on a single hourly exposure was soon questioned. At the time though, this was the most feasible time frame on which to base a standard. However, the technology of instrumentation soon advanced to where there were few such limitations. Precise concentrations could be measured over any time frame, any number of exposure regimes could be tested. Furthermore, by the 1970s, gas exchange by the plants themselves could be monitored conveniently and reliably. Any effects of pollutants on the vigour, health and productivity of plants, even in the absence of any visible expression, were studied.

The most standardised and elaborate studies were initiated through the National Crop Loss Assessment Network (NCLAN) (Heck *et al.*, 1982) to determine the exposure concentrations and durations that had the greatest effect on growth and production of such major crops as corn, wheat, soybeans, alfalfa and cotton. Most significantly, identical studies were replicated at several locations throughout the USA.

The results of these studies showed that even ozone concentrations below the established air quality standard of 120 ppb, if sustained for a few hours, could suppress growth and production of a number of crops (Lefohn *et al.*, 1988). Even a 7-hour seasonal mean ozone concentration of 40 ppb could cause yield reductions up to 28%. Such findings are only now being considered for incorporation into new, updated Air Quality Standards.

It should be noted that studies in Europe have led to conclusions that, unlike for ambient levels of other pollutants, in the case of O_3 the peak concentrations are more important than the long-term mean and thus the UNECE Critical Level is now based on accumulated dose above a threshold concentration.

Acid rain

Early in the 1970s, just when air pollution biologists thought they had a handle on SO_2, fluoride and photochemical oxidants, an apparently new set of symptoms, or syndrome, came to light. Forest decline and 'acid rain' entered the pollution lexicon. Oden in Sweden used Smith's 1872 term 'acid rain' in 1970 to denote the changing acidity of air and precipitation over North America and much of Europe. Initially the threat focused on aquatic ecosystems, but potential effects on terrestrial systems were also considered (Oden, 1976).

The decline of forests, the gradual demise of the trees, was nothing new, but details of this current type appeared to be unique. The earliest symptoms were the yellowing of the older leaves of conifers. This was followed progressively by the yellowing and dropping of younger leaves followed by dieback of the shoots that resulted in a thinner canopy and reduced radial growth. More specific symptoms varied with the tree species and geographic regions (Wellburn, 1988).

None of the classic disease agents: fungi, insects, viruses or traditional soil factors or other abiotic stresses could fully explain the new disorder. By 1980, the disease had become serious over Europe from the Scandinavian countries to France. Various species of spruce, fir and pine were affected as well as beech and oaks. Forest decline affected some 14.2 million hectares by 1987. The decline of a large number of different plant species over broad areas suggested that some climatic factors might be involved. Decline was also reported from the USA, particularly on red spruce in the Adirondack mountains of New York and in New England (Wellburn, 1988).

Various hypotheses arose for the cause of decline. Climatic stresses, most notably drought and soil factors, were studied intensively and were somehow involved, but there were often exceptions. The problems caught the public's attention, and the study of this forest decline provided a classic example of how politics and the communication media influence the thinking and study of a major disease complex. The public concern that was stimulated had a major influence on generating funds for studying the disorder.

There was no question that a major new or 'novel' forest decline had been serious in recent decades, but extensive studies in Europe and North American failed to disclose a clear relation solely between acid deposition and tree growth or vigour of the conifers studied.

As noted by Godbold and Hüttermann (1994), the major influence of acid deposition was indirect, mediated by changes caused by soil chemistry that influenced other processes. Most likely the decline was associated with various stresses, alone or in combination, that might differ with the region involved. This was also consistent with the reality that the precise decline symptoms were not exactly the same in all areas. Thus, ozone might be the principal incitant in some areas, possibly mostly where it occurred in combination with nitrogen oxides, and/or sulphur dioxide.

The situation became still more complex when it was learned that these pollutants could act synergistically when present in combination (Guderian *et al.*, 1988). Sulphur dioxide could be most critical in some instances, while soil disturbances such as heavy metal accumulation, or nitrogen imbalance and climatic factors such as drought could be most critical in other areas.

Industrialising countries

Often pollution is thought to be a problem associated largely with already industrialised countries, but this view is too narrow. Impacts from air pollutants in industrialised countries may appear at first glance to be more serious than in the industrialising or developing countries. However, this may be illusory and related mostly to public awareness and perception, as well as the greater research capability and monies the industrialised countries devote to the problems. The priorities to achieve control are less immediate in developing countries seeking industrialisation, the relative cost of controls greater, and pollution control may have a lower priority.

The problem of pollution is also aggravated when industries based in industrialised countries move their more polluting operations to developing countries having more lax regulations or when entire smelters that may have obsolete control equipment are sold to developing countries. Air pollution damage to vegetation continues to be serious in such situations.

An apparent anomaly in urban regions of developing countries is the numbers of cars. It is amazing how many vehicles crowd onto the limited roads. Even more than in developed countries, high traffic flow has not kept pace with the increasing number of vehicles adding to the total emissions. This is further aggravated by the general absence of adequate emission controls. Thus the ozone concentrations in areas such as Mexico City or Sao Paulo, both high in solar, UV radiation, are as bad as any place in the world.

Emissions from the large number of vehicles in larger cities in China have added to the already dense brown haze generated by the burning of soft coal for heat and cooking. In response to this, many, more polluting industries are being shut down and coal is slowly being replaced by natural gas. But releases of CO_2 remain especially critical.

Highlights of stages in air pollution study

Sulphur dioxide

From the ancient furnaces of the Middle East to the smelters of the industrial revolution, no effort was made to control the sulphur, arsenic, or heavy metal fumes. But extensive damage to forest surrounding large smelting operations in Central Europe ultimately fostered research that led to improved control technology. The major disciplines of forestry and plant pathology were emerging during this period and applied to understanding the nature of the decline and introduced the first research in air pollution biology.

The research methods set the standard of the day but were slow to spread to other problem areas due to limited communication. The consequence of this was that as the same problems arose in other areas, research was not always based on previous work.

Thus, rather than building on previous work, each new research group often began anew.

Oxidant pollution

When A.J. Haagen-Smit showed that the strong oxidising effect of the Los Angeles atmosphere was due to abnormally high concentrations of ozone and peroxides, a new era in air pollution effects was born.

The recognition in the 1950s that oxidising pollutants in the atmosphere were harmful to vegetation took the air pollution field beyond such recognised components as sulphur dioxide, fluoride, ammonia, and carbon monoxide. The new complex was characteristic of modern cities where the economy was based upon the use of petroleum fuels.

Refinements in understanding the etiology of the injury symptoms on plants came rapidly, and by 1959, Heggestad and Middleton (1959) had established that ozone formed by the action of sunlight on nitrogen oxides, was the principal phytotoxicant responsible for the flecking symptom. By 1961, Stephens *et al.*, had isolated PAN and produced the typical glazing oxidant damage under controlled conditions.

Fumigation chambers

The chambers or glass houses in which plants are exposed to a pollutant have undergone incredible changes since their early use in the 1930s. Two improvements especially are noteworthy: first, their numbers. In the early days, the numbers of plants in a single chamber were considered to be the replicates. Ultimately it was recognised that the chambers themselves did not provide a statistically valid sample. It was necessary to replicate the numbers of chambers (Oshima and Bennett, 1978; Sokal and Rohlf, 1969).

Secondly, since the chamber environment was not necessarily the same as in the field, a more representative controlled environment was sought. After much experimenting, an open-top design was accepted and became widely utilised (Mandl *et al.*, 1973). An upward positive air flow was maintained through the chamber to minimise the incursion of ambient air through the open top while allowing more natural light and moisture conditions to prevail in the chamber. Ambient or filtered air was introduced through openings in a large tube situated around the base of the chamber. The pollutant being studied could be introduced into the filtered air stream of the replicated chambers. Subsequently, chamberless outdoor fumigation (e.g. McLeod, 1995) and air filtration (Olszyk *et al.*, 1986) systems were introduced. These have the advantage of not modifying ambient environmental conditions, but are very expensive and subject to various technical problems.

Information age

The much improved communications in the 1940s encouraged the sharing of information with the result that the collective understanding of the new type of pollution, largely associated with urbanisation and motor vehicles, progressed rapidly. Information was shared largely through meetings and conferences, but for many years they were not centred since the air pollution biologists came from many different disciplines ranging from engineering and meteorology to biological sciences. Not everyone studying air pollutant effects attended the conferences of the Air Pollution Control Association, the primary forum of the day in North America that overlapped many disciplines.

Consequently, papers and information presented often had a limited audience and distribution.

Further exchanges of information were provided through national and international conferences sponsored by the government and addressing solely air pollution issues. These had their inception to a large measure with the Los Angeles County Air Pollution Control District Technical and Administrative Reports beginning in 1949 (LA Air Pollution Control District, 1949–1950).

The need for close co-operation and sharing information in such a diverse field was recognised by air pollution biologists when they organised the first Air Pollution Workshop, which was held at Pennsylvania State University in 1963. This informal workshop fostered sessions in which everything said was 'off the record' and nothing was to be published. Since that meeting, the workshops have been held each year, alternately in western and eastern North America with participants from around the world. The emphasis in pollutant types has shifted, but the workshops with their informality have persisted.

In Europe regular workshops and conferences have also brought together scientists in a similar manner, often also informally. Thus in the UK, the Natural Environment Research Council's Committee on Air Pollution Effects Research (CAPER) has been meeting regularly since 1974, with invited speakers from abroad to give keynote addresses. From the 1980s onwards two international fora have been available for exchange of ideas and discussion of joint research, firstly under the auspices of Directorate General XII of the European Commission and, more recently, in the form of the ICP-crops and ICP-forests programmes of the UNECE.

Air quality standards

Development and implementation of air quality criteria and standards afforded a major stage in the story of air pollution biology. Following the precedent of criteria development in the fields of water quality, sewage, occupational exposure and radiation exposures, it was natural to discover some realistic values in air quality criteria toward which attainment might be sought.

Impetus was given to the development of air quality standards in 1958 when it was realised that photochemical problems could not be resolved without control of motor vehicle emissions. In 1959, California took the lead in developing and publishing standards for the quality of air directed toward human health and damage to vegetation and interference with visibility (California State Dept. of Public Health, 1959). Three levels of standards were decided on: adverse, indicating the earliest outward symptoms or discomfort: serious, where insidious or chronic changes might be apparent in sensitive populations; and the emergency level, where acute sickness or death might occur!

An 'oxidant' standard of 0.1 ppm for one hour not to be exceeded more than 7 days in any 90-day period, and an SO_2 standard of 0.5 ppm for one-hour, were proposed.

Research findings have made it possible to move away from the original one-hour standards to more meaningful standards based on the duration of exposure as well as concentrations.

Meteorological knowledge

When ozone injury was recognised in rural areas along the Atlantic seaboard in the 1960s, the question arose as to where the ozone might have originated. Heggestad and Middleton (1959) showed how air movement extended from such urban centres as Washington DC and Philadelphia well into the surrounding agricultural areas. It also was recognised that more ozone was formed by photochemical reactions of auto exhaust type olefines and oxides of nitrogen during transport (Stephens *et al.*, 1956).

Air pollution and its consequences had originally been considered to be relatively local phenomena. Even the classic Los Angeles smog was largely limited by the surrounding mountains. But as the effects and subsequent measurements soon showed, the pollutants readily extended beyond local areas. During the 1970s and 1980s, movement of air masses, and ozone dispersal, have been followed over distances of hundreds of miles, in both North America and Europe.

Instrumentation

Mention must be made of the role instrumentation has played in facilitating measurements and enabling the continuous monitoring of any number of air pollutants. Quality control of these devices has not only given reliability to the results but has made it possible to compare validly results from around the world. Instrumentation for measuring gas exchange has made it ever more feasible and quantitative to monitor the impact of pollutants on photosynthesis and other untoward effects. The application of computer technology to the advances in instrumentation has, of course, made it more feasible to record, integrate and keep pace with the data obtained.

Interactions

It has long been understood that environmental factors can modify markedly the response of plants to air pollution, with temperature, humidity, and light intensity being studied in some depth for certain pollutant/species combinations, while other interactions, such as with wind, have received less attention. However, it is only really since the late 1970s/early 1980s that there has come the growing realisation that air pollution can modify the response of plants to other more widely recognised environmental stresses, such as frost, drought, salinity, pathogens (fungal, bacterial and viral) and herbivorous insects. There is now an abundant body of evidence that such interactions can occur in the presence of levels of pollutants which are widespread in the real world. Thus studies such as NCLAN may be grossly underestimating the true impact of pollutants, as in nearly all cases experimental procedures have been designed to eliminate these natural stresses. This is a particularly important topic in terms of control policy, as experience leads us to believe that air pollution remains unrecognised by policy makers and others concerned with managing and protecting forests, agriculture and the natural environment, but who willingly embark on major expensive programmes of research and action in alleviating the adverse effects of these natural stresses.

Changing concerns

On a careful reflection of the potential key milestones on the route from the earliest times to our current knowledge on air pollution effects on vegetation, we feel that the really important stages were the points usually not defined clearly, when 'new' pollutants were identified. John Evelyn did not have any comprehension of the components of coal smoke, but recognised that they impacted severely on vegetation. By the latter part of the nineteenth century, it became clear that SO_2 was the major culprit, but the role of particulates was also recognised – and perhaps remains one of the most enigmatic of all phytotoxic pollutants to the present day. As discussed earlier in this chapter, the arrival of photochemical smogs in Los Angeles and their subsequent spread over a large part of the world, together with elevated O_3 levels over vast areas of countryside, raised the subject of this book to an unparalleled level. Thus air pollution was not just associated with industry, but also with the everyday activities of people in the form of the use of motor vehicles. It took a long time for this understanding to penetrate outside North America. Concurrently with the latter, came the growing realisation in the 1970s that NO_2 was a serious pollutant in its own right. Thus from the pioneering work of Menser and Heggestad (1966), who demonstrated synergistic interactions between SO_2 and O_3 in damaging tobacco plants, came the understanding of the importance of considering the effects of the ambient pollutant mix. This was highlighted in studies in the UK in the 1970s which clearly indicated that ambient levels of relevant pollutants were far more phytotoxic than when they were administered from commercially purchased cylinders (Bell, 1984). Thus we came to understand the complexity of such pollutant interactions, where synergistic, antagonistic and additive effects have all been demonstrated. This remains one of the most problematical of all areas within the remit of this book.

The late 1970s also led to a growing concern over both wet and dry deposited acidity, acting via the soil or possibly the foliage. This resulted in substantial research programmes to determine impacts of different levels of acidity in simulated acid rain and later mists/fogs/clouds. Initially this was devoted to crops (Evans, 1989), but the concerns over forest decline led to a major shift to trees. One of the most interesting developments during the 1980s was the realisation that increased N deposition was not only causing acidification problems, but also eutrophication of terrestrial ecosystems, which appeared to be leading to loss of low-N species over large areas. The role of air pollution in forest decline and in the adverse effects of eutrophication remain one of the most difficult areas covered by this book to understand.

Conclusions

Air pollution biology has come far since the early days when forests surrounding the nineteenth-century heap-roasting smelters were killed. We have progressed from identifying the pollutants responsible to measuring their concentration and largely controlling them from most point sources. We have identified still other pollutants emitted from power generating plants and other sources and largely brought them under control.

Today's major pollutants from multiple vehicular sources have been identified and monitored. Air quality standards have been established and the technology to achieve their goals is moving towards achievement.

Yet, with increasing population density and global industrialisation, air pollution problems remain a challenge. Emissions once not considered pollutants are now of concern. Research has shifted toward learning more about the impacts of NO_x and CO_2 and their combined effect with other atmospheric components. The potential relationship and problems associated with CO_2 and global warming have become and continue to be of particular concern.

References

Adams, D.F., Mayhew, D.J., Gnagy, R.M., Richey, E.P., Koppe, R.R. and Allen, L.W. (1952). Atmospheric pollution in the ponderosa pine blight area. *Ind. Eng. Chem.* **44**, 1356–1365.

Ashmore, M.R. and Marshall, F.M. (1999). Ozone impacts on agriculture: an issue of global concern. *Advances in Botanical Research* **29**, 31–52.

Atkisson, A. and Gaines R.S. (eds). (1970). *Development of Air Quality Standards*. Charles E. Merrill Publ. Co., Columbus, Ohio.

Bell, J.N.B. (1984). SO_2 effects on the productivity of grass species. In Winner, W.E., Mooney, H.A. and Goldstein, R.A. (eds). *The Effects of SO_2 on Plant Productivity*, pp. 209–266. Stanford University Press, Stanford.

Bell, J.N.B. and Clough, W.S. (1973). Depression of yield in ryegrass exposed to sulphur dioxide. *Nature* (London), **241**, 47–49.

Bell, J.N.B. and Cox, R.A. (1975). Atmospheric ozone and plant damage in the United Kingdom. *Environ. Pollut.* **8**, 163–170.

Benton, J., Führer, B.S., Gimeno, B.S., Skärby, L. and Sanders, G. (1995). Results from the UN/ECE ICP-Crops indicate the extent of exceedance of the critical levels of ozone in Europe. *Water, Air and Soil Pollution* **85**, 1473–1478.

Bradley, C.E. and Haagen-Smit, A.J. (1950). The application of rubber in the quantitative determination of ozone. *Rubber Chem. Technol.* **24**, 750.

Brimblecome, P. (1987). *The Big Smoke*. Methuen, London.

California State Department of Public Health (1959). *Health and Safety Code*, Section 426.1.

Chameides, W.L., Xingsheng, L., Xiaoyan, T., Xiuji, Z., Chao, L., Kiang, C.S., St. John, J., Saylor, R.D., Liu, S.C., Lam, K.S., Wang, T. and Giorgi, F. (1999). Is ozone affecting crop yields in China? *Geophysical Research Letters* **26**, 867–870.

Cohen, J.B. and Ruston, A.G. (1925). *Smoke, a Study of Town Air*, Edward Arnold, London.

Committee on Trail Smelter Smoke (1939). *Effect of Sulphur Dioxide on Vegetation*. Nat. Res. Council of Canada, O'Hara, 44.

Evans, L.S. (1989). Effects of acidic precipitation on crops. In Adriano, D.C. and Salomons, W. (eds). *Acidic Precipitation. Vol. 2: Biological and Ecological Effects*, pp. 29–60. Springer-Verlag, New York.

Evelyn, J. (1661). *Fumifugium: or the Inconvenience of the Aer and Smoake of London Dissipated*. W. Godbid, London.

Godbold, D.L. and Hüttermann, A. (eds) (1994). *Effects of Acid Rain on Forest Processes*. John Wiley & Sons, New York.

Guderian, R. (1977). Air pollution. In *Ecological Studies*, 22, p. 127. Springer-Verlag. Berlin.

Guderian, R., Klumpp, G. and Klumpp, A. (1988). Effects of SO_2, O_3, NO_2, singly and in combination on forest species. In M.A. Ozturk (ed.). *International Symposium On Plants and Pollutants in Developed and Developing Countries*. E.U. Press, Ege University, Izmir, pp. 231–268.

Haagen-Smit, A.J., Darley, E.F., Zaitlin, M., Hull, H. and Noble, W. (1952). Investigation on injury to plants from air pollution in the Los Angeles area. *Plant Physiol.* **27**, 18–34.

Haagen-Smit, A.J. and Fox, M.M. (1956). Ozone formation in photochemical oxidation of organic substances. *Ind. and Eng. Chem.* **48**, 1484–1487.

Haywood, J.K. (1910). Injury to vegetation and animal life by smelter wastes. US Dept. Agric. Bur. Chem. No. 113.

Hartig, R. (1894). *Textbook of the Diseases of Trees*. Translated by W. Somerville. MacMillan and Co. Ltd, London.

Hasselhoff, E. and Lindau, G. (1903). *Handbuchs zur Erkennung und Beurteilung von Rauchgeschäden*. Gebrüde Borntraeger, Berlin.

Heck, W.W., Taylor, O.C., Adams, R., Bingham, G., Miller, J., Preston, E. and Weinstein, L. (1982). Assessment of crop loss to ozone. *J. Air Poll. Contr. Assoc.* **32**, 353–361.

Heggestad, H.E. and Middleton J.T. (1959). Ozone in high concentrations as a cause of tobacco injury, *Science* **129**, 208–210.

Jäger, H-J., Unsworth, M., De Temmerman, L. and Mathy, P. (eds). (1993). Effects of air pollution on agricultural crops in Europe. *Air Pollution Report 46*, CEC, Brussels.

Katz, M. (1961). Some aspects of the physical and chemical nature of air pollution, pp. 97–158. In *Air Pollution*, World Health Org., Columbia University Press. NY.

Los Angeles County Air Pollution Control District. (1950). *Technical and Administrative Report on Air Pollution Control in L.A. County*, 1949–1950.

Lefohn, A.S., Laurence, J.A. and Kohut, A.J. (1988). A comparison of indices that describe the relationship between exposure to ozone and reduction in the yield of agricultural crops, *Atmos. Environ.* **22**, 1229–1240.

Mandl, R.H., Weinstein, L.H., McCune, D.B. and Keveny, M. (1973). A cylindrical, open-top chamber for the exposure of plants to air pollutants in the field. *J. Environ. Qual.* **2**, 371–376.

Mayrhofer, J. (1893). Uber Pflanzen beschädigung, veranlosst durch der Betrieb einer Superphosphate fabrik, *Free Ver. Bayer Vertreter Angew. Chemie Ber.* **102**, 127–129.

McLeod, A.R. (1995). An open-air system for exposure of young forest trees to sulphur dioxide and ozone. *Plant, Cell and Environ.* **18**, 215–225.

Menser, H.A. and Heggestad, H.E. (1966). Ozone and sulfur dioxide synergism: Injury to tobacco plants. *Science* **153**, 424–425.

Middleton, J.T., Kendrick, J.B. Jr. and Schwalm, H.W. (1950). Injury to herbaceous plants by smog or air pollution. *Plant Dis. Reptr.* **34**, 245–252.

Middleton, J.T., Kendrick, J.B. Jr. and Darley, E.F. (1955). Airborne oxidants as plant damaging agents. *Proc. 3rd Nat. Air Poll. Symp*. pp. 191–198.

Middleton, J.T., Crafts, A.S., Brewer, R.F. and Taylor, O.C. (1956). Plant damage by air pollution. *Calif. Agric.* June, pp. 9–12.

Oden, S. (1976). The acidity problem: an outline of concepts. In L.S. Dochinger and T.A. Seliga (eds), *Proc. 1st Inter Symp. on Acid Precip. and the Forest Ecosystem*, pp. 1–42. NE Forest Exp. Sta. Rep. NE-23.

O'Gara, P.S. (1922). Sulfur dioxide and fume problems and their solutions. *Ind. Eng. Chem.* **14**, 744–745.

Olsyzk, D.M., Katz, G., Dawson, T.J., Bytnerowicz, A., Wolf, J. and Thompson, C.R. (1986). Characteristics of air exclusion systems for field air pollution studies. *Journal of Environmental Quality* **15**, 326–334.

Oshima, R.J. and Bennett, J.P. (1978). Experimental design and analysis. In W.W. Heck, S.V. Krupa and S.N. Linzon (eds), *Methodology for the Assessment of Air Pollution Effects on Vegetation*, Ch. 4, pp. 1–23, Air Poll. Cont. Assoc. Pittsburgh.

Richards, B.L., Middleton, J.T. and Hewitt, H.B. (1958). Air pollution with relation to agronomic crops V Oxidant stipple of grape. *Agron. J.* **50**, 559–561.

Sanders, G.A., Skärby, L., Ashmore, M.R. and Fuhrer, J. (1995). Establishing critical levels for the effects of air pollution on vegetation. *Water, Air and Soil Pollution* **85**, 189–200.

Schroeder, von J. and Ruess, C. (1883). *Die Beschädegung der Vegetation durch Rauch und die Oberharzer Hüttenrauchschäden*. Parey, Berlin.

Smith, R.A. (1872). *Air and Rain: The Beginnings of a Chemical Climatology*. Longmans, Green and Co., London.

Sokal, R.R. and Rohlf, J. (1969). *Biometry*. Freeman, San Francisco.

Sorauer, P. (1886). *Handbuch der Pflanzenkrankheiten*, 2nd ed., Parey, Berlin.

Sorauer, P. (1914). Non-parasitic diseases. In *Manual of Plant Diseases, Vol. 1*, Record Press. Wilkes Barre, PA.

Stoklasa, J. (1923). *Die Beschädigung der Vegetation durch Rauchgase und Fabriksexhalation*. Urban and Schwarzenberg, Munich.

Stephens, E.R., Hanst, P.L., Doerr, R.C. and Schott, W.E. (1956). Reactions of NO_2 and organic compounds in air. *Ind. Eng. Chem.* **48**, 1498–1504.

Thomas, M.D. and Hill G.R., Jr. (1935). Absorption of sulfur dioxide by alfalfa and its relation to leaf injury. *Plant Physiol.* **10**, 291–307.

Treshow, M. and Anderson F.K. (1989). *Plant Stress from Air Pollution*. Wiley, NY.

Wanta, R.C., Moreland, U.B. and Heggestad, H.E. (1961). Tropospheric ozone: an air pollution problem arising in the Washington, D.C. metropolitan area. *Monthly Weather Review.* **89**, 289–296.

Wellburn, A. (1988). *Air Pollution and Acid Rain*. Longman Scientific and Technical, Harlow.

World Health Organisation (2000). *Air Quality Guidelines for Europe*, 2nd edition, WHO Regional Publications European Series No. 91, Copenhagen.

3 Emissions, dispersion and atmospheric transformation

R.N. COLVILE

Overview

The first three parts of this section aim to provide the reader with an overview of all sources of air pollution. This is a starting point for the classification of individual pollution sources of importance to plant life leading to an assessment of their nature and significance. The emphasis is on breadth rather than depth of detail, but should be sufficient for items of special interest to individual readers to be researched further by reference to other texts. This leads on to a similarly broad overview of methods of reducing emissions. This complements Chapters 2 and 22, in which the emphasis is on how pollution impacts on plant life have driven legislation but without consideration of how the necessary emissions reductions can be achieved. The section closes with an introduction of how it is possible to make quantitative estimates of emissions of air pollution which will form a vital part of the foundation of any assessment of impacts on plant life. The emphasis here is on continental scale, although national and urban inventories are also included.

The remainder of the chapter is then intended to provide the basis of an understanding of how we can quantify air pollution at any receptor and relate its concentration and composition to emissions. Thus it considers the transport and dilution of pollutants after emission to the atmosphere, which is most important close to a source of emissions. At longer distances, the conversion of primary emissions to secondary pollutants becomes important, and these atmospheric transformations are covered later. Quantification of the relationships between emissions and concentrations involves two steps: identifying what are the most important atmospheric processes, then representing these mathematically in a model. The modelling techniques considered are Gaussian Plume models and Lagrangian or Eulerian techniques. After understanding this chapter, the reader should thus be able to move on to any of Chapters 5 onwards and appreciate how the various impacts of air pollution on plant life are related to emissions of the pollutants concerned.

Air Pollution and Plant Life, second edition. Edited by J.N.B. Bell and M. Treshow. ISBN 0 471 49090 3 (HB), 0 471 49091 1 (PB).

Sources of air pollution

Combustion of fuel

Complete combustion of a pure hydrocarbon fuel in pure oxygen produces carbon dioxide and water resulting in global emissions fluxes of order 10^{12} to 10^{13} kg per year of each. Natural fluxes of these gases are even greater than this, but the long life of carbon dioxide in the atmosphere/surface ocean system allows the anthropogenic perturbation to become significant over centuries. These greenhouse gas emissions, however, are beyond the scope of this book as outlined in Chapter 1, so will not be discussed further and are included here only for completeness and for comparison with other air pollutants.

No fuel currently burnt on earth is a completely pure hydrocarbon. The most abundant impurity in fossil fuels is sulphur, present in very small quantities in natural gas, a few tenths of a per cent to 2% in crude oil and from about 1% to 6% by weight in coal. During combustion, sulphur in fuels is oxidised to sulphur dioxide (SO_2). Anthropogenic emissions of SO_2, unlike CO_2, are larger than natural fluxes of oxidised or reduced sulphur.

Chlorine is an impurity present in coal at an average concentration of 0.15%, combustion of which leads to emissions of hydrogen chloride (HCl) gas. Incineration of plastics produces much larger quantities of HCl. (It could be considered that incineration of waste, whether to dispose of the waste or to generate power, is equivalent to using a fuel that is composed entirely of impurities!) A dozen or more other elements are present as impurities in coal and oil at levels from parts per billion to parts per thousand. Organic nitrogen compounds present in fuel can be oxidised to form nitric oxide. Some impurities are deliberately added to fuels during processing, notably lead which has now been eliminated from petrol (gasolene) in most developed countries but which is still used as an additive in developing countries.

Impurities participate in the combustion process not only from the fuel but also from the atmosphere which, of course, is not pure oxygen. During combustion, abundant atmospheric nitrogen (N_2) is oxidised to NO_x, which is a mixture of nitric oxide (NO) and nitrogen dioxide (NO_2). The amount of NO_x produced is typically comparable with the amount of SO_2 obtained from a relatively high-sulphur fuel, so for low-sulphur fuels NO_x is the most abundant pollutant produced after CO_2 and H_2O.

The final factor that determines what air pollutants are present in combustion exhaust gases is the fact that combustion is usually incomplete. Unburned material may include black soot, ash from solid fuel, white liquid droplets of unburned hydrocarbon fuel or lubrication oil, polycyclic aromatic hydrocarbons from low-temperature combustion especially from wood burning, and carbon monoxide especially from non lean-burn petrol engines that do not have an excess of air in the combustion mixture.

Chemical processes

Two chemical processes are of special interest to the study of air pollution and plant life, since they are sources of fluoride, which will be covered in detail in Chapter 9. Hydrogen fluoride receives much less attention than many other pollutants, and is not mentioned at all in many standard air pollution textbooks, probably because its occurrence is restricted to near its sources (one exception is Brimblecombe (1986), which provides more detail on sources of air pollution following a similar structure to what is adopted here).

The largest source of fluoride is the smelting of aluminium. This process is also notorious in requiring massive amounts of energy. Most aluminium is therefore produced in Norwegian and Canadian plants located close to sources of abundant hydroelectricity, which tend to be surrounded by forest in mountainous terrain where impacts of atmospheric emissions on trees can be enhanced by topography. Fluoride is emitted from aluminium smelting since molten $AlF_3 \cdot 3NaF$ cryolite is used as a solvent in the high-temperature electrolysis process. Similar emissions may arise during smelting of uranium, or from the fluoride flux that is used in steel making. The second significant source of fluoride in the atmosphere is brick making. The bricks are heated to drive off water from a number of clay minerals. Some of these minerals contain fluorine in place of hydroxyl groups, so that hydrogen fluoride is driven off instead of water. Similar emissions can arise from the manufacture of glass and ceramics. Both these processes require the input of large amounts of energy. Unless this is from a hydroelectric or nuclear power plant, pollutants from the chemical process will be in addition to those discussed. A third source of fluoride is a by-product of fertiliser manufacture.

For a wider perspective, other major chemical sources of air pollution are listed in Table 3.1, including low-temperature manufacture of sodium sulphate alongside the other high-temperature processes; in the nineteenth century, this was a large source of hydrochloric acid which resulted in destruction of vast amounts of vegetation. Nowadays, the hydrochloric acid is usually scrubbed into aqueous solution as a valuable by-product, although the collapse of Communism in Eastern Europe has at the time of writing temporarily left some examples of unabated acid emissions from chemical works. In all the examples listed in the table, the purpose of the process is the chemical change in the reagents, any energy generated being coincidental. Incineration of waste could be included in this category, although the degradation of unwanted material and the generation of energy are nowadays increasingly being combined in a single process.

Other emissions

A wide range of sources of air pollution do not fit into either of the two categories above. Volcanoes emit around 10^{10} kg of SO_2 into the atmosphere globally per year, compared with around 10^{11} kg a^{-1} from anthropogenic sources. Emissions of reduced sulphur compounds from marine phytoplankton lie somewhere between these two figures, and can be oxidised in the atmosphere to form SO_2 along with smaller quantities from terrestrial biota. Some trees and other plants are the dominant sources of certain organic gases, especially terpenes. Lightning is responsible for about 10^{10} kg per year of NO_x compared with anthropogenic emissions of around two to four times that amount. Particles are released into the atmosphere by sea surface bubble bursting processes releasing salt crystals from evaporation of seawater, and dust from erosion of soil by wind. Disturbing the soil during agricultural activity is another source of mineral dust. Any storage of ash or mechanical process such as handling powders or drilling

Table 3.1 Major chemical sources of atmospheric pollution

Chemical process	Reagents	Reaction (example)	Pollutants emitted (in addition to pollutants produced by burning fuel)
Cement manufacture	Clay, limestone	$CaAl_2Si_2O_8 + 6CaCO_3 = 2Ca_2SiO_4 + 6CO_2$	Carbon dioxide, dust
Production and refining of metals by smelting and roasting	Oxide ore	$Fe_2O_3 + 2CO = 2Fe + 2CO_2$	Carbon dioxide, particles including metals
	Sulphide ore	$Ni_2S + 4O_2 = 2NiO + 3SO_2$	Sulphur dioxide
Food processing	Especially brewing, meat	Various	Mercaptans (odour)
Incineration	Waste	Oxidation	Metals, chlorides, organic compounds
Manufacture of sodium sulphate	Salt and sulphuric acid	$2NaCl + H_2SO_4 = Na_2SO_4 + 2HCl$	Hydrochloric acid

or cutting materials similarly can produce fugitive dust emissions. Agriculture is also a major source of ammonia both from animal manure and the application of chemical fertilisers. Pesticides are a peculiar pollutant in that their polluting effects are the same as their intended use. Any pesticide spray that escapes before reaching the intended target, or evaporation of stored volatile chemicals can be a source of air pollution that is especially damaging to plant life. Paints during storage and use cause evaporative losses of volatile solvents. The lighter fractions of hydrocarbon fuels also evaporate during handling and storage, especially natural gas from leaks and losses during repair and ignition. Other sources of methane include the anaerobic decomposition of waste by bacteria in land-fill dumping sites.

Trends in emissions

The impact of any pollutant can be reduced in one of four ways:

1. Dilute its concentration before it reaches any sensitive receptor.
2. Collect the pollutant and dispose of it in a way that prevents it from reaching the most sensitive receptors.
3. Collect and recycle the pollutant for some useful purpose.
4. Reduce or prevent altogether the emissions from occurring in the first place.

In years gone by, it was uncommon to progress beyond the first of these, the only widely applied emissions abatement technology being to build a tall chimney to reduce local ground-level concentrations. In the 1970s, however, it became clear that the capacity of the atmosphere to dilute pollution is not unlimited, but that tall chimneys can only remove a problem to a longer distance away. This has led to the adoption of progressively more effective technology to reduce emissions, while the potential demand for many polluting activities has tended to increase.

There are many ways a pollutant can be collected before emission to the atmosphere. Such 'end-of-pipe' technologies include cyclones, water sprays, filtration and electrostatic precipitators to remove particles. Chemicals added to remove gases include crushed limestone ($CaCO_3$) to trap sulphur dioxide. Catalysis can reduce NO_x to N_2 and oxidise hydrocarbons to CO_2 and H_2O, on large stationary and small mobile sources. Micro-organisms can be used in biological emissions abatement supported on a filter of peat moss mixed with twigs and small branches, to remove specific pollutants including odours. Some authors have suggested planting roadside trees can soak up vehicle exhaust, but the effect is rather small and the main benefit of trees is their uptake of CO_2 when grown in large quantities.

As the obvious major energy consuming combustion processes are steadily tackled with emissions abatement measures, other sources become more significant and also require attention. Space here permits only a brief list of a few: evaporative losses of vehicle fuel during filling can now be reduced by recovering vapour from the vicinity of the nozzle of a modern petrol pump; agricultural emissions of ammonia can be reduced by banning the spreading of animal manure on the surface of field for its disposal as well as fertiliser use, requiring instead that it be injected into the soil;

chain saws and lawn mowers tend to lack the emissions abatement technology fitted to large mobile sources. Some sources are likely to remain resistant to control because people simply like causing pollution, for example barbecues and fireworks.

End-of-pipe technology can also produce new pollutants that replace the ones it is intended to remove. Electrostatic precipitators produce large amounts of fine ash; catalysts on cars can constitute the dominant source of ammonia in urban areas. Some pollutants that are removed from exhaust gases can usefully be re-used. A prime example of this is the gypsum produced by flue-gas desulphurisation of coal-fired power plants which is used as a building material.

Removal of pollutants from exhaust gases is nearly always applied in conjunction with initiatives to reduce or prevent the production of the pollution at source. For example, levels of NO_x produced can be reduced by controlling the combustion conditions and sulphur is increasingly removed from oil during refining. Evaporative losses are controlled by reducing volatility of liquid fuels at the same time as selecting hydrocarbons that produce less ozone in the atmosphere. Diesel-engined vehicles can be converted to run on compressed natural gas, alcohols and vegetable oils resulting in immediate emissions reductions or changes in emissions composition without replacing the vehicle or its engine. Wind, hydroelectric, wave, geothermal and biomass combustion are sources of energy that are gaining favour as pressure to reduce CO_2 emissions increases. Economic and political factors can favour cleaner technology, for example the change from coal-fired room heating to central heating later fired by natural gas that ended the famous London smogs of the 1950s, followed 35 years later by a similar shift from coal to natural gas following the privatisation of the UK electricity generation industry; also the adoption of nuclear power in France. Economic decline can reduce demand for polluting activity or can have the opposite effect where it impedes the adoption of cleaner technology. In some regions, taxation is being transferred away from economic activity towards production of pollution. Most recent developments include reallocation of road space from private cars to bicycles, buses and pedestrians; geographical location of industries in close proximity to take and recycle each other's waste streams ('industrial ecology'); management systems with employee involvement so that a business automatically adopts best environmental practice. These are far removed from the old 'tall chimney' policies, and represent a revolution in the way people interact with their environment.

Some of the air pollution emissions impact of technological and socio-economic changes combined (or opposing), is shown for the UK in Figure 3.1.

Emissions inventories

The data in Figure 3.1 come from an inventory of emissions. Such an inventory is a valuable and increasingly widely used tool for gaining a first impression of the relative significance of different sources of pollution, and the effects of policies to influence trends over periods of time. As well as showing time series of emissions, most inventories include some spatial information allowing maps to be produced such as the example in Figure 3.2. Here, the importance of the road network for emissions of NO_x compared with large point sources for SO_2 is clearly indicated.

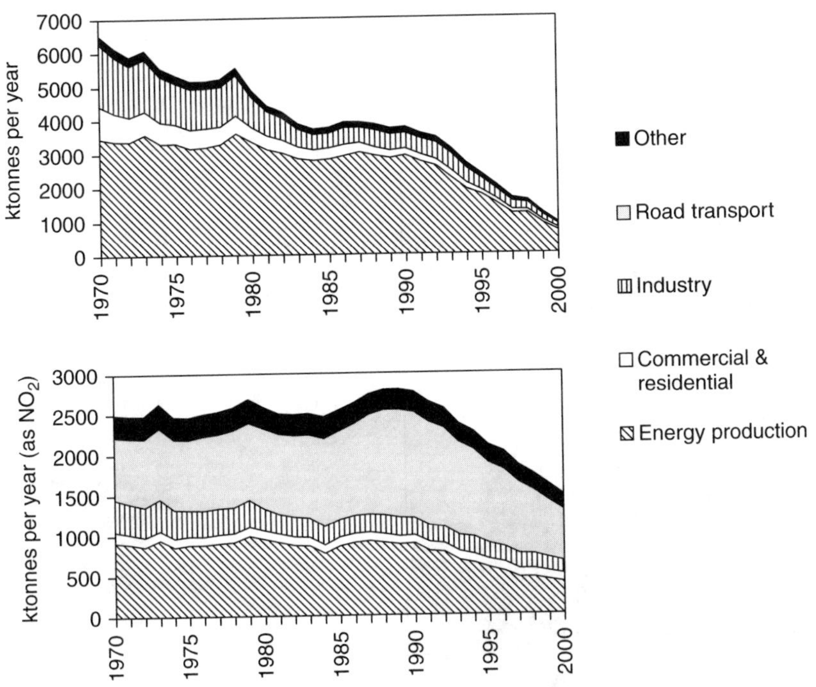

Figure 3.1 Annual total UK atmospheric emissions of oxides of nitrogen NO_x (bottom) and sulphur dioxide (top) by sector from 1970 to 1996 (Goodwin *et al.*, 1999)

Ideally, we would like hourly measurements of the rate of emission of each of the pollutants discussed above from every individual source, alongside data describing the location of each source and other relevant parameters such as temperature, height and exit velocity of the exhaust gases. For some pollutants and certain classes of large point sources, such data are increasingly coming into existence, but there remains a large number of smaller sources for which it is unlikely ever to be practical to obtain such detailed information. The biggest example of this is road traffic. For mobile sources, emissions data are now usually collected using what is called a 'bottom-up' methodology. Information on traffic flow volume and speed, and fleet composition, is collected for each road link. This is combined with emissions factors describing the mass of pollutant emitted by each vehicle type per unit distance travelled under those flow conditions. It is important to measure emissions factors representative of the whole fleet, as a few poorly maintained vehicles can be responsible for a high proportion of the emissions (Pierson *et al.*, 1996), and also to include emissions from cold, recently started engines for which current abatement technology is ineffective. To obtain the total emissions from a whole city, it is necessary to add up the contributions from every road, using a combination of measured and modelled traffic flow data. For some other sources, a 'top-down' methodology is more convenient. For example fuel use data for domestic heating is typically available as totals over large geographical areas. Finer spatial resolution is then obtained by disaggregating the data spatially based on

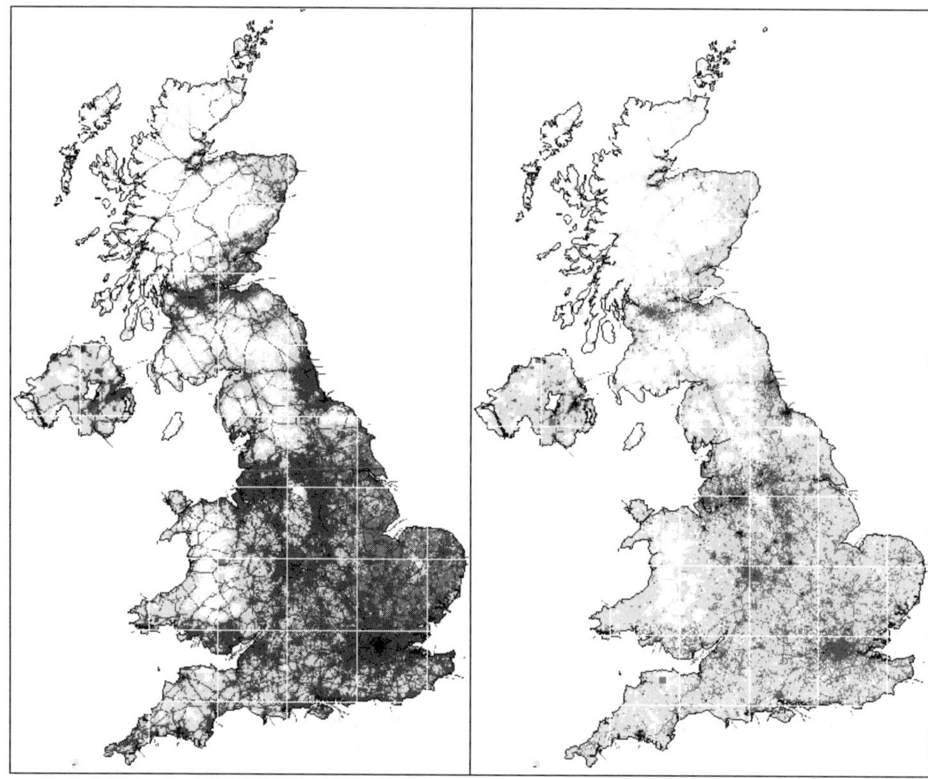

Figure 3.2 UK atmospheric emissions of oxides of nitrogen NO_x (left) and sulphur dioxide (right) in 1998 mapped on a 1 km × 1 km grid (Goodwin *et al.*, 1999)

an assumption that the local emissions rate is proportional to population density. Most emissions inventories only include annual average data, although temporal disaggregation can be applied for example by considering diurnal, weekly and seasonal cycles in road traffic flows.

In Europe, two important emissions inventories are CORINAIR (CORINAIR, 1996) and EMEP (MSC-W, 1996). Both contain information on emissions of CO, NO_x, SO_2, CH_4, CO_2, NH_3 and non-methane hydrocarbons. Methodologies are harmonised between individual EU member states for both inventories, and further harmonisation between these and the international greenhouse gas emissions inventories (IPCC, 1994) is under way. A convenient summary of the current state of these inventories may be found in Friedrich and Schwartz (1999). EMEP has a spatial resolution of 50 km, using the same grid as the modelling that was carried out initially for the development of policy to control acid damage to plant life (see Chapter 22). CORINAIR disaggregates emissions on a national basis and down to smaller administrative units, prepared for the European Environment Agency. The inventory is updated periodically, and is currently being extended to include some additional pollutants and to cover the Eastern European countries that are currently undergoing preparation for joining the

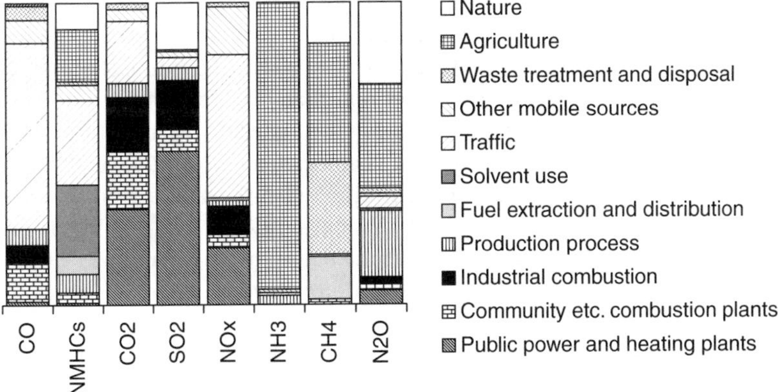

Figure 3.3 Relative contribution from the main sectors in per cent of the overall recorded European emissions in 1994 (CORINAIR, 1996)

European Union. Recently, emissions inventories have tended to become available with finer spatial resolution, for example the UK National Atmospheric Emissions Inventory (Goodwin *et al.*, 1999) increasing from 10 km to 1 km national coverage, matching the finer resolution that was previously available only for major cities such as London (Buckingham *et al.*, 1997).

An example of CORINAIR emissions data is shown in Figure 3.3, where emissions have been summed over all 25 European countries covered but the distribution between the 11 main sectors of emissions type has been retained and expressed as percentages. The 11 main sectors summarised here are divided further into 270 sub-sectors in the inventory.

Atmospheric dispersion

In this section we will focus on 'short-range' transport of air pollution from its source, which in the atmosphere means to receptors within about 100 km from the source. We will assume the pollutant travels from source to receptor without undergoing any change – atmospheric chemistry will then be considered separately.

Advection and turbulence

The concentration of a pollutant down-wind of a source, relative to its concentration at the point of emission, is determined largely by two processes. Advection is the movement of the pollution carried by the wind, averaged over the time taken for it to blow from the source to any given receptor. Stronger wind causes more rapid dilution at the point of emission; the highest concentrations are found when it is calm. Turbulence is the instantaneous random motion of the air, which causes the pollution to spread out perpendicular to the wind direction. Turbulence is initiated in the atmosphere by the drag of the movement of the lower layers of air over a rough surface, a process that is closely related to deposition (see Chapter 4). If that surface is warmer than the air above, the vertical transfer of heat will enhance the amount of turbulence present. When this heating effect is marked, the atmosphere is described as unstable and the large overturning motions that can be induced sometimes lead to the formation of thunderstorms if the amount of energy and moisture involved are sufficiently great. Conversely, if the surface is colder than the air above, the atmosphere tends to form stable layers between which turbulent mixing is inhibited, for example on a still, cloudless, misty night in winter. When the sky is covered by cloud and the wind is strong, the turbulence generated mechanically by the roughness of the surface dominates over either thermal effect and the atmosphere is described as neutral.

The Gaussian plume model

Even though the details of atmospheric turbulence and boundary layer structure are complicated, it is possible to calculate the concentration at a certain distance in a given direction downwind of a source of pollution using a rather simple mathematical model that takes account of the influence of stable and convective atmospheric conditions on dispersion discussed in the previous section.

Since the process of dispersion of pollution in a turbulent atmosphere is a random one, we can make a first guess that the profile of concentration along a cross-section through a plume will be normally distributed. Field observation indeed indicates that the Gaussian formula is a sufficiently good description of how pollutant concentration varies in cross-section through a real plume averaged over an hour or so. Plume behaviour from an elevated source is shown schematically in Figure 3.4, along with Gaussian plume parameters, for the three main atmospheric stability types. Similar nomograms can be used to look up values of the horizontal plume spread parameter σ_y, which is similar to σ_z in that at any distance from the source it is largest in unstable conditions and smallest in stable conditions. σ_y is larger than σ_z especially for stable conditions because the plume is able to meander laterally when its vertical dispersion is inhibited by the layered structure of a stable atmosphere. Table 3.2 shows how the full range of atmospheric stability classes from very stable to very unstable were defined by Pasquill (1961) as a function of wind speed and cloud cover.

Different weather conditions thus produce the maximum ground-level concentration for different types of source. Elevated point sources contribute little to ground-level air

Table 3.2 Pasquill–Gifford stability classes (according to Turner, 1969)

Surface wind speed at 10 m s^{-1}	Day-time: Solar radiation (W m^{-2})			Night-time: Cloud cover	
	>700	350 to 700	<350	≥4/8	≤3/8
<2	A	A to B	B		
2 to 3	A to B	B	C	E	F
3 to 5	B	B to C	C	D	E
5 to 6	C	C to D	D	D	D
>6	C	D	D	D	D

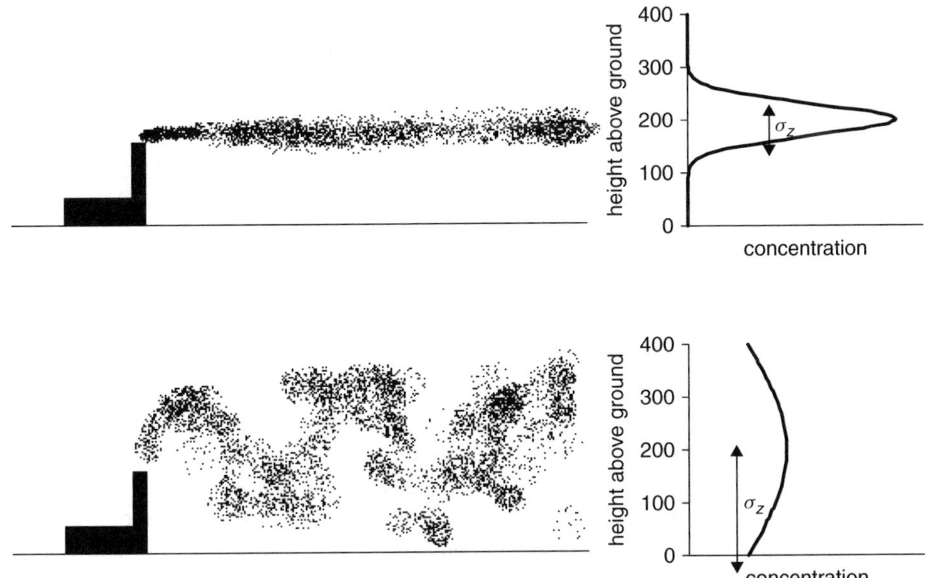

Figure 3.4 Instantaneous images of plume undergoing dispersion from an elevated point source in stable and unstable atmospheric conditions, with Gaussian plume vertical cross-sections of ensemble average concentration through the plume centreline and vertical plume spread parameters σ_z

quality in flat countryside in stable conditions, because the plume remains above the ground even though pollutant concentrations in it remain high due to inefficient dilution. The maximum ground-level concentration from a tall chimney in flat countryside will be found in unstable meteorological conditions, when the plume is being diluted rapidly but is also being brought down to the ground very rapidly. Ground-level sources (of which road traffic is the main example) behave quite differently to elevated sources, giving the lowest ground-level concentrations in unstable conditions and the highest in stable conditions, especially where a temperature inversion perhaps as low as 100 m above the ground traps pollution in an urban area. In complex terrain, however, where hills rise above the height of the top of the plume, ground-level and elevated sources alike tend to give rise to the highest ground-level air pollution concentrations under stable conditions; how to model these is discussed briefly at the end of this section.

In recent years, look-up tables and nomograms such as Table 3.2 and Figure 3.4 (for example, Clarke, 1979) have been largely replaced by computer programs such as those that are freely available from the US Environmental Protection Agency (USEPA, 1978), plus more advanced models offering improved performance under certain conditions (Carruthers *et al.*, 1994; Cimorelli *et al.*, 1998). Most usefully, with such software it is relatively straightforward to calculate ground-level pollutant concentrations on a grid of receptors for each set of meteorological conditions that are observed in a year, and then to generate from the model output weighted averages according to the number of hours per year that each meteorological situation occurs. Statistics such as accumulated total over a certain threshold can thus be calculated, which can be related to damage to plant life, converting a local emissions map into a pollutant concentration or potential pollutant impact map. At the same time, advances in our understanding of the atmospheric boundary layer have allowed some of the most serious failings of the basic Gaussian plume model to be corrected for, for example in unstable conditions more accurate results can be obtained by assuming an asymmetric vertical profile of concentration. At the time of writing, leading examples of such so-called second-generation Gaussian-type models are UK-ADMS.

Despite the widespread usefulness of Gaussian plume models and their newer derivatives, there is a wide range of sites and conditions where such models are inapplicable. Furthermore, several of these may be of special interest to the effects of pollution on plant life, for example where forests are commonly planted on hills. For relatively gentle slopes, it is possible to make a number of corrections to the basic model depending on how accurate a solution is required and how high the hills are compared with the source, including estimates of changes in wind speed using mathematical methods also applied to studies of wind damage to trees (Inglis *et al.*, 1994). Under the most complex topographical conditions, however, and also in locations such as valleys in continental interiors where calm conditions occur frequently, it is necessary to abandon the Gaussian plume model altogether and invest in a full numerical solution of equations describing the mean flow and turbulence throughout a three-dimensional grid, or to carry out laboratory measurements in a wind tunnel. Now that powerful computers are widely available, such computational fluid dynamics solutions to environmental fluid flow problems are becoming increasingly popular. Highly distorted terrain, flow and dispersion patterns do not occur only in natural mountainous areas but also in the built environment of a city. Road traffic emissions commonly become trapped in so-called street canyons where the carriageway is hemmed in on both sides by tall buildings, and any vegetation sharing that space with the vehicles is likely to be exposed to very high pollutant concentrations as the sheltering effect of the buildings prevents the emissions from being blown away and diluted effectively. The recent growth of interest in urban air quality has led to the development of a number of semi-empirical models for simple street canyon situations, for example Berkowicz *et al.* (1997), although still for any specific site such as a small urban garden surrounded by buildings a wind tunnel, computational fluid dynamics or full-scale measurement is required to characterise the air pollution microclimate. Finally, as was mentioned, we are confining ourselves here to pollutants that do not change in the atmosphere; attenuation of the plume by reactive decay or indeed by deposition can be modelled simply by an exponential factor in a Gaussian plume model, but such an approach is totally inapplicable to any more than the simplest chemistry or to a secondary pollutant such as ozone.

Atmospheric transformation

Along with wet and dry deposition (Chapter 4), atmospheric chemistry is the major process that can remove pollutants from the atmosphere. Here, however, we will consider its importance in forming new, secondary pollutants that are able to harm plant life.

Photooxidant formation

The most important example of a secondary pollutant, in terms of effect of air pollution on plant life, is ozone (O_3). This gas is produced in the stratosphere (around 20 km above the ground) by photolysis of oxygen (O_2) molecules. Closer to the ground, however, in the troposphere, ultra-violet sunlight of sufficiently short wavelength is not present for this to occur, so it is principally nitrogen dioxide NO_2 that is photolysed to generate O_3. On its own, this is unable to maintain the significant concentrations of ground-level O_3 that are observed, because the photoreduction of NO_2 is rapidly reversible, consuming the O_3 almost as soon as it is formed. Furthermore, O_3 deposits rapidly to solid surfaces and is also destroyed by photolysis. It is this last, chemical sink of O_3, however, that initiates cycles of reactions, the final result of which is net production of O_3.

The action of sunlight on a molecule of O_3 liberates an excited singlet oxygen atom which sometimes reacts with water vapour to form a hydroxyl radical ($OH^•$). Reaction with $OH^•$ is the main removal mechanism for many hydrocarbons in the atmosphere, including alkanes, alkenes and some aromatic species such as toluene (methyl benzene). Attack of an organic molecule by $OH^•$ initiates a sequence of reactions in which NO also participates, and the net result of these is oxidation of typically two or three but sometimes more molecules of NO to NO_2. The $OH^•$ is regenerated, so that it is able to participate in further cycles. Hydrogen peroxide (H_2O_2) can also be formed, which we will meet again when we consider strong acid formation. But if we shift our attention now from the $OH^•$ radical to the oxides of nitrogen, we find a cycle of reactions of NO_x that generate the observed concentrations of ground-level O_3.

Most of the NO_2 in the atmosphere is produced by oxidation of NO by O_3, so the photolysis of NO_2 by sunlight merely regenerates the O_3 that was consumed initially. But if some NO_2 is formed as part of the degradation of a hydrocarbon following $OH^•$ attack, this does not consume any O_3, so the subsequent photolysis of this NO_2 generates new O_3. In several hours, the NO_x in a plume of polluted air might thus cycle between NO and NO_2 several times, resulting in an O_3 concentration in excess of that of NO_x.

NO_x alone will not produce any O_3, and nor will any hydrocarbon without NO_x. It is together that these two pollutants, in the presence of sunlight, undergo cycles of reactions that result in the production of ground-level O_3. Different hydrocarbons lead to the production of varying amounts of O_3, for example the photochemical oxidant production potential of ethane is 1.5 times that of ethene, and o-xylene produces twice as much O_3 as ethane. Simplistically, those hydrocarbons that react most quickly and participate in cycles of reactions that consume more NO have a high photochemical oxidant production potential, but other more complex issues need to be taken into account as well. For example, the atmospheric degradation of some hydrocarbons results in an increase in the number of reactive free radicals in the atmosphere and can thus amplify the amount of O_3 produced by the degradation of all hydrocarbons. Others, including terpenes that are emitted by trees, probably have the opposite effect of suppressing the general photochemical reactivity of the atmosphere locally, but may form stable intermediate compounds that then release oxides of nitrogen and generate O_3 many thousands of kilometres from the source of emissions. Controlling rural levels of O_3 is therefore not straightforward, as O_3 responds well to reductions in NO_x in some regions but in others reduction of hydrocarbon emissions is more effective. In urban areas, air that is trapped by local topography and meteorological effects, similarly can cook under the action of sunlight to produce the highly toxic mixture of hydrocarbon droplets and ozone that is called photochemical smog.

Strong acid formation

Sulphur dioxide, like many hydrocarbons, is oxidised by hydroxyl radicals in the gas phase to form sulphuric acid (H_2SO_4). At a typical $OH^\bullet$ concentration of 10^6 molecules per cm^3 of air, SO_2 has a half-life of around 100 hours in clear air. This is long enough for oxidation not to be the dominant sink for SO_2, with greater amounts being removed by dry deposition and even more during scavenging by drizzle, rain or snow, especially if the snow is wet. A much more effective way of oxidising SO_2 is to dissolve it with the oxidant in water. In a clean cloud, O_3 and SO_2 are drawn into solution by their reaction to form H_2SO_4. This continues until the acidity produced reduces the solubility of the gases in the water, unless ammonia is also present to buffer the solution. In an acidic cloud, SO_2 undergoes acid catalysed reaction with H_2O_2 instead of with O_3, as long as H_2O_2 is available. In clean conditions, the amount of H_2O_2 available may exceed that of SO_2, such that all the SO_2 is oxidised in a matter of minutes. In polluted conditions without ammonia to allow the O_3 reaction to continue, the reaction stops with excess SO_2 remaining as soon as the H_2O_2 is used up. The amount of H_2SO_4 produced and the acidity of the cloud and any rain that falls from it are therefore often rather insensitive to the concentration of SO_2.

Some of the impact of this chemistry and the distribution of emissions on the geographical and seasonal distribution of sulphuric acid in air near ground level may be seen in Figure 3.5. This maxima of emissions over major industrial areas such as Eastern Europe and China show clearly. Unoxidised SO_2 concentrations (not shown) are higher in winter than summer, because of increased winter fuel use and the lack of extensive convective mixing in winter. However, the model used to produce these maps takes account of how the availability of oxidising agents H_2O_2 and OH is much less in winter than in summer, and this is greater than the change in emissions. Concentrations of sulphate are therefore higher in summer, when there is plenty of photochemical activity, than in winter. (Note the reversal of the difference between northern and southern hemispheres from one season to the other.) Actually, the model is showing the seasonal difference too clearly, possibly because it does not consider the oxidation of SO_2 by O_3 in the presence of NH_3, which could be more important in winter when the other oxidants are much less abundant. This is because complexities and large uncertainties in NH_3 sources and sinks make it very difficult to model this process within the constraints of this kind of model.

Until recently, sulphuric acid was the most abundant strong acid produced in the atmosphere by the removal of pollution. Since the raised awareness of the impact of H_2SO_4 deposition on plants in the 1970s, however (Chapter 2), and the resulting action to cut emissions of SO_2 (Chapter 22), coupled with the general decline of heavy industry in Europe and North America accompanied by rapid increase in road traffic,

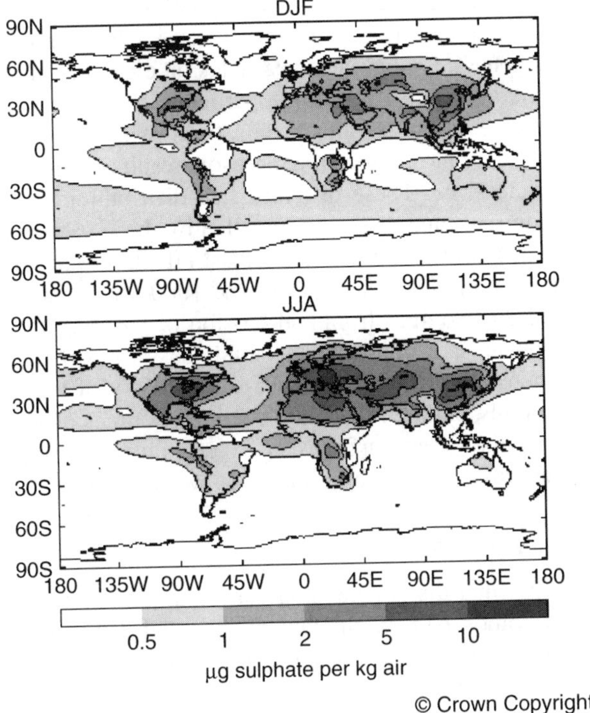

Figure 3.5 Global three-month average sulphate mass mixing ratios approximately 10 m above the ground in 1985 modelled for December–February (top) and June–August (bottom) (Jones *et al.*, 2002)

H_2SO_4 in the 1990s begun to be overtaken by nitric acid (HNO_3) as the largest source of atmospheric acidity.

We have already seen how oxides of nitrogen are predominantly emitted as nitric oxide NO. This is much more reactive as a reducing agent than SO_2, and it is oxidised within minutes to nitrogen dioxide (NO_2) by ozone in the gas phase. NO_2 is less reactive than NO but still more reactive to oxidation than SO_2, having a half-life of about a day to gas-phase attack by several oxidising agents including hydroxyl radicals, ultimately producing nitric acid HNO_3. The greater chemical reactivity but reduced aqueous solubility of NO_x relative to SO_2 means that mixed-phase chemistry of NO_x is certainly less important for NO_x than for SO_2, and remains rather poorly understood.

Particulate air pollution

The different pathways for production of sulphuric and nitric acid in the atmosphere influence the physical state of the acid or its salt in the atmosphere. An outline of how the relevant chemistry influences this is given here, with more detail to follow. Processes that occur in the gas phase, for example the oxidation of SO_2 by $OH^{\bullet}$ produce very small particles of liquid or solid pollutant a few nanometres in size. Combustion processes also generate such small particles containing soot and hydrocarbons. These coagulate to form mixed particles a fraction of a micrometre in size. Cloud processes generate rather larger particles on evaporation of the cloud, from 0.1 μm to several micrometres. The largest particles in the atmosphere are generated by physical processes such as salt crystals from evaporation of sea spray. The condensation of nitric acid onto pre-existing aerosols results in nitrate distributed over a wide range of particle sizes. These different size distributions can have a marked effect on the deposition rates and distances travelled by these particles, as discussed in Chapter 4. The 0.2 μm particles are deposited least effectively of all and so are most effective in transporting pollution long distances of 1000 km or more. In remote areas (unless there are local dust sources from soil erosion or the sea), these transboundary particles, rich in sulphate, can dominate the total mass of particulate matter in the air. In cities, superimposed on this one finds a wide variety of particles from human activity. In the most highly polluted roadside locations, sources of vehicle exhaust and resuspended dust from the street surface can exceed imported sulphate, nitrate and secondary organic particulate matter by a factor of two or larger.

Modelling secondary pollutants

It was mentioned earlier that the simple Gaussian plume model is unable to represent atmospheric chemistry other than the simplest exponential decay of a reactive primary pollutant with distance downwind of a source. For atmospheric chemistry and aerosol evolution, it is therefore necessary to focus on solving the equations describing these processes in one of two ways, nearly always sacrificing much of the detailed description of near-source dispersion that the Gaussian plume model handles best. These two main approaches are known as Eulerian and Lagrangian.

In a Eulerian model, the mathematical equations describing the chemical reactions are solved on a fixed three-dimensional grid. It is therefore necessary to include terms to describe the advection of pollutants from one grid cell to the next, mostly horizontally but also with some vertical exchange between layers of grid cells. Such a Eulerian framework is also used for computational fluid dynamics models of environmental flows in the urban environment or in complex terrain, as well as for meteorological (including weather forecasting) models at regional, national, continental or global scale.

The significant computational demands of an atmospheric chemistry model can be reduced by choosing a Lagrangian framework instead of a Eulerian one. A Lagrangian model has a co-ordinate system that moves with a volume of air as it is blown from source to receptor. For the fastest calculation, a single trajectory can be followed, using wind speed and direction data either from a weather forecasting model or from a network of measurements. It can be followed in three dimensions, including information about vertical movement of air, or only horizontal motion can be followed with a column of air. A Lagrangian column model can assume the air is well mixed vertically throughout the atmospheric boundary layer, or divide this into two layers with a movable interface between the two, or consider a vertical stack of cells with a parameterisation for the turbulent vertical exchange of air between levels in the column.

A major advantage of a Lagrangian framework is that concentrations of a primary or secondary pollutant at a receptor location can easily be attributed to the emissions at each source location. For this reason, a Lagrangian model has been used for many years to study the oxidation of sulphur in Europe and to inform the development of optimised emissions abatement policies to protect plant life from acidification (Barrett and Seland, 1995), as will be discussed in Chapter 22. Recent developments, however, requiring increased emphasis on oxides of nitrogen and ozone, have brought about a gradual change to Eulerian modelling, and this is also the approach that is being adopted where the methodology is exported to Asia.

These atmospheric chemistry models of the European Monitoring and Evaluation Programme take their wind-field input from European-scale weather forecasting models, typically the Norwegian Meteorological Institute limited area model (Grønås and Hellevik, 1982). The global sulphate fields that were shown in Figure 3.5 are quite different, in that they were not generated by an atmospheric chemistry model, but were produced as part of the development of the parameterisation of sulphuric acid production within the UK's leading global climate model (Williams *et al.*, 2001). That model remains primarily very much a model of atmospheric physics, with only half a dozen or so chemical species typically used in any one model run, to avoid increasing computing time too much. The atmospheric photochemistry required to produce fields of spatially and seasonally varying hydroxyl radical and hydrogen peroxide concentrations is therefore modelled off-line by a separate stochastic Lagrangian model (Johnson *et al.*, 1999). At the time of writing, however, the coupling between the two is becoming closer, as the two models are becoming able to run almost simultaneously, exchanging data between each other using massively parallel supercomputer processing. We can expect to see similar developments occurring not only at global scale in climate modelling, but also down to regional and urban scale in air pollution modelling, including not only photo-oxidant chemistry but also the surface and solution chemistry and physics of the aerosol particles that are suspended in the air. The challenge then will be to integrate the application of such comprehensive models with the maintenance of sufficiently detailed emissions inventories and information about the impacts of pollution on receptors of interest, including plants, so that policy responses to remaining air pollution problems can be supported effectively by the exciting achievements in the science of atmospheric physics and chemistry that are already being made.

Closing remarks and conclusions

This chapter has considered air pollution from a wide range of sources, including measures that can be taken to reduce emissions which can be read in conjunction with Chapter 2 to assess past developments and indicate what future trends might be. The process of environmental improvement in Western Europe and North America has been step-wise, starting with end-of-pipe emissions abatement technology. Impacts of pollutants have been assessed one at a time, starting with the primary pollutants that are easiest to model, then moving on to consider a few secondary pollutants one at a time. In some respects, rapidly developing countries in Asia, Africa and South America, where local air pollution problems are often still increasing in acuteness, and the Eastern

European economies in transition from Communism, might simply follow several years behind the developments that have occurred in Western Europe and North America. We do now, however, have a global opportunity to make progress more effectively, learning from the experience of the wealthier countries. Clean technology solutions can be applied at the same time as end-of-pipe clean-up measures. Assessment tools are now becoming able to consider multiple impacts of primary and secondary pollutants simultaneously, combining modelling of physical and meteorological processes with gas-phase and cloud droplet chemistry. The challenge to preserve the remaining natural environments of the globe and to develop sustainable agriculture to feed a growing population is an exciting one for science and society to tackle together, with the impact of air pollution on plant life likely to play a small but increasingly prominent role during the coming 10 to 20 years.

References

Barrett, K. and Seland, O. *European transboundary air pollution, 10 years calculated fields and budgets to the end of the First Sulphur Protocol*, Report 1/95, EMEP/MSC-W, Norwegian Meteorological Institute, N-0313 Oslo 3, Norway, 1995.

Berkowicz, R., Sørensen, N.N. and Michelsen, J.A. Modelling air pollution from traffic in urban areas. In R.J. Perkins, and S.E. Belcher, (Eds.), *Flow and Dispersion through Groups of Obstacles* Oxford, Clarendon Press, 1997.

Brimblecombe, P., *Air Composition and Chemistry*. Cambridge University Press, 1986.

Buckingham, C., Clewley, L., Hutchinson, D., Sadler, L. and Shah, S. *London Atmospheric Emissions Inventory*. London Research Centre, 1997.

Carruthers, D.J., Holroyd, R.J., Hunt, J.C.R., Weng, W.S., Robins, A.G., Apsley, D.D., Thompson, D.J. and Smith, F.B. UK-ADMS, a new approach to modelling dispersion in the earth's atmospheric boundary layer. *Journal of Wind Engineering and Industrial Aerodynamics* **52**, 139–153, 1994.

Cimorelli, A.J., Perry, S.G., Venkatram, A., Weil, J.C., Paine, R.J., Wilson, R.B., Lee, R.F. and Peters, W.D. AERMOD, Description of Model, USEPA web site (http//www.epa.gov/scram001), 1998.

Clarke, R.H. *A Model for Short and Medium Range Dispersion of Radionuclides Released to the Atmosphere*. National Radiological Protection Board, NRPB R91, Harwell, UK, 1979.

CORINAIR, *Atmospheric Emission Inventory Guidebook*, European Environment Agency, DK-1050 Copenhagen, Denmark, 1996.

Friedrich, R. and Schwartz, U.-B. Emission inventories, Chapter 6 in Fenger, J., Hertel, O., and Palmgren F. (Eds.) *Urban Air Pollution – European Aspects*, Dordrecht, Kluwer, 1999.

Goodwin, J.W.L., Salway, A.G., Murrells, T.P., Dore, C.J. and Eggleton, H.S. *UK Emissions of Air Pollutants 1970 to 1997*, Report to the Department of the Environment, Transport and the Regions, National Environmental Technology Centre, Abingdon, 1999.

Grønås, S. and Hellevik, O., *A Limited Area Prediction Model at the Norwegian Meteorological Institute*. NMI Technical Report No. 66. Norwegian Meteorological Institute, Oslo, Norway, 1982.

Inglis, D.W.F., Choularton, T.W., Stromberg, I., Gardiner, B. and Hill, M. Testing of a linear airflow model for flow over complex terrain and subject to stable, structured stratification. In J. Grace and M.P. Coutts (Eds.) *Wind and Wind-Related Damage to Trees*. Cambridge University Press, 1994.

IPCC Intergovernmental Panel on Climate Change, Guidelines for National Greenhouse Gas Inventories (3 volumes), Cambridge University Press, 1994.

Johnson, C.E., Collins, W.J., Stevenson, D.S. and Derwent, R.G. Relative roles of climate and emissions changes on future tropospheric oxidant concentrations, *Journal of Geophysical Research*, **104**, 18631–18645, 1999.

Jones, A., Roberts, D.L., Woodage, M.J. and Johnson, C.E. Indirect sulphate aerosol forcing in a climate model with an interactive sulphur cycle, *Journal of Geophysical Research,* 2002 (in press).

MSC-W, *MSC-W Status Report 1996 Part One*, The Norwegian Meteorological Institute, N-0313 Oslo 3, Norway, 1996.

Pasquill, F., The estimation of the dispersion of windborne material, *Meteorol. Magazine* **90**, 33–49, 1961.

Pierson, W.R., Gertler, A.W., Robinson, N.F., Sagebiel, J.C., Zielinska, B., Bishop, G.A., Stedman, D.H., Zweidinger, R.B., and Ray, W.D. Real-world automotive emissions – summary of studies in the Fort McHenry and Tuscarora Mountain Tunnel. *Atmospheric Environment* **30**, 2233–2256, 1996.

Turner, D.B., *Workbook of Atmospheric Diffusion Estimates*, US Environmental Protection Agency Report 999-AP-26, Washington, DC, 1969.

USEPA (United States Environmental Protection Agency), *Guideline on Air Quality Models*. Appendix-W to Part 51 of 40CFR, 1978. (Revised edition of 1 July 1999 available at http//www.epa.gov/ttn/scram/-guidance/guide/appw_99.pdf.)

Williams, K.D., Jones, A., Roberts, D.L., Senior, C.A. and Woodage, M.J. The response of the climate system to the indirect effect of anthropogenic sulfate aerosol. *Climate Dynamics* **17**(11), 845–856, 2001.

4 Pollutant deposition and uptake by vegetation

D. FOWLER

Introduction

This chapter considers the exposure of vegetation to pollutants and methods of quantifying exposure. The closely related processes of deposition and uptake are then presented for the major gaseous and particulate pollutants and the major ions in precipitation and cloud. For some of the major gaseous pollutants including SO_2, and NO_2, the underlying mechanisms which control rates of deposition and the fluxes in field conditions, are known. For other equally important pollutants, including O_3, while we know from experimental studies the overall deposition rates and the relationship between simple indices of exposure and effects on dry matter yield (e.g. for wheat, Fuhrer *et al.*, 1997), the chemical and biochemical mechanisms which determine the overall flux and the fate of deposited O_3 molecules remain obscure. It is necessary therefore to describe the range of indices used to quantify exposure of vegetation to pollutants, and the current state of understanding of the processes which regulate fluxes and the fate of deposited pollutant. There is no single index of exposure that is appropriate for all pollutants or purposes. A knowledge of the deposition flux in rain and by dry deposition and their spatial distribution form an essential prerequisite to any control strategy.

The exposure of vegetation to air pollutants has been expressed using a range of different methods (Table 4.1). The simplest measure, the average concentration to which plants are exposed, conceals a range of complicating factors. First, the mean concentration of reactive pollutants varies with height above ground and through canopies of vegetation, due to deposition. Thus the concentration must be referenced to a specific height as different parts of a crop canopy experience a different concentration. The sensitivity of vegetation to pollutant also varies with time, phenology and a wide range of environmental variables (e.g. vapour pressure deficit, solar radiation and temperature). For many pollutant–plant interactions it is the peak concentrations that cause the majority of the injury, especially in the case of acute damage. Thus a measure of the peak concentrations during a specific exposure period has proved valuable as a measure of exposure. Recent developments of the concept of exposure of vegetation to peak concentrations as the most appropriate measure of pollution risk are commonly applied for ozone. In this case the product of concentration and time above

Air Pollution and Plant Life, second edition. Edited by J.N.B. Bell and M. Treshow. ISBN 0 471 49090 3 (HB), 0 471 49091 1 (PB).

Table 4.1 Measures of exposure and dose of pollutants to vegetation

		Units	Typical averaging period	
1.	Mean concentration	$\mu g\ m^{-3}$ or ppbV	Seasonal or annual	
2.	Peak concentration	$\mu g\ m^{-3}$ or ppbV	Seasonal or annual or 99%ile	
3.	Accumulated exposure to potentially damaging concentration	AOT_{40} (ppbh > 40pph), SUM06, SUM40	Growing Season	Cereals April–June April–Sept Trees
4.	Total acidifying input $\sum SO_x + NO_y + NH_x$	k eq H^{-1} ha^{-1}	Annual	
5.	Deposited flux of pollutant	kg ha^{-1} year Nitrogen, eutrophication	Year	

a threshold concentration has been used to assess risk of any damage. This measure of dose, with units of ppb.hours is dimensionally peculiar and has been referenced to a specific season and time of day to focus the measure on the period of greatest sensitivity of the vegetation.

The deposited flux of pollutant over a specific time (e.g. kg ha^{-1} $year^{-1}$) represents a more fundamental measure of dose and may be expressed per unit ground area or per unit area of foliage.

The major pollutants differ greatly in their chemical properties and their effects on plants. The different chemical reactivities and physical forms of the pollutants provide a convenient method of classifying their interactions with plant canopies for presentation in this chapter.

Plant atmosphere exchange processes

Atmospheric pollutants are transported to vegetation from their source by wind and turbulence. Wind advects pollutants over the landscape and transports pollutants from sources and source areas. The mean speed of transport in the planetary boundary layer varies between regions, but is typically in the range 5 to 10 ms^{-1}, thus horizontal transport over a day is typically 500 to 1000 km. During atmospheric transport the pollutants disperse under the action of turbulence, and chemical processes (mostly oxidation) transform the primary gaseous pollutants into the secondary pollutants, and often into aerosols. The transformation in chemical and physical form greatly influences rates of removal from the atmosphere by precipitation as *wet deposition*, and by direct deposition as gases and aerosols to terrestrial and marine surfaces, a process known as *dry deposition*.

The transport of gases and aerosols from the atmosphere to terrestrial surfaces is by turbulent transfer, which is generated by frictional drag by terrestrial surfaces on the wind. Thus the nature of the surface strongly influences rates of transfer. The aerodynamically rough surfaces of forests and woodland generate much greater frictional drag on airflow and, as a consequence, rates of transport of pollutants from the free atmosphere to the surface are much greater over forests than over short vegetation (e.g. grassland).

The turbulence generated by interaction of the wind with vegetation may be quantified by a turbulent diffusivity for momentum (K_m) or sensible heat (K_h), which may be estimated from direct measurements of the turbulent fluxes within the surface boundary layer above vegetation. The turbulent diffusivities vary with wind velocity and atmospheric stability as well as surface roughness (Z_0). The rates of deposition of pollutant gases and particles depend on both the turbulent transfer to the surface and processes at the surface which determine the uptake of gases or capture of particles (Figure 4.1). For gases which are highly reactive, such as nitric acid (HNO_3), reaction with the vegetation surfaces are rapid and the gas is deposited to terrestrial surfaces as quickly as atmospheric turbulence can deliver the molecules to absorbing surfaces. In these conditions the surface is considered a 'perfect sink' for the pollutant, and rates of deposition are determined solely by turbulent transfer.

Deposition rates are commonly quantified using a deposition velocity (V_d) in which

$$V_{d(z)} = \frac{\text{Flux } (f)}{\text{Concentrations } \chi_{(z)}} \text{ ms}^{-1} \tag{1}$$

The vertical flux (f) is constant with height within the surface boundary layer that develops over uniform areas of vegetation within a layer which is typically 1/100 of the horizontal extent of the uniform vegetation. Thus a rectangular field with the long axis of 300 m would have a constant flux layer approximately 3 m deep above the vegetation at the downward edge, provided that there were no major discontinuities in height at the field boundary (Thom, 1975). The concentration (χ) is referred to a specific height (z), because concentration within the constant flux layer declines

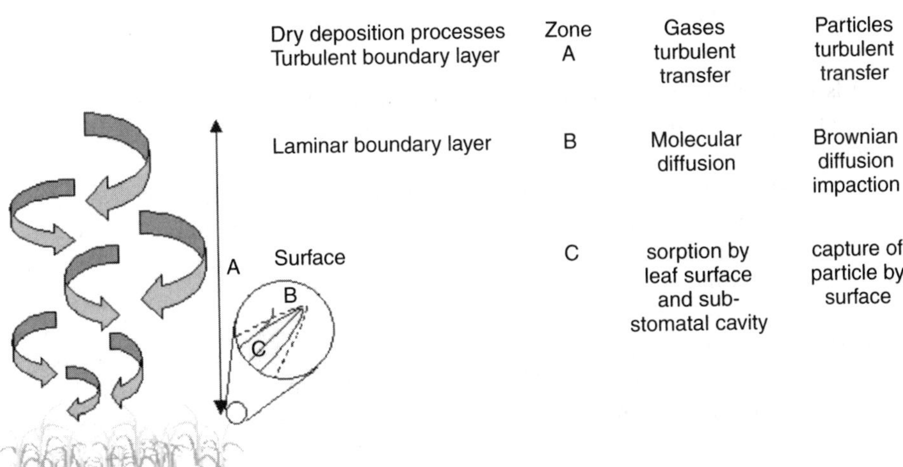

Figure 4.1 Transfer processes for gaseous and particulate pollutants from the free atmosphere to terrestrial surfaces

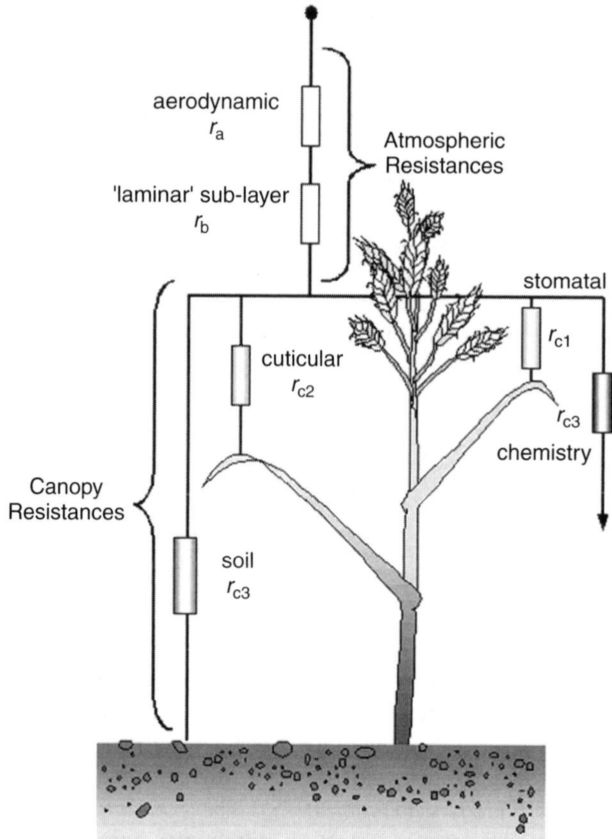

Figure 4.2 A simple resistance analogue to separate the effects of atmospheric and surface processes on rates of pollutant deposition to terrestrial surfaces

towards the surface as a result of deposition. The deposition velocity therefore varies with height and must be referenced to a specific height at which concentration is known.

Deposition velocity is commonly used to quantify rates of deposition and the deposited flux is simply the product of deposition velocity and ambient concentration from Equation (1). The underlying processes regulating deposition rates are commonly parameterised using a resistance analogy (Monteith and Unsworth, 1990). The total resistance (r_t) to transfer of pollutants from a reference height in the atmosphere to the surface is the reciprocal of the $V_{d(z)}$ and can be subdivided into an atmospheric resistance ($r_{a(z)}$) which quantifies the turbulent transfer through the turbulent boundary layer from a reference height (z) to the viscous sub-layer over leaf surfaces, r_b quantifies the viscous sub-layer resistance over leaf and other surfaces, and r_c the surface or canopy resistance quantifies the efficiency of the uptake at the surface (Figure 4.2). In the case of reactive gases which are readily absorbed by terrestrial surfaces, the canopy

Table 4.2 Potential effects of forests on deposition fluxes

		Moorland	Forest	% increase forest/moor
Canopy height	h (m)	0.15	10	–
Zero plane displacement	d (m)	0.1	7	–
Roughness length	z_o (m)	0.01	1.0	–
Friction velocity	$u*$ (m s^{-1})	0.32	0.82	156
Momentum flux	ϑ (N m^{-2})	131	840	541
Maximum deposition velocity for SO_2	V_{max} SO_2 (mm s^{-1})	18.6	35.1	89
Maximum deposition velocity for NO_2	V_{max} NO_2 (mm s^{-1})	20.0	43.5	118
Maximum deposition velocity for NH_3	V_{max} NH_3 (mm s^{-1})	21.4	55.5	160

or surface resistance is zero. For gases which are absorbed by stomata, and also deposit to external surfaces of vegetation, the canopy resistance may be subdivided to quantify each of the component terms illustrated in Figure 4.2. The challenge in this approach is in quantifying each component of the resistance network and their dependence on surface and atmospheric conditions.

The effect of different canopies of vegetation on the atmospheric components of deposition velocity may be illustrated by comparing deposition velocities calculated for short vegetation (e.g. moorland or grassland 0.15 m high) with those for a young forest 10 m in height (Table 4.2) in which canopy resistance is set to zero when the surface is considered a 'perfect sink'.

The maximum deposition velocities vary between the three gases due to differences in their molecular diffusivities, and hence r_b. However, all gases deposit rapidly (18 mm s^{-1} to 55 mm s^{-1}), and deposition rates onto forest are typically a factor of 2 larger than these onto short vegetation. Such large deposition velocities rapidly remove pollutants from the boundary layer. If for example we assume a planetary boundary layer 1 km deep, then deposition velocities of 18 mm s^{-1} to 55 mm s^{-1} would lead to lifetimes of the pollutant in the boundary layer of 15 hours and 5 hours, respectively.

It is important to note that turbulent transfer processes provide the vertical diffusion of all gaseous pollutants, and rates of molecular diffusion only become a significant contributor in the deposition process very close to absorbing surfaces (fractions of a mm) where viscous forces suppress turbulence. Thus in the control of r_b (Figure 4.2), molecular diffusivity is generally included in standard formulations (Monteith and Unsworth, 1990).

For particles, while turbulent transfer provides the bulk of the vertical transfer, gravitational sedimentation becomes progressively more important for particles larger than 5 μm in diameter (see later).

The pollutants

This chapter is focused on the major gaseous and particulate pollutants. However, no attempt has been made to be comprehensive in this treatment, as there are large numbers of more exotic, especially organic pollutants for which little is currently known about their deposition rates. The gases occur in their emitted state as primary pollutants including sulphur dioxide (SO_2), ammonia (NH_3), hydrogen chloride (HCl) and nitric oxide (NO). The oxidation of primary pollutants leads to aerosol, in the case of SO_2, as SO_4^{2-} either as the acid H_2SO_4, partially neutralised as NH_4HSO_4 or fully neutralised as $(NH_4)_2SO_4$. The emitted NO is readily oxidised by ozone to form NO_2. Other oxidation products of the atmospheric chemistry of NO_x include HONO, gaseous HNO_3 and aerosol phase NH_4NO_3 or HNO_3. Thus some pollutant gases are forced in the atmosphere through chemical processing of their primary pollutant precursors and are referred to as secondary pollutants.

The photochemical oxidants, and ozone (O_3) in particular, are secondary pollutants, formed through oxidation of volatile organic compounds (VOC) in the presence of the oxides of nitrogen NO and NO_2 (collectively NO_x). The photochemical production of O_3 in polluted air is described in detail by Wayne (2000) and in summer northern European boundary layer conditions, typical daily rates of ozone production lead to an increase in ambient concentrations of 10 to 30 ppb O_3. In slowly moving air advected away from source regions, photochemical oxidant episodes lasting 3 days may therefore lead to 30 ppb to 90 ppb of ozone in addition to the background concentration, which is generally in the range 20 to 30 ppb (PORG, 1997). Other pollutant gases produced by the photochemical degradation of VOC include hydrogen peroxide (H_2O_2) and peroxyacetyl nitrate (PAN). The volatile organic compounds include pollutants from industrial and motor vehicle sources and biogenic compounds from a wide range of vegetation.

The pollutant VOCs are numerous including alkanes, alkenes, aldehydes and aromatics which vary in reactivity. The major oxidant responsible for their removal from the atmosphere is the hydroxyl radical (OH) (PORG, 1994). However, these gases are generally deposited only very slowly at terrestrial surfaces, so that their lifetime in the atmosphere and fate are determined by chemical processing, and dry deposition represents only a minor sink. Similarly the biogenic gases including terpenes and isoprene are not subsequently deposited at significant rates. These organic compounds are not considered further in this chapter.

Particles

The secondary pollutant aerosols containing SO_4^{2-} NO_3^- and NH_4^+ comprise the majority of the sulphur and nitrogen exchanged over international borders. Thus the transformation to aerosol is of both political and scientific interest. The mass of these aerosols is primarily found in the size range 0.1 μm to 1.0 μm, and many field measurements report mass median diameters of approximately 0.5 μm for aerosol SO_4^{2-}. However, a large population of much smaller particles, the fine mode, is also present, contributing little total mass, and a small population of coarse aerosols, often of resuspended material from the surface (Figure 4.3). The bulk of the aerosol NH_4^+ and NO_3^- is found in the same sub-micron size range as the SO_4. The phase change from gas to aerosols for sulphur and nitrogen compounds forms highly efficient condensation nuclei for cloud droplet formation and are efficiently removed from the atmosphere by precipitation. However, these particles deposit quite slowly to terrestrial surfaces, so that in the absence of precipitation they have relatively long atmospheric lifetimes, extending to several days.

Among the atmospheric pollutants found largely in the particle phase are the heavy metals including Pb, Cd, Zn, Cu and the base cations Ca, Mg, Na, K. The potentially toxic heavy metals are emitted largely from metal processing industries and large combustion sources while the base cations derive from combustion sources and have substantial natural sources in the resuspension of soil particles by wind.

The toxic and combustion derived metals are primarily found in the sub-micron size range with mass median diameters between 0.1 μm and 0.5 μm while much of the resuspended particulate material is in the larger size range, 0.5 μm to 5.0 μm and consequently has a much shorter atmospheric lifetime and travel distance. This chapter is primarily devoted to the deposition of pollutants onto vegetation, but the spectrum of atmospheric pollutants extends from the gases and particles with short atmospheric lifetimes of hours to days to gases with very long lifetimes, including methane (CH_4) which has a lifetime of approximately 10 years. Vegetation plays an important role in CH_4 emission from anaerobic soils, including paddy rice and high latitude wetlands. In this case the role of the vegetation is as a conduit for CH_4 from the anaerobic zone of soil to the atmosphere, conducted through the specialised gas transport aerenchyma of vegetation, while oxygen flows in the opposite direction to maintain respiration masses in the plant root system. As in the case of the reactive short lived atmospheric pollutants, it is in regulating the rates of exchange of these gases at the interface between atmosphere and terrestrial surfaces, that vegetation exerts its main effect on the pollutants.

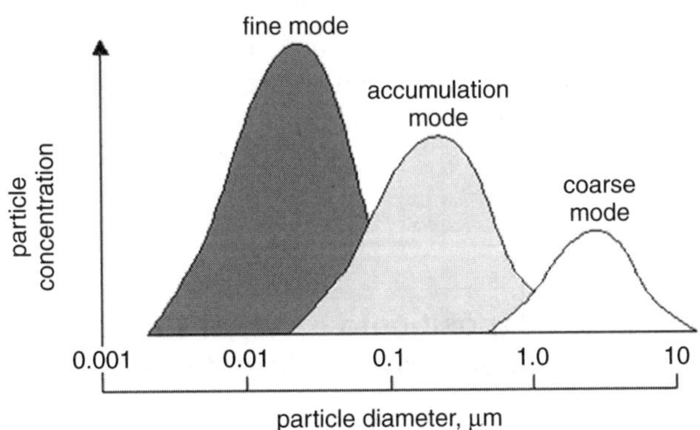

Figure 4.3 Approximate size ranges for the three aerosol modes

Highly reactive pollutant gases (HNO_3, HCl, HF)

The published literature on these gases is limited, but from the available field data, rates of deposition are controlled almost entirely by transfer from the atmosphere to foliar surfaces and hence turbulence (Huebert and Roberts, 1995). The canopy, or surface resistance for HNO_3, HCl and HF may be therefore considered negligible for most practical purposes and the surface 'perfect sinks'. With this property of vegetation, the calculation of rates of deposition, is greatly simplified because the relationships between the atmospheric resistances r_a and r_b and wind speed and height of vegetation are well documented (Monteith and Unsworth, 1990), and

$$V_{d(max)} = (r_a + r_b)^{-1} \tag{2}$$

The maximum value for deposition velocity for a given wind speed and vegetation height ($V_{d(max)}$) is given for moorland and forests in Table 4.2, showing the effect of canopy height (and therefore aerodynamic roughness) on deposition rates of these reactive gases. With a rate of deposition of 40 mm s^{-1} onto a woodland or forest, and a boundary layer 1000 m deep, the atmospheric lifetime for HNO_3, HCl or HF would be just 7 hours. The rapid deposition to vegetation, efficient uptake by precipitation and efficient gas to particle formation in the presence of NH_3, keep the ambient concentration of gases small. Concentrations of HNO_3 in Europe vary between a few ng m^{-3} and a few μg m^{-3} with a pronounced seasonal cycle and summer maximum (Erisman *et al.*, 1998). There is also a clear geographical distribution in Europe with smallest concentrations in the Atlantic coastline and northerly countries and largest concentrations in southern Europe. The large deposition velocities for HNO_3 result in significant inputs of nitrogen to forest even in regions with ambient concentrations of 1 μg HNO_3 m^{-3} such as central Holland. The deposited flux is simply the product of deposition velocity and ambient concentration, in appropriate units. Thus annual inputs of HNO_3 to forest in the Netherlands, with concentrations of about 1 μg m^{-3} HNO_3 would be of the order 3 kg N ha^{-1}, whereas for much shorter vegetation (e.g. *Calluna*) the input would be about 1 kg HNO_3-N ha^{-1} annually. The dependence of deposition rates of HNO_3, HCl and HF on turbulence leads to marked variation of deposition rate with wind speed, vegetation height and atmospheric stability as illustrated in Table 4.2, showing deposition velocity increasing by about a factor of 2 for an increase in vegetation height from 30 cm to 10 m. For short vegetation, an increase of wind speed from 1 m s^{-1} to 4 m s^{-1} is required to double the deposition velocity.

Pollutant gases which deposit onto external surfaces of vegetation and are taken up or emitted through stomata (SO_2, NH_3, O_3)

Three of the most important pollutant gases, SO_2, NH_3 and O_3 react with the external surfaces of vegetation and are also exchanged with the internal surfaces of foliage following diffusion through stomata. As such, all three gases exhibit a canopy resistance, and for most surface and atmospheric conditions the canopy resistance (r_c) exceeds the aerodynamic and leaf boundary layer resistances, r_a and r_b, respectively. Thus, the overall rates of deposition are substantially smaller than those of HNO_3 or HCl, and are controlled mainly by processes at the surface and are much less sensitive to wind speed or vegetation height. The three gases differ considerably in their chemical properties and interactions with vegetation and are therefore considered separately.

SO_2

Among the major pollutant gases the deposition rates of SO_2 onto vegetation have attracted most interest. Early studies of ambient concentrations and the mass balance of sulphur compounds over the UK by Meetham (1950), showed that dry deposition of SO_2 to the landscape must be a major removal process. From this analysis a deposition velocity of about 1.5 cm s^{-1} may be deduced. Direct measurements of SO_2 deposition to grassland (Garland *et al.*, 1973), cereal crops (Fowler and Unsworth, 1979) and forest (Erisman *et al.*, 1994) have shown the major characteristics of the deposition process. The measurements show that rates of SO_2 deposition to canopies of vegetation vary from 1 mm s^{-1} to 20 mm s^{-1} and comprise two major sinks. Stomata, which absorb SO_2 at rates consistent with those expected from measurements of stomatal conductance to water vapour, allowing for the different molecular diffusivities of SO_2 and H_2O, at least in field conditions, comprise the first sink. The rate of stomatal uptake of SO_2 can therefore be calculated from stomatal conductance and ambient SO_2 concentration.

The second sink for SO_2 is the external surface of vegetation consisting of epicuticular wax, surface debris and very frequently a layer of water. The water layer may be just a few molecules deep but is sufficient to regulate the rate of reaction of SO_2 with leaf surfaces. The variability in quantity and composition of the surface water film leads to huge variability in the rate of SO_2 uptake into foliar surfaces. The fraction of the time that foliar surfaces are wet with rain or dew in the climate of NW Europe varies between 60% and 80% depending in part on the characteristics of the vegetation (tall vegetation dries faster than short vegetation given a similar quantity of intercepted water). The depth and leaf area index and distribution also influence the 'wetness' of the vegetation. The ionic composition of the surface water then determines the solubility and fate of the dissolved SO_2.

The processes of turbulent transport to the vegetation, solution and reaction and fate of deported SO_4^{2-} have been simulated using a process-based model by Flechard *et al.* (1999). The model, illustrated schematically in Figure 4.4 shows the deposition of SO_2 to be regulated in part by the presence of ambient NH_3 and NH_4 in solution and by solution pH. Incorporating the major ion chemistry of the liquid film the model demands a thorough knowledge of leaf surface chemistry to initialise. This is provided by measuring the ionic composition of precipitation events and assuming that at the time precipitation ceases, the ionic composition of water film on vegetation is identical to that measured in an adjacent precipitation collector. The dynamic model simulates continuous fluxes and concentrations of SO_2 and related species for several days. The comparison between measured and modelled fluxes at Auchencorth Moss in the Scottish borders from the work of Flechard *et al.* (2000) shows good agreement in both the mean rates of exchange and its temporal variability. These recent data confirm earlier work which suggested a role for NH_3 in regulating SO_2 deposition (van Hove, 1989) based on laboratory studies, but which has been difficult to demonstrate in the field mainly due to the difficulty in measuring NH_3 concentrations and fluxes.

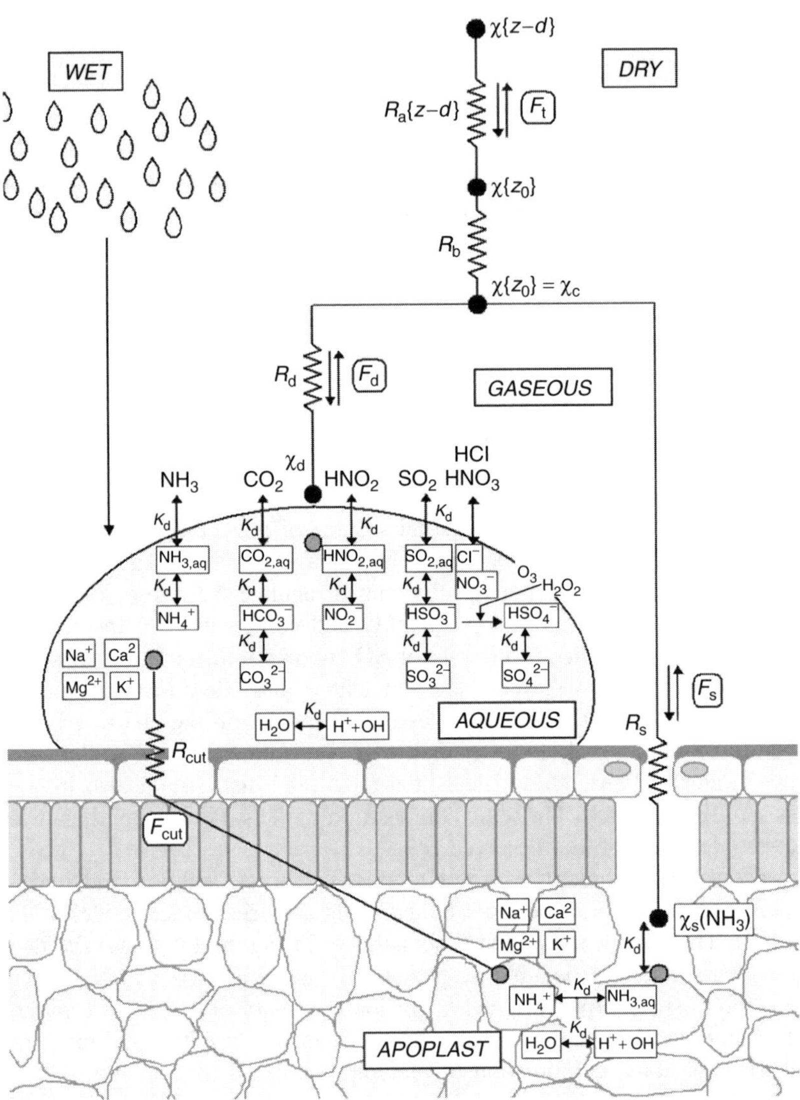

Figure 4.4 A schematic representation of the exchange of NH_3 and SO_2 between the atmosphere and vegetation including leaf surface chemical processes and stomatal uptake (from Flechard *et al.*, 2000)

The control of SO_2 deposition to vegetation by NH_3 and the composition of surface water film have several important implications. First, the adoption of simple surface resistance parameters to model dry deposition over country or regional scales is only appropriate if the ambient concentrations of the major gases which influence r_c are reasonably constant. Clearly with very different characteristics of SO_2 and NH_3 sources, the former from large combustion sources including power stations and refineries while the latter is largely emitted from livestock farms, the concentration ratio NH_3/SO_2 exhibits large spatial variability. It is clear that spatial patterns in the average deposition velocity for SO_2 in vegetation as a consequence of different relative concentrations of SO_2 and NH_3 are present, and that accurate estimates of SO_2 dry deposition require these effects to be quantified explicitly. A further and important consequence of the effect of NH_3 on SO_2 deposition (which has been widely referred to as co-deposition) is that the deposition velocity to the countryside has changed as the emissions and concentrations of SO_2 have declined in the UK and over large parts of polluted northern Europe and North America. As an example, in the East Midlands of the UK in the mid-1970s the average canopy resistance to SO_2 deposition on cereal cropland was 130 s m^{-1} in an area experiencing annual mean SO_2 concentrations of 20 ppb. By 1998 and 1999, concentrations had declined to between 1 ppb and 2 ppb of SO_2 and canopy resistance had declined to 80 s m^{-1}, increasing the deposition velocity by 40% (Fowler *et al.*, in press).

The underlying exchange processes for SO_2 are therefore reasonably well understood and given adequate data to define the concentration fields for SO_2 and the major gases which influence its exchange rate especially (NH_3), and the frequency and composition of surface water on vegetation, the calculation of deposition velocity and deposition fluxes is straightforward.

Process-based models of sulphur deposition of varying complexity form the basis for SO_2 deposition estimates at the regional scale in Europe (EMEP, 1998) and North America (Finkelstein, in press). These are generally satisfactory to define the large-scale average deposition rate, but they do not allow for differences in deposition velocity due to smaller scale variability in canopy resistance. Such variability occurs at a range of scales from tens of metres to 100 km, but at the smaller, country scale, e.g. Netherlands or UK, it has proved necessary to include land use specific canopy resistance formulations to obtain a satisfactory estimate of SO_2 deposition (Erisman *et al.*, in press; Smith *et al.*, 2000).

Ambient concentrations of SO_2 in Europe and North America have declined substantially during the last three decades. In the UK for example large areas of central England experienced annual average concentrations in excess of 50 μg m^{-3} throughout the 1950s and 1960s and concentrations in these areas are now typically 5 μg m^{-3} (Figure 4.5). The main interests in these concentrations were in effects on human health, and damage to vegetation and buildings. The deposition of atmospheric sulphur to cropland also provided the nutritional requirements for the majority of crops. With the decline in concentrations and deposition of atmospheric sulphur, the potential for effects on higher plants has largely disappeared, and the requirement for fertiliser application of sulphur to sustain crop yield and quality has emerged.

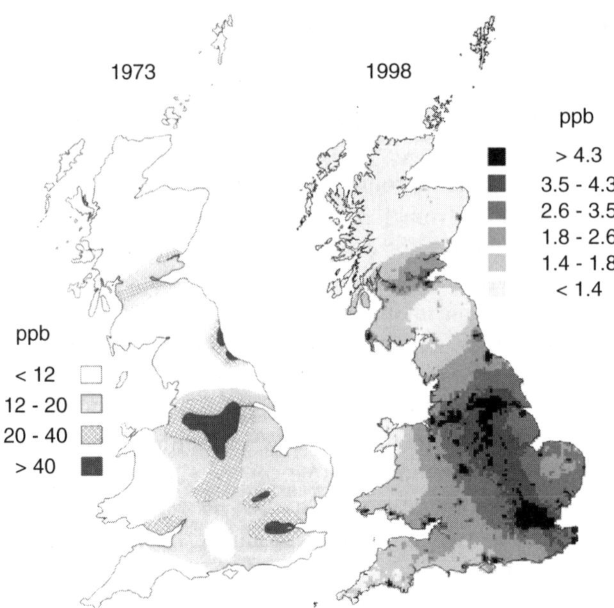

Figure 4.5 Rural concentrations of SO_2 in Britain in 1973 and 1998 showing the large decline in values over the 25-year period

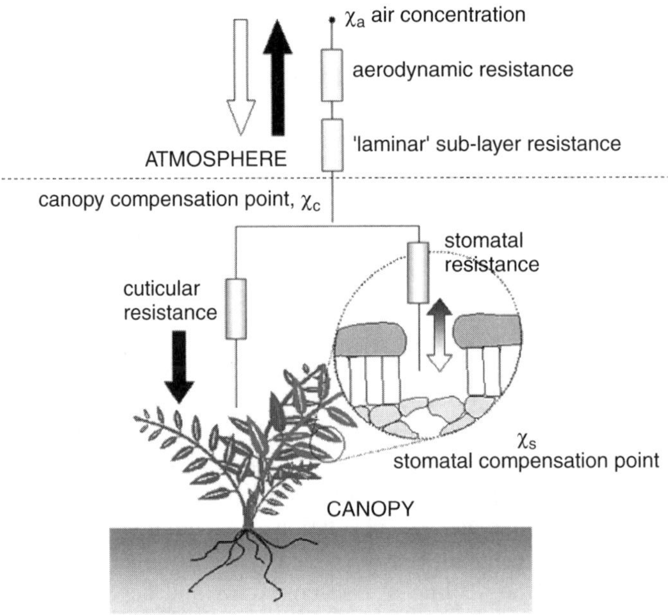

Figure 4.6 A canopy compensation point model used to simulate NH_3 exchange between vegetation and the atmosphere

Land-atmosphere exchange of NH_3

Atmospheric ammonia (NH_3) occurs as a consequence of natural processes following the emission from animal waste and from vegetation (Sutton *et al.*, 1995b). Unlike SO_2, vegetation may be either a source or sink of atmospheric NH_3 depending on the difference between ambient NH_3 concentration and the equilibrium NH_3 concentration within sub-stomatal cavities due to the presence of NH_4^+in the liquid phase leaf intercellular apoplast. The exchange of NH_3 between vegetation and the atmosphere may be regarded as analogous to that of CO_2 exchange in which a compensation point defines the concentration at which no net exchange occurs. For atmospheric concentrations in excess of the compensation point, NH_3 is deposited and at smaller concentrations NH_3 is emitted. The process is illustrated in Figure 4.6, and is shown to contain two compensation points. The first, a stomatal compensation point quantifies the net exchange through the stomatal aperture. However, the molecules of NH_3 emitted through a stomatal aperture may be absorbed by the external surfaces of vegetation, especially if wet, and so that the NH_3 emitted may not be emitted from the vegetation canopy. Thus a second compensation point is defined as the canopy compensation point to quantify the ambient concentration above which NH_3 is emitted into the planetary boundary layer. By combining the compensation point concentrations with resistance parameters for aerodynamic and boundary layer terms (Figure 4.2) the processes may be finalised in a simple big-leaf model (Sutton *et al.*, 1995a). The canopy compensation point χ_c may be written

$$\chi_{\mathrm{c}} = \frac{\chi\{z-d\}/(r_{\mathrm{a}}\{z-d\}+r_{\mathrm{b}}) + \chi_{\mathrm{s}}/r_{\mathrm{s}} + \chi_{\mathrm{d}}/r_{\mathrm{d}}}{(r_{\mathrm{a}}\{z-d\}+r_{\mathrm{b}})^{-1} + r_{\mathrm{s}}^{-1} + r_{\mathrm{d}}^{-1}} \tag{3}$$

in which most terms are defined earlier or within Figure 4.4 with the exception of r_{d}, the cuticular adsorption resistance (Flechard *et al.*, 1999).

The net exchange of NH_3 is also strongly influenced by the physical climate. In particular the partitioning between liquid phase NH_4^+ in apoplast solution and the equilibrium gas phase NH_3 in the atmosphere changes with temperature. The effect is illustrated by some data for barley in which the net NH_3 exchange measured in controlled conditions is plotted against leaf temperature (Figure 4.7). The leaf is a sink for NH_3 at temperatures below 20 °C and is a source at higher temperatures. This very strong dependence of the compensation points on temperature, means that field crops may be absorbing NH_3 at night and early in the morning and emitting NH_3 in the warmer afternoon hours. Such variations with temperature are readily simulated but again illustrate the quickly changing dynamics of this gas.

The simple canopy compensation point models have been shown to simulate emission from and deposition of NH_3 to plant canopies (Sutton *et al.*, 1995a) and to be preferable to simple resistance models of the type shown in Figure 4.2, which are unable to simulate NH_3 emission. However, even the single layer canopy compensation model shown in Figure 4.6 is necessarily simplistic. In practice conditions within-plant canopies vary with height above the ground and multi-layer approaches have been developed to simulate within-canopy and canopy–atmosphere exchange processes for field crops. Measurements and model simulations using a multi-layer

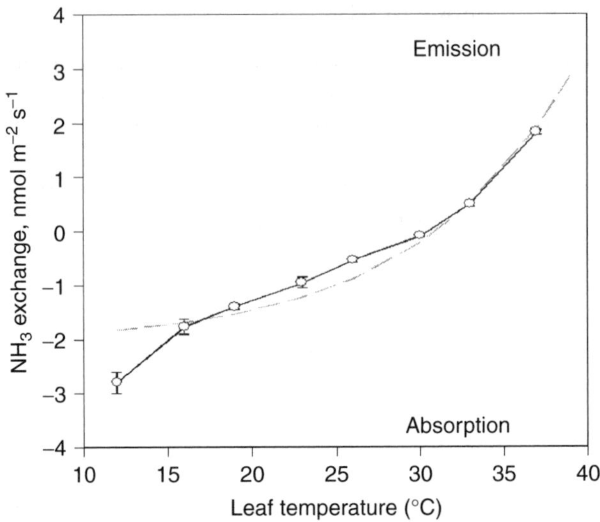

Figure 4.7 The effect of leaf temperature on net vegetation–atmosphere exchange of NH_3 from the work of Schjorring *et al.* (1998)

approach by Nemitz *et al.* (2001) show that, for an oil-seed rape canopy, within-canopy cycling of NH_3 may be an important feature of agricultural crops. The development of multi-layer approaches illustrate the natural development of models to simulate the complexity of NH_3 exchange in the field. The development of a detailed chemical treatment of NH_3 interaction with water on the external surfaces of vegetation (Flechard *et al.*, 1999) represents another important development in understanding. Both new approaches described briefly here to simulate NH_3 exchange illustrate the complexity of the processes and the requirement for a very detailed knowledge of the plant canopy, chemical composition of the boundary layer and the surface water on vegetation to simulate the net exchange of NH_3. For extrapolation to larger scales the information required for very detailed model insulation is not available. The models used to estimate the net vegetation–atmosphere exchange of NH_3 at the UK, Europe or North America scale are therefore much simpler. To date the only country using canopy compensation point models to quantify country scale NH_3 exchange is the UK (Smith *et al.*, 2000), although experimental studies have been reported for a range of other European countries (Spindler *et al.*, 2001).

As concentrations and deposition of sulphur dioxide and oxides of nitrogen decline due to control measures, NH_3 is becoming more important as a contributor to eutrophication and in some areas, of acidifying inputs to soil. In the UK for example the deposition, wet and dry, of reduced nitrogen as NH_3 and NH_4^+ represents 60% of the total countrywide annual nitrogen deposition (Figure 4.8), even though it represents only 27% of the total emissions of nitrogen during the period 1992 to 1994 (Figure 4.8). The deposition of NH_3 is therefore a very important component of the fixed nitrogen input to vegetation. The current interest in effects of deposited atmospheric nitrogen

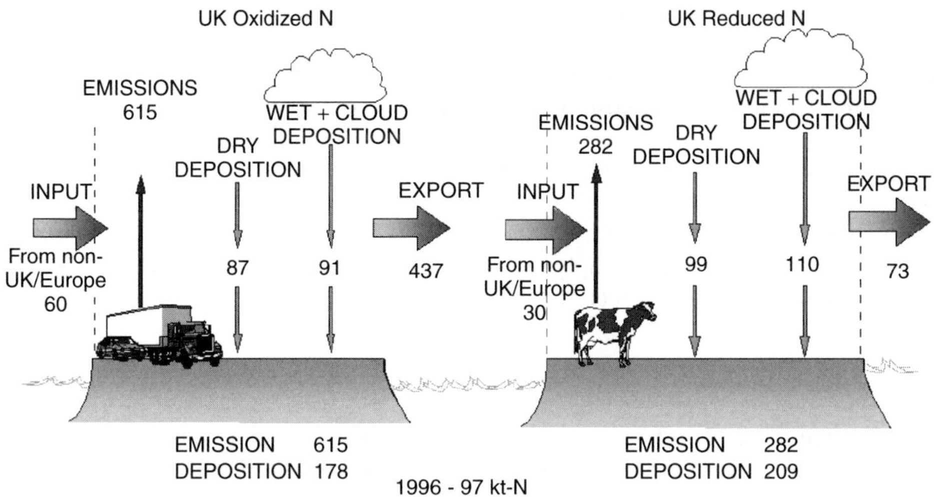

Figure 4.8 The annual mass balance for oxidised and reduced nitrogen within the boundary layer over the UK averaged over the period 1996 to 1997, all values in kton–N

Table 4.3 Nitrogen deposition to the major land uses in the UK

Land use	Area 10^6 ha	(Gg^9) Total deposited nitrogen	Kg N ha^{-1} Annual N deposit	% NH_x
Moorland/heath/bog	7.9	124	16	65
Forest	2.0	68	33	78
Grassland (fertilised)	6.5	98	15	56
Arable crops	7.9	124	16	65

on vegetation is driven mainly by the potential for changes in species composition of semi-natural plant communities and of the loss of species which are adapted to a small supply of fixed atmospheric nitrogen, often substantially less than 5 hg N ha^{-1} annually (see Chapter 12). The semi-natural plant communities, which are not fertilised and have smaller compensation point NH_3 concentrations therefore receive a disproportionately large fraction of the deposited nitrogen. In the largely arable landscape of the East Midlands for example fields of arable crops may receive a small fraction of the nitrogen deposition into shelterbelt woodland or a nature reserve in the same region. The differences in NH_3 deposition to different species and land uses depend on the compensation point and the local ambient concentration, but 'average' deposition to 5 km × 5 km grid squares of the landscape can be misleading in the assessment of efforts. A guide to the magnitude of nitrogen deposition on the major land uses in the UK and the relative importance of reduced and oxidised nitrogen is provided in Table 4.3.

The mechanisms controlling NH_3 deposition onto vegetation are reasonably well described, but like SO_2, the detailed modelling requires much more information than is generally available to quantify very fine scale deposition (~1 km). The current spatial resolution of NH_3 deposition maps in the UK is 5 km × 5 km, and even at that scale many general assumptions are required for simulation. In particular the very large number of sources of NH_3 from agricultural activities creates a concentration field for NH_3 which is very spatially variable (Sutton *et al.*, 1998). With this scale of variability, the monitoring networks (70 sites in the UK) are used to validate concentration fields provided by models, there is simply too much spatial variability to produce satisfactory maps from direct measurements.

Ozone deposition

The removal of ozone (O_3) from the boundary layer by dry deposition to the ground represents the major process of removal of O_3 from the boundary layer. Rates of deposition range from 1 mm s^{-1} to 20 mm s^{-1} and strong diurnal and seasonal cycles in deposition rate are observed (PORG, 1997). Effects of ozone on vegetation have been largely focused on ambient concentration and a wide range of indices of ozone exposure have been developed. These usually include a threshold which is considered useful in expressing the potential phytotoxicity of this gas. Examples of simple indices of O_3 exposure include the AOT40 (accumulated time over a threshold of 40 ppb) which has units of ppb.hours, and SUM06 or SUM60 in which the threshold is 60 ppb. There is usually a time window included to focus the unit on a physiologically appropriate period of the day or growing season. The major problem with these indices is that the vegetation exposed to O_3 may or may not absorb the gas depending on stomatal conductance. In fact, during periods of large O_3 concentrations the surface concentrations may be large partly as a consequence of reduced stomatal conductance. The flux of ozone through stomata is therefore a much better measure of dose.

In addition to stomatal uptake, O_3 also reacts with the external surfaces of vegetation, although unlike the case of SO_2 and NH_3, water layers do not enhance deposition rates. The deposition pathway for O_3 therefore has two routes, one which is very important for effects on the growth and physiology through stomatal uptake and a second onto external surfaces for vegetation. Stomatal uptake of O_3 at the canopy level may be readily estimated from measurements of stomatal conductance for water vapour assuming the absence of a significant internal or mesophyll resistance. Recent work to develop a flux-based approach to quantify effects of O_3 on cups and semi-natural vegetation (Emberson *et al.*, 2000) has quantified stomatal fluxes of ozone for crops throughout Europe.

Measurement of O_3 fluxes in the field by micrometeorological methods is a straightforward task (Fowler and Duyzer, 1989). However, separation of the two major O_3 deposition pathways requires independent measurements of the water vapour flux, and therefore stomatal conductance.

An example of the partitioning of stomatal and non-stomatal O_3 flux is provided by Fowler *et al.* (1999), for long-term flux measurements over moorland vegetation. The deposited flux may at its simplest, be crudely separated by assuming that nocturnal deposition is entirely non-stomatal to quantify the non-stomatal component of the resistance network, r_{c2} illustrated in Figure 4.2. Applying this value of r_{c2} to the entire day provides a very simple method of quantifying the relative importance of these two sinks in vegetation. The data (Figure 4.9) from the measurements over moorland at Auchencorth Moss in southern Scotland shows over a 2-year period that non-stomatal deposition represents the majority (65%) of the total deposited flux.

Until very recently the importance of non-stomatal O_3 deposition for effects in vegetation and for determining the atmospheric lifetime of this important photochemical oxidant have been ignored. However, if as seems very likely from the measurements reported over Auchencorth moss, the non-stomatal O_3 deposition is generally a large, if not dominant term in the deposition flux, then this term is very important whether or not the deposited O_3 has any phytotoxic effect. Strictly, the latter topic lies outside the scope of this chapter, but it seems probable that large long-term ozone fluxes to the epicuticular wax will be one of the processes which regulate the lifetime of this interface between plants and the atmosphere. The precise chemistry of O_3 interaction with external surfaces of vegetation is unknown and is a current focus for research. From recent measurements, Fowler *et al.* (2001) show that the resistance components r_{c2} decreases with ambient temperature and solar radiation. This effect has also been observed by Coe *et al.* (1995) and by Rondon *et al.* (1993), who speculatively suggest a photochemical O_3 destruction at foliar surfaces. However, from a simple analysis of the data from Auchencorth Moss, the reduction in r_{c2} with increasing temperature may be regarded as evidence of thermal decomposition of O_3 at the leaf surfaces which therefore increases with leaf surface temperature with an activation energy of 50 kJ mol^{-1} as shown in Figure 4.10, (Fowler *et al.*, 2001).

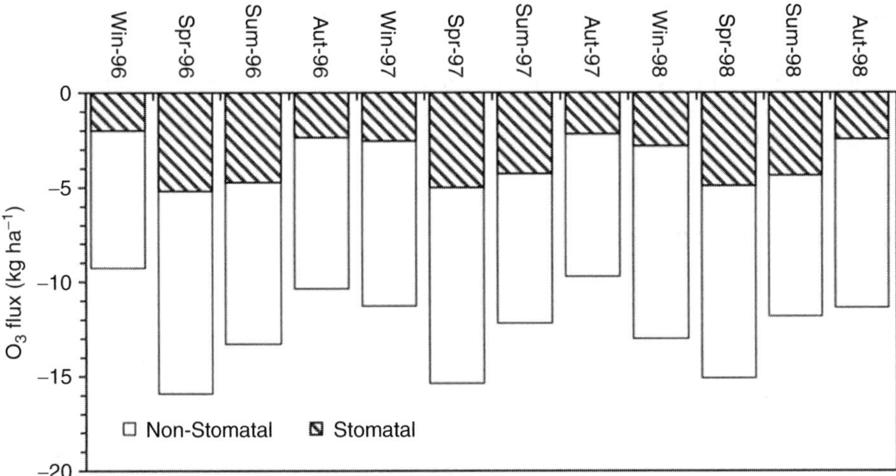

Figure 4.9 The measured ozone flux to moorland partitioned between stomatal and non-stomatal sinks from work of Fowler *et al.* (1999)

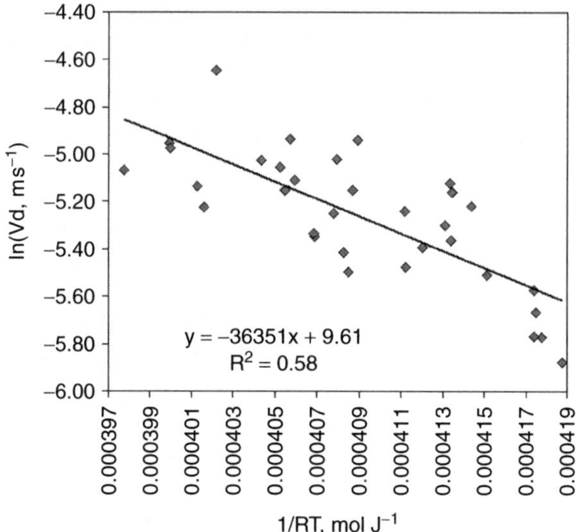

Figure 4.10 The effect of temperature on deposition velocity, illustrated using an Arrhenius analysis which implied an activation energy for O_3 thermal decomposition terrestrial surfaces of approximately 50 kJ mol^{-1} (from Fowler *et al.*, 2001)

Ozone is not conserved following deposition, and because it is not emitted there is generally no reason to present a mass balance over extended regions or time. However, it is valuable to provide a general context for the magnitude of the fluxes of O_3 relative to those of other major pollutants. The O_3 flux deposited to the UK amounts to 30 Gmol (cf. 1484 kT O_3) annually and compares with sulphur deposition of 14 Gmol (448 kT) and with total nitrogen deposition of 27 Gmol (378 kT). Tropospheric O_3 concentrations throughout the mid-northern latitude are predicted to rise through the coming decades as precursors throughout developing regions increase emissions of O_3 precursors (VOC and NO_x) (Stevenson *et al.*, 1998). The response of the processes at the surface to increasing concentrations may represent an important feedback in the atmospheric lifetime of O_3.

NO_2 deposition to vegetation

Emissions of NO_x, largely as NO from combustion are readily oxidised to NO_2 in the boundary layer by ozone. The NO_2 formed deposits onto vegetation, following uptake by stomata. Thus the resistance network for NO_2 is simple and follows Figure 4.2, with stomata representing the only significant sink within vegetation. Field data by Hanson and Lindberg (1991) show that NO_2 deposition to a range of tree species is regulated largely by stomatal conductance, that there is no uptake onto external surfaces of vegetation and that there is no internal resistance within the vegetation. Thus a knowledge of stomatal conductance (or resistance) and ambient NO_2 concentration is sufficient to quantify deposition velocities for NO_2. Similar conclusions were reached by Hargreaves *et al.* (1992) for a mixed species composition grassland. These field data are well supported by laboratory measurements showing NO_2 deposition is largely under stomatal control. Thus NO_2 deposition is a relatively simple process and provided that NO_2 concentration and the bulk canopy resistance (or conductance) for NO_2 can be estimated then good long-term estimates of NO_2 deposition can be provided.

The greatest complication in quantifying NO_x exchange between vegetation and the atmosphere arises through the role of nitric oxide NO, which is emitted from soils, especially agricultural soils and the exchange between NO and NO_2 within plant canopies.

The release of NO from soil has been widely measured and results from nitrification and denifrication processes in soil (Skiba *et al.*, 1992). Following release from soil into the plant canopy NO reacts with O_3 to form NO_2 which is then absorbed by stomata (Figure 4.11). This closed cycle does not lead to net release from the canopy, but does interact with NO_2 transfer from the free atmosphere to the surface, in generating an effective canopy compensation point. At very clean air locations with small ambient NO_2 concentrations, net NO_x emission is occasionally observed during the warmest hours of the day when the emission of NO from soil may exceed the capacity of the canopy to oxidise and absorb the NO_2. Extrapolation of NO_2 deposition to regional and annual scales provides annual inputs of the order 0.2 to 2 kg NO_2-N ha^{-1} and generally of a similar order to those of wet NO_3^--N deposition. Other reactive oxides of nitrogen may contribute to the input to vegetation and include gases such as PAN (peroxyacetyl nitrate) and nitrous acid (HONO) but in the case of the former, ambient concentrations are very small and deposition velocities are also small and dry deposition is therefore of little importance. For HONO, exchange at terrestrial surfaces may be

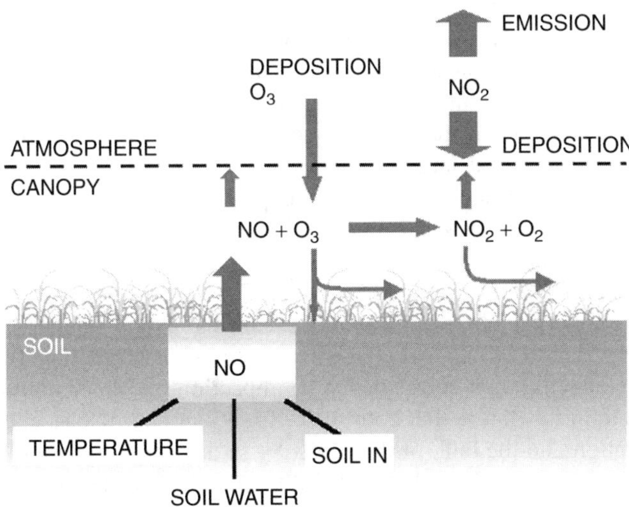

Figure 4.11 The sources and sinks for NO and NO_2 within crop canopies and interactions with O_3

much more important, but so far it is unclear whether terrestrial surfaces represent a significant sink or source due to HONO formation at the surface from NO_2 (Harrison *et al.*, 1994).

Aerosol deposition to vegetation

A substantial proportion of the primary gaseous pollutants including SO_2 and NH_3 are transformed to particle phase secondary pollutants prior to removal from the atmosphere. The phase change has profound effects on rates of removal by precipitation (wet deposition) and on the dry deposition process. The majority of long-range transport of sulphur and its atmospheric exchange between countries occurs as $SO_4{}^{2-}$ aerosol. Similarly, for reduced nitrogen the long-range transport occurs mainly as $NH_4{}^+$ in aerosols. The main reason for the different behaviour of aerosol and gaseous sulphur and reduced nitrogen is that the aerosols have longer atmospheric lifetimes than their gaseous precursors, due to slower rates of dry deposition. The underlying differences in rates of dry deposition are subtle and vary considerably with land use.

Aerosols contain not only sulphur and nitrogen but most of the heavy metals (Pb, Zn, Cu, Mn, Cr) and base cations Ca^{2+} Mg^{2+} Na^+ present in the atmosphere. Aerosols have become a very important focus of interest in climate change through their role in determining the albedo of the atmosphere and in influencing the droplet number concentration in cloud and thus their radiative properties (IPCC, 1997). Lastly aerosols have been identified as the major component of urban air pollutants associated with human health effects in Europe and North America.

It is not surprising therefore that interest in the deposition rate of aerosols onto vegetation is growing rapidly. The aerosols present in the atmosphere vary in size from a few nm to 10 μm in diameter (Figure 4.3). The bulk of the mass, especially of sulphur and nitrogen compounds is contained in the size range 0.1 to 1.0 μm, often referred to as the accumulation mode and for $SO_4{}^{2-}$, $NH_4{}^+$ and aerosol $NO_3{}^-$ the mass median diameter is typically about 0.5 μm (dia). The heavy metals are generally found in smaller size range (<0.1 μm dia) and referred to as the fine fraction, while the base cations are in the coarse mode (<1 μm), as illustrated in Figure 4.3. Conventional wisdom, based largely on the measurements of particle deposition on vegetation in wind tunnels shows aerosol dry deposition on vegetation to vary with the size of the aerosol with the aerodynamic properties of vegetation and with turbulence (Chamberlain, 1991).

The minimum deposition velocity for aerosols in the deposition velocity–aerosol size relationship for short vegetation is at about 0.5 μm (dia) and at smaller sizes deposition velocity increases due to diffusional effects, while for sizes larger than 0.5 μm, aerosol deposition rates increase as a consequence of impaction and interception as shown in Figure 4.12. Note in this figure that both scales for V_d and particle size are logarithmic. For short vegetation and the bulk of the aerosol size distribution, deposition velocity is therefore small, (<1 mm s^{-1}) and substantially smaller than deposition velocities for the precursor gases SO_2, NO_2 and NH_3.

For aerodynamically rough vegetation such as hedgerows, isolated trees, shelter belt, woodland and even for continuous forest cover, recent field measurements show

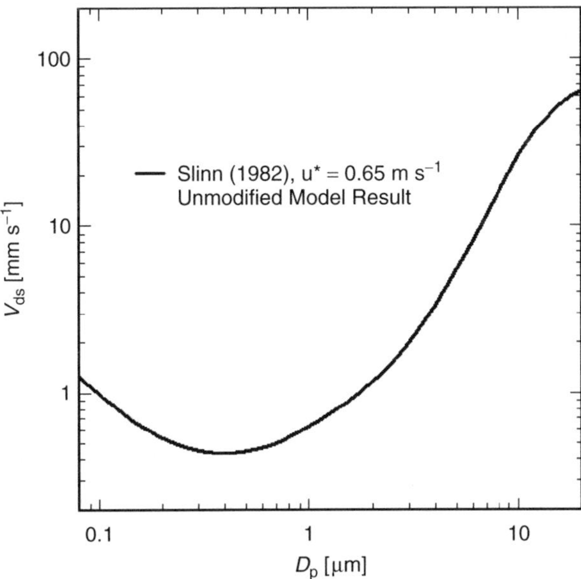

Figure 4.12 The relationship between deposition velocity for aerosol (v_{ds}) and particle size (D_p) from the Slinn (1982) model (with $U_* = 0.65$ m s^{-1})

deposition rates of a range of aerosols onto forests which are substantially larger than onto short vegetation. The comparison shows reasonable agreement at 0.1 μm and in the size range 3 to 10 μm dia, while for the size range 0.3 μm to 2 μm, field data suggest deposition velocities about an order of magnitude larger than those of the Slinn model (1982). To date, there has been no mechanistic basis provided in support of the larger values and review papers of the data suggest measurement artefacts for the large values (Garland, in press). It remains unclear why the earlier wind-tunnel values on which the Slinn model is largely based are so much smaller than the field data. It seems, however, that the weight of empirical evidence is in favour of larger deposition velocities for trees, especially isolated or hedgerow trees. An important implication of this result is that urban trees play a much more important role in removal of aerosols from the atmosphere than has earlier been suggested.

Cloud droplet deposition

The only clouds which deposit directly are fogs and orographic clouds which envelop uplands in the cool moist climate of modern Europe, and the UK and Ireland in particular. These orographic clouds are characterised by droplets in the size range 4 μm to 15 μm dia. The droplets are very efficiently deposited onto vegetation with deposition velocities very close to V_{max} and thus have similar rates of deposition to those of HNO_3 or HCl. An important consequence of this feature lies in upland areas which encounter frequent hill cloud. Taking as an example the Pennine hills in northern

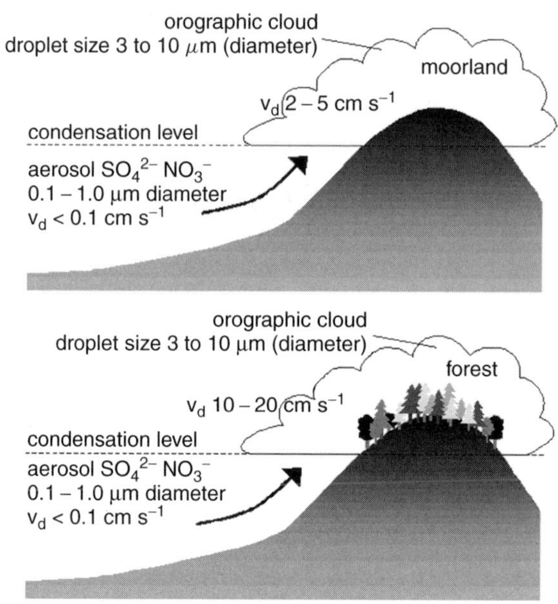

Figure 4.13 The effect of land use on cloud droplet deposition

England, this range of hills with heights in the range 600 m to 900 m is enveloped in cloud for up to 2000 hours annually and in this windy, wet climate cloud deposition makes a significant direct contribution to the deposition of sulphur, nitrogen and acidity. The large deposition rate and lack of a surface or canopy resistance also means that deposition rates are very sensitive to the aerodynamic roughness of the vegetation. In the case of the Pennine hills at Great Dun Fell it has been estimated that afforestation of the upper levels of this range of hills would increase the deposition of acidity from sulphur and nitrogen by a factor of 4, as illustrated in Figure 4.13.

Vegetation and especially forest have been shown to scavenge cloud droplets in the tropics with 'cloud forests' and in coastal areas exposed to sea fog (Hori, 1953).

For vegetation the efficient capture of cloud droplets presents a hazard to tree health where the cloud droplets form on pollutant derived aerosols. In the Appalachian mountains for example, red spruce has been shown to be very sensitive to simulated cloud water containing SO_4^{2-} and H^+, with substantial increases in frost sensitivity when exposed to concentrations of SO_4^{2-} and H^{-1} commonly observed in orographic cloud in these regions (Fowler *et al.*, 1989, Cape *et al.*, 1991).

References

Cape, J.N., Leith, I.D., Fowler, D., Murray, M.B., Sheppard, L.J., Eamus, D. and Wilson, R.H.F. (1991). Sulphate and ammonium in mist impair the frost hardening of red spruce seedlings. *New Phytologist*, **118**, 119–126.

Chamberlain, A.C. (1991). *Radioactive Aerosols*. Cambridge University Press, Cambridge.

Coe, H., Gallagher, M.W., Choularton, T.W. and Dore, C. (1995). Canopy scale measurements of stomatal and cuticular O_3 uptake by Sitka spruce. *Atmos. Environ.*, **29**, 1413–1423.

Emberson, L.D., Ashmore, M.R., Cambridge, H.M., Simpson, D. and Tuovinen, J.P. (2000). Modelling stomatal ozone flux across Europe. *Environ. Pollut.*, **109**, 403–413.

EMEP (1998). *Transboundary Photo-oxidant Air Pollution in Europe: Calculations of Tropospheric Ozone and Comparison with Observations*. EMEP MSC-W Status Report. The Norwegian Meteorological Institute, Research Report No. 67.

Erisman, J.W., Hensen, A., Fowler, D., Flechard, C.R., Grüner, A., Spindler, G., Duyzer, J.H., Weststrate, H., Römer, F., Vonk, A.W. and Jaarsveld, H.v. In press. Dry deposition monitoring in Europe. *Water Air and Soil Pollution*.

Erisman, J.W., Mennen, M., Fowler, D., Flechard, C.R., Spindler, G., Gruner, G., Duyzer, J.H., Ruigrok, W. and Wyers, G.P. (1998), Deposition monitoring in Europe. *Environ. Monit. Assess.* **53**, 279–295.

Erisman, J.W., Pul, A. van and Wyers, P. (1994). Parameterization of dry deposition mechanisms for the quantification of atmospheric input to ecosystems, *Atmos. Environ.*, **28**, 2595–2607.

Finkelstein, P.L. In press. Deposition velocities of O_3 and SO_2 over agricultural and forest ecosystems. *Water Air and Soil Pollution*.

Flechard, C.R., Fowler, D., Sutton, M.A. and Cape, J.N. (2000). A dynamic chemical model of bi-directional ammonia exchange between semi-natural vegetation and the atmosphere. *Q.J.R. Meteorol. Soc.* **125**, 2611–2641.

Fowler, D., Cape, J.N., Coyle, M., Smith, R.I., Hellbrekke, A-G., Simpson, D., Derwent, R.D. and Johnson, C. (1999). Modelling photochemical oxidant formation, transport, deposition and exposure of terrestrial ecosystems. *Environ. Pollut.* **100**, 43–55.

Fowler, D., Cape, J.N., Deans, J.D., Leith, I.D., Murray, M.B., Smith, R.I., Sheppard, L.J. and Unsworth, M.H. (1989). Effects of acid mist on the frost hardiness of red spruce seedlings. *New Phytol*. **113**, 321–335.

Fowler, D., Cape, J.N. and Unsworth, M.H. (1989). Deposition of atmospheric pollutants on forests. *Phil. Trans. R. Soc. B*., **324**, 247–265.

Fowler, D. and Duyzer, J.H. (1989). Micrometeorological techniques for the measurement of trace gas exchange. In: *Exchange of Trace Gases between Terrestrial Ecosystems and the Atmosphere*, eds. M.O. Andreae and D.S. Schimel, pp. 189–207. Chichester: Wiley.

Fowler, D., Flechard, C., Cape, J.N., Storeton-West, R.L. and Coyle, M. (2001). Measurements of ozone deposition to vegetation quantifying the flux, the stomatal and non-stomatal components. *Water Air and Soil Pollution*, **130**, 63–74.

Fowler, D., Flechard, C., Skiba, U., Coyle, M. and Cape, J.N. (1998). The atmospheric budget of oxidized nitrogen and its role in ozone formation and deposition. *New Phytol*. **139**, 11–23.

Fowler, D., Sutton, M.A., Flechard, C., Cape, J.N., Storeton-West, R., Coyle, M. and Smith, R.I. In press. The control of SO_2 dry deposition on to natural surfaces and its effects on regional deposition. *Water Air and Soil Pollution*.

Fowler, D. and Unsworth, M.H. (1979). Turbulent transfer of sulphur dioxide to a wheat crop. *Q. Jl. R. Met. Soc*. **105**, 767–783.

Fuhrer, J., Skarby, L. and Ashmore, M.R. (1997). Critical levels for ozone effects on vegetation in Europe. *Environ Pollut* **97**, 91–106.

Garland, J. In press. On the size dependence of particle deposition. *Water Air and Soil Pollution*.

Garland, J.A, Clough, W.S. and Fowler, D. (1973). Deposition of sulphur dioxide on grass. *Nature,* **242**, 256–257.

Hanson, P.J. and Lindberg, S.E. (1991). Dry deposition of reactive nitrogen compounds: a review of leaf, canopy and non-foliar measurements. *Atmos. Environ.* **25A**, 1615–1634.

Hargreaves, K.J., Fowler, D., Storeton-West, R.L. and Duyzer, J.H. (1992). The exchange of nitric oxide, nitrogen dioxide and ozone between pasture and the atmosphere. *Environ. Pollut.*, **75**, 53–59.

Harrison, R.M. and Kitto, A.M.N. (1994). Evidence for a surface source of atmospheric nitrous acid. *Atmos. Environ.* **28**, 1089–1094.

Hori, T. (1953). *Studies on Fogs*. Sapporo: Tanne Trading Co.

Huebert, B.J. and Roberts, C.H. (1985). The dry deposition of nitric acid to grass. *J. Geophys. Res.* **90**, 2085–2090.

IPCC (1997). *Greenhouse Gas Inventory Workbook.* Revised 1996 guidelines for national gas inventories. Vol 2. IPCC, Bracknell.

Meetham, A.R. (1950). Natural removal of pollution from the atmosphere. *Quart. J.R. Met. Soc.*, **76**, 359–371.

Monteith, JL, Unsworth, KH. (1990). *Principles of Environmental Physics*. Edward Arnold, London.

Nemitz, E., Milford, C. and Sutton, M.A. (2001). A two-layer canopy compensation point model for describing bi-directional biosphere-atmosphere exchange of ammonia. *Quarterly Journal Royal Meteorological Soc.* **127**, 815–833.

PORG. (1994). *Ozone in the UK 1993.* Third report of the United Kingdom. London: Department of the Environment.

PORG. (1997). *Ozone in the UK*. Fourth report of the Photochemical Oxidants Review Group. DETR (ITE Edinburgh).

Rondon, A. (1993). Atmospheric-surface exchange of nitrogen oxides and ozone. PhD thesis University of Stockholm.

Schjorring, J.K., Husted, S. and Mattsson, M. (1998). 'Physiological parameters controlling plant-atmosphere ammonia exchange'. *Atmospheric Environment*, **32**, 491–498.

Skiba, U., Hargreaves, K.J., Fowler, D. and Smith, K.A. (1992). Fluxes of nitric and nitrous oxides from agricultural soils in a cool temperate climate. *Atmos. Environ.* **26A**, 2477–2488.

Slinn, W.G.N. (1982). Prediction for particle deposition to vegetative canopies. *Atmos. Environ.*, **16**, 1785–1794.

Smith, R I., Fowler, D., Sutton, M.A., Flechard, C.R. and Coyle, M. (2000). A model for regional estimates of sulphur dioxide, nitrogen dioxide and ammonia dry deposition in the UK. *Atmos. Environ.* **34**, 3757–3777.

Spindler, G.,Teichmann, U. and Sutton, M.A. (2001). Ammonia dry deposition over grassland-micrometeorological flux-gradient measurements and bidirectional flux calculations using an inferential model. *Quarterly Journal of the Royal Meteorological Society* **127**, 795–815.

Stevenson, D.S., Johnson, C.E., Collins, W.J., Derwent, R.G., Shine, K.P. and Edwards, J.M. (1998). Evolution of tropospheric ozone radiative forcing. *Geophysical Research Letters* **25**, 3819–3822.

Sutton, M.A., Milford, C., Dragosits, U., Place, C.J., Singles, R.J., Smith, R.I., Pitcairn, C.E.R., Fowler, D., Hill, J., ApSimon, H.M., Ross, C., Hill, R., Jarvis, S.C., Pain, B.F., Phillips, V.C., Harrison, R., Moss, D., Webb, J., Espenhahn, S.E., Lee, D.S., Hornung, M., Ullyett, J., Bull, K.R., Emmett, B.A., Lowe, J. and Wyers, G.P. (1998). Dispersion, deposition and impacts of atmospheric ammonia: quantifying local budgets and spatial variability. *Environ. Pollut*. (Nitrogen Conference Special Issue). **102**, S1, 349–361.

Sutton, M.A., Place, C.J., Eager, M., Fowler, D. and Smith, R.I. (1995a). Assessment of the magnitude of ammonia emissions in the United Kingdom. *Atmos. Environ*. **29**, 1393–1411.

Sutton, M.A., Pitcairn, C.E.R. and Fowler, D. (1993). The exchange of ammonia between the atmosphere and plant communities. *Advances in Ecological Research*, **24**, 301–393.

Sutton, M.A., Schorring, J.K. and Wyers, G.P. (1995b). Plant-atmosphere exchange of ammonia. *Philosophical Transactions of the Royal Society London A.* **351**, 261–278.

Thom, AS. (1975). Momentum, mass and heat exchange of plant communities. In: Monteith, J.L (ed.) *Vegetation and Atmosphere*, pp. 57–109. London: Academic Press.

Van Hove, L.W.A., Adema, E.H., Vredenberg, W.J. and Pieters, G.A. (1989). A study of the adsorption of NH_3 and SO_2 on leaf surfaces. *Atmos. Environ.* **23**, 1479–1486.

Wayne, R.P., (2000). *Chemistry of Atmospheres*. Oxford University Press. Oxford.

5 Effects of oxidants at the biochemical, cell and physiological levels, with particular reference to ozone

S.P. LONG AND S.L. NAIDU

Introduction

Although there are many oxidizing pollutants in the atmosphere (Cape, 1997), ozone is by far the most significant for plants today because of its toxicity, widespread occurrence and increasing incidence (Fishman, 1991). This chapter focuses on ozone (O_3), however, many of the responses are common to those of other oxidizing pollutants. Ozone exposure induces a range of effects in plants depending upon duration and magnitude of exposure. Two types of exposure are defined, since their impacts on plants can differ. (1) Acute – exposure to high concentrations, 120–500 nmol [O_3] mol[air]$^{-1}$, for hours, as may occur at the most polluted sites. (2) Chronic – exposure to an elevated background concentration with peak daily concentrations in the range of 40–120 nmol mol^{-1} over several days in the growing season. Acute exposure may induce the uncontrolled death of cells, most commonly in the leaf mesophyll. It may also induce longer term changes, which in part mimic the hypersensitive response of plants to pathogens. These changes include programmed cell death, accelerated senescence, induction of biochemical systems for scavenging active oxygen species (AOS) and changes in cell wall structure. Some of these changes increase the resistance of the leaf to subsequent O_3 exposure, others decrease photosynthetic and productive capacity. Chronic exposure may induce some of the same visible symptoms, including small necrotic or chlorotic lesions. Often there are no visible symptoms, but a series of biochemical changes occur, including increases in AOS scavenging systems, decrease in photosynthetic capacity and accelerated senescence (Pell *et al.*, 1997).

The following examines how O_3, and other oxidizing pollutants, can gain access to the sites of plant metabolism, the changes induced within the cell, and how this affects physiological function, in particular photosynthesis.

Gaining entry

Although O_3 is highly reactive, primary damage is largely confined to leaves and then primarily the mesophyll; the only significant exception appears to be points of

Air Pollution and Plant Life, second edition. Edited by J.N.B. Bell and M. Treshow. ISBN 0 471 49090 3 (HB), 0 471 49091 1 (PB).

pollen reception (Black *et al.*, 2000). Why these tissues? Although O_3 attacks lipids by breaking unsaturated carbon–carbon bonds, there is little evidence that at current levels of pollution it can disrupt the cuticle to the extent that it may damage the underlying epidermal cells. Exposure to O_3 induces formation of additional epicuticular wax and causes changes in composition, in particular a shorter chain length of the lipid components. These changes result predominantly from synthesis rather than direct interaction of the lipids with O_3 (Kerfourn and Garrec, 1992). This suggests that epidermal cells sense and respond to the reactions of O_3 with the cuticle. Despite its ability to react with cuticles, available evidence suggests that the cuticle is an effective barrier against O_3 and other oxidizing air pollutants.

Decrease in O_3 concentration in air passing over a leaf is closely related to the degree to which the stomata of the leaf are open. The ease with which gases may gain access into the leaf is expressed by the stomatal conductance (g_s). This is the rate of diffusion of gas through the stomata that would occur if the concentration gradient was unity. O_3 uptake is diminished if exposure of leaves occurs under environmental conditions that cause decreased stomatal conductance, e.g. elevated CO_2 (McKee *et al.*, 1995). Deposition of O_3 in forests is also closely correlated with leaf stomatal conductance (Pilegaard *et al.*, 1995). The measured rate of O_3 uptake from air passing over a wheat leaf in a glass chamber was accurately predicted by assuming that the rate of uptake would be equal to the product of stomatal conductance and the O_3 concentration (Farage *et al.*, 1991). This prediction could only be possible if uptake through the stomata accounted for most of the O_3 lost as the air passed over the leaf (see Chapter 4).

Almost all other plant surfaces are protected by a continuous cuticle, the only exceptions are the stigmata of the flower, which catch the pollen, and the tubes of the pollen germinating on the stigmata. Most effects of O_3 on plants are consistent with O_3 entering the leaf and being absorbed by the wet, uncuticularized, exposed surfaces of

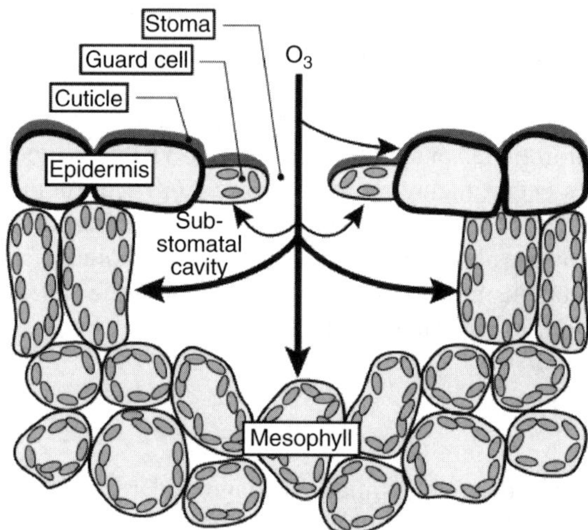

Figure 5.1 Targets of ozone damage on the leaf surface (cuticle) and within the leaf mesophyll. Small ovals represent chloroplasts

the mesophyll cells, as indicated in Figure 5.1; the one exception being decreased seed set (Black *et al.*, 2000; McKee *et al.*, 1997).

In conclusion, O_3 affects plants primarily by gaining access, via the stomata and mesophyll air space, to the exposed cell wall surfaces (Figure 5.1). The following section examines the likely chemical reactions as O_3 dissolves into the wet surface of these exposed cell walls. This region of the cell, outside the plasmalemma, is termed the apoplast.

Reactions of ozone and other oxidizing pollutants in the apoplast

Ozone is one of the most powerful oxidants known, with only a slightly lower oxidation potential than fluorine. The solubility of O_3 in pure water is 0.29 $m^3\ m^{-3}$ at 25 °C and, despite its oxidation potential of 2.07 eV in pure water with a pH equivalent to that of the apoplast (5.0–6.5), its half-life exceeds 1 hour at 25 °C (Horvath *et al.*, 1985; Moldau, 1998). Reactions of O_3 with water and solutes in the apoplast lead to the formation of free radicals. These are short-lived, highly reactive molecular fragments that contain one or more unpaired electrons; formed by the splitting of a molecular bond. Ozone reacts with the hydroxyl ions of water to yield the hydroperoxide ($\bullet O_2H$) and superoxide ($\bullet {O_2}^-$) radicals:

$$HO^- + O_3 \longrightarrow \bullet O_2H + \bullet {O_2}^-$$

Hydroperoxide radicals can combine to form hydrogen peroxide and are in equilibrium with superoxide and protons (pK 4.8):

$$\bullet O_2H + \bullet O_2H \longrightarrow H_2O_2 + O_2 \qquad \bullet O_2H \longleftrightarrow \bullet {O_2}^- + H^+$$

The pH of the mesophyll apoplast limits hydroxyl ion supply (Moldau, 1998). However, this slow reaction initiates a cyclic chain mechanism of O_3 decomposition in water by providing superoxide (Glaze, 1987) (Figure 5.2). Superoxide will react rapidly with O_3. This reaction and the further reactions of the cycle up to the formation of the hydroxyl radical ($\bullet OH$) all have very high rate constants ($1.5 \times 10^5 - 5 \times 10^{10}$ M s^{-1}). The $\bullet OH$ can in turn react rapidly with O_3 to produce further hydroperoxide and superoxide, completing and propagating the cycle (Bahnemann and Hart, 1982). The limitation on this cycle is the supply of superoxide.

Substances in the apoplast have the potential both to accelerate and quench this cycle (Figure 5.2). Electron donors, such as phenolic compounds in the cell wall, will accelerate the formation of ozonide radicals. Addition of caffeic and ferulic acid to ozonated water greatly accelerates $\bullet OH$ production (Moldau, 1998). Transition metal ions in the apoplast accelerate $\bullet OH$ production from hydrogen peroxide via the Fenton and Haber–Weiss reactions (Bielski and Cabelli, 1995). These sources of $\bullet OH$ will increase the probability of further superoxide formation to propagate the cycle (Figure 5.2). Ascorbate concentrations of 0.5 mM are found in the mesophyll apoplast. Ascorbate will react with $\bullet OH$ to form the ascorbate free radical and with O_3 and hydrogen

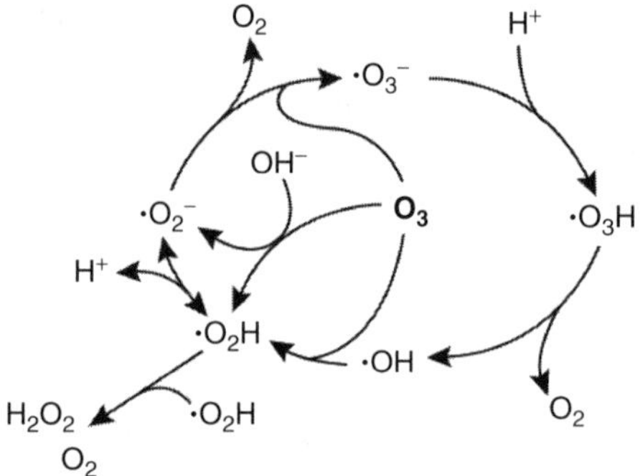

Figure 5.2 The cyclic chain mechanism of O_3 decomposition in water, as may occur in the cell wall. It is primed by the formation of superoxide

peroxide to form dehydroascorbates MDHA and DHA, discussed later. Its capacity to remove the radicals during O_3 exposure will depend on continued regeneration of ascorbate via reactions in the cell or on the plasmalemma (Moldau, 1998). Whilst reaction of •OH with ascorbate will decrease the probability of superoxide formation, protection against •OH is not possible since it will react with the nearest surface or solute. Superoxide dismutase in the apoplast accelerates removal of superoxide through conversion to hydrogen peroxide and oxygen. Hydrogen peroxide may be removed, and the cycle quenched, by ascorbate peroxidase or cell wall peroxidases. Again the phenoxyl radicals would need to be reduced, probably by ascorbate, to provide protection against continued exposure to O_3 (Moldau, 1998).

What is the probability of O_3 reaching the plasmalemma of the mesophyll cells? From the rate constants of reaction, Moldau (1998) calculated that, assuming 0.5 mM ascorbate in a typical mesophyll cell wall thickness of 0.4 μm, the probability of an O_3 molecule penetrating to the plasmalemma would be about 40%. Inability of even these high levels of ascorbate to screen the apoplast from oxidizing radicals has been confirmed experimentally by fluorescent dyes within the leaf (Jakob and Heber, 1998). The probability of reaching the plasmalemma would increase at high O_3 concentrations if the capacity for regeneration of ascorbate and other reductants is overwhelmed. These considerations suggest that O_3, and therefore the active oxygen species (AOS) that it forms, have the potential to react with the plasmalemma.

Whilst several stresses are capable of elevating concentrations of AOS, O_3 stress differs in also causing unique reactions called ozonolysis. This may explain why O_3 is more toxic than equivalent concentrations of AOS alone (Wellburn, 1994). What are the direct reactions of O_3 with the plasmalemma, and with other organic molecules in the apoplast? Carbon–carbon double bonds are subject to ozonolysis, forming molozonide (Figure 5.3). A series of further reactions rearrange bonding so that the three oxygen atoms are completely inserted between the two carbons to form an ozonide. In aqueous solution, the ozonide is hydrolyzed yielding two aldehyde residues and hydrogen peroxide (Figure 5.3). A similar series of reactions rupture carbon–carbon triple bonds (Horvath *et al.*, 1985). Unsaturated lipids are therefore the obvious targets

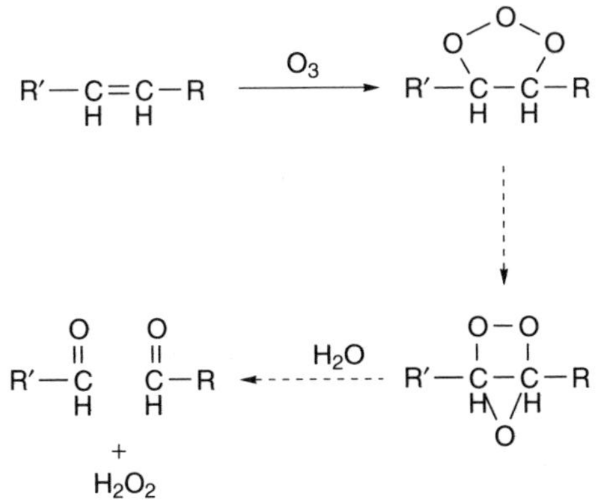

Figure 5.3 The reaction sequence in the ozonolysis of carbon–carbon double bonds in an aqueous medium

for O_3. However, proteins in the apoplast and in the outer surface of the plasmalemma may be damaged even when lipids are unaffected (Wellburn, 1994; Mudd *et al.*, 1997a). Three amino acid residues are particularly sensitive to O_3. Ozonolysis will open up the pyrrol ring of tryptophan and oxidize the sulphydryl groups (–SH) of cysteine and methionine to form disulphide bridges (–S–S–) or sulphoxides (Mudd *et al.*, 1997b; Wellburn, 1994).

In addition to direct attack, AOS formed from O_3 (Figure 5.2), in particular H_2O_2 and $\bullet O_2H$, can cause lipid peroxidation, whilst the hydroxyl radical may attack all components of the plasmalemma. Of the radicals generated by this chain of reactions (Figure 5.2), •OH is by far the most potent and will immediately oxidize almost all organic molecules. With a rate constant in excess of 100 M s^{-1}; the rate of reaction of •OH will be governed only by diffusion. Given its reactivity, •OH can only cause damage at its site of formation. At carbon–hydrogen bonds, •OH will remove the hydrogen atom to form water, leaving behind an unpaired electron on the carbon atom. This new carbon radical will react with molecular oxygen and induce autoxidation of the membrane. With unsaturated carbon bonds, •OH will add to the chain, facilitating hydrolysis of the bond or otherwise drastically altering the molecular structure. The hydroxyl radical can attack amino acid residues, yielding a decarboxylated radical (Fossey *et al.*, 1995). Many of the plant defence mechanisms for removing AOS (reviewed later) may be explained as a means of minimizing •OH formation in reference to Figure 5.2.

Do O_3 or the reactive radicals, pass beyond the plasmalemma? The reactivity of O_3 and its more damaging radicals suggests that this is highly unlikely. If it did happen, then by the same reasoning O_3 could penetrate the chloroplast. Chloroplast membranes can often be seen in cell sections viewed under the electron microscope to abut the plasmalemma on the portion of the mesophyll cells adjacent to the intercellular air space. Therefore the diffusion pathway is short, and considerably less than across the cell wall. If O_3 could penetrate membranes it would reach the thylakoid membrane and a rapid loss of function would follow. The dark-adapted ratio of variable to maximum chlorophyll fluorescence (F_v/F_m) of leaves provides a simple and rapid *in vivo* assay of photosystem II capacity. This capacity is highly dependent on membrane integrity. In a study of the sequence of changes in photosynthesis during and following exposure to high O_3 doses (400 nmol mol^{-1}), Farage *et al.* (1991) showed that despite large losses in capacity for carbon metabolism and decrease in stomatal conductance within hours, there was no significant change in F_v/F_m for 16 h. This suggests that there could have been no direct O_3 damage to the thylakoid membrane, and that subsequent loss of photosystem II capacity was a secondary or tertiary response.

A common response to O_3 is an increase in capacity for scavenging AOS, both inside and outside the cell. Why inside, if O_3 and radicals derived from O_3 do not enter the cell? As explained in the subsequent sections, stress can increase radical generation by the cell itself, in particular within the chloroplast. In this respect plant responses to O_3 are similar to those induced by other phytotoxic gaseous pollutants and by pathogens. Given that O_3 does not enter the cell, what signals this challenge on the outside to the cell?

Sensing what is happening on the outside

The most extreme acute exposures to O_3 will damage the plasmalemma to the extent that the cell will be unable to maintain its ion balance and cell death will follow. Less extreme acute and chronic exposures will induce an ordered response, up- and down-regulating various genes, which may cause specific cell death, accelerated senescence, and a range of events that may be interpreted as protective, particularly an increased capacity for AOS scavenging (Pell *et al.*, 1997). These effects may be induced by the interaction of O_3 and AOS with elicitor molecules on the plasmalemma or chemical interactions in the apoplast, for example increases in H_2O_2 or ascorbate consumption. Figure 5.4 outlines the major signal cascades thought to play a role. Of these, increased ethylene production is highly correlated with O_3 effects.

As with many plant stressors, the production of ethylene in response to O_3 exposure is an early event that is closely correlated to several responses within the cell (Tingey *et al.*, 1976; Mehlhorn and Wellburn, 1987). The plant hormone ethylene (C_2H_4), performs several functions, including promoting leaf abscission and leaf senescence, and inhibiting cell elongation (Abeles *et al.*, 1992). Methionine is the precursor to ethylene, reacting with ATP to produce S-adenosyl-L-methionine (AdoMet, also called SAM) via the enzyme AdoMet synthetase (Figure 5.5). AdoMet then reacts to form 1-aminocyclopropane-1-carboxylic acid (ACC) via ACC synthase. Finally, ACC is oxidized via ACC oxidase to ethylene, CO_2 and hydrogen cyanide (Kende, 1993). The rate-limiting step in ethylene formation is the formation of ACC by ACC synthase (Adams and Yang, 1979), which can be stimulated by various conditions including wounding, drought, and air pollution.

On exposure of *Arabidopsis thaliana* to 350 nmol mol^{-1} O_3, ethylene release rose rapidly to a peak at 1.5 h. What triggers this release of ethylene? A family of at least five genes in *A. thaliana* code for ACC synthase. Only one of these (*ACS6*) was up-regulated in this treatment with increased levels of the mRNA transcript at 30 min,

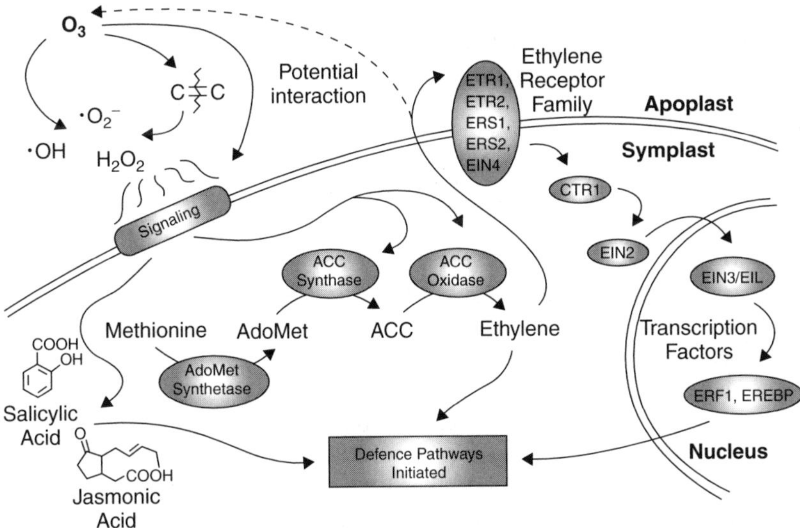

Figure 5.4 Ozone detection and signal transduction within plant cells. Abbreviations for genes, enzymes and molecules are presented in the text

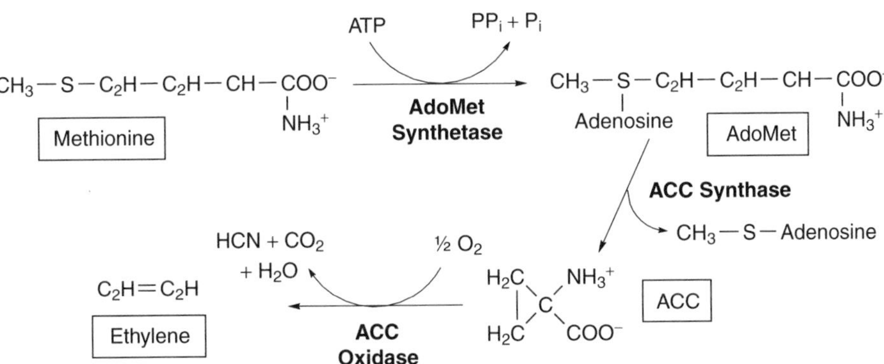

Figure 5.5 Ethylene biosynthesis pathway. Enzymes are shown in bold and key reactants and products are boxed

followed by increased ACC synthase activity (Vahala *et al.*, 1998). This explains the increase in ethylene, but how is the increase perceived? Molecular genetic studies are revealing the genes involved in the ethylene response pathway (Figure 5.4). The *ETR1* gene has been isolated from ethylene resistant mutants (Bleecker *et al.*, 1988). It codes for the ETR1 protein, which is similar to two-component histidine kinases found in bacterial systems and is assumed to be an ethylene receptor (Kieber, 1997). Subsequently, two subfamilies of genes coding for ethylene receptors have been identified: *ETR1* and *ERS1*; and *ETR2*, *EIN4*, and *ERS2* (Hua *et al.*, 1998). A further gene in the ethylene response pathway is *CTR1*, which appears to code for a serine/threonine protein kinase that restricts ethylene production (Kieber, 1997). Mutations in *CTR1* remove this restriction, causing ethylene to be constitutively produced. Kieber (1997) proposed that ethylene causes the ETR1 protein to alter its conformation to a form that blocks the kinase activity of the CTR1 protein (see also McGrath and Ecker, 1998). With CTR1 inactive, numerous downstream components, such as EIN2 are active and trigger various responses to ethylene. DNA-binding proteins, named ethylene-responsive element-binding proteins (EREBPs), have been identified in tobacco that bind to specific promoter regions on genes that are activated when plants are exposed to ethylene (Ohme-Takagi and Shinshi, 1995). Other downstream elements that may function as transcription factors (*EIN3, EIL1, EIL2, ERF1*) have also been identified (Solano *et al.*, 1998). Although first discovered in *Arabidopsis thaliana*, homologous genes have been isolated in soybean and tomato.

Ethylene induces changes which lead to increased expression of senescence associated genes (SAGs) and decreased expression of genes coding for photosynthetic proteins, in particular *rbcS* which codes for the small sub-unit of ribulose 1:5 bisphosphate carboxylase/oxygenase (Rubisco) (Miller *et al.*, 1999). The function of stress ethylene is not completely understood (Bleecker and Kende, 2000), but by accelerating leaf senescence, reserves for storage or seed production could be mobilized before a stress kills the leaf or a pathogen removes its reserves. Whilst increased ethylene is clearly associated with changes during and following O_3 exposure, it is unclear which of three potential mechanisms of interaction is of most importance (Pell *et al.*, 1997; Heath, 1994) (Figure 5.6).

The common coupling of damage and ethylene production in response to O_3 exposure suggests that ethylene is directly involved in the damage-induction pathway. Strongly supporting this hypothesis are the many studies in which the ACC synthase inhibitor aminoethoxyvinyl glycine [AVG] is used to inhibit the formation of ethylene and successfully decrease, although not eliminate, O_3-induced damage (e.g. Mehlhorn

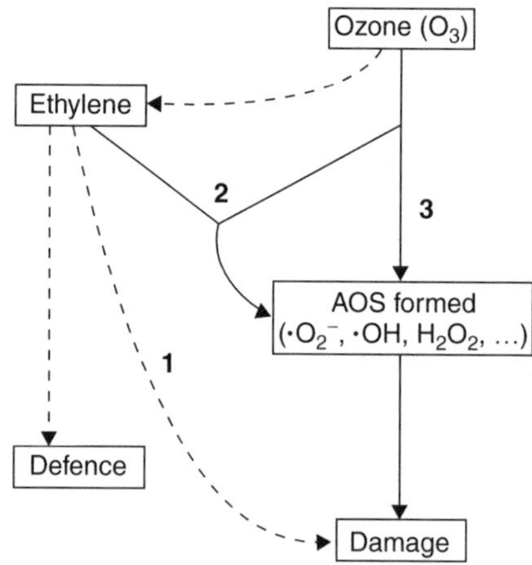

Figure 5.6 Alternative pathways (1, 2, and 3 – see text) for the role of ethylene in ozone-induced phytotoxicity. Dashed arrows indicate an induction pathway and solid arrows indicate a direct effect

et al., 1991; Wenzel *et al.*, 1995). Still, the mechanism of the interaction between O_3, ethylene, and damage remains unclear. Increased ethylene production in response to O_3 may be responsible for inducing damage via a signal transduction pathway with ethylene the primary (Bae *et al.*, 1996) or secondary signal molecule (Sandermann Jr. *et al.*, 1998) (Figure 5.4). Alternatively, ozonolysis of the ethylene molecule (Figure 5.3) may produce toxic organic radicals (AOS, Figure 5.6) as the possible primary cause of damage (Lucas and Wolfenden, 1996; Heath and Taylor, 1997).

Whether ethylene induces damage directly or produces damaging AOS via interaction with O_3, evidence for the direct involvement of ethylene in O_3 phytotoxicity from inhibition studies is compelling. It is important to note that the application of AVG in the presence of O_3 does not always prevent damage (Zilinskas *et al.*, 1990) or ethylene formation. It is possible that, in at least some species, AVG either does not completely block ethylene production (de Vries *et al.*, 1996) or that ethylene is simply not involved in the damage induction pathway. Furthermore, the inhibitory action of AVG is not specific to the ethylene production pathway; therefore, it cannot be determined conclusively from application of AVG (or any other non-specific inhibitor) that the absence of ethylene alone is responsible for reduced damage in the presence of O_3.

Alternatively, it is possible that the ethylene response is simply a parallel pathway, i.e., AOS produced directly by the interaction of O_3 with biomolecules are themselves signal molecules or are primarily responsible for damage (Figure 5.6, **3**) (Salter and Hewitt, 1992; Heath, 1987). Beside promoting senescence, ethylene may also be linked with salicylic acid and possibly jasmonate to improving defence against ozone, in particular increased capacity for AOS scavenging.

Defending the cell and apoplast against oxidants

Active oxygen species are produced during normal physiological processes, particularly the light-dependent photosynthetic reactions in plants (Foyer *et al.*, 1994), as well as during response to various stresses, including O_3, drought, temperature, and pathogen attack (Polle and Rennenberg, 1993). These AOS are produced by metabolism both within the cell and in the apoplast (Figure 5.2). Some of the AOS produced in the apoplast and all of the AOS produced in the cell in response to O_3 result from stress metabolism and are not chemically derived from the O_3 itself. This 'oxidative burst' in response to stress is thought to serve as a first line of defence against pathogens, triggering a signal cascade to promote various defence pathways (Figure 5.4) (Lamb and Dixon, 1997).

Despite this positive aspect of AOS induction, these molecules are highly reactive and damaging to living cells (Gille and Sigler, 1995). To prevent or minimize damage from AOS, plants produce a variety of antioxidant enzymes (mainly superoxide dismutase, ascorbate peroxidase, and glutathione reductase) and non-enzymatic low-molecular weight antioxidant molecules (ascorbate, glutathione, α-tocopherol) as well as certain secondary metabolites (phenolics, flavonoids, carotenoids) which can directly scavenge AOS, i.e. chemically reduce and so neutralize these potentially damaging oxidants (Figure 5.7) (Srivastava, 1999; Noctor and Foyer, 1998).

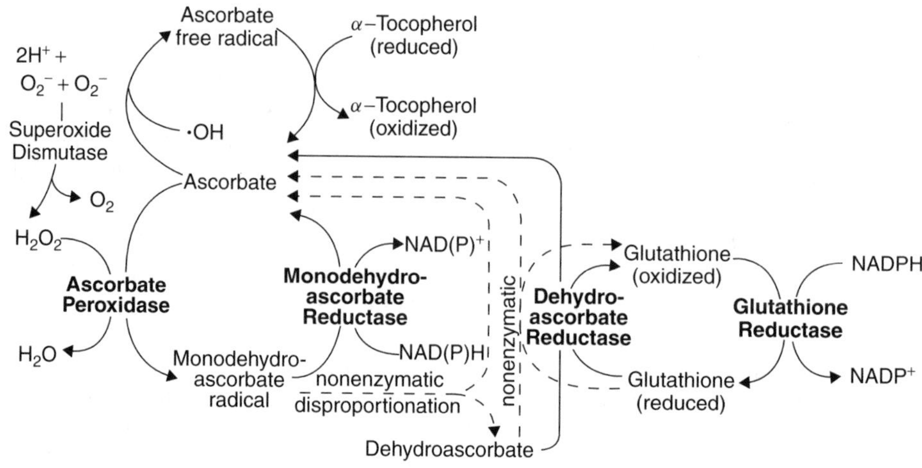

Figure 5.7 Antioxidant scavenging reactions. Solid lines are enzyme-catalyzed reactions and dashed lines are non-enzyme-catalyzed reactions. Modified from Jahnke *et al.* (1991)

Since O_3 and AOS derived from O_3 (Figure 5.2) will attack the plasmalemma via the mesophyll cell wall, it is not surprising that some protective compounds which act directly as radical scavengers (such as ascorbate, polyamines, and tocopherol) as well as some detoxifying enzymes (such as peroxidases and SOD) have been found in the cell apoplast or at the plasmalemma surface (Vanacker *et al.*, 1998; Srivastava, 1999). These constitute a first line of defence that is correlated, at least in part with genetic variation in O_3 tolerance (Turcsányi *et al.*, 2000).

In addition to certain protective molecules and enzymes that may be constitutively present in some plants, the production of some free radical scavengers and various antioxidant enzymes are induced or up-regulated by O_3 exposure. Examples are polyamines, glutathione, peroxidases, catalase, and superoxide dismutase (Srivastava, 1999; Sharma and Davis, 1997). The production of other enzymes responsible for the synthesis of various general plant protective compounds (including phenylpropanoid, flavonoids, and phytoalexines) and certain cell wall compounds (lignin, extensins, callose, etc.) that create barriers to pathogens, and perhaps O_3 diffusion, may also be induced by O_3 (Sandermann Jr. *et al.*, 1998; Srivastava, 1999). This may, in part, be a general stress response since *A. thaliana* plants pre-treated with O_3 showed increased resistance to bacterial pathogens (Sharma *et al.*, 1996).

Interspecific differences in tolerance to O_3 may involve the induction of a variety of metabolic defence pathways. An understanding of such differences will help in the identification and development of O_3-resistant plant cultivars. An increasing number of investigators have begun using transgenic plants, engineered to overproduce various enzymes of the antioxidant pathways, in an attempt to confer resistance to O_3 damage. However, data indicate that single enzyme changes are often ineffective towards increasing tolerance to O_3; a coordinated increase in multiple metabolites and enzymes may be required to confer tolerance (reviewed in Srivastava, 1999).

Increase in AOS scavenging capacity requires change within the cell. How is this triggered by O_3 at the cell surface? Ethylene may play a part (Figure 5.4), however in *A. thaliana*, salicylic acid is clearly essential. Transgenic plants expressing salicylate hydroxylase to prevent the accumulation of salicylic acid were unable to acquire increased O_3 tolerance associated with increased AOS scavenging capacity following exposure to O_3, in contrast to wild-type plants (Sharma *et al.*, 1996). The following section outlines what is known about the induction of defence pathways by O_3. While there is evidence that numerous signalling molecules and pathways are involved, the potential interactions among these various defence pathways is unclear.

Induction of defence pathways

In addition to antioxidants, O_3 is known to trigger the induction of general plant defence pathways. For example, a variety of proteins, known collectively as pathogenesis-related (PR) proteins, are synthesized in plants in response to pathogen exposure. These proteins include β-1,3-glucanases, chitinases, and proteinase inhibitors. Several studies have demonstrated the induction of PR genes by O_3 (reviewed in Kangasjärvi *et al.*, 1994; Sandermann Jr., 1996; Sharma and Davis, 1997; Pell *et al.*, 1997). These PR proteins are produced as part of two general pathways triggered by pathogens and by O_3 (among other stresses): the hypersensitive response (HR), which results in cell death, halting the spread of pathogens; and systemic acquired resistance (SAR), which helps to minimize secondary pathogen infections (reviewed in Sandermann Jr. *et al.*, 1998).

The signal transduction pathway by which O_3 triggers gene expression is not known, but likely uses components of these other, more general, stress response pathways in a process termed cross induction (Sandermann Jr., 1996). Many diverse stresses trigger parts of the same response pathways as plants challenged by one stress can often become resistant, or 'cross-tolerant' to others (Bowler and Fluhr, 2000). Speculation as to the primary signal molecule(s) in these stress pathways has lately focused on the AOS themselves (Sandermann Jr., 2000; Noctor and Foyer, 1998). In particular, H_2O_2 appears to function as a primary signal molecule (Lamb and Dixon, 1997) in a transduction pathway that includes protein phosphorylation via a mitogen-activated protein kinase cascade (Kovtun *et al.*, 2000). However, H_2O_2 formed from O_3 (Figure 5.4) is assumed not to enter the cell, so any response to H_2O_2 would be to that formed within the cell in response to changes on the outside. The apoplastic ascorbate–dehydroascorbate redox couple is linked to the cytoplasmic ascorbate–dehydroascorbate redox couple by specific transporters for either or both metabolites (Horemans *et al.*, 2000). The reduction state of this redox couple within the cell is one way in which changes on the outside may be signalled to the inside of the cell. However, neither of these primary signals could elicit a response to O_3 that could differ from other stresses inducing increased AOS. The involvement of ethylene, jasmonate, and salicylic acid as secondary signal molecules is clearer (e.g., Dong, 1998; Koch *et al.*, 2000). In the specific case of O_3-induced damage, complex interaction between these pathways is likely (Rao *et al.*, 2000). The composition of proteins in the plasmalemma and even a possible new protein has been detected following O_3 exposure, and could therefore provide another potential primary signal (TokarskaSchlattner *et al.*, 1997).

The evidence that ethylene plays a key role in O_3-induced damage is discussed above, but it is likely that ethylene is central to the induction of defence pathways as well (Figure 5.4). It has been demonstrated that various plant defence genes are regulated by ethylene (Boller, 1991; Broglie and Broglie, 1991) and there is evidence that applying ethylene prior to O_3 exposure can increase protection (Mehlhorn *et al.*, 1991).

Photosynthesis

Both theory and measurement suggest it is highly unlikely that O_3 or the AOS it directly gives rise to can cause direct damage to the photosynthetic apparatus of intact leaves. Yet, photosynthesis is an early casualty of O_3 exposure, and sometimes the only physiological symptom of damage during chronic exposure of leaves. In wheat leaves in 200–400 nmol mol^{-1} [O_3], photosynthetic capacity, independent of stomatal conductance, declined significantly within 30 min (Farage *et al.*, 1991). In leaves grown with chronic elevations of 70 nmol mol^{-1}, loss of photosynthetic capacity was, again, an early symptom evident before any visual change (McKee *et al.*, 1995; Farage and Long, 1999). O_3 has been shown capable of damaging or inhibiting almost every step of the photosynthetic process from light capture to starch accumulation (Farage *et al.*, 1991). Restriction of any step within the apparatus will decrease the ability of the photosynthetic membrane to utilize the light energy that it intercepts, increasing the potential for formation of oxidizing radicals within the chloroplast, causing photo-inhibition and photo-oxidation. This complicates analysis of cause and effect. Farage *et al.* (1991) conducted an *in vivo* analysis, of change in the photosynthetic apparatus of wheat during acute O_3 exposure, and McKee *et al.* (1995) conducted a similar analysis of flag leaves that formed and senesced in a chronic O_3 treatment (Figure 5.8). The first significant change in both studies was a loss of Rubisco activity, and Rubisco protein. Photosynthetic rate declined in direct proportion to *in vivo* Rubisco activity. Maximum quantum efficiency of CO_2 uptake and capacity for regeneration of RubP (ribulose 1:5 bisphosphate) remained high. This could only be possible if the photosynthetic membranes retained their integrity, and if the activity of the steps in electron transport and carbon metabolism controlling regeneration of RubP was unaffected. Only after 16 h at 400 nmol mol^{-1} and the loss of 75% of Rubisco activity was there any effect on the maximum efficiency of photosystem II (Figure 5.8a). These results suggest a sequence of change in which Rubisco is affected ahead of all other changes in the photosynthetic apparatus. The activation of Rubisco is either unaffected or rises (Figure 5.8b), suggesting that decreased carboxylation results from loss of Rubisco protein and not its activation (Pelloux *et al.*, 2001, McKee *et al.*, 1995). Studies of Rubisco chemistry from O_3 treated leaves suggest oxidative damage (Junqua *et al.*, 2000), followed by rapid protease digestion (Pell *et al.*, 1997). Decrease in mRNA transcripts for *rbcS* are commonly observed and could contribute to the lower levels of Rubisco in leaves that develop under chronic O_3 pollution. However, Pell *et al.* (1997)

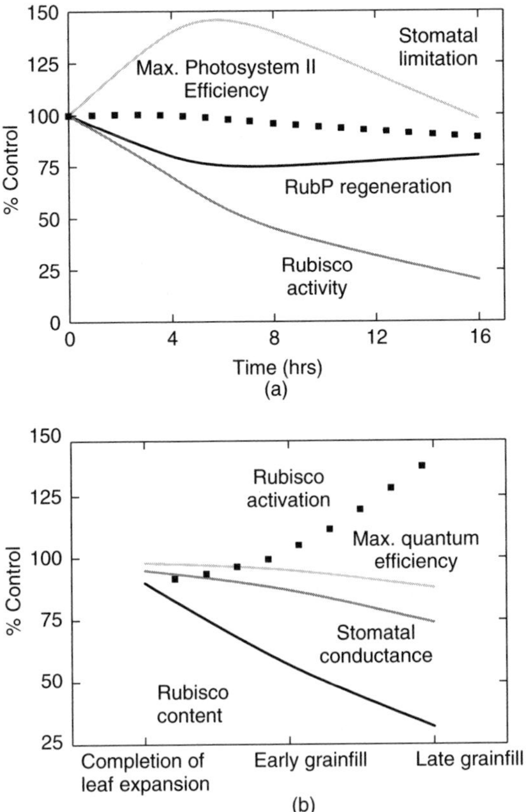

Figure 5.8 Kinetics of change in photosynthesis of wheat leaves after (a) over 16 h of acute (400 nmol mol^{-1}) ozone exposure (change over hours, redrawn from Farage *et al.*, 1991) and (b) over about 40 days of chronic (70 nmol mol^{-1}) ozone exposure of the flag leaf (redrawn from McKee *et al.*, 1995). All processes are described as a % of the measure for parallel control plants in ozone-free air. In (a) stomatal limitation is calculated as described in the text and maximum efficiency of photosystem II is determined by chlorophyll fluorescence. In (b) stomatal conductance was measured directly rather than calculated as a limitation. Rubisco activation represents the proportion of the Rubisco present that was active. Maximum quantum efficiency is the ratio of CO_2 assimilation to absorbed photons, in limiting light

notes that poplar grown in elevated O_3 had lower Rubisco content throughout the life of the leaf, although Rubisco synthesis was unaffected, suggesting that damage to existing Rubisco rather than decreased synthesis is the dominant cause of lower Rubisco activity. Loss appears to depend on the production of oxidizing radicals within the chloroplast, which, by damaging Rubisco, may increase potential for photo-oxidative damage leading to more widespread changes within the chloroplast (Figure 5.9). The primary cause of oxidative damage to Rubisco is unclear. However, ethylene signalling will, via transcription, increase lipase and lipoxygenase. This corresponds to a decrease

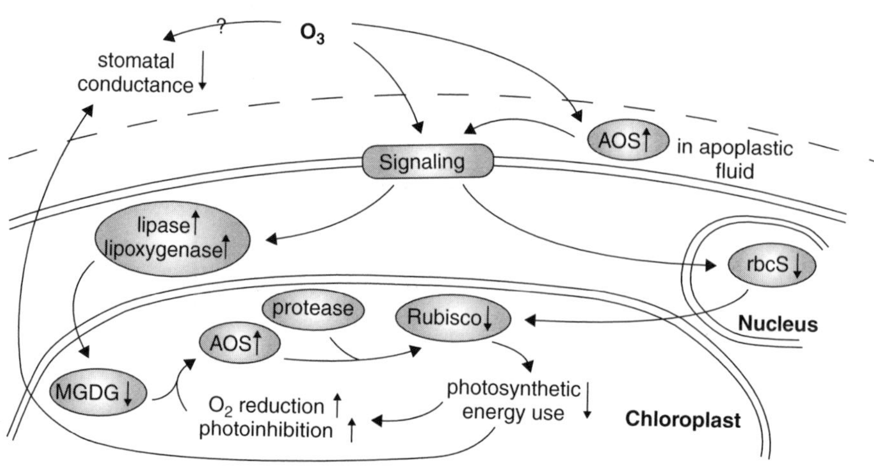

Figure 5.9 Interactions and mechanisms by which photosynthesis is reduced in plants exposed to ozone. Abbreviations are presented in the text

in chloroplast mongalactosyl diglycerides (MGDG) which may be digested to free fatty acids by lipase, followed by oxidation with molecular oxygen via lipoxygenase, to form oxidizing radicals (Pell *et al.*, 1997).

Stomata control access of carbon dioxide and O_3 to photosynthetic cells. The guard cells of the stomata are not fully cuticularized, and will be exposed to the highest concentrations along the diffusion gradient into the leaf (Figure 5.1). Following exposure to O_3 there is a decline in stomatal conductance (Figure 5.8b). But stomatal conductance is closely linked with photosynthetic activity, and in healthy leaves it is closely linked to photosynthetic rate (Long, 1985). Stomatal conductance could therefore decline as an indirect result of O_3 damage to photosynthetic capacity within the mesophyll or as a direct response to O_3. Ozone has been shown to impair the stomatal function of epidermal strips (1000 nmol $[O_3]$ mol^{-1}) and damage the K^+ uptake channels of protoplasts (10 000–30 000 nmol $[O_3]$ mol^{-1}) made from guard cells (Torsethaugen *et al.*, 1999). Whilst treatment at these extremely high concentrations demonstrates a potential for direct effects of O_3, do they have relevance to pollution levels in the field? Does decline in stomatal conductance during and following O_3 exposure decrease photosynthesis? Farquhar and Sharkey (1982) suggested a simple means of separating stomatal from non-stomatal limitation. Photosynthesis (A) is measured at the normal ambient CO_2 concentration (C_a), and then the CO_2 concentration around the leaf is elevated until the concentration inside the leaf reaches C_a. The photosynthetic rate at this point (A_0) will be that which the leaf would achieve if there were no stomatal restriction to diffusion of CO_2 into the leaf. The ratio $(A_0 - A)/A_0$ provides a simple measure of stomatal limitation (l). Farage *et al.* (1991) and Farage and Long (1999) used this approach to separate stomatal and non-stomatal effects in wheat leaves, and found either no increase or a decrease in stomatal limitation in both acute and chronic exposures (Figure 5.8a). However, Torsethaugen *et al.* (1999) found that if leaves of broad bean (*Vicia faba*) were exposed to 100 ppb $[O_3]$ from the start of the photoperiod, stomatal limitation was higher for the first 3 h of fumigation. By 4 h the difference was lost. If O_3 was introduced during the photoperiod, there was no effect on stomatal limitation, even during the first 3 h. Since O_3 will typically be minimal around dawn, it is unlikely that stomatal opening will occur in $[O_3]$ levels as high as those used by Torsethaugen *et al.* (1999). The result does indicate that elevated $[O_3]$ may alter the speed of stomatal response in other situations that do have relevance to the field: for example, opening following a period of shade, as might occur when there is a sudden increase in photon flux on a partially cloudy day or as a leaf enters a sunfleck. Using a modelling approach Martin *et al.* (2000) showed that the observed kinetics of decline in stomatal conductance could be explained solely from the dependence of g_s on A, where decline in A was due to loss of Rubisco activity. Whilst this does not rule out a direct effect on the stomatal apparatus, it shows that the indirect effect via the mesophyll dominates the response, and that any direct effect on the stomata does not affect photosynthetic rate.

Concluding comments

Ozone attacks the plant primarily by gaining access to the leaf intercellular air space and entering the apoplast of the exposed mesophyll cells. Here it forms an array of AOS (Figure 5.2). Although ascorbate and other scavenging systems will remove O_3 and AOS, O_3 will reach and attack the outer proteins and lipids of the plasmalemma. At high concentrations, damage will lead to uncontrolled cell and tissue death. At the lower concentrations now typical of rural areas in much of the industrialized and industrializing regions, O_3 in the apoplast will trigger two categories of events within the cell:

1. A loss of photosynthetic capacity, associated with a loss of Rubisco. In part this is damage induced in Rubisco and in part it is accelerated senescence, with a down-regulation of photosynthetic genes and an up-regulation of genes involved in programmed cell death and/or tissue senescence. Although high levels of O_3 can damage stomatal function, low level O_3 appears to decrease stomatal opening only indirectly through decreased photosynthesis. Decreased photosynthetic capacity and accelerated loss of leaf area will depress both plant productivity, explaining the losses in yield in crops, and plant fitness in natural vegetation observed for plants grown in current polluted air relative to charcoal filtered air (see Chapter 6).
2. Protection, by up-regulation of AOS scavenging systems both in the apoplast and within the cell; and also by changes within the cell wall that would increase diffusion resistance and removal of O_3 and AOS.

The primary signals driving these changes are uncertain, although likely to involve either redox changes in the apoplast that are linked to the cell, or alteration of plasmalemma proteins. Ethylene, jasmonate and salycilic acid are key secondary signals driving up- and down-regulation of key genes (Figure 5.4). The ever-growing range of characterized *A. thaliana* mutants and gene-chip arrays for this species, is allowing rapid discovery of the genes that are up- and down-regulated by O_3, and identification of those that are essential to different aspects of the response. Unfortunately, most responses in this species have been reported for very high concentrations, often >300 nmol mol^{-1}. Given that plant responses to such rare acute pollution events are very different from responses to the widespread chronic pollution, the relevance of *A. thaliana* as a model for understanding molecular mechanisms in crops and natural vegetation has to be questioned. Relevance of these findings will depend on the development of parallel tools in more relevant systems.

Finally, whilst it is easy to understand why induced resistance will benefit the plant, the significance of programmed cell death and senescence is more obscure. Why do cells or whole leaves accelerate their demise in response to the presence of O_3 outside of the cell, can this have any benefit to the productivity or fitness of the individual? In evolutionary terms, surface O_3 concentration above a few nmol mol^{-1} is a new phenomenon. Plant responses, and genetic variation in response, most probably derive from mechanisms that have developed in acclimating to other stresses. The similarity of O_3 damage to the hypersensitive response of plants to pathogens, suggests that O_3 may be perceived as a pathogen, inducing cell and organ death to prevent the spread of the invader. Accelerated senescence may be a more generic response to stress. The plant senses the onset of a stress that will prevent further production, such as a developing drought or a pest attack. To avoid losing limiting resources, such as nitrogen, senescence provides an ordered retreat, allowing the plant to mobilize and retranslocate the reserves to seeds or storage organs before the stress reaches a level that will cause uncontrolled cell and leaf death. However, under chronic O_3 pollution there will probably be no benefit to production or fitness in this response, since in most areas O_3 will not rise to levels that would cause uncontrolled cell and tissue death. Mutation and genetic manipulation could remove the response and probably increase production under chronic O_3 pollution. But the cost could be decreased disease resistance and ability to deal with other stresses. New genetic tools coupled with new open-air field fumigation systems provide the means to address these issues.

There have been many thorough reviews of the different areas of biochemical, molecular and physiological responses of plants to ozone. Space prevents us from more than outlining their findings. Those that have been particularly valuable in developing this outline include: chemistry of forms of AOS (Bielski and Cabelli, 1995), reproductive development (Black *et al.*, 2000), mechanisms underlying accelerated senescence and increased defence (Pell *et al.*, 1997), ozone scavenging in the cell wall (Moldau, 1998), and molecular mechanisms responses of AOS scavenging systems (Sharma and Davis, 1997).

References

Abeles, F.B., Morgan, P.W. and Saltveit Jr., M.E. (1992). *Ethylene in Plant Biology*, Academic Press, San Diego.

Adams, D.O. and Yang, S.F. (1979). Ethylene biosynthesis: identification of 1-aminocyclopropane-1-carboxylic acid as an intermediate in the conversion of methionine to ethylene, *Proceedings of the National Academy of Sciences USA*, **76**, 170–174.

Bae, G., Nakajima, N., Ishizuka, K. and Kondo, N. (1996). The role in ozone phytotoxicity of the evolution of ethylene upon induction of 1-aminocyclopropane-1-carboxylic acid synthase by ozone fumigation in tomato plants, *Plant Cell Physiology*, **37**, 129–134.

Bahnemann, D. and Hart, E.J. (1982). Rate constants of the reaction of the hydrated electron and hydroxyl radical with ozone in aqueous solution, *Journal of Physical Chemistry*, **86**, 252–255.

Bielski, B.H.J. and Cabelli, D.E. (1995). Superoxide and hydroxyl radical chemistry in aqueous solution, In: *Active Oxygen in Chemistry*, Eds., Foote, C.S., Valentine, J.S., Greenberg, A. and Liebman, J.F., Blackie A&P, London, pp. 66–104.

Black, V.J., Black, C.R., Roberts, J.A. and Stewart, C.A. (2000). Impact of ozone on the reproductive development of plants, *New Phytologist*, **147**, 421–447.

Bleecker, A.B., Estelle, M.A., Somerville, C. and Kende, H. (1988). Insensitivity to ethylene conferred by a dominant mutation in *Arabidopsis thaliana*, *Science*, **241**, 1086–1089.

Bleecker, A.B. and Kende, H. (2000). Ethylene: a gaseous signal molecule in plants, *Annual Review of Cell and Developmental Biology*, **16**, 1–18.

Boller, T. (1991). Ethylene in pathogenesis and disease resistance, In: *The Plant Hormone Ethylene*, Eds., Mattoo, A.K. and Suttle, J.C., CRC Press, Boca Raton, pp. 293–314.

Bowler, C. and Fluhr, R. (2000). The role of calcium and activated oxygens as signals for controlling cross-tolerance, *Trends in Plant Science*, **5**, 241–246.

Broglie, R. and Broglie, K. (1991). Ethylene and gene expression, In: *The Plant Hormone Ethylene*, Eds., Mattoo, A.K. and Suttle, J.C., CRC Press, Boca Raton, pp. 101–113.

Cape, J.N. (1997). Photochemical oxidants – What else is in the atmosphere besides ozone? *Phyton-Annales Rei Botanicae*, **37**, 45–57.

de Vries, H.S.M., Martis, A.A.E., Reuss, J., Parker, D.H., Petruzzelli, L. and Harren, F.J.M. (1996). Ethylene evolution of germinating peas exposed to ozone, *Progress in Natural Science*, supplement to **6**, s550–s553.

Dong, X. (1998). SA, JA, ethylene and disease resistance, *Current Opinion in Plant Biology*, **1**, 316–323.

Farage, P., Long, S., Lechner, E. and Baker, N. (1991). The sequence of change within the photosynthetic apparatus of wheat following short-term exposure to ozone, *Plant Physiology*, **95**, 529–535.

Farage, P.K. and Long, S.P. (1999). The effects of O_3 fumigation during leaf development on photosynthesis of wheat and pea: An *in vivo* analysis, *Photosynthesis Research*, **59**, 1–7.

Farquhar, G.D. and Sharkey, T.D. (1982). Stomatal conductance and photosynthesis, *Annual Reviews of Plant Physiology*, **33**, 317–345.

Fishman, J. (1991). The global consequences of increasing tropospheric ozone concentrations, *Chemosphere*, **22**, 685–695.

Fossey, J., Lefort, D. and Sorba, J. (1995). *Free Radicals in Organic Chemistry*, Wiley, Chichester.

Foyer, C., Lelandais, M. and Kunert, K. (1994). Photooxidative stress in plants, *Physiologia Plantarum*, **92**, 696–717.

Gille, G. and Sigler, K. (1995). Oxidative stress and living cells, *Folia Microbiologica*, **40**, 131–152.

Glaze, W. (1987). Drinking-water treatment with ozone, *Environmental Science and Technology*, **21**, 224–230.

Heath, R. (1987). The biochemistry of ozone attack on the plasma membrane of plant cells, *Recent Advances in Phytochemistry*, **21**, 29–54.

Heath, R. (1994). Possible mechanisms for the inhibition of photosynthesis by ozone, *Photosynthesis Research*, **39**, 439–451.

Heath, R.L. and Taylor, G.E. (1997). Physiological processes and plant responses to ozone exposure, *Ecological Studies*, **127**, 317–368.

Horemans, N., Foyer, C.H. and Asard, H. (2000). Transport and action of ascorbate at the plant plasma membrane, *Trends in Plant Science*, **5**, 263–267.

Horvath, M., Bilitzky, L. and Huttner, J. (1985). *Ozone*, Elsevier, Amsterdam.

Hua, J., Sakai, H., Nourizadeh, S., Chen, Q.G., Bleecker, A.B., Ecker, J.R. and Meyerowitz, E.M. (1998). *EIN4* and *ERS2* are members of the putative ethylene receptor gene family in *Arabidopsis*, *The Plant Cell*, **10**, 1321–1332.

Jahnke, L.S., Hull, M.R. and Long, S.P. (1991). Chilling stress and oxygen metabolizing enzymes in *Zea mays* and *Zea diploperennis*, *Plant, Cell and Environment*, **14**, 97–104.

Jakob, B. and Heber, U. (1998). Apoplastic ascorbate does not prevent the oxidation of fluorescent amphiphilic dyes by ambient and elevated concentrations of ozone in leaves, *Plant and Cell Physiology*, **39**, 313–322.

Junqua, M., Biolley, J.P., Pie, S., Kanoun, M., Duran, R. and Goulas, P. (2000). In vivo occurrence of carbonyl residues in *Phaseolus vulgaris* proteins as a direct consequence of a chronic ozone stress, *Plant Physiology and Biochemistry*, **38**, 853–861.

Kangasjärvi, J., Talvinen, J., Utriainen, M. and Karjalainen, R. (1994). Plant defence systems induced by ozone, *Plant, Cell and Environment*, **17**, 783–794.

Kende, H. (1993). Ethylene biosynthesis, *Annual Review of Plant Physiology and Plant Molecular Biology*, **44**, 283–307.

Kerfourn, C. and Garrec, J.P. (1992). Modifications in the alkane composition of cuticular waxes from spruce needles (*Picea abies*) and ivy leaves (*Hedera helix*) exposed to ozone fumigation and acid fog – Comparison with needles from declining spruce trees, *Canadian Journal of Botany–Revue Canadienne De Botanique*, **70**, 861–869.

Kieber, J.J. (1997). The ethylene response pathway in *Arabidopsis*, *Annual Review of Plant Physiology and Plant Molecular Biology*, **48**, 277–296.

Koch, J.R., Creelman, R.A., Eshita, S.M., Seskar, M., Mullet, J.E. and Davis, K.R. (2000). Ozone sensitivity in hybrid poplar correlates with insensitivity to both salicylic acid and jasmonic acid. The role of programmed cell death in lesion formation, *Plant Physiology*, **123**, 487–496.

Kovtun, Y., Chiu, W., L Tena, G. and Sheen, J. (2000). Functional analysis of oxidative stress-activated mitogen-activated protein kinase cascade in plants, *Proceedings of the National Academy of Sciences USA*, **97**, 2940–2945.

Lamb, C. and Dixon, R.A. (1997). The oxidative burst in plant disease resistance, *Annual Review of Plant Physiology and Plant Molecular Biology*, **48**, 251–275.

Long, S.P. (1985). Leaf gas exchange, In: *Photosynthetic Mechanisms and the Environment*, Eds., Barber, J. and Baker, N.R., Elsevier, Amsterdam, pp. 455–499.

Lucas, P. and Wolfenden, J. (1996). The role of plant hormones as modifiers of sensitivity to air pollutants, *Phyton*, **36**, 51–56.

Martin, M.J., Farage, P.K., Humphries, S.W. and Long, S.P. (2000). Can the stomatal changes caused by acute ozone exposure be predicted by changes occurring in the mesophyll? A simplification for models of vegetation response to the global increase in tropospheric elevated ozone episodes, *Australian Journal of Plant Physiology*, **27**, 211–219.

McGrath, R. and Ecker, J. (1998). Ethylene signaling in *Arabidopsis*: events from the membrane to the nucleus, *Plant Physiology and Biochemistry*, **36**, 103–113.

McKee, I.F., Bullimore, J.F. and Long, S.P. (1997). Will elevated CO_2 concentrations protect the yield of wheat from O_3 damage?, *Plant, Cell and Environment*, **20**, 77–84.

McKee, I.F., Farage, P.K. and Long, S.P. (1995). The interactive effects of elevated CO_2 and O_3 concentration on photosynthesis in spring wheat, *Photosynthesis Research*, **45**, 111–119.

Mehlhorn, H., O'Shea, J. and Wellburn, A. (1991). Atmospheric ozone interacts with stress ethylene formation by plants to cause visible plant injury, *Journal of Experimental Botany*, **42**, 17–24.

Mehlhorn, H. and Wellburn, A. (1987). Stress ethylene formation determines plant sensitivity to ozone, *Nature*, **327**, 417–418.

Miller, J.D., Arteca, R.N. and Pell, E.J. (1999). Senescence-associated gene expression during ozone-induced leaf senescence in *Arabidopsis*, *Plant Physiology*, **120**, 1015–1023.

Moldau, H. (1998). Hierarchy of ozone scavenging reactions in the plant cell wall, *Physiologia Plantarum*, **104**, 617–622.

Mudd, J.B., Dawson, P.J. and Santrock, J. (1997a). Ozone does not react with human erythrocyte membrane lipids, *Archives of Biochemistry and Biophysics*, **341**, 251–258.

Mudd, J.B., Dawson, P.J., Tseng, S. and Liu, F.P. (1997b). Reaction of ozone with protein tryptophans: Band III, serum albumin, and cytochrome c, *Archives of Biochemistry and Biophysics*, **338**, 143–149.

Noctor, G. and Foyer, C.H. (1998). Ascorbate and glutathione: keeping active oxygen under control, *Annual Review of Plant Physiology and Plant Molecular Biology*, **49**, 249–279.

Ohme-Takagi, M. and Shinshi, H. (1995). Ethylene-inducible DNA binding proteins that interact with an ethylene-responsive element, *The Plant Cell*, **7**, 173–182.

Pell, E.J., Schlagnhaufer, C.D. and Arteca, R.N. (1997). Ozone-induced oxidative stress: Mechanisms of action and reaction, *Physiologia Plantarum*, **100**, 264–273.

Pelloux, J., Jolivet, Y., Fontaine, V., Banvoy, J. and Dizengremel, P. (2001). Changes in Rubisco and Rubisco activase gene expression and polypeptide content in *Pinus halepensis M.* subjected to ozone and drought, *Plant, Cell and Environment*, **24**, 123–131.

Pilegaard, K., Jensen, N.O. and Hummelshoj, P. (1995). Seasonal and diurnal variation in the deposition velocity of ozone over a spruce forest in Denmark, *Water Air and Soil Pollution*, **85**, 2223–2228.

Polle, A. and Rennenberg, H. (1993). Significance of antioxidants in plant adaptation to environmental stress, In: *Plant Adaptation to Environmental Stress*, Eds., Fowden, L., Mansfield, T. and Stoddart, J., Chapman and Hall, London, UK, pp. 263–271.

Rao, M.V., Lee, H., Creelman, R.A., Mullet, J.E. and Davis, K.R. (2000). Jasmonic acid signaling modulates ozone-induced hypersensitive cell death, *Plant Cell*, **12**, 1633–1646.

Salter, L. and Hewitt, N. (1992). Ozone-hydrocarbon interactions in plants, *Phytochemistry*, **31**, 4045–4050.

Sandermann, Jr., H. (1996). Ozone and plant health, *Annual Review of Phytopathology*, **34**, 347–366.

Sandermann, Jr., H. (2000). Active oxygen species as mediators of plant immunity: Three case studies, *Biological Chemistry*, **381**, 649–653.

Sandermann, Jr., H., Ernst, D., Heller, W. and Langebartels, C. (1998). Ozone: an abiotic elicitor of plant defence reactions, *Trends in Plant Science*, **3**, 47–50.

Sharma, Y. and Davis, K. (1997). The effects of ozone on antioxidant responses in plants, *Free Radical Biology and Medicine*, **23**, 480–488.

Sharma, Y., Leòn, J., Raskin, I. and Davis, K. (1996). Ozone-induced responses in *Arabidopsis thaliana*: The role of salicylic acid in the accumulation of defense-related transcripts and induced resistance, *Proceedings of the National Academy of Sciences USA*, **93**, 5099–5104.

Solano, R., Stepanova, A., Chao, Q. and Ecker, J. (1998). Nuclear events in ethylene signaling: A transcriptional cascade mediated by ETHYLENE-INSENSITIVE3 and ETHYLENE-RESPONSE-FACTOR1, *Genes and Development*, **12**, 3701–3714.

Srivastava, H.S. (1999). Biochemical defence mechanisms of plants to increased levels of ozone and other atmospheric pollutants, *Current Science*, **76**, 525–533.

Tingey, D.T., Standley, C. and Field, R.W. (1976). Stress ethylene evolution: A measure of ozone effects on plants, *Atmospheric Environment*, **10**, 969–974.

TokarskaSchlattner, M., Fink, A., Castillo, F.J., Crespi, P., Crevecoeur, M., Greppin, H. and Tacchini, P. (1997). Effects of ozone on the plasma membrane proteins in *Arabidopsis thaliana (L.)* leaves, *Plant, Cell and Environment*, **20**, 1205–1211.

Torsethaugen, G., Pell, E.J. and Assmann, S.M. (1999). Ozone inhibits guard cell K+ channels implicated in stomatal opening, *Proceedings of the National Academy of Sciences USA*, **96**, 13577–13582.

Turcsányi, E., Lyons, T., Plöchl, M. and Barnes, J. (2000). Does ascorbate in the mesophyll cell walls form the first line of defence against ozone? Testing the concept using broad bean (*Vicia faba L.*), *Journal of Experimental Botany*, **51**, 901–910.

Vahala, J., Schlagnhaufer, C.D. and Pell, E.J. (1998). Induction of an ACC synthase cDNA by ozone in light-grown *Arabidopsis thaliana* leaves, *Physiologia Plantarum*, **103**, 45–50.

Vanacker, H., Harbinson, J., Ruisch, J., Carver, T.L.W. and Foyer, C.H. (1998). Antioxidant defences of the apoplast, *Protoplasma*, **205**, 129–140.

Wellburn, A. (1994). *Air Pollution and Climate Change: The Biological Impact*, Longman, Harlow.

Wenzel, A.A., Schlautmann, H., Jones, C.A., Küppers, K. and Mehlhorn, H. (1995). Aminoethoxyvinylglycine, cobalt and ascorbic acid all reduce ozone toxicity in mung beans by inhibition of ethylene biosynthesis, *Physiologia Plantarum*, **93**, 286–290.

Zilinskas, B.A., Greenhalgh-Weidman, B. and Brennan, E. (1990). The relationship between EDU pre-treatment and C_2H_4 evolution in ozonated pea plants, *Environmental Pollution*, **65**, 241–249.

6 Effects of oxidants at the whole plant and community level

M.R. ASHMORE

Introduction

The effects of oxidants on vegetation were first demonstrated in the 1950s – much later than those, for example, of sulphur dioxide. Since oxidants are not emitted directly into the atmosphere, but are formed by a series of complex reactions, some of which are photochemical in character, localised hot-spots of intense plant damage close to sources are not typical of ozone; rather, high concentrations of oxidants are a regional problem. During a photochemical smog episode, a range of chemical species may be formed, but there is little doubt that ozone (O_3) is by far the most important in terms of effects on vegetation. Although evidence of the effects of other oxidants, such as peroxyacetyl nitrate (PAN) and hydrogen peroxide (H_2O_2), on vegetation exists (e.g. Temple and Taylor, 1983; Terry *et al.*, 1995), these are of minor regional significance compared with ozone.

This chapter therefore is focussed on the effects of ozone. It first discusses the effects of acute ozone exposures, which may result in direct foliar injury, and then considers in qualitative terms the longer-term effects of chronic ozone exposure on growth, yield and community composition. This is followed by sections reviewing the quantitative relationships between ozone exposure and these responses and considering the significance of ozone effects on vegetation in the context of methods developed to assess its regional impacts. The chapter is based primarily on the extensive field and experimental evidence from both Europe and North America over the last three decades. However, there is increasing recognition of ozone as a global problem, and its effects outside these regions are considered briefly in the final section of the chapter. The content of this chapter is directly linked to that of Chapter 5, which described the biochemical and physiological impacts of oxidants.

Visible injury

One characteristic feature of the effects of high ozone concentrations on sensitive species is the appearance of visible leaf injury. Ozone injury most commonly takes the form of small necrotic flecks or stipples on the upper leaf surface, which may

Air Pollution and Plant Life, second edition. Edited by J.N.B. Bell and M. Treshow. ISBN 0 471 49090 3 (HB), 0 471 49091 1 (PB).

be brown, white or black. Detailed compendia have been produced to illustrate the range of symptoms produced by ozone on different species (e.g. Krupa *et al.*, 1998). Although such symptoms may be widespread geographically, demonstrating that they are due to ozone requires the use of techniques to reduce exposure to the pollutant or to reproduce the symptoms by experimental fumigation. One of the first confirmed reports of widespread foliar injury which could be attributed to ozone was so-called 'weather fleck' of tobacco in the eastern United States (Heggestad and Middleton, 1959); this symptom had previously been thought to be caused by climatic stress. Over the subsequent decades, many other reports of visible injury to crops have been made in North America.

Visible injury due to ozone to species such as morning glory, taro and peanut have been commonly reported in Japan, especially in the regions around Tokyo, from the mid-1970s. In Europe, the first reports of visible injury to crop species also appeared in the 1970s. For example, Ashmore *et al.* (1980) reported visible injury to radish and peas following an episode of high ozone concentrations, which was verified by experiments to demonstrate the protective effect of filtration. Visible ozone damage to crops is now most regularly reported in Mediterranean areas of Europe; for example, Velissariou *et al.* (1996) documented instances of visible foliar injury to 21 crop species in Italy, Spain and Greece. Table 6.1 lists 22 agricultural and horticultural crops for which visible damage to commercial fields have been reported in Europe in ozone episodes.

In North America, several studies have been conducted to assess the extent of visible injury symptoms on forest trees, especially in the Appalachian mountain region. In the 1980s, ozone-like foliar symptoms were observed on several species in this area, with experimental studies using charcoal-filtered air confirming the role of ozone for three species – yellow-poplar (*Liriodendron tulipifera*), green ash (*Fraxinus pennsylvanica*) and sweetgum (*Liquidambar styraciflua*) (Duchelle *et al.*, 1982). More

Table 6.1 List of commercial agricultural and horticultural crops injured by ambient ozone episodes in Europe (from Mills *et al.*, 2000). Reproduced by permission of Gina Mills, Chairperson, ICP vegetation

Agricultural crops		Horticultural crops	
Bean	*Phaseolus vulgaris*	Courgette	*Cucurbita pepo*
Clover	*Trifolium repens*	Chicory	*Chicorium endiva*
Corn	*Zea mays*	Lettuce	*Lactuca sativa*
Grape-vine	*Vitis vinifera*	Muskmelon	*Cucumis melo*
Peanut	*Arachis hypogea*	Onion	*Allium cepa*
Potato	*Solanum tuberosum*	Parsley	*Petroselinum sativum*
Soybean	*Glycine max*	Peach	*Prunus persica*
Tobacco	*Nicotiana tabacum*	Pepper	*Capiscum anuum*
Wheat	*Triticum aestivum*	Radish	*Raphanus sativus*
	Triticum durum	Red beetroot	*Beta vulgaris*
		Spinach	*Spinacea oleracea*
		Tomato	*Lycopersicon esculentum*
		Watermelon	*Citrullus lanatus*

recently, surveys of the sensitive species black cherry (*Prunus serotina*) revealed that injury symptoms were present on almost 50% of individual trees in the Great Smoky Mountains National Park (Chappelka *et al.*, 1997). The evidence of visible damage to tree species is more limited in Europe, although this may reflect the lack of systematic surveys of injury. For example, recent field surveys in southern Switzerland and in Spain, regions with relatively high ozone exposures, have revealed ozone-like symptoms on a large number of shrub and deciduous tree species, with experimental studies confirming the diagnosis in several species (Skelly *et al.*, 1999).

Acute injury to leaves of species with a market value dependent on their visible appearance, such as many horticultural crops, can cause an obvious and immediate loss of economic value. For example, an ozone episode north of Athens in 1998 caused such severe reddening and necrosis on chicory and lettuce that local crops could not be sold (Vellisariou, 1999), with severe implications for local producers. However, the significance of an acute ozone episode causing visible injury for longer-term growth responses depends on a range of factors, including changes in resource allocation to support new leaf growth, which are discussed further below.

Historically, the extent of visible injury has been used to classify the sensitivity of species to ozone. However, sensitivity rankings based on visible injury may not be a useful guide to sensitivity rankings in terms of growth responses. For example, Taylor (1994), reviewing studies of loblolly pine, concluded that in most, but not all, studies, sensitivity rankings based on visible injury differed significantly from those based on growth responses.

In addition to the characteristic visible symptoms, exposure to ozone can also result in more general symptoms such as chlorosis, and there is extensive evidence from both crop and tree species that ozone can accelerate leaf senescence. This is often an important factor in its long-term effects on growth and yield. It is important to distinguish the process of ozone-induced leaf senescence from an accelerated natural senescence, in which cellular constituents are remobilised before leaf abscission. For example, Lippert *et al.* (1996) demonstrated for beech (*Fagus sylvatica*) that, while nitrogen is retranslocated before leaf fall in natural autumn senescence, ozone not only accelerated leaf loss but also inhibited translocation, leading to significant losses of nitrogen.

Biomonitoring of oxidants

The presence of characteristic foliar injury symptoms has been used as a semi-quantitative measure of the degree of phytotoxic ozone in the atmosphere. This method of biomonitoring was developed in the US in the 1960s, through the identification of the super-sensitive tobacco cultivar Bel-W3, and the less sensitive cultivars Bel-B and Bel-C (Heggestad, 1991). The use of these three differentially sensitive cultivars can enhance the capacity to identify different levels of ozone in regional surveys.

Tobacco is by no means the only species which has been used successfully as a biomonitor of ozone, and its rapid growth rate and susceptibility to other diseases, together with changes in ozone sensitivity with continued exposure, can limit its long-term application. One method of improving the standardisation of tobacco plants used in regional biomonitoring is the use of a miniaturised kit using two-week old seedlings grown in filtered air environments (Lorenzini *et al.*, 1995). Assessment of leaf injury on local vegetation, rather than transplantation of standard plants raised from seed, has also been used in biomonitoring of ozone. For example, surveys of foliar injury to sensitive tree species such as black cherry have been used to assess the extent of elevated ozone levels in remote forest areas of the US. However, such *in situ* surveys may be confounded by variations in site conditions, such as soil moisture levels, which influence ozone injury.

Effects on agricultural production

The metabolic and physiological effects of ozone which may lead to reductions in agricultural yields were fully described in Chapter 5. Nonetheless, the processes by which the effects of ozone at its initial points of impact in the leaf lead to reductions in agricultural yields are complex and not always fully understood. For example, translocation patterns to, and hence the growth rate of, different plant organs may depend as much on sink activities as on source strength. Thus, a reduction in the assimilate pool, or in carbon export from leaves, induced by ozone may affect carbon translocation in different ways, depending on partitioning priorities at the time of ozone exposure. For such reasons, it is important to appreciate the dynamic interactions between crop development and ozone exposure.

Sellden and Pleijel (1995) speculated on the role of the balance between source activity, translocation and sink activity, especially in the context that modern crop breeding has mainly stimulated grain production in cereals through changes in harvest index, the proportion of assimilate invested in the grain, as opposed to straw. They suggest that the lower sensitivity to ozone of barley yield, compared with wheat yield, is due primarily to the larger potential surplus of carbohydrate for grain filling in barley, which means that grain yield is less sensitive to reductions in photosynthesis.

The significance of the balance between source activity, sink activity and translocation clearly means that there is a potential for the same ozone episode to have different effects on crop yield depending on when it occurs. However, there are only a limited number of studies in which a crop has been exposed to the same ozone exposure at different points during its development and the effect on final yield compared. For determinant crops in particular, the growth stage at which ozone exposure occurs can be significant. In soybean, kidney bean and dry bean (e.g. Younglove *et al.*, 1994), exposure during pod filling has a greater effect than exposure during earlier growth stages, although studies with bush bean have not demonstrated the same effect. In tomato, the period between flowering and fruit set was most sensitive to ozone. In wheat, it appears that the period between anthesis and grain filling is the most sensitive in terms of yield loss (e.g. Pleijel *et al.*, 1998). The key effect of ozone on wheat and other cereals appears to be an acceleration of leaf ageing, particularly of the flag leaf. For example, Pleijel *et al.* (1997) showed a strong positive association between the effect

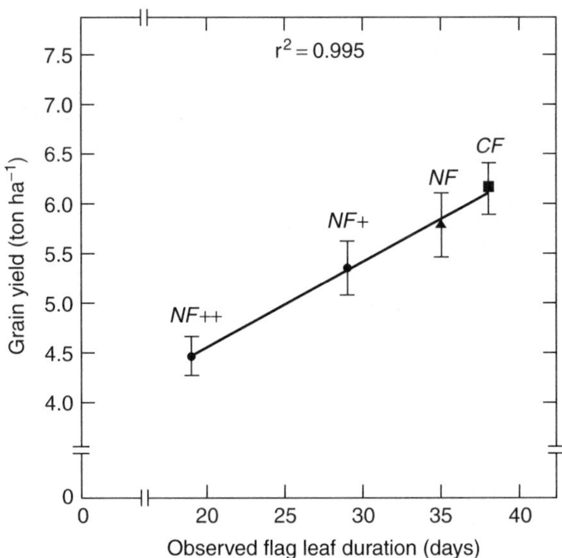

Figure 6.1 Grain yield of wheat plotted against the longevity of the flag leaf in different ozone treatments (CF: charcoal-filtered air; NF: non-filtered air with approximately ambient ozone; NF+ and NF++: non-filtered air with additional 25 ppb and 35 ppb ozone, respectively). From Pleijel *et al.* (1997)

of ozone on the number of days with a green flag leaf and its effect on grain yield (Figure 6.1)

It is widely assumed that effects of ozone on reproductive organs are mediated through reduced carbon allocation. However, it is also possible, although poorly researched, that ozone may have direct effects on reproductive structures. Bosac *et al.* (1994) developed a method of exposing reproductive structures to ozone independently of the vegetative parts, and were able to show that exposing flowering racemes of *Brassica napus* to ozone caused bud abortion and abscission, while Stewart *et al.* (1996), using a similar approach, showed that multiple exposures to ozone caused seed abortion and reduced mature seed number per pod in *Brassica campestris.*

It is also assumed that the main effect of ozone is on crop yield, but the effects on crop quality, although poorly researched, could also be significant. Studies of fruit crops have produced little clear evidence of effects on fruit quality, although a study of grape by Soja *et al.* (1997) indicated that ozone exposure over 2 years caused a large decrease in fruit sugar content. A recent study of the impacts of elevated ozone on oilseed rape in the UK by Ollerenshaw *et al.* (1999) showed a significant reduction in both seed yield and the oil content of the harvested seeds. Since contract price in the UK is based on oil content, this effect of ozone would be an additional economic loss to the producer.

The effects of ozone on fruit species have been poorly studied compared with arable and horticultural crops, despite their economic significance in many areas. The effects of ozone on such species need to be considered over several years, and could be cumulative. For example, Soja *et al.* (1997) found greater effects on grape yield in the second year of exposure, which they attributed either to reduced carbohydrate storage in the stems after the first season's exposure, or to the greater yield in the second year producing a greater assimilate demand. However, other studies have reported no cumulative effects of ozone; for example, Retzlaff *et al.* (1997) reported similar yield reductions in plum trees over four consecutive years of fumigation.

The effects of ozone on managed pastures have also received relatively little attention compared with effects on arable crops, despite their economic importance. Several studies in both Europe (e.g. Fuhrer *et al.*, 1994) and North America (e.g. Montes *et al.*, 1982) have demonstrated that increased ozone exposure reduces clover biomass in grass-clover swards, with this being accompanied in some, but not all, cases by an increase in the biomass of the grass component. There is evidence from studies of managed swards that regular cutting enhances this process, probably because of the reduced capacity for regrowth of clover exposed to ozone. The reduction of the clover component will reduce nitrogen fixation, and the overall forage quality in such systems, to an extent which depends on the proportion of clover in the forage mix.

Most experimental studies have grown crops under conditions which prevent the occurrence of pests and diseases. While this is understandable, given the need to prevent confounding effects of uncontrolled variables, it does mean that the possible role of indirect effects of ozone have been studied to a limited extent. Furthermore, many commercial fungicides and pesticides have been shown to provide significant protection against ozone injury (e.g. Taylor and Rich, 1973), and this is an important factor both in interpreting experimental studies and assessing the extent of damage to crops in the field. Interactions with pests and diseases are considered in more detail in Chapter 20 of this book.

Effects on forest vitality and production

As for crop species, the effects of ozone on forests can be explained through effects on photosynthesis, translocation and sink activity. There is evidence for several tree species that ozone can increase carbon retention in the leaves, either because of higher demand for assimilate for repair processes or because of decreased phloem loading, and can reduce carbon allocation to the roots. There is also evidence (e.g. Andersen *et al.*, 1997) that ozone exposure can reduce root carbohydrate content and new root growth in the spring following a season of ozone exposure. In certain species, such as poplar, premature abscission of older leaves can be partly compensated by increased production of new leaves (e.g. Woodbury *et al.*, 1994), but this may be at the expense of carbon partitioning to the stem and root. The key difference between forests and arable crops is the longevity of tree species, which means that the secondary effects of ozone exposure on resource allocation need to be considered over a much longer time frame. Furthermore, nutrient and water supply may be critical factors for forest vitality and growth, while small changes in crown or root development can eventually lead to changes in competitive balance between or within species. It is therefore essential to consider ozone impacts on forests in this wider context.

Where ozone is present in high concentrations, the evidence linking cause and effect in the field is strong, and can provide insights into the impacts of ozone on forest systems over decades. The area where the impacts of ozone stress on forest community composition has been most intensively studied is in the San Bernardino mountains, which surround the city of Los Angeles (Miller and McBride, 1999). Effects of ozone pollution, generated from pollutant emissions in the city below, began to be observed on the native forest community in the 1960s. The most dominant species of these mixed-conifer forests prior to European settlement were ponderosa pine (*Pinus ponderosa*) and Jeffrey pine (*Pinus jeffreyi*), due to their tolerance of the frequent wildfires; however, these have also proved to be the most sensitive species to ozone. In many areas, both species have shown severe foliar injury and reduced needle longevity. These symptoms are associated with reduced radial growth, or even years with missing growth rings. Trees affected by ozone are more susceptible to attack by bark beetles, which are often the direct cause of mortality. Outbreaks of bark beetles are associated with drought years, as well as high ozone concentrations. Regeneration in these forests is greater for trees such as white fir or cedar species, which are more resistant to ozone, although at some higher elevation sites, these, and other, conifer species do not naturally regenerate and the area may become dominated by shrubs. These patterns of change in community composition are confounded by the role of fire, since current fire exclusion policies favour replacement of ponderosa and Jeffrey pine by more fire-sensitive species which also happen to be more ozone-tolerant.

In areas where ozone is a less significant stress, a limited number of field studies have tried to identify temporal or spatial associations between ozone exposure and the growth of mature trees. Mclaughlin and Downing (1995) made short-term measurements of stem growth of loblolly pine (*Pinus taeda*) in the eastern United States over a period of 5 years, using a sensitive dendrometer, and related these to records of weekly variations in ozone concentrations, climate and soil moisture stress. The results showed that the strongest predictor of short-term radial growth was the interaction between ozone and soil moisture, with the short-term ozone effect being greater in a moist year than a dry year. In contrast to these very short-term variations in growth, Peterson *et al.* (1995) examined annual radial growth patterns in bigcone Douglas fir (*Pseudotsuga macrocarpa*) in the San Bernardino mountains of southern California since 1950. Although this species is less sensitive than ponderosa pine and Jeffrey pine, which are extensively damaged in these mountains, there was evidence of long-term reductions in growth compared with the period prior to 1950 over the whole range of sites; furthermore, those with higher ozone exposures had shorter needle retention and lower radial growth rates. Over a shorter time period, Braun *et al.* (1999) examined the diameter growth of beech trees at 57 permanent plots in Switzerland over a 4-year period in relation to the estimated ozone exposures at these sites, and a range of other site and climatic variables. A significant negative association between radial growth and ozone exposure was found, with a slope which was greater than that determined in earlier experimental studies with beech seedlings (Braun and Fluckiger, 1995).

This difference illustrates the problems of relating the extensive data from short-term chamber studies with young trees to long-term effects on mature forest stands (Kelly *et al.*, 1995; Samuelson and Kelly, 2001). One possible approach is to combine experimental and field approaches to strengthen the interpretation of field data. For example, Karnosky *et al.* (1999) investigated the effects of ozone on three clones of trembling aspen (*Populus tremuloides*), whose relative sensitivity had been determined in previous chamber studies. The results demonstrated effects of ozone on foliar injury and epicuticular wax degradation at three field sites with differing ozone exposures which were consistent with those of the earlier chamber studies and with an open-air fumigation experiment.

Other studies have directly compared responses in seedlings and mature trees experimentally. These studies show variable results, with the mature trees appearing to be more sensitive in certain studies, but less sensitive in others. Such differences reflect the interplay of a number of factors (Matyssek and Innes, 1999). These include carbon allocation patterns, the higher respiratory demand relative to green biomass in mature trees, differences in stomatal conductance due to the much higher hydraulic resistance of mature trees, the effect of the varied contribution of shade leaves (which may differ in ozone sensitivity and exposure from sun leaves) in mature crowns, and differences in hormonal signalling.

The use of physiological growth models with ozone exposure data is another possible method of assessing effects on mature trees. Figure 6.2 provides a simplified illustration of how one model integrates data on ozone impacts at the leaf level into a broader model of canopy water relations and the tree's carbon budget (Ollinger *et al.*, 1997). Such models may also be used to explore the implications of different

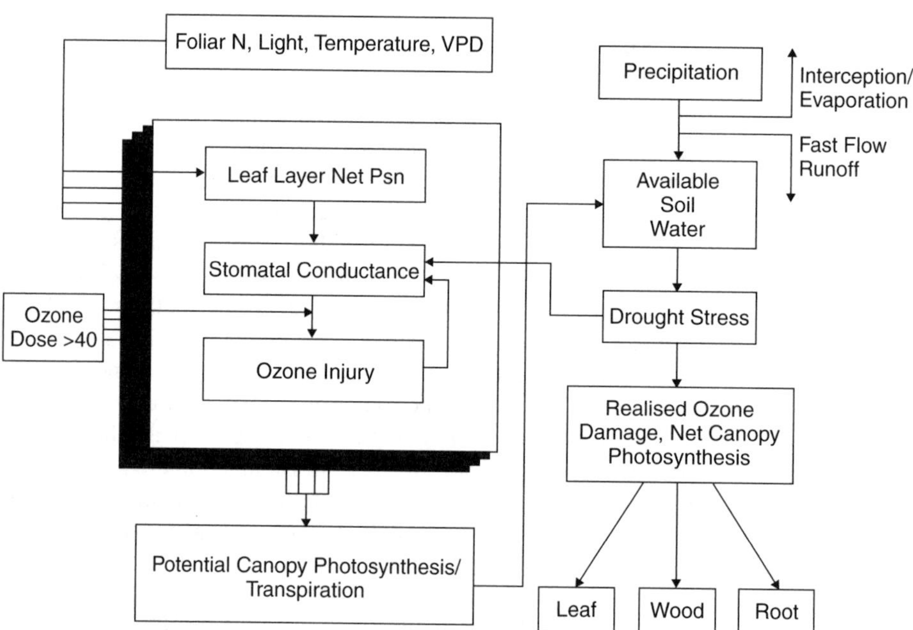

Figure 6.2 Simplified flow diagram of processes included in the PnET ozone response model. From Ollinger *et al.* (1997). Reproduced by permission of Ecological Society of America

morphology and physiology for ozone responses. For example, Constable and Taylor (1997) used a model to demonstrate that very different effects of ozone will occur on two varieties of *Pinus ponderosa*, with different stomatal and photosynthetic characteristics, and different needle morphology and tree architecture, assuming a fixed relationship between the maximum photosynthetic rate and the amount of ozone absorbed, based on experimental data. Weinstein *et al.* (1998) used a modelling approach to compare the response of seedlings and mature trees of red oak, based on a three-year experimental study. In this particular case, the mature trees were more sensitive, primarily because of their higher stomatal conductance and ozone flux, and the model predicted, because of the assumed hierarchy of carbon allocation, much bigger effects of ozone on root growth in mature trees. There are limited data on root responses to ozone in mature trees in the field, although Grulke *et al.* (1998) reported much lower root biomass of *Pinus ponderosa* in the spring at sites with higher ozone exposures in the San Bernardino mountains. Such effects may be crucial to long-term forest vitality at sites where water availability and nutrient supply are limited.

Table 6.2 summarises the results of a number of assessments of the impact of ambient ozone on forest growth in the eastern US, drawing on studies of both seedlings and mature trees, and illustrating the range of techniques which have been used (Mclaughlin and Percy, 1999). The estimates of annual growth reductions for seedlings are in the range 0–13%, and those for mature trees in the range 0–30%, depending on species, assessment techniques, ozone exposure and climatic conditions. This illustrates the current difficulties of reaching any firm conclusions about the scale of impact of ozone on mature forests.

Table 6.2 Estimates of the effects of ozone on annual growth of forest tree species in the eastern US, based on studies with seedlings and with mature trees, modified from Mclaughlin and Percy (1999)

Species	Basis of estimate	Growth reduction
Seedling studies		
Southern pines	Synthesis of field chamber studies	2–5%
Lobolly pine	Synthesis of field chamber studies	
	Mean response	0–3%
	Sensitive family response	1–10%
Hardwoods	Values derived from response surfaces	13%
Conifers	Values derived from response surfaces	3%
Mature trees		
Loblolly pine	Whole-tree model based on branch chamber data	2–9%
Loblolly pine	Analysis of short-term changes in stem growth	
	Year of low water stress	0.5%
	Year of moderate water stress	0–30%
Loblolly pine	Regional forest simulation model based on small tree exposures	5–12%
Shortleaf pine	Empirical estimate from analysis of annual growth variation	1%
Hardwoods	Regional canopy-stand simulation model	2–17%

Effects on biodiversity

The effects of ozone on wild herbaceous or shrub species have received much less attention than those on agricultural crops or major forest species. When the concern is ozone effects on biodiversity, or on the survival of individual wild species with different ecological strategies, the traits which are examined should be those that relate most directly to ecological fitness, rather than visible symptoms or vegetative growth (Davison and Barnes, 1998). Seed output is of undoubted ecological significance for ruderal species, but the extent to which seed output can be reduced by ozone exposure without affecting population viability is unclear. For competitor species, the capacity to alter resource allocation in response to competitive pressure is important; where competition is primarily for below-ground resources, it may be the responses of root morphology to ozone that are critical. The critical response for perennial stress-tolerant species is more difficult to identify, but may depend on the extent to which ozone alters sensitivity to specific stress factors.

The effects of ozone on resource allocation may be more important than those on photosynthetic capacity for many wild species. The effects of ozone on resource allocation have been studied infrequently in wild species, but the available data indicate that there is considerable variation between species in effects on resource allocation. For example, Bergmann *et al.* (1996) showed that species can differ widely in their relative allocation of resources to vegetative growth and seed or flower production (Figure 6.3). Of 17 species exposed to ozone from seedling stage to flowering, 12 showed comparable reductions in both vegetative and reproductive biomass. However, five species were clear outliers. In two species (*Papaver dubium* and *Trifolium arvense*) resources

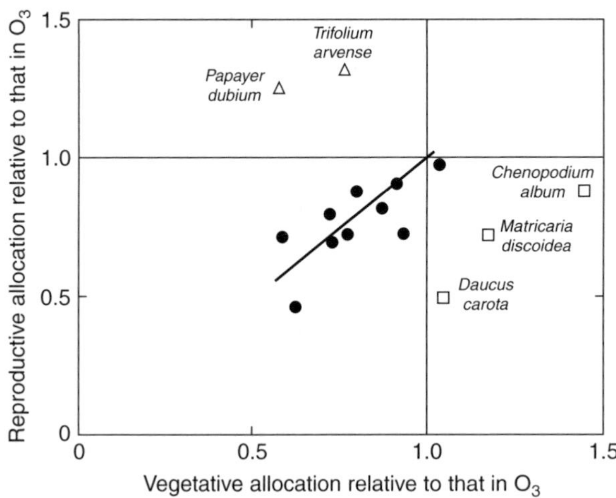

Figure 6.3 Effects of ozone on the relative allocation of resources to vegetative and reproductive organs of 17 wild annual species. Allocation is expressed as the ratio of that in ozone to that in charcoal-filtered air. Solid symbols represent species which show similar changes in allocation to vegetative and reproductive growth, open symbols represent species which show a difference in allocation to growth. From Bergmann *et al.* (1996)

were switched to reproductive growth, leading to an increased reproductive biomass in ozone-treated plants. In contrast, in three species (*Chenopodium album, Matricaria discoides* and *Daucus carota*), resources were switched to vegetative growth, leading to an increased vegetative biomass in ozone. Similarly, Warwick and Taylor (1995) showed that the responses of different calcareous grassland species in terms of root responses could be very different from that of above-ground responses; furthermore, the species differed in the nature of the root response.

The ozone response of well over 100 species of semi-natural communities has been investigated and it is clear that there are many wild species that are, when grown as individual plants, as susceptible as the most sensitive crops (Ashmore and Davison, 1996). Some attention has been paid in recent years to the question of whether there are ecological or physiological characteristics of species, as identified, for example, in Grime's CSR growth strategy, which tend to be associated with ozone sensitivity. The most comprehensive analysis of this issue is that of Franzaring *et al.* (1999), who compiled a database of 131 European species in terms of visible injury and growth responses to ozone. Their results suggested a slight association with CSR strategies, with stress tolerators tending to be somewhat less sensitive and ruderals tending to be somewhat more sensitive. However, it may not be meaningful to compare species with very different ecological strategies on the basis of growth and visible injury alone.

Franzaring *et al.* (1999) also examined the association with relative growth rate (RGR) and specific leaf area ratio (SLA), using data taken from literature compilations rather than the actual experiments. The results (Figures 6.4(a) and 6.4(b)) show a weak negative association with both RGR and SLA. The relationship to SLA might be explained by the fact that leaves with high SLA are thin and have a high surface to volume ratio. The RGR relationship is consistent with the weak positive relationship between ozone sensitivity and RGR reported by Reiling and Davison (1992). One reason for the association between high RGR and high ozone sensitivity may be that high growth rates are associated with high stomatal conductance and hence high ozone flux; positive associations between stomatal conductance and visible injury in different species have been reported, e.g. by Bungener *et al.* (1999).

However, it is unclear how well the responses of plant communities to ozone can be predicted from our knowledge of the responses of individual plants. The few studies in which artificial species mixtures have been exposed to ozone indicate that the performance of the more sensitive species tends to be reduced further by ozone in competition compared with monoculture (Davison and Barnes, 1998). However, the empirical evidence from the rather few studies which have examined the effects of ozone on intact communities in their natural habitat suggests that it is not so easy to predict the effects of ozone on mixtures or communities. Thus, Evans and Ashmore (1992) found that the effects of filtration on an acid grassland community were the opposite of those predicted from the responses of the individual species, with ozone-sensitive forbs showing a greater cover in unfiltered air. They hypothesised that this was due to these species responding to changes in the cover of the dominant grass species, rather than directly to ozone. Barbo *et al.* (1998) examined the impacts of a range of ozone concentrations on early succession communities in a pine forest. Ozone caused a significant decrease in species richness, species diversity and species

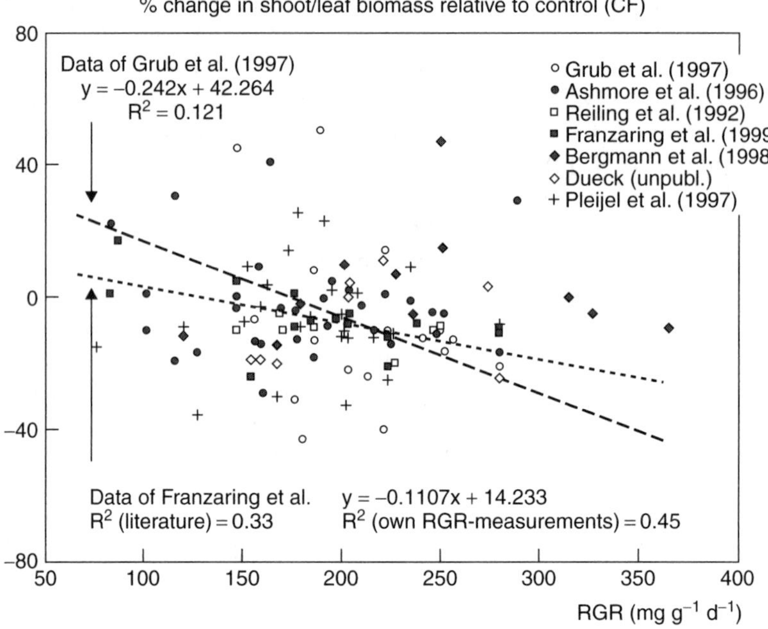

(a)

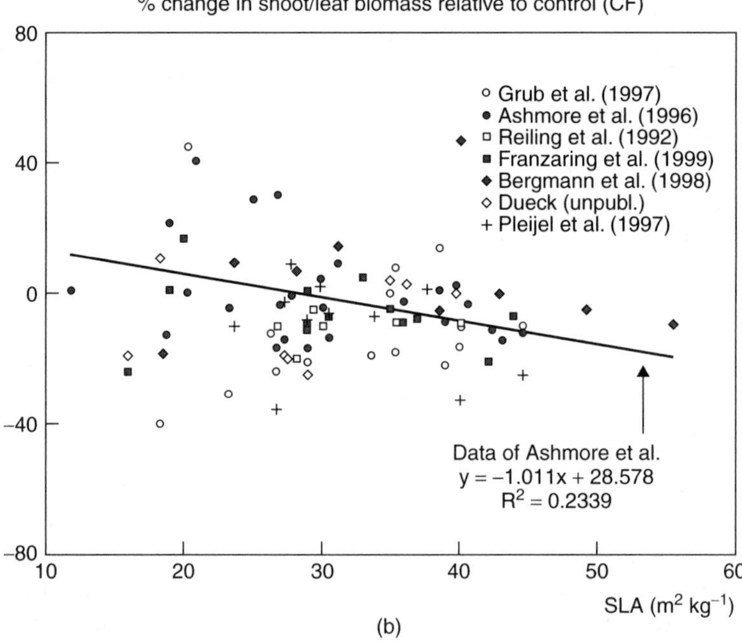

(b)

Figure 6.4 Relationship between growth reduction in ozone and relative growth rate (RGR) (a) and specific leaf area ratio (SLA) (b) in wild annual species. From Franzaring *et al.* (1999)

evenness in the first year of fumigation, although the effect was less clear in the second year of treatment, when, however, there was a stronger effect on canopy cover. These results imply that ozone may cause less diverse plant communities, but effects on individual species may be more difficult to predict. Thus, blackberry (*Rubus cuneifolius*) increased in cover in the highest ozone treatment, despite its known ozone sensitivity and despite showing clear visible symptoms of damage. This again illustrates that release of competitive stress under ozone may cause secondary effects on community composition which cannot readily be predicted from the known sensitivity of individual species.

These experiments were of very limited duration and provide little guidance as to whether, for example, long-term ozone exposure could result in species loss from particular communities. Furthermore, almost all experimental studies to date describe effects on relatively common species; the effects of ozone on rare or threatened species are almost unknown. The approach of using spatial variation in the field to evaluate the impacts of ozone on communities has only been adopted in one study, that of Westman (1979), who compared cover and species richness at 67 sites in California with coastal shrub vegetation with a range of variables, including mean oxidant concentration. The results showed a significant negative correlation between mean oxidant concentration and complexity of community structure, in terms of percentage cover and species richness.

Factors such as water and nutrient supply have a strong influence on inter-specific competition and may interact with the effects of ozone; these interactions are considered in Chapter 18. Nutrient supply may be a particularly important factor, as most natural vegetation has limited supplies of at least one major nutrient. Studies on crop species, which have very different nutrient demands, are of little relevance, but there have been very few studies of wild plants. Interestingly, however, there is evidence (Whitfield *et al.*, 1998) that nutrient limited *Plantago major* plants show a greater reduction in seed production when exposed to ozone than do well-fertilised plants. This is in contrast to the results with young birch trees (Maurer *et al.*, 1997), in which low nutrition enhanced the antioxidative defence capacity and delayed leaf loss.

Little is known about the impact of ozone on lichens and bryophytes. However, adverse effects of moderate ozone exposure have been reported in certain studies (e.g. Eversman and Sigal, 1987; Potter *et al.*, 1996), and further work is needed (see Chapter 17).

Intra-specific variations in sensitivity

The discussion of species responses in the previous sections has largely ignored the fact that there is considerable variation in sensitivity to ozone between different cultivars of crop species, and within tree species and wild herbaceous species. There is evidence that there is a genetic basis to such differences in sensitivity, for example, in terms of visible injury between crop cultivars (e.g. Mebrahtu *et al.*, 1990). In some cases, active breeding programmes have been adopted to improve ozone resistance; one such successful programme was adopted to reduce the widespread field damage to tobacco crops in the eastern USA (Menser and Hodges, 1972). It is unclear for crop species to what extent the difference between cultivars reflects inadvertent selection in the breeding process, or by farmers themselves, in areas with high ozone exposures. There is, however, no evidence that modern crop cultivars, bred under greater ozone concentrations than those of decades ago, are more tolerant of ozone. Indeed, modern Greek cultivars of wheat (Barnes *et al.*, 1990) are actually more sensitive to ozone than older cultivars, suggesting that selection for higher yield has also led to selection for characteristics associated with a lower ozone resistance.

Outside the agronomic environment, natural selection may be operating to favour more ozone-tolerant genotypes in areas with higher ozone exposures. The evolution of tolerance to other pollutants, such as toxic metals and sulphur dioxide is well established. However, there is less empirical evidence for ozone, partly because of its regional distribution, which means that spatial variation in ozone exposures can be confounded by variation in climate. The most convincing studies are probably those on trembling aspen (*Populus tremuloides*) in the US and on common plantain (*Plantago major*) in the UK. The work of Berrang *et al.* (1989) with trembling aspen involved vegetative propagation from collections of roots made in five national parks with different exposure, followed by a field trial at a single location with high ozone exposures. There was wide variation in visible injury between clones within each of the populations and each population contained clones which showed no injury. However, the most sensitive clones from the parks with lower ozone exposures had more leaf injury than the most sensitive clones from parks with higher exposures, suggesting that such genotypes are eliminated either directly or indirectly, e.g. through intraspecific competition for light, at sites with higher ozone.

In the case of *Plantago major*, differences in the sensitivity of populations collected from different locations within the British Isles have been found to relate to the estimated ozone exposure at the point of collection. This relationship has subsequently been extended by the inclusion of continental populations (Lyons *et al.*, 1997), which provide a more comprehensive dataset showing a significant positive correlation between ozone exposure and the ozone resistance of the population (Figure 6.5). However, it is difficult to be completely confident that ozone, rather than other climatic variables such as sunshine hours or temperature, which correlate spatially with ozone, are the causal factors. Davison and Reiling (1995), showed that *Plantago major* plants grown from seed collected at two locations in 1985, after several years of relatively low ozone exposures, were more sensitive than plants grown from seed collected from the same locations in 1991, after two years of high ozone exposures. This observation lends support to the argument that ozone is the direct cause of the spatial differences in ozone resistance. However, other studies have failed to relate intraspecific variation in sensitivity to ozone exposure; for example, Lee *et al.* (1999) found large differences in ozone sensitivity of black cherry clones which were unrelated to their geographical origin.

The implications of this variation in ozone sensitivity were considered by Taylor (1994), who summarised the role of genotype in the responses to ozone of one of the most intensively studied North American tree species, loblolly pine (*Pinus taeda*). In most, but not all, experimental studies of this species there was a statistically significant

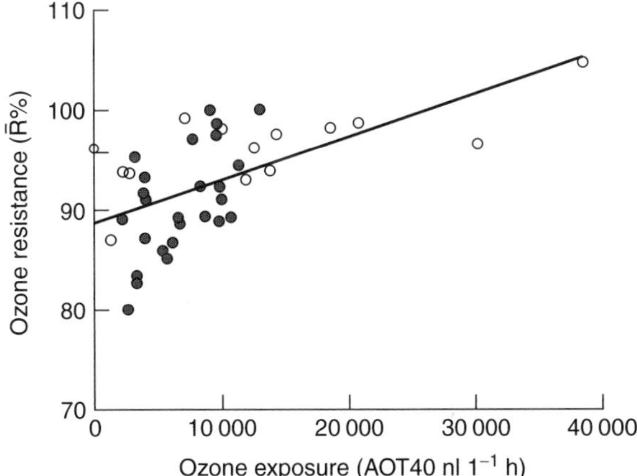

Figure 6.5 Ozone resistance of seed-grown *Plantago major* populations plotted against ozone exposure at the sites of collection. Filled symbols indicate UK sites, open symbols indicate continental European sites. From Davison and Barnes (1998). Reproduced by permission of Davison, A.W. and Barnes, J.D., Effects of ozone on wild plants, in *New Phytologist*, Blackwell Science

difference in response between families. There is evidence of a genetic component of this variation which differs with response variable, suggesting a different genetic and mechanistic basis for resistance to visible injury and growth responses. The family ozone interaction term in different studies was then used to estimate the difference in response between the mean of studied populations and the most sensitive cohorts. As shown in Figure 6.6, there is a large difference in the responses; while the threshold for the mean population is comparable to current ozone exposures in the parts of the southern US where this species is grown commercially, that for the sensitive cohorts is well below current exposures and comparable to those at the beginning of the last century.

Exposure–response relationships

Thus far, this chapter has considered the effects of ozone in a generic manner, with no reference to the ozone exposures used in specific studies. Nevertheless, it is clearly of considerable practical importance to define the concentrations of ozone which produce a given level of response in different species, and to compare these with ambient exposures. Studies which generate exposure–response relationships by including a range of ozone treatments within a single experiment are most useful in determining the forms of the relationships and possible thresholds for adverse effects.

The first major programme to determine exposure–response relationships for crops was the US National Crop Loss Assessment Network (NCLAN), which was established in the late 1970s, with five experimental sites chosen to reflect the variation in climatic conditions and cropping systems across the country (Heck *et al.*, 1988). The

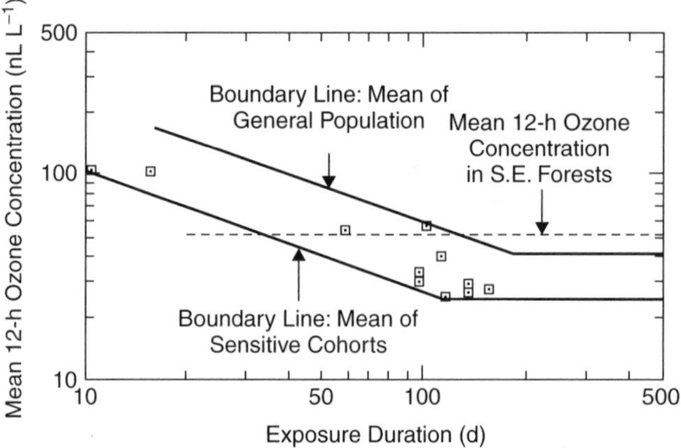

Figure 6.6 Minimum ozone exposure (expressed as a combination of 12 h mean ozone exposure and exposure duration) for effects on the growth of loblolly pine seedlings. The area between the lines represents the difference between the mean population response and that of the most sensitive cohort. From Taylor (1994). Reproduced by permission of American Society of Agronomy, Soil Science Society of America, Crop Science Society of America

experimental studies involved the use of a standard design of open-top chambers, which were placed over the crop and provided growth conditions close to those outside. Ozone exposure was expressed as the seasonal mean ozone concentration for the 7 or 12 hours (depending on the experimental design) during which ozone was added. Ten crops were examined, representing 85% of the cropped area of the US. The exposure–response relationships allowed the yield losses for different crops to be estimated for a given ozone exposure. For a seasonal 7 h mean ozone concentration of 50 ppb, estimated yield reductions for the most sensitive crops, such as soybean and cotton, exceeded 10%, for winter wheat were slightly under 10%, and for the most tolerant crops, such as rice, barley and sorghum, were below 5% (Adams *et al.*, 1988).

During the 1980s, an attempt was made by the European Commission to develop a similar coordinated assessment of air pollutant impacts on agriculture (Jäger *et al.*, 1993). This programme did not use such a standardised experimental approach, and focussed on a more limited range of crops, but has also been used to derive exposure–response relationships. Rather than using the 7 h seasonal mean ozone concentration, a new method of summarising the cumulative seasonal ozone exposure was developed, termed the AOT40 (Accumulated exposure Over a Threshold concentration of 40 ppb) index (Fuhrer *et al.*, 1997). The use of this index allowed a linear relationship to be established with wheat yield reductions in a pooled dataset covering 17 experiments in six different countries (see Figure 22.2 of Chapter 22).

Figure 6.7 shows a recent compilation of experimental data from the EU and the USA for six crops, using the 7 h mean ozone concentration, which indicates considerable consistency in exposure–response relationships from the two continents (Mills *et al.*, 2000). Soybean and wheat are the most sensitive crops in terms of yield,

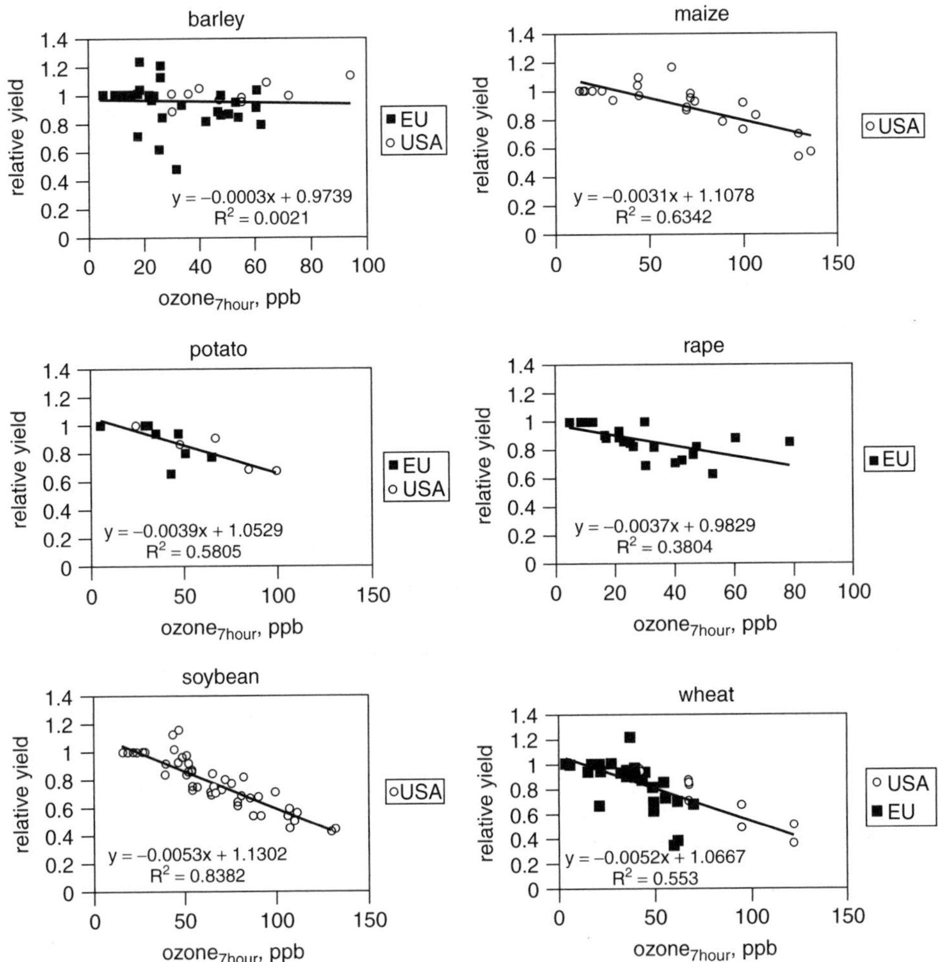

Figure 6.7 Yield response functions plotted from data presented in the literature on the response of yield of six crop species to 7 h mean ozone concentrations, expressed relative to the control treatment. Closed symbols represent data from studies in Europe, open symbols represent studies from USA. From Mills *et al.* (2000). Reproduced by permission of Gina Mills, Chairperson, ICP Vegetation

followed by potato and rape, with barley showing no significant effect even at high exposures.

However, considerable caution needs to be exercised in using such exposure–response relationships to assess the regional impacts of ozone. In typical exposure–response experiments, ozone is added simultaneously in all treatments, i.e. under the same atmospheric and micrometeorological conditions. When comparing the impacts of ozone in different locations, or in different years, it is important to appreciate that the different climates may lead to variable impacts of ozone. For example,

Grunehage *et al.* (1997) showed that in the field the highest ozone concentrations tend to occur under meteorological conditions which limit the dose of ozone absorbed by the plant. This is because of the high resistance to ozone flux across the atmospheric boundary layer to the vegetation and because such concentrations often occur with high vapour pressure deficits, which lead to low values of stomatal conductance. These high resistances can cause large vertical gradients in ozone concentrations over crops which lead to large differences between cumulative ozone exposures between normal measurement height and the crop surface, a factor which needs to be considered when assessing the likely impact of ozone concentrations measured in national monitoring networks (Figure 6.8). In summary, empirical exposure–response relationships for crops derived from open-top chamber studies cannot be applied with any confidence to value the relative benefits of policies to reduce ozone exposures at different locations.

Fewer exposure–response relationships comparing different species exist for trees. One of the most comprehensive studies is the work of Matsumura and Kohno (1999), who exposed seedlings of 17 Japanese coniferous and deciduous tree species to ozone over three growing seasons, and measured effects on dry weight. For most species, significant linear relationships were found with the AOT40 index. There was an association between low leaf life-span and high ozone sensitivity; other short-term studies have also reported that deciduous species are generally more sensitive to ozone than evergreen species. However, chamber studies, which are limited to studies of seedlings or young saplings lasting at most a few years, can provide only limited information on the long-term effects of ozone on mature trees. There is also an important statistical issue, as even with relatively small effects of ozone on growth rate, the cumulative effects over decades of tree growth could be substantial. This is of practical importance in interpretation of standard chamber experiments which typically can detect differences between treatments in the range of 10%.

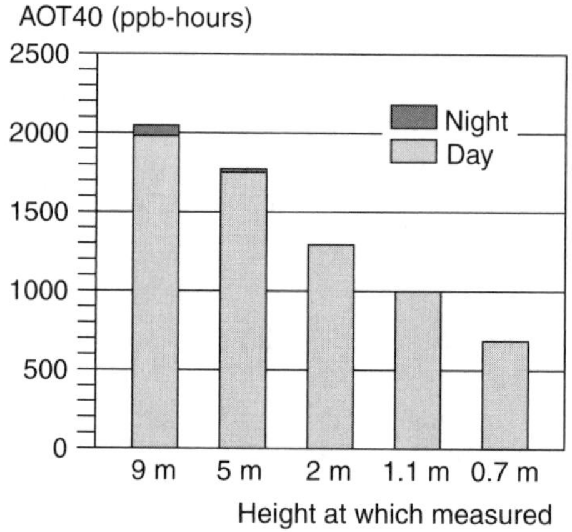

Figure 6.8 Seasonal ozone exposure measured at different heights above a barley crop 1 m high. From Pleijel *et al.* (1995)

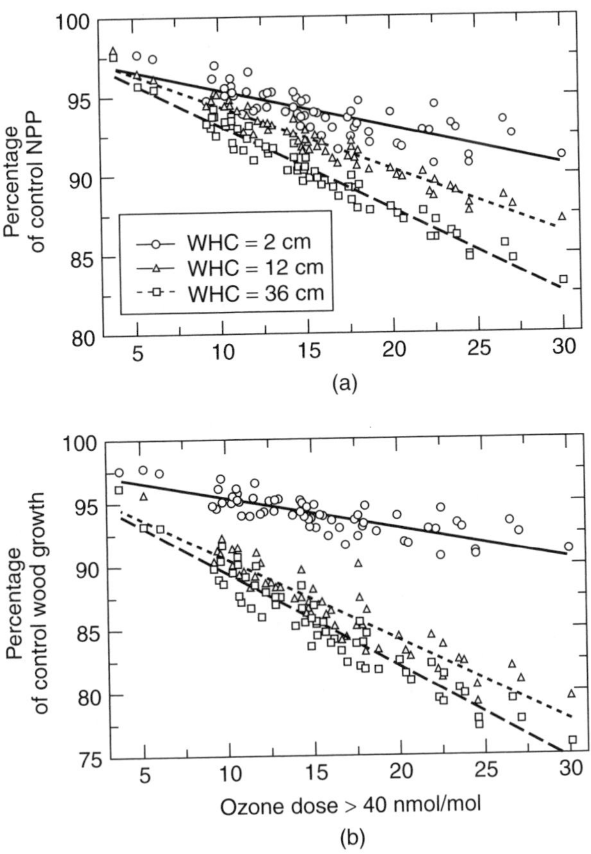

Figure 6.9 Modelled reduction by ozone of mean annual wood production of hardwood species at 64 sites across the northeastern USA, plotted against mean ozone exposures for the period 1987–1992. Predictions are shown for three levels of soil water-holding capacity (WHC) ranging from 36 cm to 2 cm. From Ollinger *et al.* (1997)

This problem raises new methodological challenges in the derivation of exposure–response relationships, which will require integrating experimental data from experiments with both young and mature trees with mechanistic models of long-term responses which are appropriate for regional climatic and edaphic conditions. For example, Ollinger *et al.* (1997) used a canopy model, parameterised with experimental data on ozone effects on stomatal conductance and photosynthetic rate to predict the impacts of ozone on the growth of hardwood forests using data from 64 sites in the northeastern US. The results (Figure 6.9) showed a strong relationship between modelled wood production and net primary productivity and seasonal ozone exposure, expressed as an index equivalent to the AOT40, with reductions in annual wood production of between 3% and 22% at the different sites. Soil water-holding capacity had a strong influence on model predictions, and the results were also found to be sensitive to model assumptions about the interactions between ozone exposure, stomatal function and water stress. Research is urgently needed to assess the impacts of ozone in mature forest stands to improve the reliability of such models.

Use of flux concept

An alternative approach which would improve the assessment of ozone impacts, and which could account to some extent for the influence of climatic and ontogenetic factors, is the use of the estimated flux of ozone into the leaf, rather than the external exposure (see Chapter 4). This involves modelling the pathway of ozone from the atmosphere through the plant canopy and the stomata to sites of damage in the leaf mesophyll. Reich (1987) was one of the first to argue that responses to ozone should be considered in terms of ozone flux, and showed that the different sensitivity of plant species could at least partly be explained by the different values of stomatal conductance. Within a species, visible injury in *Phaseolus vulgaris* was demonstrated by Amiro *et al.* (1984) to relate closely to the estimated ozone flux in laboratory experiments.

The value of using ozone flux is well illustrated by a recent analysis of experimental data for wheat collected over several seasons in southern Sweden by Pleijel *et al.* (2000). The yield data were first related to the ozone exposure over the period of grain filling (Figure 6.10(b)). This showed a wide variation in the slope of the relationships in different years, with a yield loss of 10% being predicted at AOT40 values which differ by an order of magnitude over the different years. Figure 6.10(a) shows the same data plotted against the modelled absorbed dose of ozone over the same period. Five of the six experiments now fall on a common line, indicating that by incorporating the effects of irradiance, temperature and vapour pressure deficit in modifying stomatal ozone uptake a stronger mechanistic relationship to yield is obtained. The outlier in Figure 6.10(a) may be explained by the fact that winter wheat, rather than spring wheat, was used in that one year. Similarly, Baumgarten *et al.* (2000) compared thresholds for visible leaf injury to mature and young beech trees at different sites in Germany. Very different thresholds were found in terms of ozone exposure expressed as AOT40, depending on age location and season, but when they were calculated as a cumulative stomatal ozone flux, very similar thresholds of around 3 mmol m^{-2} were found in all cases.

Progress has been made in developing such ozone flux models for application on the regional scales required to assess the impacts of ozone. For example, Emberson *et al.* (2000) present a comparative analysis of spatial and temporal patterns of AOT40 and modelled ozone flux in Europe, by linking an ozone flux model to a photochemical ozone transport model which predicts ozone concentrations at a height of 50 m in 150 km grid squares across Europe. Whereas the highest AOT40 exposures were in central and southern Europe, high ozone fluxes were modelled in parts of northern and western Europe. Calculations of cumulative stomatal ozone dose over the growing season for wheat and beech for four different grid squares (in Sweden, UK, Czech Republic and Spain), which experienced quite different AOT40 values, showed very little difference in the modelled cumulative stomatal dose, primarily

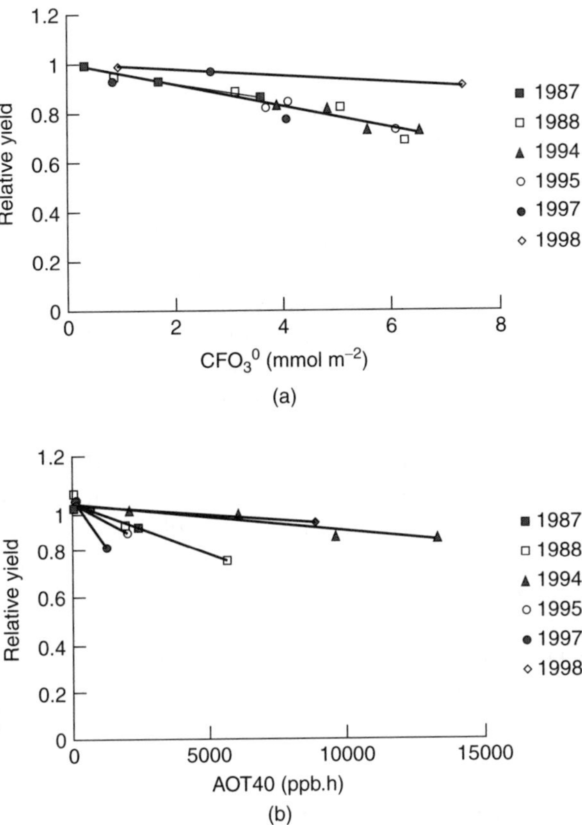

Figure 6.10 Linear regression of relative yield of wheat in experiments in Sweden in six different years against (a) modelled cumulative ozone uptake by flag leaves and (b) ozone exposure during the grain filling period. From Pleijel *et al.* (2000). Reprinted from *Environmental Pollution*, vol. 109, Pleijel *et al.*, figures 1 and 2 in 'An ozone flux-response relationship for wheat', pp. 453–462 (2000)

because of the effects of differences in phenology and of modelled vapour pressure deficit.

Massman *et al.* (2000) proposed extending this approach through a conceptual model of plant response linking estimates of ozone dose through the stomata to the capacity of defence mechanisms to detoxify the incoming ozone flux. Models are currently being developed of these detoxification mechanisms; for example, Plochl *et al.* (2000) modelled the extent to which incoming ozone flux can be detoxified by reactions with ascorbate in the apoplast. However, much more work is needed before unified mechanistic models of ozone response based on these different components of the ozone pathway from the bulk troposphere to the site of damage inside the leaf can be developed. Furthermore, such models do not consider the secondary effects of the modelled impacts at the initial site of damage in the leaf, as mediated, for example, by changes in resource allocation within the plant.

Economic effects of ozone

There is increasing interest in conducting cost-benefit analyses of the efficacy of measures to reduce ambient concentrations of pollutants such as ozone. This raises the challenge of using knowledge of plant responses to ozone to assess the economic benefits of measures to reduce the exposure to ozone of crops and forests. Ozone episodes have caused widespread visible injury on sensitive crops, such as spinach or tobacco, for which leaf appearance is a significant factor in their value, leading to major economic losses. However, quantifying the regional economic impacts of ozone on agricultural or forest production is a more complex task.

In theory, exposure–response relationships, such as those shown in Figure 6.7 could be integrated with databases on crop distributions and ozone concentrations to estimate the regional or national impacts of ozone, or rather to assess the benefits of measures to reduce ozone concentrations. For example, Adams *et al.* (1988) estimated, using a model incorporating demand curves to determine effects on crop prices, that a national decrease in ozone concentrations in the US of 40% would provide a net annual economic benefit of $3000 million, or 2.8% of national production. However, such national or international estimates may disguise the much greater losses in particular crops and regions; for example, Olszyk *et al.* (1988) estimated yield losses due to ozone in California to be in the range 20–25% for cotton, bean and onions. A recent evaluation, in the context of the development of EU emission control policy, estimated the annual benefit of reducing ozone exposure by about 30% across the member states as about 2000 million Euro; interestingly, this value is consistent in scale with the US estimate a decade earlier. The benefits in the EU study in terms of forestry were an order of magnitude lower, but only considered the effects on timber production, and not the wider societal value of healthy forests.

Such estimates clearly have many uncertainties. These include, for example, the representivity of the exposure–response functions over wide areas, the validity of the extrapolation from chamber to field conditions, and the need to group crops into different sensitivity categories. On the economic side, the simpler models assume that prices are not altered by the changes in production and that ozone does not cause shifts to less sensitive cultivars or crops. When changes in price occur through increased or decreased production, then individual growers or consumers may both suffer or benefit from changing ozone levels (Adams *et al.*, 1988). The assessments also ignore the possible indirect effects, through altered pest and pathogen damage, of ozone exposure. Thus, such estimates can only provide a rough guide to the possible regional economic significance of ozone as a pollutant, and are not reliable quantitative values.

One alternative approach to assessing whether ozone has significant economic impacts is to use farm economic data directly. This approach was used by Garcia *et al.* (1986), who showed that profits on Illinois farms growing soybean and corn were significantly negatively related to the estimated 7 h mean ozone concentration, when other factors influencing profit margins were taken into account. The estimated coefficient implied that a 10% increase in seasonal mean ozone concentrations would cause a 6% fall in profits, a result which is consistent with the biological exposure–response relationships. However, this type of approach has not been applied widely.

Impacts of ozone outside Europe and North America

This chapter has concentrated on the impacts of ozone in Europe and in North America, where the great majority of studies have been based. There is a paucity of field and experimental data from other parts of the world. However, the limited data which are available suggest that there may be significant effects of ozone in areas of Asia, Latin America and Africa with high local or regional emissions of ozone precursors (see also Chapter 21).

There have been several reports of visible injury caused by ozone outside North America and western Europe. The role of ozone in causing visible injury to tobacco crops in Taiwan was established in the early 1970s, while more recent studies have reported injury to a wide range of crops, including sweet potato. Visible injury to a potato crop was reported in the Punjab region of India in the 1980s and the diagnosis confirmed using chemical protectants (Bambawale, 1986). Hassan *et al.* (1995) reported visible symptoms of ozone damage on experimental plants of radish and turnip in Egypt, with the diagnosis again being confirmed by chemical protectants. Ozone symptoms on bean cultivars, eucalyptus and pine species were first reported in the areas around Mexico City in the early 1980s, and confirmed by fumigation or filtration experiments. Thus, there is little doubt that visible injury symptoms characteristic of ozone are found in many areas of the world; furthermore experience in both North America and western Europe suggests that once systematic surveys are carried out, such visible symptoms are found to be much more common than was previously thought.

The evidence of the impacts of ozone on crop yield or forest growth in these regions is less substantial. In Pakistan, studies by Wahid *et al.* (1995) have demonstrated that filtering ambient air at a site at the edge of the city of Lahore results in yield increases in local varieties of wheat and rice of about 40%, at a site where SO_2 concentrations are negligible but there are significant concentrations of NO_2 and ozone. Subsequent controlled fumigation studies with these rice varieties has demonstrated that, whereas NO_2 at the concentrations found at the Lahore site had no effect on growth or yield, ozone at these concentrations had very substantial effects (Maggs and Ashmore, 1998). Experiments with ozone protectant chemicals have also indicated that ozone can cause significant effects on the yield of tomato in and around New Delhi, on radish and turnip yields at a rural site in Egypt, and on bean yields in the Valley of Mexico (Ashmore and Marshall, 1998). Although no direct evidence of adverse effects of ozone on crops in mainland China is available, recent data demonstrate that rural ozone levels are high enough potentially to affect yields of winter wheat (Chameides *et al.*, 1999) and other crops (Zheng *et al.*, 1998), while Aunan *et al.* (2000) estimated that projected increases in ozone precursors are likely to lead to significant national yield losses in wheat, soybean and maize by 2020.

Evidence of effects of ozone on forests outside Europe and North America is also very limited. The most extensively studied area is the forests in mountain areas outside Mexico City, where some of the highest ozone concentrations in the world are now recorded. Miller *et al.* (1994) reported that ozone exposures in these forests were comparable to those in the San Bernadino mountains outside Los Angeles where extensive ozone damage has been documented; annual 7 h mean concentrations in the Mexican forests were approximately 75 ppb. Visible foliar symptoms are commonly found in these areas on pine species such as *P. hartwegii* and *P. montezumae*, on sacred fir (*Abies religiosa*), and on black cherry, while dendrochronological analysis has shown clear evidence of a growth decline since the early 1970s (Alvarado *et al.*, 1993).

Future impacts of ozone

There are now good observational data to suggest that global background tropospheric concentrations are increasing as a result of human activities, and in particular increased emissions of nitrogen oxides. Several studies have used global models to examine the impacts of continued increased emissions of nitrogen oxides on future ozone concentrations (e.g. Collins *et al.*, 2000). These in turn may be linked to effects on crop production using exposure–response relationships. For example, Chameides *et al.* (1994) used economic projections which suggested that global NO_x emissions would increase from 110 kT day^{-1} in 1985 to between 150 and 180 kT day^{-1} in 2025. Their model predicted that the proportion of the world's cereal crop exposed to ozone levels above an assumed threshold for significant (i.e. $>10\%$) effects on yield would increase from 9–35% in 1985 to 30–75% in 2025. Much of this increased area of cereal production at risk was in Asia. Although this model can provide no more than a crude estimate of the effects of continued increases in global NO_x emissions, it does indicate the potential severity of the problem.

Measures in North America and western Europe to reduce emissions of ozone precursors are expected to lead to future reductions in peak ozone concentrations, and indeed there is already evidence of such a trend in long-term monitoring records (e.g. Gardner and Dorling, 2000). However, the effect of these changes may be offset by the predicted increases in global tropospheric background ozone concentrations. The trends of increasing emissions of ozone precursors in parts of Asia, Latin America and Africa, combined with the predicted increase in global background concentrations, suggest that future ozone impacts on crops and forests in these areas may be very significant. The economic and social implications of widespread loss of yield of staple crops, in regions where there are problems in maintaining food supplies in the face of rapidly increasing populations and loss of productive land, could be very serious (Ashmore and Marshall, 1998).

The significance of these trends needs to be considered in the context of other predicted changes in global atmospheric composition and climate (see Chapter 23). Experimental studies suggest that the ambient CO_2 concentrations expected by the end of the next century may provide some, but not complete, protection against the effects of ozone. For example, Mckee *et al.* (1997) found that elevated CO_2 prevented any adverse effect of ozone on vegetative growth of winter wheat, but that there was an additional effect of ozone on grain yield on which elevated CO_2 appeared to have no effect. Changes in water availability, temperature and nutrient cycling may also interact with changing ozone exposure. It is thus essential that further research is conducted to elucidate the mechanisms of long-term ozone impacts on individual plants and on plant communities, to provide a sound basis for assessing the global implications of future changes in ozone exposure of vegetation.

References

Adams, R.M., Glyer, J.D. and McCarl, B.A. (1988). The NCLAN economic assessment: approach, findings and implications. In: *Assessment of Crop Loss from Air Pollutants: Proceedings of an International Conference* (W.W. Heck, O.C. Taylor and D.T. Tingey, eds.), pp. 473–504. Elsevier Applied Science, New York.

Alvarado, D., Bauer, L.I. de and Galindo, J. (1993). Decline of sacred fir (*Abies religiosa*) in a forest park south of Mexico City. *Environmental Pollution*, **80**, 115–121.

Amiro, B.D., Gillespie, T.J. and Thurtell, G.W. (1984). Injury response of *Phaseolus vulgaris* to ozone flux density. *Atmospheric Environment*, **18**, 1207–1215.

Andersen, C.P., Wilson, R., Plocher, M. and Hogsett, W.E. (1997). Carry-over effects of ozone on root growth and carbohydrate concentrations of ponderosa pine seedlings. *Tree Physiology*, **17**, 805–811.

Ashmore, M.R. and Davison, A.W. (1996). Towards a critical level of ozone for natural vegetation. In: *Critical Levels for Ozone in Europe* (L. Karenlampi and L. Skarby, eds.), pp. 58–71. University of Kuopio, Finland.

Ashmore, M.R. and Marshall, F.M. (1998). Ozone impacts on agriculture: an issue of global concern. *Advances in Botanical Research*, **29**, 31–52.

Ashmore, M.R., Bell, J.N.B., Dalpra and Runeckles, V.C. (1980). Visible injury to crop species by ozone in the United Kingdom. *Environmental Pollution (Series A)*, **21**, 209–215.

Aunan, K., Bernsten, T.K. and Seip, H.M. (2000). Surface ozone in China and its possible impact on agricultural crop yields. *Ambio*, **29**, 294–301.

Bambawale, O.M. (1986). Evidence of ozone injury to a crop plant in India. *Atmospheric Environment*, **20**, 1501–1503.

Barbo, D.N., Chappelka, A.H., Somers, G.L., Miller-Goodman, M.S. and Stolte, K. (1998). Diversity of an early successional community as influenced by ozone. *New Phytologist*, **138**, 653–662.

Barnes, J.D., Velissariou, D., Davison, A.W. and Holevas, C.D. (1990). Comparative ozone sensitivity of old and modern Greek cultivars of spring wheat. *New Phytologist*, **116**, 707–714.

Baumgarten, M., Werner, H., Haberle, K-H, Emberson, L.D., Fabian, P. and Matyssek, R. (2000). Seasonal ozone response of mature beech trees (*Fagus sylvatica*) growing at high altitude in the Bavarian Forest (Germany) in comparison with young beech trees grown in the field and in phytotrons. *Environmental Pollution*, **109**, 431–442.

Bergmann, E., Bender, J. and Weigel, H. (1996). Effects of chronic ozone stress on growth and reproduction capacity of native herbaceous plants. In: *Excedance of Critical Loads and Critical Levels* (M. Knoflacher, J. Schneider and G. Soja, eds.), pp. 177–185. Federal Ministry for Environment, Youth and the Family, Vienna.

Berrang, P., Larnosky, D.F. and Bennett, J.P. (1989). Natural selection for ozone tolerance in *Populus tremuloides:* field verification. *Canadian Journal of Forest Research*, **19**, 519–522.

Bosac, C., Roberts, J.A., Black, V.J. and Black, C.R. (1994). Impact of O_3 and SO_2 on reproductive development of oilseed rape (*Brassica napus* L.). II. Reproductive site losses. *New Phytologist*, **126**, 71–79.

Braun, S. and Fluckiger, W. (1995). Effects of ambient ozone on seedlings of *Fagus sylvatica* L and *Picea abies* (L.) Karst. *New Phytologist*, **129**, 33–44.

Braun, S., Rihm, B., Schindler, C. and Fluckiger, W. (1999). Growth of mature beech in relation to ozone and nitrogen deposition: an epidemiological approach. *Water, Air and Soil Pollution*, **116**, 357–364.

Bungener, P., Ball, G.R., Nussbaum, S., Geissman, M., Grub, A. and Fuhrer, J. (1999). Leaf injury characteristics of grassland species exposed to ozone in relation to soil moisture condition and vapour pressure deficit. *New Phytologist*, **142**, 271–282.

Chameides, W.L., Kasibhatla, P.S., Yienger, J. and Levy, II H (1994). Growth of continental-scale metro-agro-plexes, regional ozone pollution and world food production. *Science*, **264**, 74–77.

Chameides, W.L., Xingsheng, L., Xiaoyan, T., Xiuji, Z., Chao, L., Kiang, C.S., St. John, J., Saylor, R.D., Liu, S.C., Lam, K.S., Wang, T. and Giorgi, F. (1999). Is ozone pollution affecting crop yields in China? *Geophysical Research Letters*, **26**, 867–870.

Chappelka, A.H., Renfro, J., Somers, G. and Nash, B. (1997). Evaluation of ozone injury on foliage of black cherry (*Prunus serotina*) and tall milkweed (*Asclepias exaltata*) in Great Smoky Mountains National Park. *Environmental Pollution*, **95**, 13–18.

Collins, W.J., Derwent, R.G., Johnson, C.E. and Stevenson, D.S. (2000). The European regional ozone distribution and its links with the global scale for the years 1992 and 2015. *Atmospheric Environment*, **34**, 255–267.

Constable, J.V.H. and Taylor, G.E. Jr (1997). Modelling the effects of elevated tropospheric ozone on two varieties of *Pinus ponderosa*. *Canadian Journal of Forest Research*, **27**, 527–537.

Davison, A.W. and Reiling, K. (1995). A rapid change in ozone resistance of *Plantago major* after summers with high ozone concentrations. *New Phytologist*, **131**, 337–344.

Davison, A.W. and Barnes, J. (1998). Effects of ozone on wild plants. *New Phytologist*, **139**, 135–151.

Duchelle, S.F., Skelly, J.M. and Chevone, B.I. (1982). Oxidant effects on forest tree seedling growth in the Appalachian Mountains. *Water, Air and Soil Pollution*, **18**, 363–373.

Emberson, L.D., Ashmore, M.R., Cambridge, H., Tuovinen, J.-P. and Simpson, D. (2000). Modelling stomatal flux across Europe. *Environmental Pollution*, **109**, 403–413.

Evans, P.A. and Ashmore, M.R. (1992). The effects of ambient air on a semi-natural grassland community. *Agriculture, Ecosystems and Environment*, **38**, 91–97.

Eversman, S. and Sigal, L. (1987) Effects of SO_2, O_3 and SO_2 and O_3 in combination on photosynthesis and ultrastructure of two lichen species. *Canadian Journal of Botany*, **65**, 1806–1818.

Franzaring, J., Dueck, T.A. and Tonneijck, A.E.G. (1999). Can plant traits be used to explain differences in ozone sensitivity between native European plant species? In: *Critical Levels for Ozone – Level II* (J. Fuhrer and B. Achermann, eds.), pp. 83–87. Swiss Agency for Environment, Forests and Landscape, Bern.

Fuhrer, J., Skarby, L. and Ashmore, M.R. (1997). Critical levels for ozone effects on vegetation in Europe. *Environmental Pollution*, **97**, 91–106.

Fuhrer, J., Shariat-Madari, H., Perler, R., Tschannen, W. and Grub, A. (1994). Effects of ozone on managed pasture: II. Yield, species composition, canopy structure and forage quality. *Environmental Pollution*, **86**, 307–314.

Garcia, P., Dixon, B.L., Mjelde, J.W. and Adams, R.M. (1986). Measuring the benefits of environmental change using a duality approach: the case of ozone and Illinois cash grain farms. *Journal of Environmental Economics and Management*, **13**, 69–80.

Gardner, M.W. and Dorling, S.R. (2000). Meteorologically adjusted trends in UK daily maximum surface ozone concentrations. *Atmospheric Environment*, **34**, 171–176.

Grunehage, L., Jäger, H.J., Haenel, H.D., Hanewald, K. and Krupa, S. (1997). PLATIN (Plant Atmosphere Interaction) II Co-occurrence of high ambient ozone concentrations and factors limiting plant absorbed dose. *Environmental Pollution*, **98**, 51–60.

Grulke, N.E., Anderson, C.P., Fenn, M.E. and Miller, P.R. (1998). Ozone exposure and nitrogen deposition lowers root biomass of ponderosa pine in the San Bernadino Mountains, California. *Environmental Pollution*, **103**, 63–73.

Hassan, I.A., Ashmore, M.R. and Bell, J.N.B. (1995). Effect of ozone on radish and turnip under Egyptian field conditions. *Environmental Pollution*, **89**, 107–114.

Heck, W.W., Taylor, O.C. and Tingey, D.T. (eds.) (1988). *Assessment of Crop Loss from Air Pollutants*. Elsevier Applied Science, London.

Heggestad, H.E. (1991). Origin of Bel-W3, Bel-C and Bel-B tobacco cultivars and their use as indicators of ozone. *Environmental Pollution*, **74**, 264–291.

Heggestad, H.E. and Middleton, J.T. (1959), Ozone in high concentrations as a cause of tobacco leaf injury. *Science*, **129**, 208–210.

Jäger, H-J, Unsworth, M., de Temmerman, L. and Mathy, P. (eds.) (1993). *Effects of Air Pollution on Agricultural Crops in Europe*. Air Pollution Report 46, Commission of the European Communities, Brussels.

Karnosky, D., Mankovka, B., Percy, K., Dickson, R.E., Podila, G.K., Sober, J., Noormets, A., Hendrey, G., Coleman, M.D., Kubiske, M., Pregitzer, K.S. and Isebrands, J.G. (1999). Effects of tropospheric O_3 on trembling aspen and interaction with CO_2: results from an O_3-gradient and a FACE experiment. *Water, Air and Soil Pollution*, **116**, 311–322.

Kelly, J.M., Samuelson, L.J., Edwards, G., Hanson, P., Kelting, D., Mays, A. and Wullschleger, S. (1995). Are seedlings reasonable surrogates for trees? An analysis of ozone impacts on *Quercus rubra*. *Water, Air and Soil Pollution*, **85**, 1317–1324.

Krupa, S.V., Tonneijck, A.E.G. and Manning, W.J. (1998). Ozone. In: *Recognition of Air Pollution Injury to Vegetation: A Pictorial Atlas* (R.B. Flager, ed.), 2nd Edition. Air and Waste Management Association, Pittsburgh.

Lee, J.C., Skelly, J.M., Steiner, K.C., Zhang, J.W. and Savage, J.E. (1999). Foliar response of black cherry (*Prunus serotina*) clones to ambient ozone exposure in central Pennsylvania. *Environmental Pollution*, **105**, 325–331.

Lippert, M., Steiner, K., Payer, H.D., Simons, S., Langebartels, C. and Sandermann, H. Jr. (1996). Assessing the impact of ozone on photosynthesis of European beech (*Fagus sylvatica* L.) in environmental chambers. *Trees*, **10**, 268–275.

Lorenzini, G., Nali, C. and Biagioni, M. (1995). Long-range transport of photochemical ozone over the Tyrrhenian Sea, demonstrated by a new miniaturised bioassay with ozone-sensitive tobacco seedlings. *Science of the Total Environment*, **166**, 193–199.

Lyons, T.M., Barnes, J.D. and Davison, A.W. (1997). Relationships between ozone resistance and climate in European populations of *Plantago major*. *New Phytologist*, **136**, 503–510.

Mckee, I.F., Bullimore, J.F. and Long, S.P. (1997). Will elevated CO_2 protect the yield of wheat from O_3 damage? *Plant Cell and Environment*, **20**, 77–84.

Mclaughlin, S.B. and Downing, D.J. (1995). Interactive effects of ambient ozone and climate measured on growth of mature forest trees. *Nature*, **374**, 252–254.

Mclaughlin, S.B. and Percy, K. (1999). Forest health in North America: some perspectives on actual and potential roles of climate and air pollution. *Water, Air and Soil Pollution*, **116**, 151–197.

Maggs, R. and Ashmore, M.R. (1998). Growth and yield responses of Pakistan rice (*Oryza sativa*) cultivars to O_3 and NO_2. *Environmental Pollution*, **88**, 147–154.

Massman, W.J., Musselman, R.C. and Lefohn, A.S. (2000). A conceptual ozone dose-response model to develop a standard to protect vegetation. *Atmospheric Environment*, **34**, 745–759.

Matsumura, H. and Kohno, Y. (1999). Impact of O_3 and/or SO_2 on the growth of young trees of 17 species: an open-top chamber study conducted in Japan. In: *Critical Levels for Ozone – Level II* (J. Fuhrer and B. Achermann, eds.), pp. 187–192. Swiss Agency for Environment, Forests and Landscape, Bern.

Matyssek, R. and Innes, J.L. (1999). Ozone – a risk factor for trees and forests in Europe? *Water, Air and Soil Pollution*, **116**, 199–226.

Maurer, S., Matyssek, R., Gunthardt-Georg, M.S., Landolt, W. and Einig, W. (1997). Nutrition and the ozone sensitivity of birch (*Betula pendula*). I. Response at the leaf level. *Trees*, **12**, 1–10.

Mehbrahtu, T., Mersie, W. and Rangpappa, M. (1990). Inheritance of ambient ozone insensitivity in common bean (*Phaseolus vulgaris* L.). *Environmental Pollution*, **67**, 79–89.

Menser, H.A. and Hodges, G.H. (1972). Oxidant injury to shade tobacco cultivars developed in Connecticut for weather fleck resistance. *Agronomy Journal*, **64**, 189–192.

Miller, P. and McBride, J. (eds.) (1999). *Oxidant Air Pollution Impacts in the Montane Forests of Southern California: The San Bernadino Case Study*. Springer-Verlag, New York.

Miller, P.R., Bauer, L.I. de, Quevedo, A. and Hernandez Tejeda, T. (1994). Comparison of ozone exposures in forested regions near Mexico City and Los Angeles. *Atmospheric Environment*, **28**, 141–148.

Mills, G., Hayes, F., Buse, A. and Reynolds, B. (2000). *Air Pollution and Vegetation*. Annual Report 1999/2000 of UN/ECE ICP Vegetation. Centre for Ecology and Hydrology, Bangor.

Montes, R.A., Blum, U. and Heagle, A.S. (1982). The effects of ozone and nitrogen fertiliser on tall fescue, ladino clover and a fescue-clover mixture. I. Growth, regrowth and forage production. *Canadian Journal of Botany*, **60**, 2745–2752.

Ollerenshaw, J., Lyons, T. and Barnes, J.D. (1999). Impacts of ozone on the growth and yield of field-grown winter oil-seed rape. *Environmental Pollution*, **104**, 53–59.

Ollinger, S.V., Aber, J.S. and Reich, P.B. (1997). Simulating ozone effects on forest productivity: interactions among leaf-, canopy-, and stand-level processes. *Ecological Applications*, **7**, 1237–1251.

Olszyk, D.M., Thompson, C.R. and Poe, M.P. (1988). Crop loss assessment for California: modelling losses with different ozone standard scenarios. *Environmental Pollution*, **53**, 303–311.

Peterson, D.L., Silsbee, D.G., Poth, M., Arbaugh, M.J. and Biles, F.E. (1995). Growth responses of bigcone Douglas fir to long-term ozone exposure in Southern California. *Journal of the Air Pollution Management Association*, **45**, 36–45.

Pleijel, H., Wallin, G., Karlsson, P., Skarby, L. and Sellden, G. (1995). Gradients of ozone at a forest site and over a field crop – consequences for the AOT40 concept of a critical level. *Water, Air and Soil Pollution*, **131**, 241–246.

Pleijel, H., Ojanpera, K., Danielsson, H., Sild, E., Gelang, J., Wallin, G., Skarby, L. and Sellden, G. (1997). Effects of ozone on leaf senescence in spring wheat – possible consequences for grain yield. *Phyton*, **37**, 227–232.

Pleijel, H., Danielsson, H., Gelang, J., Sild, E. and Sellden, G. (1998). Growth stage dependence of the grain yield response to ozone in spring wheat (*Triticum aestivum* L.). *Agriculture, Ecosystems and Environment*, **70**, 61–68.

Pleijel, H., Danielsson, H., Karlsson, G.P., Gelang, J., Karlsson, P.E. and Sellden, G. (2000). An ozone flux–response relationship for wheat. *Environmental Pollution*, **109**, 453–462.

Plochl, M., Lyons, T., Ollerenshaw, J. and Barnes, J.D. (2000). Simulating ozone detoxification in the leaf apoplast through the direct reaction with ascorbate. *Planta*, **210**, 454–462.

Potter, J., Foot, J.P., Caporn, S.J.M. and Lee, J.A. (1996). The effects of long-term elevated ozone concentrations on the growth and photosynthesis of *Sphagnum recurvum* and *Polytrichum commune*. *New Phytologist*, **134**, 649–656.

Reich, P.B. (1987). Quantifying plant response to ozone: a unifying theory. *Tree Physiology*, **3**, 63–91.

Reiling, K. and Davison, A.W. (1992). The response of native, herbaceous species to ozone: growth and fluorescence screening. *New Phytologist*, **120**, 29–37.

Retzlaff, W.A., Williams, L.E. and DeJong (1997). Growth and yield response of commercial bearing-age 'Casselman' plum trees to various ozone partial pressures. *Journal of Environmental Quality*, **26**, 858–865.

Samuelson, L. and Kelly, J.M. (2001). Scaling ozone effects from seedlings to mature trees. *New Phytologist*, **149**, 21–41.

Sellden, G. and Pleijel, H. (1995). Photochemical oxidant effects on vegetation – response in relation to plant strategy. *Water, Air and Soil Pollution*, **85**, 111–122.

Skelly, J.M., Innes, J.L., Savage, J.E., Snyder, K.R., Vanderheyden, D., Zhang, J. and Sanz, M.J. (1999). Observation and confirmation of foliar ozone symptoms on native plant species of Switzerland and southern Spain. *Water, Air and Soil Pollution*, **116**, 227–234.

Soja, G., Eid, M., Gangl, H. and Redl, H. (1997). Ozone sensitivity of grapevine (*Vitis vinifera* L): evidence for a memory effect in a perennial crop plant? *Phyton*, **37**, 265–270.

Stewart, C.A., Black, V.J., Black, C.R. and Roberts, J.A. (1996). Direct effects of ozone on the reproductive development of *Brassica* species. *Journal of Plant Physiology*, **148**, 172–178.

Taylor, G.E. Jr. (1994). Role of genotype in the response of loblolly pine to tropospheric ozone: effects at the whole-tree, stand and regional level. *Journal of Environmental Quality*, **23**, 63–82.

Taylor, G.S. and Rich, S. (1973). Ozone injury to tobacco in the field modified by soil treatments with benomyl and carboxin. *Phytopathology*, **64**, 814–817.

Temple, P.J. and Taylor, O.C. (1983), World wide ambient measurements of peroxyacetyl nitrate (PAN) and implications for plant injury. *Atmospheric Environment*, **17**, 1583–1587.

Terry, G.M., Stokes, N.J., Lucas, P.W. and Hewitt, C.N. (1995). Effects of reactive hydrocarbons and hydrogen peroxide on antioxidant activity in cherry leaves. *Environmental Pollution*, **88**, 19–26.

Vellissariou, D. (1999). Toxic effects and losses of commercial value of lettuce and other vegetables due to photochemical air pollution in agricultural areas of Attica, Greece. In: *Critical Levels for Ozone – Level II* (J. Fuhrer and B. Achermann, eds.), pp. 253–256. Swiss Agency for Environment, Forest and Landscape, Bern.

Velissariou, D., Gimeno, B., Badiani, M., Fumigalli, I. and Davison, A.W. (1996). Records of visible injury in the ECE Mediterranean region. In: *Critical Levels for Ozone in Europe* (L. Karenlampi and L. Skarby, eds.), pp. 343–350. University of Kuopio, Finland.

Wahid, A., Maggs, R., Shamsi, S.R.A., Bell, J.N.B. and Ashmore, M.R. (1995). Effects of air pollution on rice yield in the Pakistan Punjab. *Environmental Pollution*, **90**, 323–329.

Warwick, K.R. and Taylor, G. (1995). Contrasting effects of tropospheric ozone on five native herbs which coexist in calcareous grassland. *Global Change Biology*, **1**, 143–151.

Weinstein, D.A., Samuelson, L.J. and Arthur, M.A. (1998). Comparison of the response of red oak (*Quercus rubra*) seedlings and mature trees to ozone exposure using simulation modelling. *Environmental Pollution*, **102**, 307–320.

Westman, W.E. (1979). Oxidant effects on California coastal sage scrub. *Science*, **205**, 1001–1003.

Whitfield, C., Davison, AW. and Ashenden, TW. (1998) The effects of nutrient limitation on the response of *Plantago major* to ozone. *New Phytologist*, **140**, 219–230.

Woodbury, P.B., Laurence, J.A. and Hudler, G.W. (1994). Chronic ozone exposure alters the growth of leaves, stems and roots of hybrid *Populus*. *Environmental Pollution*, **85**, 103–108.

Younglove, T., McCool, P.M., Musselmann, R.C. and Kahl, M.E. (1994). Growth-stage dependent crop yield response to ozone exposure. *Environmental Pollution*, **86**, 287–295.

Zheng, Y., Stevenson, K.J., Barrowcliffe, R., Chen, S., Wang, H. and Barnes, J.D. (1998). Ozone levels in Chongqing: a potential threat to crop plants commonly grown in the region? *Environmental Pollution*, **99**, 299–308.

7 Nitrogen oxides: old problems and new challenges

T.A. MANSFIELD

Introduction

NO_x is the term used to describe air pollution that is now abundant in most urban areas in the world, the 'x' indicating the uncertainty about the relative amounts of NO and NO_2, which vary considerably over time and between locations. This variation leads to difficulties in assessing the phytotoxicity of NO_x because the two gases have some contrasting chemical properties, and they may have quite different impacts at the cellular level. Much of the experimental evidence available to us concerns the impact of NO_2, even though many plants grow in situations where NO is the primary component of the NO_x to which they are exposed. In fact most of the oxidised nitrogen in the atmosphere is emitted as NO, which can then be converted to NO_2, slowly in the absence of O_3 but in as little as 100s in the presence of 40 ppb O_3 (Neubert *et al.*, 1993). The atmospheric chemistry of NO_x is of key importance in the production and removal of O_3, the formation of photochemical oxidants, and tropospheric cycles involving nitrous and nitric acids. The participation of plants in these processes is of great significance, because they act as receptors of much of the deposited NO_x some of which can be emitted again during cellular processes or as a result of microbial activity in soil. Indeed the role of plants and microbes is so important that the resolution of many aspects of tropospheric chemistry requires a detailed knowledge of impacts upon them. Thus even if the economic outcome of the phytotoxicity of NO_x were unimportant, we would still need to explore plant processes in detail in order to identify significant controls over the atmospheric cycling of oxidised N.

The impacts of NO_x have been most frequently studied in the context of urban air pollution. Only a few decades ago the most important phytotoxic component of urban air was SO_2, although the presence of NO_2 was seen to be of consequence because it could interact with SO_2 to cause greater damage than the sum of the independent effects of the two gases (Mansfield and Freer-Smith, 1981). (See Chapter 14.) There are major problems in defining the impacts of NO_x and SO_2 because both can stimulate plant growth at very low doses, whether they are present separately or in combination. The switch to toxicity does, however, occur rapidly as the dose increases, and the nature of the dose-response curve is strongly influenced by the environmental conditions (Mansfield, 1999). We now know that there are even greater difficulties in presenting

Air Pollution and Plant Life, second edition. Edited by J.N.B. Bell and M. Treshow. ISBN 0 471 49090 3 (HB), 0 471 49091 1 (PB).

reliable dose-response data for NO_x than for SO_2 because the impacts of NO and NO_2 may be fundamentally different at the cellular level.

Uptake of NO_x

The uptake by terrestrial plants of some aerial pollutants, including NO and NO_2, is largely determined by stomatal conductance. Compared with the amounts diffusing through stomata, deposition onto leaf surfaces and passage though the cuticle are relatively small (Kerstiens, 1996). Studies of the gas exchange of individual plants show clearly that the uptake of NO and NO_2 is linearly dependent on stomatal opening (Neubert *et al.*, 1993). The uptake of NO_x from the atmosphere by leaves has been distinguished using a ^{15}N dilution technique which was first devised for estimating N_2 fixation by legumes (e.g. Fried and Middelboe, 1977). Okano *et al.* (1986) grew sunflower and maize plants in media containing ^{15}N-labelled potassium nitrate, then exposed them to controlled levels of unlabelled NO_2 pollution, and calculated the extent to which the ^{15}N in the plant tissue had been diluted. For both species the rate of absorption of NO_2 per unit leaf area increased linearly as the atmospheric concentration increased from 0 to 1000 ppb. Okano and Totsuka (1986) found that changes in nitrate supply to the roots of sunflower did not affect the rate of absorption of NO_2 at an atmospheric concentration of 300 ppb, but the NO_2 did act as a useful source of nutrient for nitrate-deficient plants. Okano *et al.* (1988) extended these ^{15}N studies to eight different species of herbaceous plants, and found that the variation in rate of absorption of NO_2 was closely dependent on stomatal conductance (Figure 7.1). They concluded that the greater susceptibility of sunflower and radish to damage in higher concentrations of NO_2 could be attributed to higher rates of uptake. Conversely, maize and sorghum, with the lowest conductances, showed a high degree of tolerance. The ^{15}N dilution technique thus provided some useful preliminary information about uptake

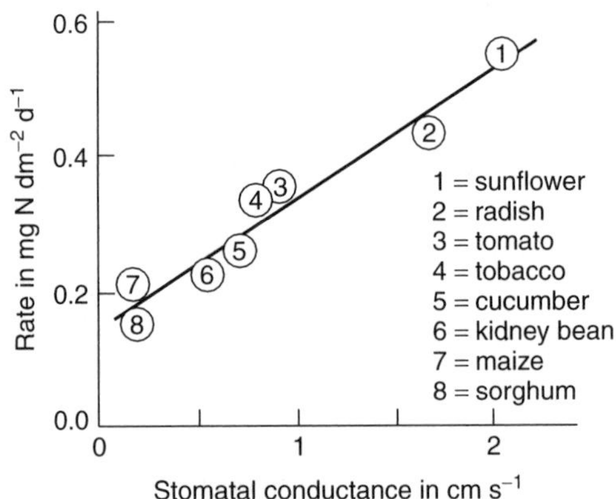

Figure 7.1 Dependence of the rate of NO_2 sorption per unit leaf area upon stomatal conductance in eight herbaceous species. From Okano, Machida and Totsuka (1988). Reproduced by permission of *The New Phytologist*, Blackwell Science

rates of NO_2 in relation to tolerance, but it was not suitable for exploring metabolic impacts in detail because it could only be used reliably with higher NO_2 concentrations than are normally found in polluted air. For such purposes as we shall see later, studies of the fate of $^{15}NO_2$ taken up through the stomata have proved invaluable.

Because of the significance of stomatal conductance, it might be assumed that uptake of NO_x would usually increase during the day and fall at night. This may be true for individual plants in moderately or severely polluted air, but it is not always true of the *net* gas exchange for a plant canopy in the field in a rural situation with low levels of NO_x pollution. Fowler *et al.* (1998) used micrometeorological methods to make detailed measurements at a moorland site in the south of Scotland, and found small fluxes of NO_2 into the canopy at night (up to about 5 ng NO_2-N m^{-2} s^{-1}). These were *reduced* during the day and, during the month of May, *emissions* were as high during the day as the maximum rates of uptake found at night. They attributed this daytime output of NO_2 not to the canopy itself, but to generation of NO in the soil beneath the canopy and its subsequent oxidation by O_3 to NO_2. Thus when there is relatively little anthropogenic pollution, there are compensatory mechanisms governing the exchange of NO_x so that direct uptake from the atmosphere by the plants is balanced by microbial processes generating NO in the soil. It is in more heavily polluted situations that a net flux into the canopy will be maintained, and then stomatal control will become a dominant component. The ability of the individual plants to tolerate a net intake of NO_x which may far exceed compensatory processes, is the issue with which we shall be principally concerned in this chapter. This will have clear implications for the responses of the ecosystems of which they are components.

Atmospheric NO_x and the apoplast

The diagram in Figure 7.2 displays the three main cellular compartments in which we need to consider the fate of the products of any NO_x which has diffused from the atmosphere through the stomata into the intercellular spaces of the leaf. Here it will make contact with the aqueous layer surrounding each cell, the apoplast, which is mostly contained within the cell wall. Although the impact on the chemical composition of the apoplast will be highly dependent on the NO/NO_2 balance in the atmosphere, we know comparatively little about the relative contribution of the two gases to the total uptake of NO_x into the plant. It is generally assumed that the rate of uptake of NO_2 must exceed that of NO because it is more reactive with water. Wellburn (1990) quoted values of solubility for the two gases which are not vastly different, but pointed out that the subsequent conversion to nitrous and nitric acid greatly increases the apparent solubility of NO_2. However, he drew attention to evidence from isotopic exchange studies that NO may be converted to nitrite and nitrate ions, though at slower rates. In Figure 7.2, the production of nitrite and nitrate ions from NO_2 is the simple assumption, with NO providing a possible additional source of these ions. The likelihood of NO *per se* crossing the apoplast and entering the cytoplasm in solution is also indicated, for it is known that its permeability coefficients in lipid bilayer

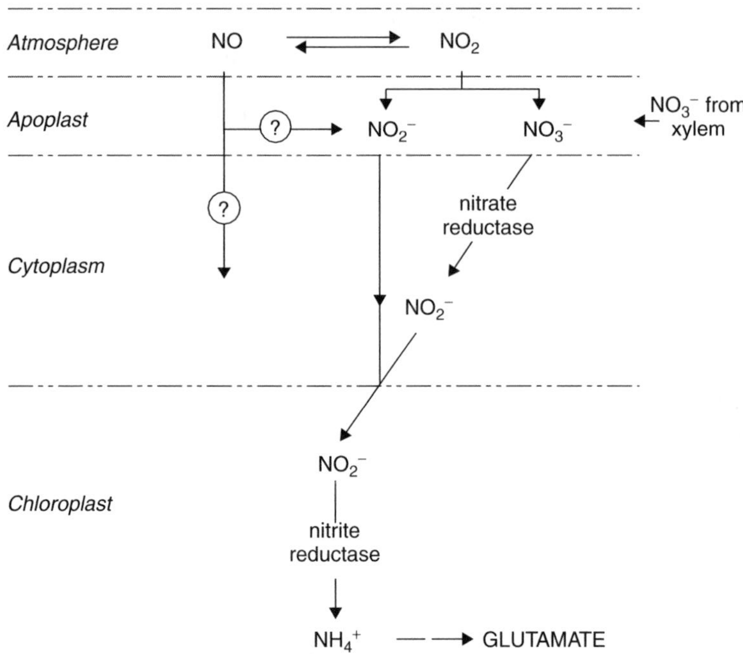

Figure 7.2 Simplified picture of the fate of NO_x when it enters a leaf cell. Nitrate is shown moving into the apoplast from the xylem to join that derived from the uptake of NO_2 from the atmosphere. The question marks indicate uncertainty about the fate of nitric oxide after it enters the apoplast

membranes are sufficiently high (Subczynski, Lomnicka and Hyde, 1996) for this to be feasible. Current understanding of the cellular impacts of NO_x suggests that we need to consider closely both the incorporation of NO_2^- and NO_3^- into amino acids, and any separate effects that may be attributable to NO entering the cytoplasm after diffusion through the apoplast and the cell membrane (Figure 7.2).

It is important to recognise that apoplastic fluid not only contains many dissolved materials, but also that its composition is to some extent controlled by the cell which it surrounds. There has been much recent interest in the constituents of the apoplast which appear to play a role in determining ozone tolerance, for example ascorbate, glutathione and superoxide dismutases (Lyons *et al*., 1999). It seems likely that there are significant apoplastic contributions to NO_2 tolerance, and it may be important to recognise that NO_2 can react with apoplastic solutes in ways which would affect the subsequent participation of its derivatives in cellular processes. Ramge *et al*. (1993) used mathematical models to explore why the simple transformation of NO_2 into the apoplastic nitrate and nitrite pools (Figure 7.2) fails to account for measured uptake characteristics. They considered two other possibilities: (1) Because NO_2 has an unpaired electron (as also does NO) it must be regarded as a free radical capable of being pro-oxidative, likely to damage cell membranes by oxidising the polyunsaturated fatty acids by hydrogen abstraction. (2) Ascorbate present in the apoplastic fluid acts to reduce NO_2 as it diffuses through the cell wall, leading to this reaction

$$Asc - OH + {}^{!}NO_2 \; Z \; Asc - O^{!} + H^+ \; NO_2^-$$

i.e. the formation of dehydroascorbate and nitrite. Their model calculations based on these possible processes/mechanisms gave unequivocal support to the second one involving ascorbate, and they were able to locate many data in the literature to support their conclusions. As we shall see below, however, some problems do arise from enzymatic studies which suggest that we should be cautious in accepting a single apoplastic mechanism as a dominant feature.

Metabolic impacts

When a plant is displaying large values of conductance associated with a high rate of growth, tolerance of NO_2 pollution is likely to depend on metabolic activity within the leaf. Because the solution of NO_2 in the apoplast is likely to produce both nitrate and nitrite ions, attention has focused mainly on the ability to reduce these to ammonium, and to the subsequent incorporation into amino acids (Figure 7.2). Nitrate reductase (NaR) and nitrite reductase (NiR) are clearly of fundamental importance, and there have been major advances in recent years in understanding the factors regulating these two enzymes. Nitrate induces the expression of both, and also of nitrate transporters. Furthermore, it induces enzymes that provide the carbon skeletons for amino acids, and represses those which divert carbon towards storage, e.g. to the synthesis of starch. It is now recognised that nitrate can act as a cellular signal, activating receptors in complex signalling networks within tissues (MacKintosh, 1998). It was originally thought that changes in the activity of NaR in plants depended on the synthesis of fresh enzyme and its subsequent breakdown, and not on the activation/inactivation of the protein (Remmler and Campbell, 1986). Recently, it has been established that NaR displays a cycle in its activity state, becoming active when photosynthesis is functioning and returning to low activity in the dark. The precise mechanisms, involving changes in the phosphorylation of NaR and a specific NaR-inhibitor protein (NIP) as regulatory components, are currently being elucidated (MacKintosh, 1998; Moorhead *et al.*, 1999). It may well be very important in relation to tolerance of NO_2 that the diurnal cycle of activity of NaR is so closely tied to photosynthesis, because this means that when stomata are wide open and allowing the most rapid entry of NO_2, NaR will be at or near the peak of its activity level. NIP is one of the 14-3-3 proteins that are currently attracting much interest as cellular regulators in plants (Chung, Sehnke and Ferl, 1999). Originally discovered in animals, these proteins occur in most eukaryotic organisms where they are essential components of signalling pathways. Under unpolluted conditions, the bulk of the nitrogen supply for land plants is acquired from the soil as a result of active transport of nitrate, involving H^+ ATPases in the plasma membranes of cells in the outer tissues of roots. In some cases the nitrate is assimilated in the roots, being converted first to nitrite and then to ammonia, but in most plants these processes take place in the leaves after nitrate has been delivered via the xylem. The activation/inactivation of NaR involving NIP is now seen as one of the key regulatory processes in plant cell metabolism. The inactivation of NaR involves two steps: first, phosphorylation takes place under the action of a protein kinase; and second, the binding onto the 14-3-3 NIP occurs, whereupon the enzyme becomes inactive or shows greatly reduced activity (Figure 7.3). The NaR kinase is Ca^{2+}-dependent through the agency of calmodulin, and the binding of Mg^{2+} to NIP is necessary before the latter can bind to the phosphorylated NaR (Chung *et al.*, 1999). Light is believed to lead to dephosphorylation of the inactive NaR, followed by the dissociation of NIP.

These intricate cellular control processes have evolved in order to regulate the assimilation of nitrate, virtually all of which is derived from the soil in non-leguminous plants. There is tight coupling with closely similar regulation of some enzymes concerned with carbon metabolism, especially sucrose phosphate synthase. In unpolluted air NO and NO_2 are normally present only in minute traces, and leaf cells in which these controls govern events do not normally encounter significant inputs of nitrate and nitrite ions directly from the atmosphere.

An important study by Rowland, Drew and Wellburn (1987) predated these recent discoveries of mechanism for the activation and deactivation of NaR, but the findings remain invaluable as an integrated picture of the effects of fumigation with NO_2 on nitrogen metabolism in the plant as a whole. They exposed hydroponically grown barley plants to 300 ppb NO_2 for 9 days, taking care to prevent the NO_2 from entering directly into the nutrient solution, and they pointed out that the fumigation therefore provided an

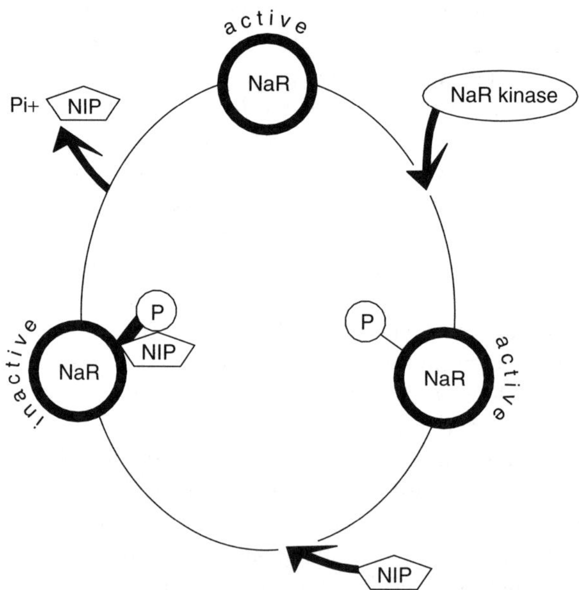

Figure 7.3 Simplified model of the regulation of the activity nitrate reductase (NaR) which involves both phosphorylation and dephosphorylation and one or several nitrate reductase inhibitor proteins (NIP). The regulation brings about activity/inactivity of the enzyme to coincide with the daily cycle of light and darkness

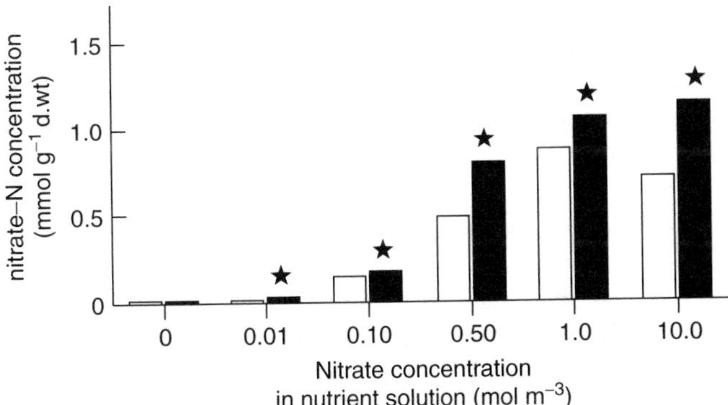

Figure 7.4 Effect of NO_2 on nitrate-N concentration in shoots of barley grown with different nitrate supplies to the roots. The unfilled columns on the left apply to the controls in clean air, and the filled columns on the right to plants grown in 300 ppb NO_2 for 9 days. Asterisks indicate a statistically significant difference as a result of the NO_2 treatment. From Rowland, Drew and Wellburn (1987). Reproduced by permission of Rowland, A.J. *et al.*, in *The New Phytologist*, Blackwell Science

opportunity to manipulate the nutritional state of the plants without changing the supply of N to the roots. They found that the response to NO_2 depended on the nitrogen status of the plants at the time of exposure. There were significant increases in *total nitrogen content* of the plants only when nitrate supplies to the roots were very low. However, even when nitrate was abundantly supplied to the roots, exposure to NO_2 did cause significant increases in the nitrate-N content of the shoots (Figure 7.4). They concluded that the solution and dissociation of NO_2 in the extracellular water (Figure 7.2) caused an increase in the nitrate pool, which stimulated the induction of extra NaR and NiR activities. Using ^{15}N as a label, they were able to show that N derived from NO_2 was translocated from the shoots to the roots, and it appeared that in the case of plants supplied with little or no nitrate to the roots, this process might alleviate the nutrient deficiency. There were clear indications that there was a reduced supply of nitrate from root to shoot because of the direct input of nitrate from NO_2 entering the leaves. There is now evidence that it is the amino compounds such as glutamine transported from the shoot to the root in the phloem which act as signals to control nitrate uptake by the roots (Rennenberg *et al.*, 1998). Nitrogen derived from $^{15}NO_2$ in the atmosphere can be quickly incorporated into amino compounds in the shoot and thereafter transferred to the roots (Nussbaum *et al.*, 1993; Weber *et al.*, 1995). Muller *et al.* (1996) examined the impact of NO_2 fumigation on nitrogen uptake by the roots of seedlings of Norway spruce. When the roots were supplied with NO_3^- and NH_4^+ as nitrogen sources, fumigation with NO_2 led to a decrease in the root uptake of NO_3^- equivalent to the amount of nitrogen gained from the deposition of NO_2 into the leaves. Fumigated plants accumulated less dry weight, especially in the roots. Less NO_3^- was reduced in the roots in plants fumigated with NO_2, while the amount of organic nitrogen transported in the phloem was increased three-fold. This study not only increased our understanding of the whole-plant regulation of the metabolism of inorganic nitrogen, but it also provided insight into basic mechanisms that are involved in the considerable tolerance of NO_x pollution displayed by some plants. However, Näsholm (1998) has pointed out that although plants may be able to regulate the root uptake of nitrogen, foliar intake of NO_x cannot be restricted in the same advantageous way. A fall in stomatal conductance to limit NO_x uptake (Figure 7.1) would also obstruct the gain of CO_2 for photosynthesis. There are also possible problems with pH regulation if foliar uptake of NO_x is excessive, because unlike roots, shoots cannot control their internal pH by exchange with the medium surrounding them.

Rowland *et al.* (1987) found that NO_2 fumigation caused an increase in NaR activity, and this response has been confirmed in several subsequent studies on a wide range of species, e.g. red spruce (Norby *et al.*, 1989), and Norway spruce (Thoene *et al.*, 1991). The increase may, however, be transient. The time course of the increases in NaR activity in response to moderate treatments with NO_2 has raised some important questions that cannot be answered satisfactorily at present. Weber *et al.* (1995) found an increase of about 30% in NaR in the leaves when wheat was exposed to 60 ppb NO_2, but after 96 hours the level had declined to that in the controls in clean air. Explanations based on mechanisms to reduce the amount of nitrate derived from NO_2 appear inadequate. There was no decline in stomatal conductance, which would have reduced uptake, and changes in ascorbate level in the apoplast which would quickly convert NO_2 to nitrite (Ramge *et al.*, 1993) seem unlikely to provide an explanation, because although NiR activity increased by more than 100%, this was also transient and after 96 hours fell to the value in the controls. Weber *et al.* suggest that nitrate and/or nitrite transport into storage pools, or export by long-distance transport may provide the answer, but we do not have evidence in support of these suggestions. The newly discovered regulation of NaR activity (Figure 7.3) may perhaps be linked to more extensive controls in leaf tissues of which we are ignorant at present. Compartmental analysis of plant cells providing information such as the distribution of nitrate ions between apoplast, cytoplasm and vacuole has only become feasible within the last few years with the development of nitrate-specific electrodes. There are fewer technical difficulties in acquiring data for the vacuole than for the cytoplasm, and when nitrate supplies are abundant vacuolar concentrations around 100 mol m^{-3} have been found. In the case of the cytoplasm the estimated values are more controversial, but the current view is that they may be as low as 3–4 mol m^{-3} (Miller and Smith, 1996; Crawford and Glass, 1998). It is likely, therefore, that there are mechanisms for maintaining the cytoplasmic concentration in this range, with controls over influx and efflux at the plasma membrane and tonoplast, and over nitrate reduction (cf. Figure 7.3). Unfortunately, most of the research into the processes involved in the transport of nitrate has been concerned with roots, and there is little specific information on leaves. The nitrate concentration in xylem sap is maintained at a high level between 5 and 40 mol m^{-3} (Crawford and Glass, 1998) and thus leaf cells are in a similar situation to those in the root, nitrate being delivered by bulk flow in both cases (in roots, in aqueous solution from the soil). It seems likely, therefore, that leaf cells will have similar nitrate-transport characteristics to those now well established for roots. Nitrate uptake into root cells is energy dependent even when the concentration in the soil solution is high, but nevertheless there is also a significant amount of efflux which, although passive, is inducible by nitrate (Aslam, Travis and Rains, 1996). The rate of efflux actually increases with higher external nitrate concentrations. If there is a similar situation in the leaf, any input of nitrate into the apoplast from NO_2 would increase efflux from the cells, tending to maintain a higher nitrate concentration in the apoplast than would be expected if active uptake across the plasma membrane were the only regulatory process. The amount of nitrate delivered to leaves in the transpiration stream is determined by xylem-loading from cortical cells within the roots, the rate of loading being determined by amino acids delivered to the roots via the phloem (Crawford and Glass, 1998). There are thus very complex regulatory processes which may be unable to react quickly when a plant is suddenly exposed to NO_2 pollution after growing in clean air, which may explain why increases in NaR activity are transient (Weber *et al.*, 1995). The intracellular signalling that achieves regulation of NaR activity (Figure 7.3) is just one of many processes that must play a part in determining a plant's responses to NO_2. Regulation of NiR activity may also require more attention than hitherto, for a study using transgenic *Arabidopsis* with higher NiR activity than the wild type suggests that NiR could be a controlling enzyme in NO_2 assimilation (Takahashi *et al.*, 2001).

Enzymatic controls, cellular influx and efflux controls at the plasma membrane and tonoplast, communications between root and shoot, co-ordination between xylem and phloem transport, and expression of nitrate transporter genes are just some of the many factors that will have to be considered in future research.

When is NO_x a damaging form of pollution?

Considerations of NO and NO_2 as phytotoxic air pollutants in their own right began in the 1950s and 1960s, the early work of J.T. Middleton and of O.C. Taylor being very influential in the USA (see Taylor *et al.*, 1975). Although leaf damage was found after fumigation with NO_2, and often described in great detail, the concentrations used were usually so high (up to 250 ppm in some cases) that the research bore little relevance to effects of polluted atmospheres then or since. Perhaps the most important outcome of the work was the finding that NO_2 *per se* was considerably less toxic than two other pollutants also being studied at the time, i.e. SO_2 and O_3. As late as 1981, it was being suggested that leaf injury in trees due to NO_2 would only be expected at doses of 1.6–2.6 ppm for periods up to 48 hours, and that a concentration of 20 ppm might be required to produce visible symptoms in just 1 hour (Smith, 1981). However, there had been early indications that NO_2 could prove very toxic in the presence of SO_2 (Tingey *et al.*, 1971; Fugiwara, 1973), and since NO_2 is very rarely likely to be encountered alone, a major part of its phytotoxicity needs to be considered in the context of interactions with other pollutants (see Chapter 14). The focus of attention in recent research has been upon the ways in which deposited NO_2 may become involved in the nitrogen metabolism of plants, and may perhaps cause disorder in some important regulatory mechanisms. In fact, low doses of NO_2 will sometimes provide a beneficial nitrogen supply for plants, particularly when there is nitrogen deficiency in the soil. Nevertheless the ecological consequences of additional nitrogen deposition (NO_x and NH_3) are complex, and are rarely considered to be advantageous in the long term (see Chapter 12).

As a result of the early research, NO was largely disregarded as a pollutant that bore any realistic direct threat to plant life (National Academy of Sciences, 1977). Now, however, we might feel justified in regarding it as the more important component of NO_x, at least with respect to its potential to disturb physiological mechanisms in plants. Certainly its impacts are likely to become a major focus of research in the future.

Compared with NO_2 and NH_3 the uptake rate of NO into plants is low, and it is not thought to have any great impact on nitrogen metabolism (Stulen *et al.*, 1998). Wellburn (1990) drew attention to studies indicating that NO may well form both nitrate and nitrite ions in solution, though the rate would be much lower than is the case with NO_2. There are many reports of effects of NO in the literature which point to distinctive impacts on plants, which are unlikely to be explained by its entering into the nitrate/nitrite reduction pathway in the same manner as NO_2. Much of the information has come from studies of plants growing in glasshouses enriched with CO_2 derived from oil or gas burners. The growth of plants in closed glasshouses in winter can be limited by a shortage of CO_2 for photosynthesis, and it has become common practice to supply additional CO_2 up to concentrations of about 1000 ppm. The least expensive way of achieving this is to use burners to generate CO_2 at the required rate, but the disadvantage is the formation of NO_x during the combustion, in amounts that exceed those in all but the most polluted situations out of doors. Figure 7.5 shows some data from a glasshouse enriched with CO_2 to around 1100 ppm using the output from a kerosene burner. Values for NO_x and for NO_2 are presented,

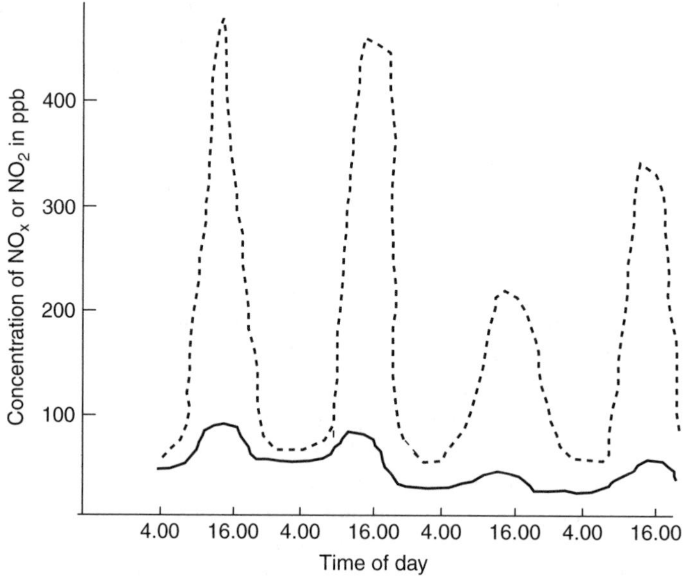

Figure 7.5 Concentrations of NO_2 (continuous line) and of total NO_x (dotted line) in a glasshouse enriched with CO_2 from a kerosene burner. Data of R.M. Law and T.A. Mansfield previously unpublished in this form

and the difference between the two curves represents the NO concentrations, which were usually 4-or 5-fold those of NO_2. NO_x concentrations between 500 and 1000 ppb have generally been found when hydrocarbons are burnt to provide CO_2 enrichment (Mortensen, 1985). Law and Mansfield (1982) found that $NO:NO_2$ mixtures in this high ratio caused marked inhibition in the growth of glasshouse crops, which was unlikely to be attributable to effects of NO_2 alone. Some important studies by Saxe and Christensen (1985) and Saxe (1986) on a range of potted ornamental plants attempted to quantify the impacts of NO_2 and NO separately, and they concluded that 1 ppm NO was, on average, four times more inhibitory to photosynthesis than NO_2. This high physiological impact of NO was not normally accompanied by any visible symptoms of injury, such as leaf scorching. Other authors have quantified the impacts of pollution during CO_2 enrichment by employing experimental fumigations with high $NO:NO_2$ ratios, usually around 4 : 1 (Mortensen, 1985; Murray and Wellburn, 1985; Caporn, 1989; Caporn *et al.*, 1991, 1994; Hufton, Besford and Wellburn, 1996), and in general responses have been found that are consistent with the view that NO acts independently from NO_2. In particular, it appears that NO can inhibit photosynthesis in the absence of any change in stomatal conductance. Studies of the effect of urban air pollution on aphid performance on several crop plants revealed that the strongest correlation was with NO content (concentration range 42–87 ppb), and the effect appeared to be mediated via the plant rather than as a result of a direct response of the aphids to NO (Houlden *et al.*, 1991). Nussbaum *et al.* (1995) found that NO concentrations in the low ppb range could cause additional stress ethylene production in spring wheat. These apparently disparate reports come together in one important respect: there appear to be distinct effects of NO on plants which can be clearly distinguished from those of NO_2.

NO is now recognised as a very important agent in signalling transduction pathways in a variety of cell functions in animals (Wink and Mitchell, 1998). Recognition of some comparable functions in plants has come very recently, and as yet there is no comprehensive picture, but it is already clear that NO has an important role in defence responses (Durner and Klessig, 1999). One aspect of defence responses against pathogens appears to involve reactive oxygen species operating with NO, for example superoxide ($O_2{}^{!-}$) reacts with NO to produce peroxynitrite ($ONOO^-$) which is believed then to kill the cells of pathogens. The enzyme nitric oxide synthase is known to occur in plants, and to show increased activity in tissues subject to pathogen attack. It is obvious that a defence mechanism dependent on free radical reactions must be both precisely located and closely controlled if it is to destroy the pathogen and not components of the host cells. It is not known whether NO pollution in the atmosphere could penetrate plant tissues and interfere with this process. Perhaps more likely would be disturbance of any role that NO might have in long-distance signalling in the plant. Durner and Klessig (1999) have postulated that because glutathione is a major component of sap in the phloem, it might become bound to NO to form nitroso glutathione, which is now thought to act as a carrier for NO both within and between cells in animals. They cited recent evidence that nitroso glutathione is very effective in inducing plant defence genes (Durner, Wendehenne and Klessig, 1998). Most of the studies of the impacts of NO as a pollutant were conducted before these revelations about the role of NO as a major component of signalling in plants.

The rates of uptake of NO *per se* from the atmosphere into plants have not been accurately determined. They will be lower than those of NO_2 because NO is not highly soluble in water, but nevertheless it is sufficiently diffusible within and between cells to function as a signalling molecule, and it has strong binding properties with many cellular constituents. When NO enters leaves via the stomata, it may well gain access to sites where it can be physiologically active. In view of the fast accumulating evidence that NO has a widespread physiological role in plants (Beligni and Lamattina, 2001) a full re-evaluation is now needed of the impacts of NO_x when NO is the principal constituent, such as in many urban environments.

Conclusions

In an influential review, Wellburn (1990) asked some critical questions relating to the action of NO_x on plants, especially concerning the factors determining whether the nitrogen input is useful nutritionally or causes damage. We can now go just a little further in answering some aspects of those questions but, on the other hand, even more areas of uncertainty have been exposed by the advances during the 1990s in our awareness of cellular and long-distance control mechanisms. The recognition that amino compounds transported from shoot to root in the phloem signal the nitrogen-demand of the shoot, and thus govern the intake of nitrate into the roots from the soil (Rennenberg *et al.*, 1998), provides a major step forward in our understanding of whether NO_x uptake into the leaves is toxic or not. If the N from NO_x is sufficient to meet the nutritional needs of the shoot, then the development of an excessive N burden could be prevented by this means. There are, however, important questions about the capacity for pH regulation in shoots deriving the bulk of their N supply from NO_x pollution. It is also clear that when the intake of NO_x exceeds the nutritional demand for N, there will be metabolic and perhaps structural costs involved in disposing of the surplus, for example transporters may be required to transfer nitrate from the apoplast and cytoplasm to cell vacuoles, and elevated levels of NiR may be needed to prevent toxic nitrite ions from accumulating in the cytoplasm. Progress in describing the diurnal controls upon NaR (Figure 7.3) are important, but although it is known that the uptake of NO_x can stimulate NaR activity, it is not known how it might impinge on these regulatory mechanisms and the way they are normally co-ordinated with the diurnal behaviour of other enzymes. The importance of considering the impact of NO_x in a broad physiological context is indicated by the work of Siegwolf *et al.* (2001), who were able to draw a distinction between potential benefits of NO_2 on hybrid poplar, in particular increased CO_2 assimilation rate and biomass, and effects that could prove detrimental, particularly reduced root:shoot biomass ratio which could impede nutrient acquisition and reduce drought tolerance.

There is great difficulty at present in assessing any impact of NO that might arise from interference with signalling mechanisms in plants. In physiological studies, controlled treatments are usually conducted by applying NO-releasing chemicals which deliver known amounts of NO into the aqueous phase around tissue sections (e.g. Ferrer and Barceló, 1999). It is not yet known to what extent NO in ppb concentrations in polluted atmospheres may penetrate tissues to cause similar effects in intact plants.

Future studies of the impacts of the dry deposition of NO_x should focus not only on the nutritional consequences of the foliar acquisition of nitrogen, but also on the action of NO which may be quite different and independent. The disproportionality in the concentrations of NO and NO_2 seen in Figure 7.5 is to be found in most urban areas, and the focus on total NO_x in many studies in the past may have concealed significant, distinctive responses to NO that must now be explored.

References

Aslam, M., Travis, R.L. and Rains, D.W. (1996). Evidence for substrate induction of a nitrate efflux system in barley roots. *Plant Physiology* **112**, 1167–1175.

Beligni, M. and Lamattina, L. (2001). Nitric oxide in plants: the history is just beginning. *Plant, Cell and Environment* **24**, 267–278.

Caporn, S.J.M. (1989). The effects of oxides of nitrogen and carbon dioxide enrichment on photosynthesis and growth of lettuce (*Lactuca sativa* L.) *New Phytologist* **111**, 473–481.

Caporn, S.J.M., Hand, D.W., Mansfield, T.A. and Wellburn, A.R. (1994). Canopy photosynthesis of CO_2- enriched lettuce (*Lactuca sativa* L.) – response to short-term changes in CO_2, temperature and oxides of nitrogen. *New Phytologist* **126**, 45–52.

Caporn, SJM, Mansfield, T.A. and Hand, D.W. (1991). Low temperature-enhanced inhibition of photosynthesis by oxides of nitrogen in lettuce (*Lactuca sativa* L.). *New Phytologist* **118**, 309–313.

Chung, H.-J., Sehnke, P.C. and Ferl, R.J. (1999). The 14-3-3 proteins: cellular regulators of plant metabolism. *Trends in Plant Science* **4**, 367–371.

Crawford, N.M. and Glass, A.D.M. (1998). Molecular and physiological aspects of nitrate uptake in plants. *Trends in Plant Science* **3**, 389–395.

Durner, J. and Klessig, D.F. (1999). Nitric oxide as a signal in plants. *Current Opinion in Plant Biology* **2**, 369–374.

Durner, J., Wendehenne, D. and Klessig, D.F. (1998). Defense gene induction in tobacco by nitric oxide, cyclic GMP, and cyclic ADP-ribose. *Proceedings of the National Academy of Sciences, USA* **95**, 10328–10333.

Ferrer, M.A. and Barceló, A.R. (1999). Differential effects of nitric oxide on peroxidase and H_2O_2 production by the xylem of *Zinnia elegans*. *Plant, Cell and Environment* **22**, 891–897.

Fowler, D., Flechard, C., Skiba, U., Coyle, M. and Cape, J.N. (1998). The atmospheric budget of oxidized nitrogen and its role in ozone formation and deposition. *New Phytologist* **139**, 11–23.

Fried, M. and Middelboe, V. (1977). Measurement of amount of nitrogen fixed by a legume crop. *Plant and Soil* **47**, 713–715.

Fugiwara, T. (1973). Effects of nitrogen oxides in the atmosphere on vegetation. *Journal of Air Pollution Control* **9**, 253–257.

Houlden, G., McNeill, S., Craske, A. and Bell, J.N.B (1991). Air pollution and agricultural aphid pests. 2. Chamber filtration experiments. *Environmental Pollution* **72**, 45–55.

Hufton, C.A., Besford, R.T. and Wellburn, A.R. (1996). Effects of NO (+NO_2) pollution on growth, nitrate reductase activities and associated protein contents in glasshouse lettuce grown hydroponically in winter with CO_2 enrichment. *New Phytologist* **133**, 495–501.

Kerstiens, G. (ed.) (1996). *Plant Cuticles – an Integrated Functional Approach*. Bios Scientific Publishers, Oxford, UK.

Law, R.M. and Mansfield, T.A. (1982). Oxides of nitrogen and the greenhouse atmosphere. In: Unsworth, M.H. and Ormrod, D.P., eds. *Effects of Gaseous Air Pollution in Agriculture and Horticulture*. Butterworths Scientific Press, London, pp. 93–112.

Lyons, T., Plöchl, M., Turcsányi, E. and Barnes, J. (1999). Extracellular antioxidants: a protective screen against ozone? In: Agrawal, S.B. and Agrawal, M., eds. *Environmental Pollution and Plant Responses*. Lewis Publishers, Boca Raton, pp. 183–201.

MacKintosh, C. (1998). Regulation of plant nitrate assimilation: from ecophysiology to brain proteins. *New Phytologist* **139**, 153–159.

Mansfield, T.A. (1999). SO_2 pollution: a bygone problem or a continuing hazard? In: Press, MC, ed. *Physiological Plant Ecology*. Blackwell Science, Oxford, UK, pp. 219–240.

Mansfield, T.A., Freer-Smith, P.H. (1981). Effects of urban air pollution on plant growth. *Biological Reviews* **56**, 343–368.

Miller, A.J. and Smith, S.J. (1996). Nitrate transport and compartmentation in cereal root cells. *Journal of Experimental Botany* **47**, 843–854.

Mortensen, L.M. (1985). Nitrogen oxides produced during CO_2 enrichment. 1. Effects on different greenhouse plants. *New Phytologist* **101**, 103–108.

Moorhead, G., Douglas, P., Cotelle, V., Harthill, J., Morrice, N., Meek, S., Deiting, U., Stitt, M., Scarabel, M., Aitken, A. and MacKintosh, C. (1999). Phosphorylation-dependent interactions between enzymes of plant metabolism and 14-3-3 proteins. *Plant Journal* **18**, 1–12.

Muller, B., Touraine, B. and Rennenberg, H. (1996). Interaction between atmospheric and pedospheric nitrogen nutrition in spruce (*Picea abies* L. Karst) seedlings. *Plant Cell and Environment* **19**, 345–355.

Murray, A.J.S and Wellburn, A.R. (1985). Differences in nitrogen metabolism between cultivars of tomato and pepper during exposure to glasshouse atmospheres containing oxides of nitrogen. *Environmental Pollution* **A 39**, 303–316.

Näsholm, T. (1998). Qualitative and quantitative changes in plant nitrogen acquisition induced by anthropogenic nitrogen deposition. *New Phytologist* **139**, 87–90.

National Academy of Sciences (1977). *Nitrogen Oxides*. National Academy of Sciences, Washington DC, pp. 147–158.

Neubert, A., Kley, D., Wildt, J., Segschneider, H.J. and Forstel, H. (1993). Uptake of NO, NO_2 and O_3 by sunflower (*Helianthus annuus* L.) and tobacco plants (*Nicotiana tabacum* L.) – dependence on stomatal conductivity. *Atmospheric Environment* **Part A 27**(14), 2137–2145.

Norby, R.J., Weerasuriya, Y. and Hanson, P.J. (1989). Induction of nitrate reductase activity in red spruce needles by NO_2 and HNO_3 vapor. *Canadian Journal of Forest Research* **19**, 889–896.

Nussbaum, S., Geissmann, M. and Fuhrer, J. (1995). Effects of nitric oxide and ozone on spring wheat (*Triticum aestivum*). *Water Air and Soil Pollution* **85**, 1449–1454.

Nussbaum, S., Vonballmoos, P., Gfeller, H., Schlunegger, U.P., Fuhrer, J., Rhodes, D. and Brunold, C. (1993). Incorporation of atmospheric $^{15}NO_2$-nitrogen into free amino-acids by Norway spruce *Picea abies* (L.) Karst. *Oecologia* **94**, 408–414.

Okano, K. and Totsuka, T. (1986). Absorption of nitrogen-dioxide by sunflower plants grown at various levels of nitrate. *New Phytologist* **102**, 551–562.

Okano, K., Fukuzawa, T., Tazaki, T. and Totsuka, T. (1986). ^{15}N dilution method for estimating the absorption of atmospheric NO_2 by plants. *New Phytologist* **102**, 73–84.

Okano, K., Machida, T. and Totsuka, T. (1988). Absorption of atmospheric NO_2 by several herbaceous species – estimation by the ^{15}N dilution method. *New Phytologist* **109**, 203–210.

Ramge, P., Badeck, F.W., Plochl, M. and Kohlmaier, G.H. (1993). Apoplastic antioxidants as decisive elimination factors within the uptake process of nitrogen-dioxide into leaf tissues. *New Phytologist* **125**, 771–785.

Remmler, J.L. and Campbell, W.H. (1986). Regulation of corn leaf nitrate reductase. II. Synthesis and turnover of the enzyme's activity and protein. *Plant Physiology* **80**, 442–447.

Rennenberg, H., Kreutzer, K., Papen, H. and Weber, P. (1998). Consequences of high loads of nitrogen for spruce (*Picea abies*) and beech (*Fagus sylvatica*) forests. *New Phytologist* **139**, 71–86.

Rowland, A.J., Drew, M.C. and Wellburn, A.R. (1987). Foliar entry and incorporation of atmospheric nitrogen dioxide into barley plants of different nitrogen status. *New Phytologist* **107**, 357–371.

Saxe, H. (1986). Effects of NO, NO_2 and CO_2 on net photosynthesis, dark respiration and transpiration of pot plants. *New Phytologist* **103**, 185–197.

Saxe, H. and Christensen, O.V. (1985). Effects of carbon dioxide with and without nitric oxide pollution on growth, morphogenesis and production time of pot plants. *Environmental Pollution* **A 38**, 159–169.

Siegwolf, R.T.W., Matyssek, R., Saurer, M., Maurer, S., Gunthardt-Goerg, M.S., Schmutz, P. and Bucher, J.B. (2001). Stable isotope analysis reveals differential effects of soil nitrogen and nitrogen dioxide on the water use efficiency of hybrid poplar leaves. *New Phytologist* **149**, 233–246.

Smith, W.H. (1981). *Air Pollution and Forests*. Springer-Verlag, New York.

Stulen, I., Perez-Soba, M., De Kok, L.J. and Van der Eerden, L. (1998). Impact of gaseous nitrogen deposition on plant functioning. *New Phytologist* **139**, 61–70.

Subczynski, W.K., Lomnicka, M. and Hyde, J.S. (1996). Permeability of nitric oxide through lipid bilayer membranes. *Free Radical Research* **24**, 343–349.

Takahashi, M., Sasaki, Y., Ida, S. and Morikawa, H. (2001). Nitrite reductase gene enrichment improves assimilation of NO_2 in *Arabidopsis*. *Plant Physiology* **126**, 731–741.

Taylor, O.C., Thompson, C.R., Tingey, D.T. and Reinert, R.A. (1975). Oxides of nitrogen. In: Mudd, J.B. and Kozlowski, T.T. (eds) *Responses of Plants to Air Pollution*. Academic Press, New York, pp. 121–139.

Thoene, B., Schröder, P., Papen, H., Egger, A. and Rennenberg, H. (1991). Absorption of atmospheric NO_2 by spruce (*Picea abies* L. Karst) trees. I. NO_2 influx and its correlation with nitrate reduction. *New Phytologist* **117**, 575–585.

Tingey, D.T., Reinert, R.A., Dunning, J.A. and Heck, W.W. (1971). Vegetation injury from the interaction of NO_2 and SO_2. *Phytopathology* **61**, 5106 –5111.

Weber, P., Nussbaum, S., Fuhrer, J., Gfeller, H., Schlunegger, U.P., Brunold, C. and Rennenberg, H. (1995). Uptake of atmospheric $^{15}NO_2$ and its incorporation into free amino acids in wheat (*Triticum aestivum*). *Physiologia Plantarum* **94**, 71–77.

Wellburn, A.R. (1990). Why are atmospheric oxides of nitrogen usually phytotoxic and not alternative fertilizers? *New Phytologist* **115**, 395–429.

Wink, D.A. and Mitchell, J.B. (1998). Chemical biology of nitric oxide: Insights into regulatory, cytotoxic, and cytoprotective mechanisms of nitric oxide. *Free Radical Biology and Medicine* **25**, 434–456.

8 Effects of sulphur dioxide

A.H. LEGGE AND S.V. KRUPA

Introduction

Sulphur (S) is an essential nutrient for normal plant growth and development. The need for S by plants has been recognized for over 200 years (Duke and Reisenauer, 1986). Although the primary source of S for plants is from the soil in the form of sulphate (SO_4^{2-}), which is taken up by the roots and translocated to the leaves where most of it is reduced and assimilated into organic S compounds (Marschner, 1995), an important secondary source of S for plants can be from the atmosphere (DeKok, 1990). Fifteen molecular species of S are known to occur in the atmosphere according to Berresheim *et al.* (1995). Plants that are unable to obtain adequate S nutrition from the soil can fulfil their S requirements by taking up sulphur dioxide (SO_2) or other volatile S compounds such as hydrogen sulphide (H_2S) when present at low concentrations in the atmosphere (DeKok, 1990; DeKok *et al.*, 1998). When more S is taken up from the atmosphere than is needed by the plants to fulfil their S requirements, plants can be adversely impacted due to the excess S availability (Rennenberg, 1984). It has been known for centuries that plants exposed to atmospheric S gases can be adversely affected (Evelyn, 1661). Among these S gases, SO_2 is considered to be the most important phytotoxic molecule (Legge *et al.*, 1998) and, as such, is the focus of this discussion. In the following sections, the discussion is directed at higher plants, although it is recognized that lichens are known to be very sensitive to SO_2 exposure (see Chapter 17).

Background SO_2 sources and ambient concentrations

A number of natural and anthropogenic sources release SO_2 into the atmosphere. The primary natural sources of SO_2 are volcanoes and biomass burning, while the main anthropogenic sources of SO_2 are stationary or point sources involved in fossil fuel combustion, metal smelting and oil and natural gas processing (Cullis and Hirschler, 1980; Brimblecombe *et al.*, 1989). Together, natural and anthropogenic global SO_2 sources emit an estimated 194 million tonnes annually, of which 83% is due to fossil fuel combustion (Watson *et al.*, 1990). Although considerable progress has been made in the development and implementation of SO_2 control technologies in North America, Europe and Japan, ambient SO_2 concentrations are still a significant problem in other

Air Pollution and Plant Life, second edition. Edited by J.N.B. Bell and M. Treshow. ISBN 0 471 49090 3 (HB), 0 471 49091 1 (PB).

parts of the world (Yunus *et al.*, 1996; Innes and Haron, 2000) (see Chapter 21). Brimblecombe *et al.* (1989) provide an overview of the influence of human activity on the global sulphur cycle.

According to Finlayson-Pitts and Pitts (1986), typical peak ambient SO_2 concentrations vary from <1 ppb (<2.6 $\mu g/m^3$) in remote areas to 1–30 ppb (2.6–78.6 $\mu g/m^3$) in rural areas to 30–200 ppb (78.6–524 $\mu g/m^3$) in moderately polluted areas and to 200–2000 ppb (524–5240 $\mu g/m^3$) in heavily polluted areas. Once emitted, SO_2 is transferred from the atmosphere onto surfaces by diffusion (dry deposition) at variable rates which are strongly influenced by meteorological conditions. Any observed foliar injury or changes in growth and productivity due to SO_2 exposures are the result of this dry deposition and plant uptake. It is also important to note that SO_2 in the atmosphere is also transformed to SO_4^{2-} at variable rates, and these SO_4^{2-} particles are deposited onto surfaces by Brownian motion (dry deposition) and by precipitation (wet deposition) (see Chapters 3 and 4 for more details).

Selected examples of ambient effects of SO_2 on plants

As noted earlier, the adverse effects of sulphur pollution on plants was documented more than 300 years ago (Evelyn, 1661; see also Chapter 2). Table 8.1 provides a selected listing of locations and source types along with some of the boreal and temperate forest and shrub species that have exhibited visible foliar injury due to SO_2 in North America, Europe and Asia. The most damage to forests has occurred from SO_2 emissions from nonferrous and ferrous metal smelters and coal-burning power plants. Classic examples include some of the subalpine regions in northern Czechoslovakia (now Czech Republic) and Poland and the former East Germany (especially the Erzgebirge), along with forests near Trail, British Columbia, Canada, and Sudbury, Ontario, Canada. Some additional examples from the 1960s and later are the Ruhr Valley (former West Germany); the Cumberland Plateau, Tennessee, USA; Wawa, Ontario, Canada; the Kola Peninsula, Russia, along with adjacent Finland and Norway, and Noril'sk, Russia.

The emphasis of investigations in the early years was on the prevalence of SO_2-induced foliar injury symptoms in the vicinity of industrial sources. Based upon the occurrence of visible foliar injury, listings of plant species sensitive to SO_2 were compiled, an example of which is provided in Table 8.2. It was common for selected plant species from these lists to be used as biological indicators of SO_2 pollution (see Legge *et al.*, 1998, for additional details). As noted by Kozlowski (1985), plant community and ecosystem level effects were studied when such impacts were visible. These types of investigations continue but have significantly decreased in number in most developed nations where regulatory policies have imposed controls on SO_2 emissions (Krupa, 1997). That being said, the occurrence of low levels of SO_2 jointly with other phytotoxic air pollutants and the prevalence of appropriate biotic and abiotic factors can lead to chronic, long-term, subtle effects on plant growth and productivity. (For additional discussion, refer to the following section and also to Chapters 14, 19 and 20.)

Table 8.1 Selected examples of boreal and temperate forest tree and shrub species that have exhibited visible foliar injury under ambient conditions due to sulphur (SO_2) emissions from industrial point or source complexes

Location/Reference	Source type	Tree and shrub species with foliar injury
North America		
Redding, CA, USA Haywood (1905)	Metal smelter	Pine (*Pinus* spp.), oak (*Quercus* spp.)
Anaconda, MT, USA Scheffer and Hedgecock (1955)	Metal smelter	Douglas fir [*Pseudotsuga menziesii* (Mirb.) Franco], lodgepole pine (*Pinus contorta* Doug. ex Loud.)
Trail, BC, Canada NRC (1939), Katz (1949), Archibold (1978), Fox *et al.* (1986)	Metal smelter	Ponderosa pine (*Pinus ponderosa* Doug. ex P and C Laws.), Douglas fir [*Pseudotsuga menziesii* (Mirb.) Franco], western larch (*Larix occidentalis* Nutt.), lodgepole pine (*Pinus contorta* Doug. ex Loud.)
Sudbury and Wawa, ON, Canada Linzon (1978), Freedman and Hutchinson (1980), Gunn (1995)	Metal smelter	White pine (*Pinus strobus* L.), white spruce [*Picea glauca* (Moench) Voss.], black spruce [*Picea mariana* (Mill.) BSP], trembling aspen (*Populus tremuloides* Michx.)
Whitecourt, AB, Canada Legge *et al.* (1977), Legge *et al.* (1981), Krupa and Legge (1998)	Sulphur recovery gas plant	Lodgepole pine x jack pine (*Pinus contorta* Doug. ex Loud. x *Pinus banksiana* Lamb.), trembling aspen (*Populus tremuloides* Michx.)
Yellowknife, NWT, Canada Hocking *et al.* (1978)	Gold smelter	Black spruce [*Picea mariana* (Mill.) BSP], white spruce [*Picea glauca* (Moench) Voss], paper birch (*Betula papyrifera* Marshall), poplar (*Populus* spp.), willow (*Salix* spp.)
Copper Basin, TN, USA Miller and McBride (1975)	Metal smelter	Oak (*Quercus* spp.), White pine (*Pinus strobus* L.)
Europe/Asia		
Ruhr Valley, Germany (former West Germany) Knabe (1970)	Mixed industrial complex	Scots pine (*Pinus sylvestris* L.)
Ore Mountains, Czech Republic (former Czechoslovakia) Materna (1973), Tichy (1996)	Industrial complex	Norway spruce [*Picea abies* (L.) Karsten]

continued overleaf

Table 8.1 *(continued)*

Location/Reference	Source type	Tree and shrub species with foliar injury
Industrial Pennines, Great Britain Farrar *et al.* (1977)	Mixed industrial complex	Scots pine (*Pinus sylvestris* L.)
Erzgebirge, Germany (former East Germany) Thomasius (1987)	Industrial complex	Spruce (*Picea* spp.)
Kola Peninsula, Russia-Eastern Lapland, Finland and Norway Kryuchkov (1993), Tikkanen and Niemelä (1995), Rigina and Kozlov (2000)	Metal smelters	Scots pine (*Pinus sylvestris* L.), Norway spruce [*Picea abies* (L.) Karsten], Downy birch (*Betula pubescens* Ehrh.), Dwarf birch (*Betula nana* L.), European aspen (*Populus tremula* L.), Willow (*Salix* spp.)
Noril'sk, Siberia, Russia Kharuk (2000), Odintsov (2000)	Metal smelters	Siberian larch [*Larix russica* (Endl.) Sab. ex Trautv.), Siberian spruce (*Picea obovata* Ledeb.), Silver birch (*Betula pendula* Roth.), Willow (*Salix* spp.)

Source: modified from Legge *et al.* (1998).

Table 8.2 Selected list of some plant species that are relatively sensitive to SO_2. Nomenclature mostly based on Wiersema and León (1999) and Brako *et al.* (1995). Refer to Legge *et al.* (1998) for a more comprehensive listing

Common name	Latin name
Alfalfa	*Medicago satavia* L.
Alsike clover	*Trifolium hybridum* L.
Austrian pine	*Pinus nigra* J.F. Arnold
Barley	*Hordeum vulgare* L.
Balsam fir	*Abies balsamea* (L.) Mill.
Bean, field	*Phaseolus vulgaris* L.
Beet, table	*Beta vulgaris* L.
Blackberry	*Rubus* spp.
Black willow	*Salix nigra* Marsh.
Blueberry	*Vaccinium* spp.
Bracken fern	*Pteridium aquilinum* (L.) Kuhn
Broccoli	*Brassica oleracea* L. var *italica* Plenck
Cabbage	*Brassica oleracea* L. var. *capitala* L.
Carrot	*Daucus carota* L.
Celery	*Apium graveolens* L.

Table 8.2 (*continued*)

Common name	Latin name
Corn	*Zea mays* L.
Cotton	*Gossypium hirsutum* L.
Cow-parsnip	*Heracleum lanatum* Michx.
Cucumber	*Cucumis sativus* L.
Eastern white pine	*Pinus strobus* L.
European white birch	*Betula pendula* Roth
Green ash	*Fraxinus pennsylvanica* Marsh.
Jack pine	*Pinus banksiana* Lamb.
Labrador tea	*Ledum groenlandicum* Oeder
Large-tooth aspen	*Populus grandidentata* Michx.
Lodgepole pine	*Pinus contorta* Doug. x Loud.
London plane-tree	*Platanus* x *acerifolia* (Aiton) Willd.
Northern red oak	*Quercus rubra* L.
Norway maple	*Acer platanoides* L.
Oats	*Avena sativa* L.
Onion	*Allium cepa* L.
Potato	*Solanum tuberosum* L.
Prickly rose	*Rosa acicularis* Lindl.
Radish	*Raphanus sativus* L.
Ragweed	*Ambrosia artemisiifolia* L.
Red maple	*Acer rubrum* L.
Rye	*Secale cereale* L.
Saskatoon serviceberry	*Amelanchier alnifolia* (Nutt.) Nutt.
Silver fir	*Abies alba* Mill., *A. amabilis* (Doug. ex J. Forbes) or *A. grandis* (Douglas ex D. Dan) Lindl.
Silver maple	*Acer saccharinum* L.
Soybean	*Glycine max.* (L.) Merr.
Spreading wood fern	*Dryopteris expansa* (K.B. Presl.) Fraser-Jenkins and Jermy
Sweet pea	*Lathyrus odoratus* L.
Tamarack	*Larix laricina* (DuRoi) K. Koch
Trembling aspen	*Populus tremuloides* Michx.
Wheat	*Triticum aestivum* L.
White birch	*Betula papyrifera* Marsh.
White fir	*Abies concolor* (Gord. and Glend.) Lindl. ex F.H. Hildebr.
Wild grape	*Vitis* spp.

There are numerous artificial exposure studies on the effects of SO_2 on crops (see Legge and Krupa, 1990). Table 8.3 provides a selected listing of locations and source types along with some of the agricultural crop species that have exhibited foliar injury due to ambient SO_2 from industrial sources in North America and Europe. In contrast to the cases for SO_2 damage to forests, however, there are only a few well-documented examples of the adverse effects of ambient SO_2 exposures on crop

Table 8.3 Selected examples of agricultural crop species that have exhibited visible foliar injury under ambient conditions due to sulphur (SO_2) contaminants from industrial point or source complexes

Location	Source	Species with acute injury
North America		
Sudbury area, ON, Canada Dreisinger and McGovern (1970)	Copper and nickel smelter complex	Buckwheat (*Fagopyrum esculentum* Moench), barley (*Hordeum vulgare* L.), red clover (*Trifolium pratense* L.), radish (*Raphanus sativus* L.), oats (*Avena sativa* L.), pea (*Pisum sativum* L.), rhubarb (*Rheum rhabarbarum* L.), timothy (*Phleum pratense* L.), Swiss chard (*Beta vulgaris* L. ssp. *cicla* (L.) Koch), bean (*Phaseolus vulgaris* L.), beet (*Beta vulgaris* L.), turnip (*Brassica rapa* L.), carrot (*Daucus carota* L.), cucumber (*Cucumis sativus* L.), lettuce (*Lactuca sativa* L.), tomato (*Lycopersicon esculentum* Miller), potato (*Solanum tuberosum* L.), raspberry (*Rubus* spp.), celery (*Apium graveolens* L.), spinach (*Spinacia oleracea* L.), cabbage (*Brassica oleracea* L. var. *capitata* L.), corn (*Zea mays* L.)
Widow's Creek, AL, USA McLaughlin and Lee (1974)	Coal-fired electricity generating plant	Soybean (*Glycine max.* (L.) Merr.)
Douglas/Hereford, AZ, USA Haase *et al.* (1980)	Copper smelter	Lettuce (*Lactuca sativa* L.), green onion (*Allium cepa* L.), cabbage (*Brassica oleracea* L. var. *capitata* L.), carrot (*Daucus carota* L.), chili pepper (*Capsicum annuum* L.), cotton (*Gossypium hirsutum* L.)
An area in southeastern Ohio and northwestern Virginia, USA Jacobson and Showman (1984)	Coal-fired electricity generating plant	Bush bean (*Phaseolus* spp.), corn (*Zea mays* L.)
Europe		
Biersdorf, former West Germany Guderian and Stratmann (1962)	Metal smelter	Winter wheat (*Triticum aestivum* L.), spring wheat (*Triticum aestivum* L.), oats (*Avena sativa* L.), winter rye (*Secale cereale* L.), potato (*Solanum tuberosum* L.)

Source: modified from Legge *et al.* (1998).

yields. A classic example of such adverse effects is from Biersdorf in the former West Germany (Guderian and Stratmann, 1962, 1968). More recently, there have been isolated but hard to access reports of SO_2 exposures and negative effects on crop productivity from rapidly developing countries such as India and the People's Republic of China (Yunus *et al.*, 1996; Chameides *et al.*, 1999). To our knowledge there may be many other similar cases, but such examples have not been fully investigated and reported.

Foliar symptoms of SO_2 exposure

Sulphur dioxide injury to plants is first seen on the foliage which is more sensitive to SO_2 exposure than stems, buds and reproductive parts. The extent to which foliage responds to SO_2 is determined by both biotic and abiotic factors as well as by the concentration, duration and frequency of SO_2 exposure. Biotic factors include genetic make-up, developmental stage of growth, plant nutrient status and insects and disease. Abiotic factors include soil moisture and nutrient status, air and soil temperature, relative humidity, radiation, precipitation and meteorological conditions as well as the presence of other air pollutants (see Chapters 18, 19, 20 and 14 for more details). Simply put, when biotic and abiotic factors are favourable for plant growth and development, there is a high probability that plants exposed to SO_2 will be adversely affected while, when one or more biotic and/or abiotic factors limit plant growth and development, there is a lowered probability that plants will be adversely affected. That being said, SO_2 concentration, as well as the duration and frequency of SO_2 exposure which are determined by meteorological conditions, play a significant role in determining the potential for an adverse response of vegetation to SO_2. This is particularly the case in the vicinity of point sources where, under meteorological conditions that promote surface level atmospheric inversions, vegetation can be exposed to relatively high SO_2 concentrations ranging in duration from a few minutes to hours (Boubel *et al.*, 1994).

Foliar SO_2 injury symptoms and the SO_2 concentrations causing these symptoms in the field are commonly characterized as being either acute or chronic. An acute SO_2 exposure is viewed as a short duration SO_2 exposure of minutes to hours that is of sufficient concentration to result in the expression of necrotic injury to the foliage within a few hours or days. On broad-leaved plants, these acute SO_2 injury symptoms commonly consist of bifacial, marginal and/or interveinal necrosis and chlorosis on leaves at the full stage of development. The necrotic areas can range in colour from white to reddish-brown to black depending on the plant species. The margins of the necrotic areas are mostly irregular and occasionally dark pigmented. Illustrations of acute foliar SO_2 injury symptoms on the following broad-leaved plant species with varying leaf forms are found in Figure 8.1a–f, respectively: alsike clover (*Trifolium hybridum* L.), alfalfa (*Medicago sativa* L.), prickly rose (*Rosa acicularis* Lindl.), cow-parsnip (*Heracleum lanatum* Michx.), Labrador tea (*Ledum groenlandicum* Oeder) and Arctic willow (*Salix arctica* Pall.). In monocotyledenous plants, acute injury symptoms start at the tip of the leaves and spread downward as necrotic and chlorotic streaks

Figure 8.1 Illustrations of acute foliar SO_2 injury on selected broad-leaved plant species. (a) Acute SO_2 injury symptoms on leaves of alsike clover (*Trifolium hybridum* L.) are seen as a whitish-tan marginal and interveinal necrosis. (Photograph courtesy Dr. A. Legge, Biosphere Solutions, Calgary, Alberta, Canada.): (b) Acute SO_2 injury symptoms on leaves of alfalfa (*Medicago sativa* L.) are seen as an apical whitish-tan necrosis with occasional dark pigmented margins spreading towards the leaf base. (Photograph courtesy Dr. A. Legge, Biosphere Solutions, Calgary, Alberta, Canada.): (c) Acute SO_2 injury symptoms on leaves of prickly rose (*Rosa acicularis* Lindl.) are seen as reddish-brown interveinal necrotic areas with dark pigmented margins. These symptoms are most pronounced on the basal leaflets. Note the presence of small water droplets on the surface of the three apical leaflets. (Photograph courtesy Dr. A. Legge, Biosphere Solutions, Calgary, Alberta, Canada.): (d) Acute SO_2 injury symptoms on a leaf of cow-parsnip (*Heracleum lanatum* Michx.) are seen as brownish-tan interveinal necrotic areas. (Photograph courtesy Dr. A. Legge, Biosphere Solutions, Calgary, Alberta, Canada.): (e) Acute SO_2 injury symptoms on leaves of Labrador tea (*Ledum groenlandicum* Oeder) are seen as brown to reddish-brown marginal and interveinal necrotic areas with red to dark pigmented margins. (Photograph courtesy Dr A. Legge, Biosphere Solutions, Calgary, Alberta, Canada.): (f) Acute SO_2 injury symptoms on leaves of Arctic willow (*Salix arctica* Pall.) are seen as reddish-brown interveinal necrotic areas with dark pigmented margins. (Photograph courtesy Dr A. Legge, Biosphere Solutions, Calgary, Alberta, Canada.)

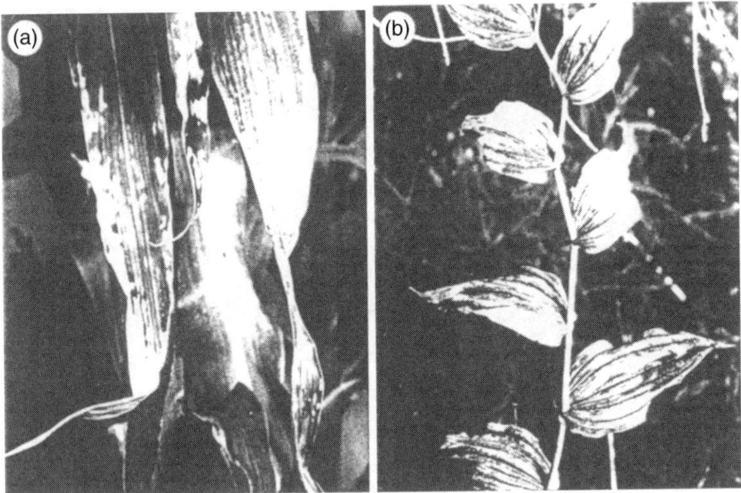

Figure 8.2 Illustrations of acute foliar SO_2 injury on selected monocotyledenous plant species. (a) Acute SO_2 injury symptoms on leaves of corn (*Zea mays* L.) are seen as a whitish-tan and a brownish-red marginal and interveinal necrosis spreading downward from the leaf tips as well as from the leaf margins towards the mid-rib. (Photograph courtesy Dr G.H.M. Krause, Landesumweltamt Nordrhein-Westfalen, Essen, Germany.): (b) Acute SO_2 injury symptoms on leaves of fairy-bells (*Disporum hookeri* (Torr.) Nicholson) are seen as marginal and interveinal necrotic areas spreading downward from the leaf tips towards the leaf base. (Photograph courtesy Dr A. Legge, Biosphere Solutions, Calgary, Alberta, Canada.)

with occasional reddish pigmentation. Illustrations of acute foliar SO_2 injury symptoms to the monocotyledenous plant species corn (*Zea mays* L.) and fairy-bells (*Disporum hookeri* (Torr.) Nicholson) are shown in Figure 8.2a–b. For conifers, acute symptoms appear on second-year needles and older and consist of a tan to reddish-brown necrosis that starts at the needle tip, spreading downward toward the base, commonly preceded by chlorosis. Premature needle drop of the older needles is also a common occurrence. Illustrations of acute foliar SO_2 injury symptoms to the coniferous species eastern white pine (*Pinus strobus* L.) and Scots pine (*Pinus sylvestris* L.) are shown in Figure 8.3a–b.

Although the above acute foliar SO_2 injury symptom examples are documented cases of acute SO_2 exposure, very similar injury symptoms can be caused by other air pollutants and biotic and abiotic stress factors (see Legge *et al.*, 1998). When leaf surfaces are wet at the time of exposure to SO_2, the SO_2 can be absorbed by the water droplets on the leaf surface. The water becomes acidic as the SO_2 which is absorbed is converted to sulphuric acid (H_2SO_4) and can cause acute foliar injury. The acute foliar injury from this 'acidic wet deposition' occurs as necrotic areas with regular margins. These regular margins reflect the outline of the acidified water droplets on the leaf surface. This is illustrated in Figure 8.4a and 4b for foliage from green alder (*Alnus crispa* (Ait.) Pursh.) exposed to SO_2 when the foliage was dry and when the foliage was wet, respectively. It is also important to note that short-term, acute foliar

Figure 8.3 Illustrations of acute foliar SO_2 injury on selected coniferous plant species. (a) Acute foliar SO_2 injury on eastern white pine (*Pinus strobus* L.) is seen as extensive premature needle drop with only one year of needles remaining attached. Although not clearly evident in this black and white photograph, there is extensive SO_2 injury to the needles which is seen as a reddish-brown tip necrosis spreading downward towards the needle base along with occasional horizontal dark pigmented bands. (Photograph courtesy Dr A. Legge, Biosphere Solutions, Calgary, Alberta, Canada.): (b) Acute SO_2 injury symptoms on second year needles of Scots pine (*Pinus sylvestris* L.) are seen as a reddish-brown tip necrosis spreading downward towards the needle base with numerous horizontal dark pigmented bands. (Photograph courtesy Dr A. Legge, Biosphere Solutions, Calgary, Alberta, Canada.)

injury symptoms may or may not lead to long-term reductions in plant growth and productivity (Smith, 1990).

Chronic SO_2 exposure is viewed as low concentration SO_2 exposures that occur during the entire growth cycle or life of a plant with periodic intermittent and random peak levels (Krupa, 1996). Such exposures may or may not result in chronic foliar injury symptoms such as marginal and/or interveinal chlorosis in broad-leaved plants, chlorosis in second-year and older conifer needles, premature fall colouration and premature leaf/needle abscission (see Legge *et al.*, 1998 for more details). As noted for acute foliar injury symptoms, however, symptoms similar to chronic injury can be caused by other air pollutants and biotic and abiotic stress factors. Chronic SO_2 exposures can lead to reductions in the rate of plant growth and productivity (biomass) as shown by Bell (1982) for grasses, which are impacted most severely under winter conditions, and by Legge *et al.* (1996) for pine. It is also important to note that reductions in plant growth and productivity from chronic exposure may occur without the development of visible chronic foliar injury symptoms. Physiological, biochemical, cellular and tissue-level markers can be used to identify the presence of chronic SO_2 stress when visible symptoms are present and/or absent (USNRC, 1989).

Acute and chronic SO_2 exposures/symptoms have been treated to this point as if they occur separately. This is not always true. In the vicinity of point sources, acute SO_2 exposures can occur on top of chronic SO_2 exposures. Further, depending on the SO_2 concentration, duration and frequency, acute and chronic SO_2 injury symptoms can co-occur on the same or different foliage on the same plant. Acute and/or chronic

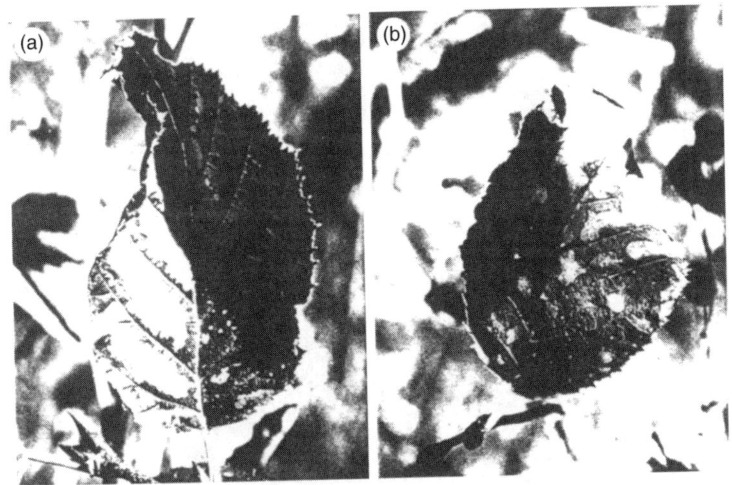

Figure 8.4 Comparison of acute foliar injury symptoms to the leaves of green alder (*Alnus crispa* (Ait.) Pursh.) caused by exposure to elevated concentrations of SO_2 when the leaf surface is dry (a) and when the leaf surface is wet (b). Acute foliar injury in (a) is seen as a predominantly interveinal necrosis with irregular margins while acute foliar injury in (b) is seen as an interveinal and marginal necrosis with regular margins. The regular margins of the necrotic areas in (b) reflect the outline of the SO_2 acidified water droplets (i.e. acidic wet deposition) which were present on the wet leaf surface. An illustration of the varied shapes of water droplets on leaf surfaces is shown in Figure 1c on the apical leaflets of prickly rose. The acute foliar injury symptoms in (a) are reddish-brown while in (b) the acute foliar injury symptoms are light brown to tan. (Photographs courtesy Dr A. Legge, Biosphere Solutions, Calgary, Alberta, Canada.)

SO_2 exposure can also result in changes in the physical nature of epicuticular waxes as illustrated in Figure 8.5a–b for Scots pine (*Pinus sylvestris* L.) as well as changes in the chemical composition of epicuticular waxes of plants (Percy *et al.*, 1994). It is important to note that the expression of acute and/or chronic SO_2 symptoms is highly variable and can vary at the genus, species, variety or cultivar, provenance and population levels (Karnosky, 1985; Tingey and Olszyk, 1985). That being said, broad-leaved plants that have satisfied their normal S requirements through uptake from the soil tend to be more sensitive to SO_2 deposition or overload (Krupa and Legge, 1999). As noted by Keller (1981), this is also true for conifers and deciduous tree species that absorb SO_2 during winter periods, leading to carry-over or latent responses during the following growth season.

Mechanisms of foliar injury and detoxification

The primary pathway of entry of SO_2 into the leaf is through the stomata. Inside the leaf SO_2 rapidly dissolves in the aqueous phase of the cell wall (apoplastic space) to form bisulphite (HSO_3^-) and sulphite (SO_3^-). Rapid conversion of toxic SO_3^- to non-toxic SO_4^- may be achieved in the apoplast (Pfanz *et al.*, 1990), leading to

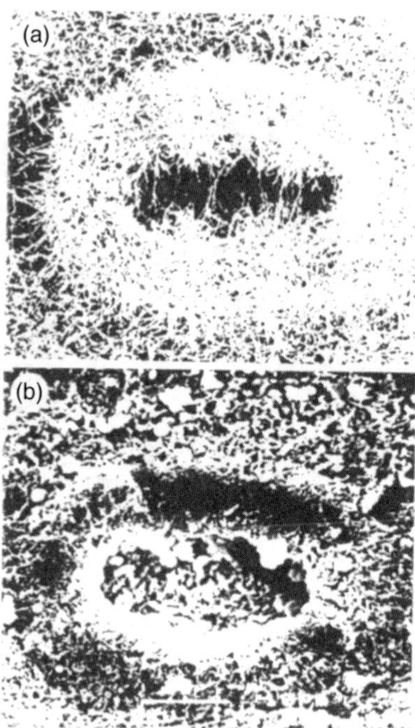

Figure 8.5 Comparison of epicuticular waxes on the surface of current year Scots pine (*Pinus sylvestris* L.) needles with no exposure to SO_2 (a) and elevated exposure to SO_2 and particulate matter (b). Note the extensive erosion of the epicuticular waxes in (b) compared to (a). (Photographs courtesy Dr Satu Huttunen, Department of Biology, University of Oulu, Oulu, Finland.)

an elevated S influx into the leaf cells through the transpiration stream (Wolfenden *et al.*, 1991; Rennenberg and Herschbach, 1996). The presence of SO_3^- is inhibitory to peroxidase activity, and oxidation of SO_3^- is competitive with the oxidation of phenolic compounds in lignin formation. Rennenberg and Polle (1994), however, calculated that the capacity of apoplastic enzymatic fluids to oxidize SO_3^- is three orders of magnitude higher than the stomatal influx of SO_2 into the leaf at 30 ppb (80 $\mu g/m^3$) SO_2. Nevertheless, depending on the rate of influx of SO_2 and the rates of the various reactions, the observed foliar necrosis is governed by the accumulation of the oxidation products of phenolic compounds.

Materna (1966) was the first to observe that plants exposed to SO_2 re-emit part of that absorbed S, predominantly as hydrogen sulphide (H_2S). This has been shown to be a general phenomenon (7–15% of the absorbed SO_2) among higher plants (Rennenberg, 1991). The conversion of SO_4^{2-} and SO_3^- to H_2S is mediated by light and is thought to occur in the chloroplast. At the metabolic level, an important consequence of the reductive conversion of sulphite to sulphide rather than the oxidative conversion to sulphate is the prevention of SO_2-derived acidification. This overall process of the loss of S into the atmosphere can be viewed as a means for the plant to reduce excess S and control its S nutrition via homeostatic regulation of the cell cysteine concentration

(Rennenberg and Herschbach, 1996). However, this capacity appears to be related to the SO_2 sensitivity of the plant species and tissue (age, maturity, etc.). Filner *et al.* (1984) noted, for example, that young leaves of cucumber (*Cucumus sativus* L. cultivar), absorbing high amounts of SO_2, were more resistant than mature leaves of cucumber absorbing smaller amounts. About 60% of the absorbed SO_2 in both young and mature leaves was oxidized to SO_4^{2-}, but the young leaves emitted H_2S at a rate of up to 100-fold higher than the mature leaves, indicating that young leaves were more able to photoreduce SO_2 than H_2S. Similarly, SO_3^--tolerant duckweed (*Lemna valdiviana* Philippi) emitted four times more H_2S than the SO_3^--susceptible duckweed (*Lemna gibba* L.) (Takemoto *et al.*, 1986).

Effects on plant physiology and metabolism

The effects of SO_2 on the stomata are too complex to allow satisfactory conclusions to be drawn, but in general it appears that short-term exposures, particularly at concentrations <50 ppb (<134 μg/m^3), often cause wider stomatal opening, while long-term exposures with higher concentrations usually cause partial stomatal closure (Mansfield and Pearson, 1996). These patterns have important consequences, first through either enhancement or depression of photosynthetic carbon dioxide uptake and transpirational water loss. Secondly, there will be an alteration in the quantity or rate at which SO_2 enters the plant and arrives at the metabolic sites in the underlying mesophyll tissue (Black, 1982). Environmental conditions (e.g., abiotic factors such as light, temperature, relative humidity, soil moisture availability, etc.) as well as other chemical constituents in the atmosphere (e.g., the concentration of CO_2) during exposure can also modify stomatal response (Mansfield, 1998, see also Chapter 19 in this book).

In general, acute and/or chronic exposures to SO_2 can result in a reduction of the photosynthetic rate (Winner *et al.*, 1985). In this regard, Pratt *et al.* (1983) showed an accumulation of total S in SO_2-fumigated soybean plants and a reduction in leaf total chlorophyll content. Although there are a few reports of short-term SO_2-induced stimulation of photosynthesis, there is some question as to whether such studies were satisfactorily able to account for CO_2 loss by photorespiration. In this context, there is no clear picture of the effects of SO_2 exposures on respiration. However, in one study, Harvey and Legge (1979) found a reduction in total ATP (adenosine triphosphate) content in pine needles subjected to chronic SO_2 exposures, suggesting a possible reduction in the rate of oxidative phosphorylation and/or increased energy consumption in stress repair. It is well known that stress due to infections by foliar pathogens can result in increases in the rate of respiration in the infected tissue (Kosuge and Kimpel, 1981).

In contrast to the effects on the primary metabolic pathways (energy metabolism), exposures to SO_2 can also alter the secondary metabolism. For example, cell wall rigidity can be induced by direct exposure to SO_2 through increased production of phenols (Hyodo *et al.*, 1978) and peroxidases (Morgan and Fowler, 1972) and consequently increased oxidation of phenols (Biggs and Fry, 1987). According to Mayo *et al.* (1992) S-gas fumigation and S-dust deposition, combined with the accumulation of phenolic compounds, were associated with increased concentrations of soluble aluminium and iron and altered elastic properties of the cell wall and water relations.

Based on other previous studies, the authors concluded that leaf diffusive resistance and photosynthesis were altered by changes in stomatal physiology and a reduction in carbon assimilation.

Accumulation and translocation of atmospheric sulphur in plants

A number of investigators have used the accumulation of total S in foliage as an indicator of plant response to atmospheric SO_2 exposures (Guderian, 1977; Linzon *et al.*, 1978; Huttunen *et al.*, 1985; Dmuchowski and Bytnerowicz, 1995; Manninen and Huttunen, 1995). In contrast to the uptake of SO_4^{2-} from the soil, once SO_2 enters the foliage, as noted earlier, it is initially converted to HSO_3^- and SO_3^- and then to SO_4^{2-}, the latter representing the accumulated S species. In contrast, since S is an essential element for plant growth, the assimilated rather than the accumulated level would be found in the organic fraction (cysteine, cystine, methionine, biotin, lipoic acid, thiols, sulphoquinovose, etc.).

Legge *et al.* (1988) measured total (S_T) and inorganic (S_I) foliar S content in mature lodgepole (*Pinus contorta* Doug. ex Loud.) x jack (*Pinus banksiana* Lamb.) pine trees exposed over a period of 11 years to atmospheric S at five locations in the vicinity of a point source near Whitecourt, Alberta, Canada. The authors calculated foliar organic S (S_O) concentrations as $S_T - S_I$. The results of the S analyses were related to the observed reductions in the photosynthetic rate of the current year's needles and a reduction in soil pH. The authors concluded that foliar S_I to S_O ratio can be used as an indicator of plants under stress from atmospheric S pollution. Manninen *et al.* (1997) made a similar observation while working with Scots pine needles in Finland but suggested that the S_I to S_O ratio works best when the trees have a low nitrogen supply from the soil. Recently Krupa and Legge (1999 and 2001) found that the S_T to S_I ratio can also be used in understanding plant foliar injury to SO_2 deposition in the broad-leaved plant Saskatoon serviceberry (*Amelanchier alnifolia* Nutt.).

It is also important to note, in addition to the foliar accumulation of S from SO_2 exposures, that absorbed S can also be translocated to roots, appearing there in various S fractions, including SO_4^{2-}. Part of the SO_2 assimilated into S_O in the shoot can also be transported as glutathione and cysteine to the roots (DeKok, 1990). Similarly, thiols are known to accumulate in measurable amounts in the roots upon exposure of the shoot to SO_2 (Herschbach *et al.*, 1995).

Dose-response relationships

There have been numerous efforts to develop quantitative, mathematical relationships between acute SO_2 exposures and the occurrence of foliar injury symptoms (O'Gara, 1922; Guderian *et al.*, 1960; Zahn, 1963). Dreisinger and McGovern (1970) developed a comprehensive summary of minimum levels of SO_2 exposures resulting in visible foliar injury on vegetation in the vicinity of nickel smelters at Sudbury, Ontario, Canada. In contrast, Guderian (1977) provided summary data on average threshold SO_2 concentrations that caused significant effects on the growth and yield of a number of plant species (Table 8.4). Note that the average SO_2 concentration threshold is much higher when only the time in which the plants were exposed to SO_2 is considered versus the total time that SO_2 was monitored.

More recently, other investigators have examined the issue of chronic exposures and changes in growth and biomass through modelling. Such efforts have been both assessive and predictive, mostly statistical or empirical by nature, while a few have been mechanistic or process oriented (for details, see Krupa and Kickert, 1987; Legge *et al.*, 1998).

A problem with many of the statistical models for chronic responses is that they are single-point models. In other words, a growth season-end measurement of a biomass variable is related to a season-end SO_2 exposure parameter, for example, S emissions (tonnes y^{-1}) or sum of average hourly concentrations. Some studies have included hourly SO_2 concentrations above a set threshold (Krupa and Kickert, 1987).

Independent of the aforementioned considerations, single-point models cannot account for the feedbacks of stress avoidance, repair and/or compensation, and the associated stochasticity or randomness of SO_2 exposures and differing plant response governed by the patterns of its physiology and phenology. As pointed out by Van Haut and Stratmann (1970), different growth stages of a given plant will respond differently to the same or differing SO_2 exposures (concentration and duration). This is further influenced, however, by the joint effects of other air pollutants, plant pathogens and insect pests (Krupa, 1996, also see Chapters 14, 19 and 20 in this book). Furthermore, a critical consideration is the variability in the essential plant growth regulating factors such as air temperature, soil moisture and nutrient availability.

Krupa and Nosal (1989) developed a multivariant, time series (multipoint) statistical model to examine the effects of ambient SO_2 exposures on alfalfa leaf biomass. This model included different growth stages of the plant within each harvest period and also growth-regulating factors such as maximum and minimum air temperature and precipitation depth. The authors found that during the first growth season, the number of SO_2 exposures and the integral of the concentrations and exposure durations were more important predictors than the peak SO_2 concentrations. During the second growth season (more mature plants that had overwintered), number of exposures and peak concentrations were more important than the integral.

Legge *et al.* (1996) also developed a multivariant statistical model to establish relationships between changes in basal area increment of mature hybrid *Pinus contorta* Doug. ex Loud. x *Pinus banksiana* Lamb. (lodgepole pine x jack pine) and chronic SO_2 exposures downwind from a point source. This study did not include growth-regulating climatic variables. However, the authors concluded that peak SO_2 concentrations or the number of episodes did not play as important a role as the integral (a product of concentration and the exposure duration).

The observed differences between the two aforementioned studies may in part be due to differences in the physiology of the two plant species examined and differences in the plant response period modelled (15-day intervals for alfalfa versus yearly intervals for the pine species). Nevertheless, results from both studies address the overall importance of the random occurrences of SO_2 exposures. All three descriptors of the SO_2 exposures (peak concentration, number of episodes and the integral of concentration and exposure

Table 8.4 Threshold values for SO_2-induced growth and yield effects for a number of forest and agricultural plant species from a field study near an iron ore smelter SO_2 source in Germany

Plant species	Threshold concentration in ppm[1]	
	$\overline{c}_i$ during exposure time t_i	$\overline{c}_m$ during monitor time t_m
Fruit trees and berry shrubs		
Gooseberry (*Ribes grossularia* L.)	0.22–0.26	0.010–0.020
Currant (*Ribes rubrum* L.)	0.22–0.26	0.010–0.020
Apple (*Malus communis* L.)	0.22–0.26	0.010–0.020
Sour cherry (*Prunus cerasus* L.)	0.22–0.44	0.010–0.083
Sweet cherry (*Prunus avium* (L.) L.)	0.26–0.44	0.020–0.083
Prune (*Prunus domestica* L.)	0.26–0.44	0.020–0.083
Forest cultures		
Spruce (*Picea abies* (L.) Karsten)	0.22–0.26	0.010–0.020
Pine (*Pinus sylvestris* L.)	0.22–0.26	0.010–0.020
Larch (*Larix europaea* Lam.)	0.22–0.26	0.010–0.020
Red beech (*Fagus sylvatica* L.)	0.22–0.26	0.010–0.020
English oak (*Quercus pedunculata* Ehrh.)	0.22–0.26	0.010–0.020
Grains and summer rape		
Winter wheat (*Triticum sativum* L.)	0.24–0.28	0.009–0.024
Winter rye (*Secale cereale* L.)	0.28–0.31	0.024–0.051
Summer wheat (*Triticum sativum* L.)	0.23–0.38	0.015–0.050
Oats (*Avena sativa* L.)	0.23–0.38	0.015–0.050
Summer rape (*Brassica napus* L.)	0.23–0.57	0.015–0.124
Truck crops		
Potato (*Solanum tuberosum* L.)	0.21–0.23	0.010–0.015
Red beet (*Beta vulgaris* L.)	0.28–0.31	0.024–0.051
Fodder plants		
Red clover (*Trifolium pratense* L.)	0.28–0.31	0.024–0.051
Alfalfa (*Medicago sativa* L.)	0.28–0.31	0.024–0.051
Oats (for green fodder)	0.23–0.38	0.015–0.050
Winter rye (for green fodder)	0.28–0.31	0.024–0.051
Summer rape (for green fodder)	0.28–0.31	0.024–0.051
Vegetables		
Spinach (*Spinacia oleracea* L.)	0.22–0.25	0.010–0.020
Carrot (*Daucus carota* L.)	0.28–0.49	0.024–0.104
Tomato (*Lycopersicon esculentum* Miller)	0.31–0.57	0.051–0.124

Source: modified from Guderian (1977).
[1] $\overline{c}_i$ = average SO_2 concentration for t_i
t_i = exposure time, calculated by summing all 10-minute time intervals with mean SO_2 concentration greater than or equal to 0.10 ppm
$\overline{c}_m$ = average SO_2 concentration for monitoring time t_m
t_m = SO_2 monitoring time of the test plants

duration) were identified as important factors in each case in the overall explanation of the plant response relationship.

As opposed to the statistical models, process-oriented models have a mechanistic basis. For example, GROW 1 (Kercher, 1978, 1982) was conceived with five main submodels to: (a) calculate light penetration into the canopy; (b) simulate soil moisture and its effects on stomatal resistance; (c) calculate internal SO_3^- concentration; (d) calculate the effects on photosynthesis; and (e) calculate carbon flow or allocation. GROW 1 has been used to examine the concept of threshold concentration for deleterious effects of SO_2 on plants.

Similarly, there are also published process models for examining responses of trees, grassland species and some other native vegetation (Krupa and Kickert, 1987). The development and application of process models are highly desirable since they can account for the feedbacks of stress avoidance, repair and compensation. It is important to note, however, that although their initial development is based on available experimental results, in many cases model validation with independent sets of data has been a limitation.

Independent of the modelling strategy used to establish cause-effect relationships, many such efforts are based on artificial SO_2 exposures. Manning and Krupa (1992) have reviewed the literature regarding various exposure methodologies. Although their review is directed to ozone, much of the information presented therein is also applicable to SO_2. An important shortcoming in many artificial exposure studies is that the SO_2 exposure protocols used have had little resemblance to the ambient patterns (Kohut, 1985; Krupa, 1996). To overcome this and other concerns, a few studies have used artificial open-air fumigation (Preston and Lee, 1982; Thompson *et al.*, 1984; McLeod, 1995) or ambient exposures in the vicinity of an SO_2 source (Guderian, 1977; Legge *et al.*, 1977; Jones *et al.*, 1987). One innovative approach is to take advantage, if appropriate, of the abundance of the occurrence of the stable isotope ^{34}S in the source emissions and its ratio to the predominant naturally occurring ^{32}S (Legge *et al.*, 1981; Krouse *et al.*, 1984; Krupa and Legge, 1998; Wadleigh and Blake, 1999). This subject is discussed further in a following section. Another approach is to couple source apportionment methods (Atmospheric Environment, 1984) to S deposition and plant response measurements in a common geographic scale (Sloof, 1995). In this case, appropriate tracer cations associated with the SO_2 emissions by the source (for example, chromium and nickel for emissions from a petroleum processing plant) are used as a source fingerprint or signature to track the spatial and temporal variations in SO_2 deposition. Subsequently, accumulation of S in the biological receptor (plants) is related to the measured effects in that receptor. However, prior to application of either approach (isotopes and signatures), initial characterization of the source emissions would be required to validate the feasibility of the method for the particular study.

Sulphur deficiency in plants

It has been demonstrated that plants growing in S-deficient soils can benefit by taking up S from the atmosphere during chronic exposures (Thomas *et al.*, 1943; Olsen,

1957; Cowling *et al.*, 1973; Noggle, 1980). According to Noggle and Jones (1979), this is species specific, for example, cotton (*Gossypium hirsutum* L.) is more efficient than tall fescue (*Festuca arundinacea* Schreb.) in accumulating atmospheric sulphur. Furthermore, the extent of the positive growth response to SO_2 will vary with the nitrogen (N) supply in the soil, being low under low N and high under N sufficiency (Cowling and Lockyer, 1978). Similar information is not readily available for forest ecosystems, although in the recent years widespread stimulation of tree growth under atmospheric nitrogen fertilization has been implicated for Europe (IPCC, 1995), while such a conclusion specifically did not consider exposures to SO_2.

From a mechanistic perspective, apparently exposure to SO_2 can be used for S nutrition and can counteract SO_4^{2-} uptake through the roots and transport to the shoots. In this manner, the negative effects of SO_2 absorption by the shoot and the consequent acidification and excess S accumulation in the foliage may be reduced (Rennenberg and Herschbach, 1996).

Effects at the plant community and ecosystem levels

Plant communities can be viewed as hierarchical entities (Allen and Starr, 1982). A response of a plant community to SO_2 exposure will reflect the response at various levels, modified by the degree of connectedness of the hierarchy within a given plant community (Lauenroth and Milchunas, 1985).

According to Guderian (1977), under SO_2 exposure, inter-species competition is altered. As a result, in his studies the primary effect of SO_2 on more susceptible members was magnified to such a degree that they could no longer compete effectively for vital growth-determining factors. Because of the changed competition in the community, the decline of the more sensitive members allowed improved growth of the more resistant species. Steubing and Fangmeier (1987) came to a similar conclusion.

In a study by Preston (1988) of the California coastal sage shrubland community dominated by black sage (*Salvia mellifera* Green), of particular ecological interest was the relative importance of plant species at different study sites in the vicinity of an oil refinery. The abundance of perennial shrub species was found to be significantly lower and the abundance of annuals higher at the most polluted (SO_2) sites compared to the reference sites. The influx of annuals and decrease in perennial cover resulted in a greater species richness and reduced the extent of dominance, similar to a situation observed in early post-fire sage scrub stands.

Although studies that focus on shifts in plant community structure under SO_2 exposure are important, it is also of importance to determine the fate and impacts of atmospheric S deposition in different compartments within a primary producer and its immediately associated ecosystem components. Legge *et al.* (1981) and Krupa and Legge (1998) have provided an integrative analysis of this subject. It was based on long-term case study investigations (1974–88) of the responses of a boreal pine forest near Whitecourt, Alberta, Canada, to SO_2 and elemental S deposition. A summary listing of ecological and plant physiological characteristics shown to change in the study area is provided in Table 8.5. This study used the relative abundance of the

Table 8.5 Changes in ecological and plant physiological characteristics observed in the West Whitecourt study area, Alberta, Canada

Parameter	Effect[1]
Count of S oxidizing bacteria	Increase
Microbial oxidation of elemental S to H_2SO_4	Increase
Soil acidification	Increase
Mineral nutrient cycling (leaching)	Increase
Litter decomposition	Decrease
Needle concentrations of S, Al, Mn and Fe in pine	Increase
Needle ratio of accumulated (SO_4-S) S to assimilated (organic) S in pine	Increase
Ratio S^{34} to S^{32} in needles, soils and surface waters	Increase
Needle greenness in pine	Decrease
Photosynthesis in pine	Decrease
Needle ATP content in pine	Decrease
Needle cell wall thickness, elasticity and water relations in pine	Change
Needle retention in pine	Decrease
Basal area increment in pine	Decrease

[1]The observed effects exhibited geographic spatial gradients with a maximum closer to the sources of S gas and S dust, reaching a no effect level at the distant regional background site(s).
Source: modified from Krupa and Legge (1998). See that article for specific references.

stable S isotope ^{34}S in the source emissions as a tracer of the source signature and the accumulation of ^{34}S in various biological components (^{34}S to ^{32}S ratio) to investigate the flow of S through the pine forest ecosystem.

Although adverse effects of atmospheric S deposition on the growth rate of a primary producer (pine species) and associated ecosystem processes were demonstrated in the West Whitecourt case study, more recent, progressive reductions in industrial S emissions in the vicinity showed that earlier adverse impacts can be overcome by ecosystem resilience. In this context, however, our knowledge of ecosystem resilience can be further improved if we understand better how ecosystems function in the absence of anthropogenic stress. The West Whitecourt case study attempted to address this by examining the background S flow within the ecosystem. However, much more work beyond that investigation is needed to unravel the complexities of the atmosphere-terrestrial ecosystem interface. Although simulation models may address this, results from such models need to be validated by independent experimental data. This need represents a significant gap in our current knowledge.

Evolution of resistance in plants

Evolution (change in the genetic constitution of a population) will occur when a population contains individuals with heritable differences in characters that affect fitness (Roose *et al.*, 1982). Here, fitness is defined as the net reproduction of offspring, including vegetative reproduction for perennial plants.

Taylor and Murdy (1975) and Taylor (1978) studied the evolution of resistance to air pollution in the populations of the winter annual Carolina crane's bill (*Geranium carolinianum* L.) from the SO_2-polluted area in the vicinity of a power plant in Georgia, USA. Populations collected from within 0.8 km of the point source sustained an average of about 20% less injury from 12 h exposures to 0.8 ppm ($\sim$2136 $\mu g/m^3$) SO_2 than populations from adjacent unpolluted areas. Visible foliar injury was observed under ambient conditions, but SO_2 concentrations were not monitored. Taylor (1978) showed that consistent differences in SO_2 resistance could be obtained in five generations by artificial selection and that these differences were inherited as a quantitative character. The power plant had been operating for 31 years when the plant populations were sampled, thus under ambient conditions evolution had occurred within 31 generations.

Evolution of resistance to SO_2 has also been reported for perennial ryegrass (*Lolium perenne* L.) in the UK (Bell and Clough, 1973). In this case, the yield of two clones indigenous to a polluted area was more resistant to SO_2 exposures than the yield of a commercial cultivar. Since the clones from the polluted area were exposed for a number of years, it is reasonable to infer that their resistance was due to selection by SO_2 and other forms of stress that were prevalent. Overall it appears that evolution of resistance to SO_2 is fairly common among a number of native plant populations in the UK (Ayazloo and Bell, 1981).

Murdy (1979) observed that plants of the annual weed peppergrass (*Lepidium virginicum* L.) from the SO_2-polluted Copper Basin, Tennessee, USA showed significantly less flower sterility after exposure of inflorescences to 0.8 ppm ($\sim$2136 $\mu g/m^3$) SO_2 for 9 h, than populations outside the basin. The populations did not differ in sterility in the absence of SO_2.

Clearly there are large variations in resistance to SO_2 and other air pollutants within and between species. As discussed in the previous section, when a variable population is stressed by SO_2 and other air pollutants, the growth of sensitive genotypes is reduced more than the resistant genotypes. Consequently, a hierarchy of size results, which leads to further reductions in the size of sensitive genotypes and their eventual death. Decreased growth and mortality of sensitive genotypes frees up resources (primarily light) for resistant genotypes, which respond by increased growth, maintaining overall yield (Roose *et al.*, 1982).

In some crops (uniformly inbred cultivars) it may be necessary to breed for resistance to SO_2. In genetically variable crop species, natural, unassisted evolution of resistance may maintain yields, since they normally experience mortality before harvest. There is evidence to show that the clones of most forest tree species possess a large amount of genetic variability for resistance to SO_2 (Scholz *et al.*, 1989). However, it is important to realize that most resistance mechanisms have associated energy costs that may reduce potential yields or food quality or alter the niche of a species. Therefore, it would be an error to assume that breeding or evolution of resistance will always compensate for stress from SO_2 and other air pollutants.

Concluding remarks

Historically among all air pollutants, the effects of SO_2 on plants have been studied the longest. Sulphur dioxide was the first to be designated as a phytotoxic air pollutant. During the middle part of the twentieth century, numerous studies were undertaken to examine the effects of SO_2 on forests. Subsequent emphasis also included crop responses. As opposed to ozone, SO_2 as in the case of fluoride, is an accumulative pollutant in plant tissue. Much knowledge has been gained on the mechanism of foliar injury and acute and chronic responses of plants to SO_2 exposures. Current information includes detoxification processes in plants to overcome SO_2 stress and evolution of resistance. A few studies have also addressed the issue of beneficial effects of SO_2 on plants growing in S-deficient soils.

From an ecological perspective, exposures to SO_2 have been shown to alter plant community structure. Functional changes in ecosystem compartments have also been demonstrated. However, in many of these studies SO_2 has been viewed as the sole factor governing plant growth and productivity. In the last decades this view has changed to include a more holistic approach, with the joint effects of multiple air pollutants, plant growth regulating climatic factors, pathogens and insect pests. To implement this philosophy and the associated tools of understanding requires interdisciplinary cooperation among scientists from multiple areas of specialization. Innovative approaches are available, such as the use of stable S isotope abundance and source apportionment techniques, coupled with computer-based routines for establishing quantitative relationships of cause and effect. Research in the future needs to exploit these and other avenues to advance further our knowledge and understanding of this subject.

Acknowledgements

The support provided by Biosphere Solutions in the preparation of this manuscript is most appreciated. The junior author was supported in kind by the University of Minnesota Agricultural Experiment Station. We thank Linda Jones and Leslie Johnson for their meticulous help in the preparation of this manuscript.

References

Allen, T.F.H. and Starr, T.B. (1982). *Hierarchy – Perspectives for Ecological Complexity*. University Chicago Press, Chicago, IL.

Archibold, O.W. (1978). Vegetation recovery following pollution control at Trail, B.C. *Can. J. Bot.* **56**, 1625–1637.

Atmospheric Environment (1984). Proceedings of the Receptor Models Workshop (Quail Roost II). *Atmos. Environ.*, **18**, 1499–1582.

Ayazloo, M. and Bell, J.N.B. (1981). Studies on the tolerance to sulphur dioxide of grass populations in polluted areas. I. Identification of tolerant populations *Dactylis glomerata, Festuca rubra, Holcus lanatus. New Phytol.*, **88**, 203–222.

Bell, J.N.B. (1982). Sulphur dioxide and growth of grasses. In *Effects of Gaseous Pollutants in Agriculture and Horticulture* (Eds. M.H. Unsworth. and D.P. Ormrod). pp. 225–246. Butterworth Scientific, London, UK.

Bell, J.N.B. and Clough, W.S. (1973). Depression of yield in ryegrass exposed to sulphur dioxide. *Nature*, **241**, 47–49.

Berresheim, H., Wine, P.H. and Davis, D.D. (1995). Sulphur in the atmosphere. In *Composition, Chemistry, and Climate of the Atmosphere* (Eds. H.B. Singh). pp. 251–307. Van Nostrand Reinhold, New York, NY.

Biggs, K. and Fry, C. (1987). Phenolic cross-linking in the cell wall. In *Physiology of Cell Expansion During Plant Growth* (Eds. D.J. Cosgrove and D.-P. Knievel). pp. 46–57. American Society of Plant Physiologists, Rockville, MD.

Black, V.J. (1982). Effects of sulphur dioxide on physiological processes in plants. In *Effects of Gaseous Air Pollution in Agriculture and Horticulture* (Eds. M.H. Unsworth and D.P. Ormrod). pp. 67–92. Butterworth Scientific, London, UK.

Boubel, R.W., Fox, D.L., Turner, D.B. and Stern, A.C. (1994). *Fundamentals of Air Pollution, Third Edition*, Academic Press, San Diego, CA.

Brako, L., Rossman, A.Y. and Farr, D.F. (1995). *Scientific and Common Names of 7,000 Vascular Plants in the United States*. American Phytopathological Society Press, St. Paul, Minnesota.

Brimblecombe, P., Hammer, C., Rodhe, H.l., Ryaboshapko, A. and Boutron, C.F. (1989). Human influence on the sulphur cycle. In *Evolution of the Global Biogeochemical Sulphur Cycle, Scope*, **39**, 77–121. (Eds. P. Brimblecombe and A.Y. Lien). John Wiley & Sons, New York, NY.

Chameides, W.L., Yu, H., Bergin, M., Zhou, X., Mearns, L., Wang, G., Kiang, C.S., Saylor, R.D., Luo, C., Huang, Y., Steiner, A. and Giorgi, F. (1999). Case study of the effects of atmospheric aerosols and regional haze on agriculture: An opportunity to enhance crop yields in China through emission controls? *Proc. Nat. Acad. Sci.* **96**(24), 13626–13633.

Cowling, D.W. and Lockyer, D.R. (1978). The effects of SO_2 on *Lolium perenne* L. grown at different levels of sulphur and nitrogen nutrition. *J. Exp. Bot.*, **29**, 257–265.

Cowling, D.W., Jones, L.H.P. and Lockyer, D.R. (1973). Increased yield through correction of sulfur deficiency in ryegrass exposed to sulfur dioxide. *Nature*, **243**, 479–480.

Cullis, C.F. and Hirschler, M.M. (1980). Atmospheric sulphur: Natural and man-made sources. *Atmos. Environ.*, **14**, 1263–1278.

DeKok, L.J. (1990). Sulphur metabolism in plants exposed to atmospheric sulphur. In *Sulphur Nutrition and Sulphur Assimilation in Higher Plants* (Eds. H. Rennenberg, C. Brunold, L.J. DeKok and I. Stulen). pp. 111–130. SPB Academic Publishers, The Hague, The Netherlands.

DeKok, L.J., Stuiver, C.E.E. and Stulen, I. (1998). Impact of H_2S on plants. In *Responses of Plant Metabolism to Air Pollutants and Global Change* (Eds. L.J. DeKok and I. Stulen). pp. 51–63. Backhuys Publishers, Leiden, The Netherlands.

Dmuchowski, W. and Bytnerowicz, A. (1995). Monitoring environmental pollution in Poland by chemical analysis of Scots pine (*Pinus sylvestris* L.) needles. *Environ. Pollut.*, **87**, 87–104.

Dreisinger, B.R. and McGovern, P.C. (1970). Monitoring atmospheric sulphur dioxide and correlating its effects on crops and forests in the Sudbury area. In *Proceedings of the Impact of Air Pollution on Vegetation*, 7–9 April 1970, Toronto (Ed. S.N. Linzon). pp. 11–28. Air Pollution Control Association, Pittsburgh, PA.

Duke, S.H. and Reisenauer, H.M. (1986). Roles and requirements of sulfur in plant nutrition. In *Sulfur in Agriculture* (Ed. M.A. Tabatabai). pp. 123–168. American Society of Agronomy, Madison, WI.

Evelyn, J. (1661). *Fumifugium or The Inconvenience of the Aer and Smoake of London Dissipated: Together with Some Remedies Humbly Proposed*. W. Godbid, London.

Farrar, J.F., Relton, J. and Rutter, A.J. (1977). Sulphur dioxide and the scarcity of *Pinus sylvestris* in the industrial Pennines. *Environ. Pollut.* **14**, 63–68.

Filner, P., Rennenberg, H., Sekiya, J., Bressan, R.A., Wilson, I.G., LeCureux, L. and Shimei, T. (1984). Biosynthesis and emission of hydrogen sulfide by higher plants. In *Gaseous Air Pollutants and Plant Metabolism* (Eds. M.J. Koziol. and F.R. Whatley). pp. 291–312. Butterworth, London.

Finlayson-Pitts, B.J. and Pitts, Jr., J.N. (1986). *Atmospheric Chemistry: Fundamentals and Experimental Techniques*, John Wiley & Sons, New York, NY.

Fox, C.A., Kincaid, W.B., Nash III, T. H., Young, D.L. and Fritts, H.C. (1986). Tree-ring variations in western larch (*Larix occidentalis*) exposed to sulfur dioxide emissions. *Can. J. For. Res.*, **16**, 283–292.

Freedman, B. and Hutchinson, T.C. (1980). Long-term effects of smelter pollution at Sudbury, Ontario, on forest community composition. *Can. J. Bot.* **58**, 2123–2140.

Guderian, R. (1977). *Air Pollution. Phytotoxicity of Acidic Gases and its Significance in Air Pollution Control*. Ecological Studies 22. Springer Verlag, New York, NY.

Guderian, R. and Stratmann, H. (1962). Freilandversuche zur Ermittlung von Schwefeldioxidwirkungen auf die Vegetation. I. Teil: Übersicht zur Versuchmethodik und Versuchsauswertung. Köln & Opladen. Forsch. Ber. d Landes Nordrhein-Westfalen, Nr. 1118, Westdeutscher Verlag, Köln, West Germany.

Guderian, R. and Stratmann, H. (1968). Freilandversuche zur Ermittlung von Schwefeldioxidwirkungen auf die Vegetation. III. Teil: Grenzwerte schädlicher SO_2-Immissionen für Obst- und Forstkulturen sowie für landwirtschaftliche und gärtnerische Pflanzenarten. Köln & Opladen. Forsch. Ber. d Landes Nordrhein-Westfalen, Nr. 1920. Westdeutscher Verlag, Köln, West Germany.

Guderian, R., Haut, H. van and Stratmann, H. (1960). Probleme der Erfassung und Beurteilung von Wirkungen gasförmiger Luftverunreinigungen auf die Vegetation. *Z. Pflanzenkrankh. Pflanzenschutz* **67**, 257–264.

Gunn, J.M. (1995). *Restoration and Recovery of an Industrial Region*. Springer-Verlag, New York, NY

Haase, E.F., Morgan, G.W. and Salem, J.A. (1980). Field surveys of sulfur dioxide injury to crops and assessment of economic damage. In *Proceedings of the 73rd Annual Meeting, Air Pollution Control Association*, Montréal, Québec, June 22–27, 1980. Paper 80–26.2, Air Pollution Control Association, Pittsburgh, PA.

Harvey, G.W. and Legge, A.H. (1979). The effect of sulfur dioxide upon the metabolic level of adenosine triphosphate. *Can. J. Bot.* **57**, 759–764.

Haywood, J.K. (1905). Injury to vegetation by smelter fumes. *U.S. Dept. Agric. Bur. Chem., Bull.* **89**, 1–23.

Herschbach, C., DeKok, L.J. and Rennenberg, H. (1995). Net uptake of sulphate and its transport to the shoot in spinach plants fumigated with H_2S or SO_2: Does atmospheric sulphur affect the 'inter-organ' regulation of sulphur nutrition? *Bot. Acta* **108**, 41–46.

Hocking, D., Kuchar, P., Plambeck, J.A. and Smith, R.A. (1978). The impact of gold smelter emissions on vegetation and soils of a sub-arctic forest-tundra ecosystem. *J. Air Pollut. Cont. Assoc.* **28**(2), 133–137.

Huttunen, S., Laine, K. and Toruela, H. (1985). Seasonal sulphur contents of pine needles as indices of air pollution. *Ann. Bot. Fenn.* **22**, 343–359.

Hyodo, H., Kuroda, H. and Yang, S.F. (1978). Induction of phenylalanine ammonia-lyase and increase in phenolics in lettuce leaves in relation to the development of russet spotting caused by ethylene. *Plant Physiol.* **62**, 31–35.

Innes, J.L. and Haron, A.H. (2000). *Air Pollution and the Forests of Developing and Rapidly Industrializing Regions*. Report No. 4 of the IUFRO Task Force on Environmental Change. CAB International, Cambridge University Press, Cambridge, UK.

IPCC (Intergovernmental Panel on Climate Change) (1995). *Radiative Forcing of Climate Change and an Evaluation of the IPCC IS92 Emissions Scenarios*. Cambridge University Press, Cambridge, UK.

Jacobson, J.S. and Showman, R.E. (1984). Field surveys of vegetation during a period of rising electric power generation in the Ohio River Valley. *J. Air Pollut. Control Assoc.* **34**, 48–51.

Jones, H.C., Noggle, J.C. and McDuffie, Jr. C. (1987). Effects of emissions from a coal-fired power plant on soybean production. *J. Environ. Qual.* **16**(4), 296–306.

Karnosky, D. (1985). Genetic variability in growth responses to SO_2. In *Sulfur Dioxide and Vegetation – Physiology, Ecology and Policy Issues* (Eds. W.E. Winner, H.A. Mooney and R.A. Goldstein). pp. 346–356. Stanford University Press, Stanford, CA.

Katz, M. (1949). Sulfur dioxide in the atmosphere and its relation to plant life. *Ind. Eng. Chem.* **41**(11), 2450–2465.

Keller, T. (1981). Winter uptake of airborne SO_2 by shoots of deciduous species. *Environ. Pollut.*, **26**, 313–317.

Kercher, J.R. (1978). A model of leaf photosynthesis and the effects of simple gaseous sulfur compounds (H_2S and SO_2). UCRL-52 643, Lawrence Livermore National Laboratory, Livermore, CA.

Kercher, J.R. (1982). An assessment of the impact on crops of effluent gases from geothermal energy development in the Imperial Valley, California. *J. Environ. Manage.*, **15**, 213–228.

Kharuk, V.I. (2000). Air pollution impacts on subarctic forests at Noril'sk, Siberia. In *Forest Dynamics in Heavily Polluted Regions* (Eds. J.L. Innes and J. Oleksyu). pp. 77–86. CAB International Publishing, Wallingford, United Kingdom.

Knabe, W. (1970). Distribution of Scots pine forest and sulfur dioxide emissions in the Ruhr area. *Staub-Reinhalt.* Luft **30**, 43–47.

Kohut, R. (1985). The effects of SO_2 and O_3 on plants. In *Sulfur Dioxide and Vegetation: Physiology, Ecology and Policy Issues* (Eds. W.E. Winner, H.A. Mooney and R.A. Goldstein). pp. 296–312. Stanford University Press, Stanford, CA.

Kosuge, T. and Kimpel, J.A. (1981). Energy use and metabolic regulation in plant-pathogen interactions. In *Effects of Disease on the Physiology of the Growing Plant* (Ed. P.G. Ayres). pp. 29–45. Cambridge University Press, Cambridge, England.

Kozlowski, T.T. (1985). SO_2 effects on plant community structure. In *Sulfur Dioxide and Vegetation: Physiology, Ecology, and Policy Issues* (Eds. W.E. Winner, H.A. Mooney and R.A. Goldstein). pp. 431–453. Stanford University Press, Stanford, CA.

Krouse, H.R., Legge, A.H. and Brown, H.M. (1984). Sulphur gas emissions in the boreal forest: the West Whitecourt case study. V. Stable sulphur isotopes. *Water Air Soil Pollut.*, **22**, 321–347.

Kryuchkov, V.V. (1993). Extreme anthropogenic loads and the northern ecosystem condition. *Ecol. Applic.*, **3**(4), 622–630.

Krupa, S.V. (1996). The role of atmospheric chemistry in the assessment of crop growth and productivity. In *Plant Response to Air Pollution* (Eds. M. Yunus and M. Iqbal). pp. 35–73. John Wiley & Sons, Chichester, England.

Krupa, S.V. (1997). *Air Pollution, People and Plants*. APS Press, St. Paul, MN.

Krupa, S.V. and Kickert, R.N. (1987). An analysis of numerical models of air pollutant exposure and vegetation response. *Environ. Pollut.*, **44**(2), 127–158.

Krupa, S.V. and Legge, A.H. (1998). Sulphur dioxide, particulate sulphur and its impacts on a boreal forest ecosystem. In *Modern Trends in Ecology and Environment* (Ed. R.S. Ambasht). pp. 285–306. Backhuys Publishers, Leiden, The Netherlands.

Krupa, S.V. and Legge, A.H. (1999). Foliar injury symptoms of Saskatoon serviceberry (*Amelanchier alnifolia* Nutt.) as a biological indicator of ambient sulfur dioxide exposures. *Environ. Pollut.*, **106**, 449–454.

Krupa, S.V. and Legge, A.H. (2001). Saskatoon serviceberry and ambient sulfur dioxide exposures: study sites re-visited, 1999. *Environ. Pollut.*, **111**, 363–365.

Krupa, S.V. and Nosal, M. (1989). A multivariate, time series model to relate alfalfa responses to chronic, ambient sulphur dioxide exposures. *Environ. Pollut.*, **61**, 3–10.

Lauenroth, W.K. and Milchunas, D.G. (1985). SO_2 effects on plant community function. In *Sulfur Dioxide and Vegetation* (Eds. W.E. Winner, H.A. Mooney and R.A. Goldstein). pp. 454–477. Stanford University Press, Stanford, CA.

Legge, A.H. and Krupa, S.V. (eds) (1990). *Acidic Deposition: Sulphur and Nitrogen Oxides*. Lewis Publishers, Chelsea, MI

Legge, A.H., Jaques, D.R., Harvey, G.W., Krouse, H.R., Brown, H.M., Rhodes, E.C., Nosal, M., Schellhase, H.U., Mayo, J., Hartgerink, A.P., Lester, P.F., Amundson, R.G. and Walker, R.B. (1981). Sulphur gas emissions in the boreal forest: The West Whitecourt Case Study I: Executive Summary. *Water Air Soil Pollut.*, **15**, 77–85.

Legge, A.H., Bogner, J.C. and Krupa, S.V. (1988). Foliar sulphur species in pine: A new indicator of a forest ecosystem under air pollution stress. *Environ. Pollut.*, **55**, 15–27.

Legge, A.H., Nosal, M. and Krupa, S.V. (1996). Modeling the numerical relationships between chronic ambient sulphur dioxide exposures and tree growth. *Can. J. Forest Res.*, **26**, 689–695.

Legge, A.H., Jäger, H.-J. and Krupa, S.V. (1998). Sulfur dioxide. In *Recognition of Air Pollution Injury to Vegetation: A Pictorial Atlas, 2nd Edition* (Ed. R.B. Flagler). pp. 3-1–3-42. Air and Waste Management Association, Pittsburgh, PA.

Legge, A.H., Jaques, D.R., Amundson, R.G. and Walker, R.B. (1977). Field studies of pine, spruce and aspen periodically subjected to sulfur gas emissions. *Water Air Soil Pollut.*, **8**, 108–129.

Linzon, S.N. (1978). Effects of airborne sulfur pollutants on plants. In *Sulfur in the Environment Part II: Ecological Impacts* (Ed. J.O. Nriagu). pp. 109–162. John Wiley & Sons, New York, NY.

Linzon, S.N., Temple, P.J. and Pearson, R.G. (1978). Sulfur concentrations in plant foliage and effects. 71st Annual Meeting Air Pollution Control Association, Houston, TX, June 25–30, 1975. Paper 78–7.4.

Manninen, S. and Huttunen, S. (1995). Scots pine needles as bioindicators of sulphur deposition. *Can. J. For. Res.*, **25**, 1559–1569.

Manninen, S., Huttunen, S. and Perämäki, P. (1997). Needle S fractions and S to N ratios as indices of SO_2 deposition. *Water Air Soil Pollut.*, **95**, 277–298.

Manning, W.J. and Krupa, S.V. (1992). Experimental methodology for studying the effects of ozone on crops and trees. In *Surface Level Ozone Exposures and Their Effects on Vegetation* (Ed. A.S. Lefohn). pp. 93–156. Lewis Publishers, Chelsea, MI.

Mansfield, T.A. (1998). Stomata and plant water relations: Does air pollution create problems? *Environ. Pollut.*, **101**, 1–11.

Mansfield, T.A. and Pearson, M. (1996). Disturbances in stomatal behavior in plants exposed to air pollution. In *Plant Response to Air Pollution* (Eds. M. Yunus and M. Iqbal). pp. 179–194. John Wiley & Sons, Chichester, UK.

Marschner, H. (1995). *Mineral Nutrition of Higher Plants, 2nd Edition*, Academic Press, San Diego, CA.

Materna, M. (1966). Die Ausscheidung des durch Fichtennadeln absorbierten Schwefeldioxids. *Arch. Forstwesen* **15**, 691–692.

Materna, J. (1973). Relationship between SO_2 concentration and damage of forest trees in the region of the Slavkov forest. *Práce Vúlhm* **43**, 169–180.

Mayo, J.M., Legge, A.H., Yeung, E.C., Krupa, S.V. and Bogner, J.C. (1992). The effects of sulphur gas and elemental sulphur dust deposition on *Pinus contorta x Pinus banksiana*: Cell walls and water relations. *Environ. Pollut.*, **76**, 43–50.

McLaughlin, S.B. and Lee, N.T. (1974). Botanical studies in the vicinity of Widow's Creek Steam Plant. Review of Air Pollution Effects Studies, 1952–72 and Results of 1973 Surveys. Tennessee Valley Authority Internal Report I-EB-74-1, Muscle Shoals, AL.

McLeod, A.R. (1995). An open-air system for exposure of young forest trees to sulphur dioxide and ozone. *Plant, Cell and Environ.*, **18**, 215–225.

Miller, P.R. and McBride, R. (1975). Effects of air pollutants on forests. In *Responses of Plants to Air Pollution* (Eds. J.B. Mudd and T.T. Kozlowski). pp. 195–235. Academic Press, New York, NY.

Morgan, P.W. and Fowler, J.L. (1972). Ethylene: Modification of peroxidase activity and isozyme complement in cotton (*Gossypium hirsutum* L.). *Plant and Cell Physiol.*, **13**, 727–736.

Murdy, W.H. (1979). Effect of SO_2 on sexual reproduction in *Lepidium virginicum* L. originating from regions with different SO_2 concentrations. *Bot. Gaz.*, **140**, 299–303.

Noggle, J.C. (1980). Sulfur accumulation by plants: the role of gaseous sulfur in crop nutrition. In *Atmospheric Sulfur Deposition: Environmental Impact and Health Effects* (Eds. D.S. Shriner, C.R. Richmond and S.E. Lindberg). pp. 289–298. Ann Arbor Science, Ann Arbor, MI.

Noggle, J.C. and Jones, H.C. (1979). *Accumulation of atmospheric sulfur by plants and sulfur-supplying capacity of soils*. EPA-600/7-79-109. Report prepared by Tennessee Valley Authority, Chattanooga, TN for US Environmental Protection Agency, Washington, DC.

NRC (National Research Council). (1939). *Effects of Sulphur Dioxide on Vegetation*. Report prepared for the Associate Committee on Trail Smelter Smoke, National Research Council of Canada, Ottawa, ON: 447 pp.

Odintsov, D.I. (2000). Forestry problems related to air pollution in central Asia. In *Air Pollution and the Forests of Developing and Rapidly Industrializing Countries* (Eds. J.L. Innes and A.H. Haron). pp. 101–120. CAB International Publishing, Wallingford, UK.

O'Gara, P.J. (1922). Sulfur dioxide and fume problems and their solutions. *J. Ind. Eng. Chem.*, **14**, 744–745.

Olsen, R.A. (1957). Absorption of sulfur dioxide from the atmosphere by cotton plants. *Soil Sci.*, **84**, 107–111.

Percy, K.E., Cape, J.N. Jagels, R. and Simpson, G.J. (1994). *Air Pollutants and the Leaf Cuticle*. Springer-Verlag, Berlin, Germany.

Pfanz, H., Dietz, K.-J., Weinerth, J. and Oppmann, B. (1990). Detoxification of sulphur dioxide by apoplastic peroxidases. In *Sulphur Nutrition and Sulphur Assimilation in Higher Plants* (Eds. H. Rennenberg, C. Brunold, L.J. DeKok and I. Stulen). pp. 229–233. SPB Academic Publishers, The Hague, The Netherlands.

Pratt, G.C., Kromroy, K.W. and Krupa, S.V. (1983). Effects of ozone and sulphur dioxide on injury and foliar concentrations of sulphur and chlorophyll in soybean *Glycine max*. *Environ. Pollut.*, **32**, 91–99.

Preston, K.P. (1988). Effects of sulphur dioxide pollution on a Californian coastal sage scrub community. *Environ. Pollut.*, **51**, 179–195.

Preston, E.M. and Lee, J.J. (1982). Design and performance of a field exposure system for evaluation of the ecological effects of SO_2 on a natural grassland. *Environ. Monit. and Assess.*, **1**, 213–228.

Rennenberg, H. (1991). The significance of higher plants in the emission of sulphur compounds from terrestrial ecosystems. In *Trace Gas Emissions by Plants* (Ed. T.D. Sharkey, E.A. Holland and H.A. Mooney). pp. 217–260. Academic Press, San Diego, CA.

Rennenberg, H. (1984). The fate of excess sulfur in higher plants. *Ann. Rev. Plant Physiol.*, **35**, 121–153.

Rennenberg, H. and Herschbach, C. (1996). Responses of plants to atmospheric sulphur. In *Plant Response to Air Pollution* (Eds. M. Yunus and M. Iqbal). pp. 285–294. John Wiley & Sons, Chichester, UK.

Rennenberg, H. and Polle, A. (1994). Metabolic consequences of atmospheric sulphur influx into plants. In *Plant Response to the Gaseous Environment* (Eds. R. Alscher and A. Wellburn). pp. 165–180. Chapman and Hall, London, UK.

Rigina, O. and Kozlov, M.V. (2000). The impacts of air pollution on the northern taiga forests of the Kola Peninsula, Russian Federation. In *Forest Dynamics in Heavily Polluted Regions* (Eds. J.L. Innes and J. Oleksyn). pp. 37–65. CAB International Publishing, Wallingford, UK.

Roose, M.L., Bradshaw, A.D. and Roberts, T.M. (1982). Evolution of resistance to gaseous air pollutants. In *Effects of Gaseous Air Pollution in Agriculture and Horticulture* (Eds. M.H. Unsworth and D.P. Ormrod). pp. 379–410. Butterworth Scientific, London, England.

Scheffer, T.C. and Hedgecock, G.C. (1955). Injury to northwestern forest tree species by sulfur dioxide from smelters. U.S. Department Agriculture, Forest Service. Tech. Bull. No. 1117, Washington, DC.

Scholz, F., Gregorius, H.-R. and Rudin, D. (eds) (1989). *Genetic Effects of Air Pollutants in Forest Tree Populations*. Springer-Verlag, Berlin, Germany.

Sloof, J.E. (1995). Pattern recognition in lichens for source apportionment. *Atmos. Environ.*, **29**, 333–343.

Smith, W.H. (1990). *Air Pollution and Forests*, *Second Edition*, Springer-Verlag, New York, NY.

Steubing, L. and Fangmeier, A. (1987). SO_2 sensitivity of plant communities in a beech forest. *Environ. Pollut.*, **44**, 297–306.

Takemoto, B.K., Noble, R.D. and Harrington, H.M. (1986). Differential sensitivity of duckweeds (Lemnaceae) to sulphite. II. Thiol production and hydrogen sulphide emission as factors influencing phytotoxicity under low and high irradiance. *New Phytol.*, **103**, 541–548.

Taylor, G.E. Jr. (1978). Genetic analysis of ecotypic differentiation with an annual plant species, *Geranium carolinianum* L. in response to sulfur dioxide. *Bot. Gaz.*, **139**, 362–368.

Taylor, G.E. Jr. and Murdy, W.H. (1975). Population differentiation of an annual plant species, *Geranium carolinianum*, in response to sulfur dioxide. *Bot. Gaz.*, **136**, 212–215.

Thomas, M.D., Hendricks, R.H., Collier, T.R. and Hill, G.R. (1943). The utilization of sulphate and sulphur dioxide for the sulphur nutrition of alfalfa. *Plant Physiol.*, **18**, 345–371.

Thomasius, H. (1987). Experience on forest regeneration of pollution-damaged forests in the GDR. In *Forest Decline and Reproduction: Regional and Global Consequences* (Eds. L. Kairkutstis, S. Nilssen and A. Staszak). pp. 605–620. International Institute for Applied Systems Analysis, Laxenburg, Austria.

Thompson, C.R., Olszyk, D.M., Kats, G., Bytnerowicz, A., Dawson, P.J. and Wolf, J.W. (1984). Effects of ozone or sulfur dioxide on annual plants of the Mojave desert. *J. Air Pollut. Contr. Assoc.*, **34**, 1017–1022.

Tichy, J. (1996). Impact of atmospheric deposition on the status of planted Norway spruce stands: a comparative study between sites in southern Sweden and the northeastern Czech Republic. *Environ. Pollut.*, **93**, 303–312.

Tikkanen, E. and Niemelä, I. (1995). Kola Peninsula Pollutants and Forest Ecosystems in Lapland. Final Report of the Lapland Forest Damage Project, Finland Ministry of Agriculture and Forestry and the Finnish Forest Research Institute, Jyväskylä, Finland.

Tingey, D.T. and Olszyk, D.M. (1985). Intraspecies variability in metabolic responses to SO_2. In *Sulfur Dioxide and Vegetation – Physiology, Ecology and Policy Issues* (Eds. W.E. Winner, H.A. Mooney and R.A. Goldstein). pp. 179–205. Stanford University Press, Stanford, CA.

USNRC (National Research Council). (1989). *Biologic Markers of Air Pollution Stress and Damage in Forests*. National Academy Press, Washington, DC.

van Haut, H. and Stratmann, H. (1970). *Farbtafelatlas über Schwefeldioxidwirkungen und Pflanzen*. Verlag W. Girardet, Essen, West Germany.

Wadleigh, M.A. and Blake, D.M. (1999). Tracing sources of atmospheric sulphur using epiphytic lichens. *Environ. Pollut.*, **106**, 265–271.

Watson, R.T., Rodhe, H., Oeschger, H. and Siegenthaler, U. (1990). Greenhouse gases and aerosols. In *Climate Change – The IPCC Scientific Assessment* (Eds. J.T. Houghton, G.J. Jenkins and J.J. Ephraums). pp. 1–40. Cambridge University Press, Cambridge, Great Britain.

Wiersema, J.H. and Leónn, B. (1999). *World Economic Plants – A Standard Reference*. CRC Press, Boca Raton, FL.

Winner, W.E., Mooney, H.A., Williams, K. and von Caemmerer, S. (1985). Measuring and assessing SO_2 effects on photosynthesis and plant growth. In *Sulfur Dioxide and Vegetation – Physiology, Ecology and Policy Issues* (Eds. W.E. Winner, H.A. Mooney and R.A. Goldstein). pp. 118–132. Stanford University Press, Stanford, CA.

Wolfenden, J., Pearson, M. and Francis, J. (1991). Effects of overwinter fumigation with sulphur and nitrogen oxides on biochemical parameters and spring growth in red spruce (*Picea rubens* Sarg.). *Plant, Cell Environ.*, **14**, 35–45.

Yunus, M., Singh, N. and Iqbal, M. (1996). Global status of air pollution: An overview. In *Plant Response to Air Pollution* (Eds. M. Yunus and M. Iqbel). pp. 1–34. John Wiley & Sons, New York, NY.

Zahn, R. (1963). Untersuchungen uber die Bedeutung kontinuierlich und intermittier-ender Schwefeldioxideinwirkung fur die Pflanzenreaktion. *Staub* **23**, 343–352.

9 Effects of fluorides

D.C. MCCUNE AND L.H. WEINSTEIN

The occurrence of fluorides

Although fluorine (F) is a common element in the earth's crust, F-containing air pollutants are not common and usually occur within a local area as emissions from certain industrial processes. Some examples are the treatment of phosphate rock with acid in fertilizer production, the firing of clays in kilns at brickworks, the use of F-containing minerals as fluxes in the smelting of aluminum or steel, and the production of hydrogen fluoride (HF) or its use in the manufacturing of glass frit or fluorocarbon compounds (National Academy of Science, 1971).

There are also natural sources of airborne F as well as a global fluorine cycle. Volcanic eruptions emit F-containing ejecta and geothermal activity, such as hot springs or fumaroles, are sources of volatile fluorides. The stratospheric degradation of fluorocarbons will add to the global deposition of F originating from natural sources. However, neither the natural nor the global aspects are of direct consequence to the effects of airborne fluorides on vegetation. Industrial emissions contain both particulate and gaseous forms of F. The former comprise a variety of particle sizes and diverse chemical compounds, such as fluorapatite, cryolite, or fluorite. Of the gaseous forms, HF may be the predominant gaseous component and of greatest importance with reference to effects on vegetation although silicon tetrafluoride (SiF_4), fluosilicic acid (H_2SiF_6), or elemental fluorine (F_2) may also be emitted.

Accumulation of fluoride

Whether in gaseous or particulate form, airborne fluorides will be dispersed from their points of emission and carried to the foliage of plants by the same atmospheric processes that transport other air pollutants. The predominant route by which gaseous fluoride enters the plant is diffusion through the stomata on the leaf; entrance through the lenticels of stems may be a minor route of uptake in some species. When diffusion into the aqueous phase of the mesophyll from the substomatal space is the rate-limiting process, HF will be absorbed at a greater rate than other gaseous pollutants owing to its lower molecular weight and greater solubility in water (Davison, 1986).

Particulate F is deposited to the surface of the leaf, and its subsequent penetration into the leaf is slow and depends upon the solubility of the material, particle size,

Air Pollution and Plant Life, second edition. Edited by J.N.B. Bell and M. Treshow. ISBN 0 471 49090 3 (HB), 0 471 49091 1 (PB).

relative humidity, and the presence of dew or other free water on the foliar surface. The superficial deposits, which can include gaseous F sorbed to the waxy cuticle of the leaf or materials residing upon it as well as F from the interior of the leaf, can be eluted from the foliar surface by precipitation or by the loss or abrasion of the cuticle.

Once F enters the aqueous phase of the mesophyll, it may passively diffuse into the cytoplasm or through cell walls and extracellular space. Within the leaf, however, the major path of F is in the transpirational stream towards the apical or marginal tissues of the leaf where F remains as an immobile, inorganic species. There is no evidence for the synthesis of an organofluorine compound, such as monofluoroacetate, from F except in one species of plant native to South America, two species in Africa, and several native species of legumes in Australia (Weinstein, 1977). Translocation from the leaf to other organs of the plant is insubstantial and the F that has been accumulated by the plant is eventually lost by abscission of the foliage or its consumption by herbivores.

The most general empirical relationship between the concentration of fluoride found in vegetation (F) and the mean concentration of HF (C) and duration of exposure (T) is given by

$$F = F_0 + kCT \tag{1}$$

where F_0 is the concentration attributable to F taken up by the roots from fertilizer or soil minerals (which usually have 20 to 500 parts per million F by weight [ppm F w/w]). Some woody species in North America and Europe can have 'background' levels of F greater than the normal range of 5 to 20 ppm F, and some species in the tea family (*Theaceae*) can naturally accumulate F from the soil to give foliar concentrations greater than 2000 ppm F.

At its simplest, the leaf or plant can be represented as a system wherein the concentration of F is a variable driven by the dynamics of two sets of processes: flux of F to or from the foliage and the generation or loss of dry matter in the leaf or plant. Consequently, both the form of the regression equation and the values of the apparent accumulation coefficient (k), can be influenced by three major sets of factors: biological, environmental, and analytical (Davison, 1986; Doley, 1986; Guderian, 1977).

Biological factors, such as species, cultivar, and stage of development, can determine the uptake of F and its concentration in the foliage by affecting stomatal conductance, surface to volume ratios of the leaf, and leaf area index of the plant. Moreover, the pattern of growth of the plant as it affects the rate of accretion (and loss) of foliar mass can also exert an effect upon the dose-response relationship. If one assumes that the uptake of F is a function of the dry matter present during exposure, then growth during or following the period of exposure would dilute concentration of F and decrease its apparent rate of accumulation.

Environmental factors, such as temperature, light intensity, photoperiod, relative humidity, and water stress can exert a direct effect by altering the stomatal conductance, which determines gaseous exchange of the leaf and thereby the uptake of pollutant. Their effects may also be expressed indirectly to the extent that they alter the rate of growth or senescence of the foliage.

Analytical factors that decrease the accuracy and precision of the methods used for monitoring the air and analyzing the tissue for F will increase bias and decrease the correlation between exposure and effect. The results will also depend upon whether the sample to be analyzed comprises the entire shoot, foliage of a certain age, or just the apical halves of leaves or needles. Often a sample of foliage is split into two aliquots and one is washed to remove superficial deposits; the difference between the two is used as a measure of the contribution of particulate F (Jacobson and Weinstein, 1977).

Metabolic responses

Exposure of a plant to HF can decrease its photosynthetic activity, as measured by the rate of carbon dioxide assimilation in the light. This response has three general characteristics: (1) a threshold concentration of HF must be exceeded; (2) the effect is reversible and returns to the level of the unfumigated plants following the exposure period if no foliar symptoms have developed; and, (3) the permanent reduction in assimilation following an acute exposure is associated with the appearance of HF-induced foliar lesions (Hill and Pack, 1983). Respiratory activity, as measured by oxygen uptake or the release of carbon dioxide in the dark, can increase, decrease, or remain unaffected depending upon the species of plant, developmental stage of the leaf or plant, mineral nutrient status, concentration of HF, and duration of exposure. The preponderance of evidence suggests that in glucose catabolism, the glycolytic pathway is of greater sensitivity to the effects of F than the pentose phosphate pathway and that this sensitivity decreases with the age of the tissue.

Many enzymes have been shown to be inhibited by the fluoride ion *in vitro* and thus there are many possible sites of action. In general, the mechanism of action appears to be by the formation of a complex of F with a divalent or trivalent metal at the active site of the enzyme as exemplified by the inhibition of enolase by fluoromagnesium phosphate. Changes in the activities of many enzymes, such as enolase, glucose-6-phosphate dehydrogenase, phosphoglucomutase, peroxidase, catalase, acid phosphatase, and succinic dehydrogenase, have been found in leaves that did not exhibit HF-induced foliar symptoms. Whether the activity increased or decreased depended upon the enzyme and isoenzyme, species of plant, and length of exposure. A change often progressively increased in magnitude during the period of exposure and then decreased or approached the same levels as the controls after exposure had ceased (Weinstein, 1977).

Changes in the levels of various metabolites – keto acids, nonvolatile organic acids, amino acids and amides, sugars, starch and non-starch polysaccharides, chlorophylls, and DNA – and RNA-phosphorus – occur in foliar tissue not only during but also following exposure to HF. Increases as well as decreases in pool size have been found depending upon the species of plant, age of leaf, concentration of HF, duration of exposure, and metabolite. For example, changes in the pool sizes of phosphate esters (including mono-, di-, and triphosphate nucleotides) were not observed in plants receiving exposures that produced changes in other metabolites. Determinations of the effect of HF on metabolites have been predominantly on pool size with little information on turnover rates although an altered incorporation of ^{32}P into phosphate esters was associated with the occurrence of chlorosis in maize leaves (Weinstein, 1977).

Effects on cellular organelles in foliar tissues include changes in the number, ultrastructure, and chlorophyll content of chloroplasts as well as changes in the structure of mitochondria, starch grains, and the endoplasmic reticulum. Some of the histological changes associated with the development of visible lesions in needles of ponderosa pine (*Pinus ponderosa*) included hypertrophy of epithelial cells of the resin duct and hypertrophy of phloem, xylem, and transfusion parenchyma. In broad-leaved plants, collapse of mesophyll and epidermal cells of the leaf and tyloses in vessel elements of the petiole can occur (Treshow, 1984).

Owing to the complexity of the metabolic pathways and their regulatory mechanisms, it is difficult to interpret a change in enzymes, metabolites, or process with reference to the action of F at a particular metabolic site. Moreover, the action of F could depend upon both the distribution of F within the cell and the metabolic pathways that are operationally significant in the cell during and after exposure of the plant to HF. Of all the diverse metabolic changes that can be produced by HF, those of greatest practical importance are changes in the constituents of the plant that could decrease its value as food, forage, or raw material for refined products.

Symptomatology

Visible symptoms

The type of foliar symptom produced by an exposure to HF depends primarily on the species of plant and secondarily on the concentration of HF and duration of the exposure (Guderian, 1977; Treshow, 1984; Weinstein, 1998).

On conifers, such as pine (*Pinus*), fir (*Abies*), spruce (*Picea*), and larch (*Larix*), symptoms may appear first as a chlorosis of the tip of the needle. With time or increasing severity, the apical tissue becomes necrotic – usually with a brown or reddish-brown color – and is delineated from the healthy tissue by a darker zone of necrotic tissue. The length of the necrosis is usually a measure of the magnitude of the exposure. The occurrence of two or more narrow, dark zones within the necrotic tissue is an indication that two or more exposures occurred, with the effect of successive exposures proceeding basipetally along the needle.

On dicotyledonous (broad-leaved) species, symptoms may occur initially as a chlorosis at the tip or along the margins of the leaf. With increasing severity, the chlorosis becomes more intense and the chlorotic area extends basipetally along the margins and medially towards the midrib; at its most severe, only the midrib and major veins remain green. The chlorotic lesions may also be associated with deformation of the leaf, such as a downward cupping or savoying owing to an inhibition of the growth of the lamina. In some species, the initial symptom may occur as a necrosis of the tip of the leaf or of a narrow zone along the margin. Successive exposures may increase the extent of the necrotic area with dark, narrow zones delimiting the increments. Over time, the necrotic tissue may be lost leaving the leaf with a notch at its tip or a scalloped margin. On St. John's wort (*Hypericum perforatum*) and slender bush-clover (*Lespedeza virginiana*), the necrotic area may proceed basipetally as a line across the entire leaf or leaflet.

On monocotyledonous (narrow-leaved) species, the typical chlorotic symptom on grasses occurs at the tip and along the margin of the leaf and decreases in width with distance from the tip. With increasing severity, the chlorotic area extends basipetally and medially toward the midrib. On maize, sorghum, and some grasses, chronic symptoms may first appear as a chlorotic mottling on middle-aged leaves, which may occur at the arch or along the apical margins and later coalesce. On *Gladiolus* and other species of the iris family (*Iridaceae*), symptoms occur as a necrosis of the apical tissue which progresses uniformly towards the base of the leaf with increasing severity. A narrow band of dark brown sharply delimits this necrosis from the healthy tissue; when necrosis is produced by several exposures, each increment may be separated from the preceding necrosis by these narrow bands.

Foliar lesions that mimic those induced by F can be caused by edaphic and climatic stresses, by pests and pathogens, and by other air pollutants. Needle tipburn in conifers can be caused by late frost in the spring, drought stress, scale insects, soil contamination or drift from highway de-icing salts, and chlorine (Cl_2), hydrogen chloride (HCl), or sulfur dioxide (SO_2). On broad-leaved plants, marginal leaf scorch can also be caused by drought, frost, salt, HCl, SO_2, and the application of some pesticides whereas marginal and interveinal chlorotic patterns resembling those induced by HF can be caused by deficiencies of manganese (Mn), iron (Fe), or zinc (Zn) in the soil, particularly in citrus and members of the *Rosaceae*, as well as by some virus disorders. A chlorotic mottle of maize leaves can be caused by mites or some pathogens.

Symptoms on flowers and fruit are less common than those on foliage. Although petals have been injured in some species under acute exposures, visible symptoms on the corollas of flowers are usually not present at the concentrations of HF that induce foliar symptoms. The bracts of gladiolus will have tip necrosis under the same exposures that injure the leaves. A premature ripening of the stylar end of cherry, peach, and pear fruit, which results in local necrosis and shrivelling, is attributed to high concentrations of HF. On peach fruit, a premature ripening of the flesh along both sides of the suture ('soft suture') can be induced by long-term exposure to HF, primarily at the pit-hardening stage, at concentrations below those associated with foliar injury on the tree. The effect of stylar-end lesions can also be produced by water stress and the soft suture syndrome can also be mimicked by the application of phenoxyactic acid herbicides.

Factors affecting sensitivity

Sensitivity is usually defined with respect to the concentration of HF (C) for a given duration of exposure (T) required to elicit a response.

One major biological factor determining the sensitivity of the plant is the stage of development of the plant and its foliage. Generally, the period of greatest foliar sensitivity occurs when the leaves or needles are elongating. Another factor is the species and cultivar of plant. The literature contains many compilations of the relative sensitivities of crops and native species, usually with reference to three classes – 'sensitive', 'intermediate', and 'tolerant' – and the difference between a sensitive and tolerant species may be equivalent to a 20-fold difference in concentration in HF (Arndt *et al.*, 1995; Horning and Mitchell, 1982).

Several environmental factors appear to affect sensitivity by their effect on the rate of gaseous exchange and thereby on the uptake of F by the leaf, such as increased sensitivity with increased light intensity, temperature, relative humidity, and wind speed and decreased sensitivity of plants growing under soil water deficits. However, in some species light may also affect the sensitivity of the foliar tissue to F: injury is produced at lower levels of foliar F when accumulated in the dark or injury occurs after the plant is transferred to the light from an exposure in the dark. On the other hand, the mineral nutrient status of the plant – with reference to N, Ca, P, K, and Mg – can affect both the accumulation of F and the sensitivity to foliar injury depending upon the species of plant and element (Weinstein, 1977). When the environmental factor is another pollutant, such as SO_2, their joint action on the growth and reproduction of the plant can be complex (Murray and Wilson, 1988).

Growth and reproduction – yield

Various measures of growth and reproduction have been studied in diverse species with the major emphasis on what could be directly evaluated as or indirectly related to a commercially relevant effect on yield (Weinstein and McCune, 1971). These have included the dry mass of shoots of romaine lettuce, chard, alfalfa, and orchard grass, the production of flowers and corms in gladiolus, effects on flower production, fruit set, or mass of fruit of tomato, bean, strawberry, sweet cherry, and citrus, the yield of seeds of sorghum and maize, cone production in spruce, linear growth in rose and citrus, as well as effects on the radial growth in the boles of conifers and citrus (Weinstein, 1977).

These studies have shown that an effect of HF on growth or reproduction can be produced if the concentration of HF is sufficiently high. Of equal importance is the evidence that the risk or magnitude of an effect on growth or yield is greatest if exposure to HF occurs during a critical developmental stage. For example, grain yield in sorghum and maize is most sensitive just before the emergence of the inflorescence and during anthesis, and winter barley is more sensitive during the two-leaf stage and its major phase of growth than in the intervening developmental phases. In beet, root development is sensitive in the very early stages of its development, that is, during the cotyledonary and first leaf stages. Corm yield in gladiolus is most sensitive to the loss of foliar tissue during flowering when corm development starts (Guderian, 1977).

Although a reduction in the photosynthetic capacity of the foliage by chlorotic or necrotic lesions may be directly related to yield in some crops and trees, the relation of the effects on growth and reproduction to the occurrence and magnitude of foliar injury is not consistent (Hill and Pack, 1983). The increased growth of stems also found indicates that allocation as well as production of dry matter may be affected. Moreover, some reproductive processes, such as flower stalk extension, pollen release and germination, and fruit maturation may be directly affected by HF.

Ecological effects

Around some industrial facilities, fluorides have altered the structure and composition of adjacent plant communities. The most striking changes occurred where the dominant overstory was killed and sensitive species were eliminated, which allowed the release of more tolerant herbaceous species. In forests where the mortality of dominant species is not present, changes in the epiphytic community may still occur, particularly with respect to the decrease in fruticose lichens, such as *Usnea* and *Alectoria* (Treshow, 1984).

Observations in the field have indicated that the incidence of some pests and diseases can be positively or negatively correlated with the incidence of F-induced injuries to vegetation. Some of these may be due to infection or infestation by secondary invaders on weakened trees or necrotic foliage. Others may result because the meteorological conditions that disperse the emissions are correlated with the climatic conditions that affect the development or dispersion of the pest (see Chapter 20).

Nevertheless, experimental studies with virus, bacterial, and fungal pathogens have shown that exposure of the plant to HF can increase or decrease the severity of the disease by several possible modes of action, which depending upon the pathogen and species of plant include: a direct effect of HF on the infectivity of the pathogen; an effect of F accumulated by the leaf on the pathogen; and, an effect of F-induced metabolic changes in the foliar tissue on the development of the disease. Similarly, experimental studies with the Mexican bean beetle (*Epilachna varivestis*) indicated that their reduced growth, fecundity, and feeding on HF-fumigated plants could be the result of both F in the leaf and F-induced changes in its metabolites (Weinstein, 1977).

Practical problems

The assessment of the effects of airborne fluorides on vegetation and the establishment of air quality standards to reduce the risk of adverse effects are the principal problems of practical concern.

A common means for determining the incidence, severity, and extent of effects on vegetation is the assessment of foliar symptoms, particularly the patterns at four spatial scales: (1) the leaf as to whether the lesions produced are of the type induced by HF; (2) the plant as to whether distribution reflects the relationships among phenology, sensitivity, and time of exposure; (3) the community with reference to interspecific differences in sensitivity and the occurrence of symptoms on indicator species; (4) the area with reference to the dispersion pattern of the emissions and the occurrence of mimicking symptoms on affected species in more remote but similar habitats. Visual inspections should also utilize the sampling and chemical analysis of the foliage for F not only as complementary information for the diagnosis of injury but also with respect to dispersion of the pollutants and possible hazards to herbivores (Treshow, 1984). The ingestion of excessive amounts of F in vegetation by livestock can lead to the development of symptoms, such as lesions of the teeth and bones, lameness and decreased appetite (Shupe and Olson, 1983).

Very sensitive species, especially those with an indeterminate pattern of growth, can be used as 'indicators' to assess the severity and extent of F-induced effects in the field. Among these are gladiolus, St. John's wort, Oregon grape, eastern redbud (*Cercis canadensis*), Italian prune, and Chinese apricot. In some surveillance programs, plants of one species or cultivar were cultured in pots and then placed at selected sites around a source as biomonitors. A standardized grass culture was used for F-accumulation and gladiolus were used for monitoring both the occurrence of foliar injury and the accumulation of F.

The establishment of air quality standards for gaseous fluoride to reduce the risk of adverse effects on vegetation has been undertaken by various jurisdictions. One approach was to establish limiting concentrations for F in vegetation, which may be averaged over several periodic sampling periods during the year or growing season (MacLean *et al.*, 1992). The rationale is that the plant serves as a biomonitor, its concentration of F is a surrogate measure of its exposure to HF, and limits on foliar F will be protective both against adverse effects on plants and against the risk that accumulated F would pose to livestock, in the form of industrial fluorosis, which results in damage to bones and teeth. In the State of Maryland, the standard specifies that the following concentrations (ppm F) shall not be exceeded: (1) deciduous trees and shrubs – 100 in washed samples of fully expanded functional leaves; (2) conifers and evergreen trees and shrubs – 50 in washed samples of fully expanded leaves or needles of the current year and 75 in leaves and needles of prior seasons; (3) field crops – 35 in washed samples of foliage at any stage of growth; (4) forage/pasture and baled hay and/or silage (unwashed samples) – 80 for the average of samples collected during any one month, 60 for the average of any two consecutive months, and 35 as a running average for 12 consecutive months.

Another approach would be to establish standards for HF in the air with shorter term (12-hour to 1-week) limits to protect against the effects of acute exposures or exposures during critical periods of development and longer term (30-day to growing season) averages to limit the frequency of occurrence of short-term exposures and to reduce the F accumulated by the foliage (Scholl, 1971). For example, acceptable limits for averaging times of 24 hours, one month, and seven months could be, respectively, 1.6, 0.4, and 0.25 $\mu gF\ m^{-3}$ for highly sensitive plants/effects and 10.0, 2.5, and 1.2 $\mu gF\ m^{-3}$ for plants of low sensitivity.

Summary

The effects of airborne fluorides span the broad scale of biological organization from molecule to ecosystem: the accumulation of F itself; changes in metabolic states and physiological processes; the production of foliar lesions; changes in growth, development, and reproduction of the plant; and, the secondary effects that F-induced changes in plants may have on other organisms in the community.

Some special features of fluorides as air pollutants and their effects on vegetation are these: they are emitted from distinct, stationary industrial facilities; F is accumulated by vegetation as a stable, inorganic chemical species and therefore the plant can be a channel for F into other compartments of the ecosystem; and, the principal toxic component – gaseous hydrogen fluoride (HF) – can be of concern at concentrations ranging from 10.0 to 0.25 $\mu gF\ m^{-3}$, which are lower than those usually associated with the effects of other major atmospheric pollutants.

References

Arndt, U., Flores, F. and Weinstein, L.H. (1995). *Fluoride Effects on Plants. Diagnosis of Injury in the Vegetation of Brazil*. Editora da Universidade, Universidade Federal do Rio Grande do Sul, Porto Alegre.

Davison, A.W. (1986). Pathways of fluoride transfer in terrestrial ecosystems. In: Coughtrey, P.J., Martin, M.H., Unsworth, M.H. (eds) *Pollutant Transport and Fate in Ecosystems*, Blackwell Scientific Publications, Oxford, pp. 193–210.

Doley, D. (1986). *Plant-Fluoride Relationships. An Analysis With Particular Reference to Australasian Vegetation*. Inkata Press, Melbourne.

Guderian, R. (1977). *Air Pollution: Phytotoxicity of Acidic Gases and its Significance in Air Pollution Control*. Springer-Verlag, Berlin.

Jacobson, J.S. and Weinstein, L.H. (1977). Sampling and analysis of fluoride: methods for ambient air, plant and animal tissues, water, soil and foods. *J. Occup. Med.* **19**: 79–87.

Hill, A.C., Pack, M.R. (1983). Effects of atmospheric fluoride on plant growth. In: Shupe, J.L., Peterson, H.B., Leone, N.C. (eds.) *Fluorides: Effects on Vegetation, Animals, and Humans, Proc. Int. Symp. Fluorides*, Logan, Utah, 1982. Paragon Press, Salt Lake City, pp. 105–120.

Horning, D.S., Jr., Mitchell, A.D. (1982). Relative resistance of Australian native plants to fluoride. In: Murray, F. (ed.) *Fluoride Emissions: Their Monitoring and Effects on Vegetation and Ecosystems, Proc. Australasian Fluoride Workshop*, 1st, Sydney, NSW, 1981. Academic Press, Sydney, pp. 157–176.

MacLean, D.C., McCune, D.C., Convey, M.P. (1992). A fluoride surveillance program for an agricultural area. *Light Metals* 1992: 289–291.

Murray, F., Wilson, S. (1988). The joint action of sulphur dioxide and hydrogen fluoride on the yield and quality of wheat and barley. *Environ. Pollut.* **55**: 239–249.

National Academy of Sciences. (1971). *Biological Effects of Atmospheric Pollutants – Fluorides*. National Academy of Sciences, Washington, DC.

Scholl, G. (1971). Die Immissionsrate von Fluor in Pflanzen als Massstab für eine Immissionsbegrenzung. (with discussions by Hill, A.C., Weinstein, L.H.) *VDI-Ber.* **164**: 39–52.

Shupe, J.L., Olson, A.E. (1983). Clinical and pathological aspects of fluoride toxicosis in animals. In: Shupe, J.L., Peterson, H.B., Leone, N.C. (eds.) *Fluorides: Effects on Vegetation, Animals, and Humans, Proc. Int. Symp. Fluorides*, Logan, Utah, 1982. Paragon Press, Salt Lake City, pp. 319–338.

Treshow, M. (1984). *Environment and Plant Response*. McGraw-Hill, New York.

Weinstein, L.H. (1977). Fluoride and plant life. *J. Occup. Med.* **19**: 49–78.

Weinstein, L.H. and McCune, D.C. (1971). Effects of fluoride on agriculture. *J. Air Pollut. Control Assoc.* **21**: 410–413.

Weinstein, L.H. and McCune, D.C. (1971). Effects of fluoride on vegetation. Chapter 7, *Biological Effects of Atmospheric Pollutants – Fluorides*. National Academy of Sciences, Washington, DC, 1971.

Weinstein, L.H., Davison, A.W. and Arndt, U. (1998). Fluoride. In: Flagler, R.B. (ed.) *Recognition of Air Pollution Injury to Vegetation: A Pictorial Atlas*, Air & Waste Management Association, Pittsburgh: Chapter 4.

10 Effects of volatile organic compounds

C.D. COLLINS AND J.N.B. BELL

Introduction

The atmosphere contains a vast range of volatile organic compounds (VOCs) of both anthropogenic as well as natural origin, the latter including emissions from plant, animal and microbial sources. So far as effects on vegetation are concerned, the overwhelming interest is in the concentrations of those VOCs which have a high potential to generate phytotoxic levels of photochemical oxidants (see Chapter 3). Thus the direct impacts of VOCs on plant life have received relatively little attention. Indeed, there is only one VOC for which there is unequivocal evidence of widespread effects in the field, that is ethylene or ethene (C_2H_4), and studies on this are relatively limited, particularly in recent years. In addition concerns have developed recently that various volatile synthetic organic compounds may play a role in forest decline (see Chapter 15). Both these issues will be reviewed in this chapter.

Ethylene

Ethylene is an extremely unusual pollutant in that it is a natural plant hormone involved in a wide range of processes, such as growth regulation, flower development, fruit ripening, senescence and abscission of organs. In fact C_2H_4 is produced internally by plants in order to effect these processes under normal circumstances, while so-called 'stress ethylene' is also produced at higher concentrations in response to a wide range of environmental stresses, including air pollutants such as O_3. A literature search on the subject of C_2H_4 and plants will throw up a vast number of references, but the great bulk of these are concerned with internal C_2H_4 or, in some cases, artificially applied C_2H_4 as a method of ripening fruit. However, there is a small but significant literature concerned with C_2H_4 as a pollutant and, also, some of the other publications have relevance to this topic. Many of the current reports on VOC concentrations in ambient air do not concern C_2H_4 due to its lack of impact on human health. Indeed, the literature on sources and ambient concentrations is somewhat scattered and in some cases contains data of sufficient variability to make generalisations difficult. One of the more recent reviews on the subject is by Sawada and Totsuka (1986), who estimated a global emission of 35.39×10^6 t y^{-1}, with 74% and 26% being from natural and anthropogenic sources, respectively (Table 10.1). The vast bulk of natural C_2H_4 arises

Air Pollution and Plant Life, second edition. Edited by J.N.B. Bell and M. Treshow. ISBN 0 471 49090 3 (HB), 0 471 49091 1 (PB).

Table 10.1 Global C_2H_4 emissions from Sawada and Totsuka, 1986 (10^6 t y^{-1})

Natural sources	
Terrestrial ecosystems	23.3
Aquatic ecosystems	2.9
Total:	26.2
Anthropogenic sources	
Coal combustion	0.42
Fuel oil combustion	1.54
Refuse incineration	0.10
Leakage from C_2H_4 industry	0.03
Biomass burning	7.10
Total:	9.19

Table 10.2 Ethylene concentrations reported worldwide (Squier *et al.*, 1985). Reprinted with permission from Squier *et al.*, in *Environmental Science and Technology*, **19**, pp. 432–437 (1985), The American Chemical Society

Environment	Continent	Mean ethylene concentration (ppb)	Maximum ethylene concentration (ppb)
Pristine	Antarctica	0.9	–
Urban/Industrial	Australia	6	–
	Europe	5	10
	Asia	16	109
	N. America	16	25
	N. America	338	641
	N. America	5	10
Rural	Europe	4	9
	N. America	0.2	–
	N. America	1.8	1.8
	N. America	0.4	–
	N. America	0.7	–
	N. America	2.0	3.4
	N. America	0.9	3.1

from emissions from soil and vegetation, while biomass burning, followed by fossil fuel combustion dominates the anthropogenic sources.

Much of the information on ethylene concentrations worldwide is published in earlier literature and its current validity may be questionable in view of increasing controls on emissions of VOCs from both mobile and stationary sources in the developed world, and rising emissions, particularly in the case of the former, in developing countries. Table 10.2 shows a range of C_2H_4 concentrations reported worldwide pre-1985. These range from means of 0.2 ppb to 338 ppb (with a maximum concentration of 641 ppb) in rural and urban areas of North America, respectively (Squier *et al.*, 1985). There are reports of much greater concentrations on a highly localised scale, such as of 4000 ppb in smoke from burning rice stubble (Abeles, 1973) and 3000 ppb in the proximity of a polyethylene plant in Texas (Hall *et al.*, 1957). In cities motor vehicles represent the major source of C_2H_4 and it has been estimated that 80 ppb may be present next to busy roads (Abeles, 1985) although catalytic convertors reduce emissions by about 80% and this figure is likely to be lower now in the developed world, as illustrated below. Not surprisingly, the diurnal pattern of concentration closely follows levels of traffic flow, with peaks in the morning and evening rush hours (Abeles, 1985).

The concentrations reported above were measured before the days of introduction of catalytic converters on motor vehicles and thus their current relevance in developed countries must be called into question. In the UK C_2H_4 is at present measured at 12 urban locations and in many of these concentrations have approximately halved between 1994 and 1998, with current means in the range 1–4 ppb and maximum hourly means of 20–60 ppb. In contrast at a roadside site in central London the mean concentration in 1998 was 16 ppb, and at a rural location 0.8 ppb.

The earliest reports of C_2H_4 damage to vegetation occurred in 1864, with descriptions of defoliation of street trees in Germany, due to escape of illuminating gas from underground pipes (Abeles, 1973), although there is the possibility that this was not entirely an air pollution problem, but might have been caused by root asphyxiation. Subsequently reports occurred elsewhere, with problems also arising in glasshouses where gas was used as fuel for lights. The damaging agent was identified as C_2H_4 by a Russian plant physiologist in 1901 (Abeles and Heggestad, 1973). Much research was conducted into the effects of C_2H_4 derived from these sources over the early parts of the twentieth century, notably at the Boyce Thompson Institute for Plant Research in the USA. After World War II interest switched to C_2H_4 as an urban traffic derived pollutant, with particular concern over damage to the cut flower industry, with problems arising from leakage largely disappearing with the switch to natural gas.

Much of this work is summarised in Abeles (1973), who noted that the dose/response relationships for different effects and species tended to follow the same curve, that is 1–10 ppb: no effect; 10–100 ppb: discernible effects; 100–1000 ppb: half maximal effect; 1000–10 000 ppb: saturation. Abeles (1973) stated that 'When reference is made to ethylene, the reader is left with the impression that it is of limited consequence and its effects are of concern primarily to orchid growers', noting that this probably resulted from its lack of human health effects or production of acute visible injury on most plant species. As might be expected of such an important plant hormone, effects of C_2H_4 as a pollutant tend to take the form of growth abnormalities and developmental changes. Common symptoms occurring at concentrations $\leq$100 ppb and thus within the ambient range are shown in Table 10.3, including inhibition of bud growth, abscission of various organs and reduced flower and fruit production. Other symptoms reported include yellowing of conifer needles, premature opening of broad-leaved tree flowers, inward rolling or twisting of leaves on monocotyledons, and chlorosis or red/purple discoloration of leaves, induction of roots from stems, bushy growth habit and premature fruit ripening in broad-leaved plants (Taylor *et al.*, 1987). Heck and Pires (1962) categorised plant sensitivity into six main classes, after standard fumigations of 89 species from 39 families, covering ornamentals, trees, cereals, and fruit, vegetable, fibre and forage crops. A very common response to C_2H_4 is epinasty,

Table 10.3 Summary of effects of ethylene at concentrations <100 ppb

Concentration (ppb)	Duration	Species	Effects
2	24 hours	*Cattleya* orchid	Injury to sepals
10	48 hours	Apple	Production of swellings on twigs
20	24 hours	*Cattleya* orchid	Inhibition of bud growth
30	48 hours	Carnation	Inhibition of flowering
100	6 hours	Carnation	Closing of flowers
100	8 hours	Pepper, tomato	Abscission of flower buds
25	46 days	Cucumber	Reduced flower production
	70 days	Marigold	Reduced flower production
	50 days	Soybean	Reduced flower and fruit production
	44 days	Kidney bean	Reduced flower and fruit production

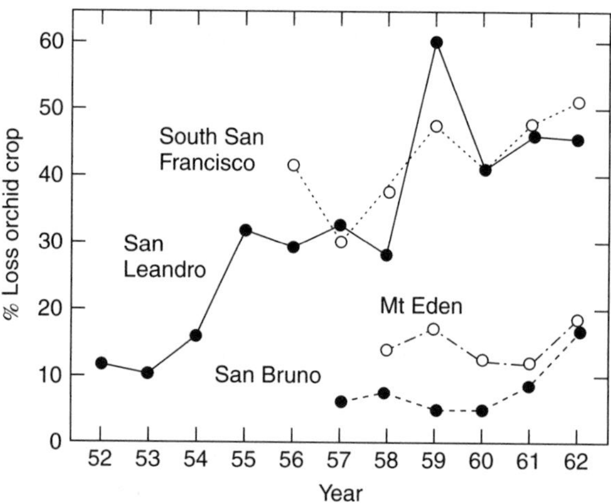

Figure 10.1 Trends in November losses of *Cattleya* orchids grown in the San Francisco area from 1952 to 1962 (Abeles, 1973). Reproduced by permission of Academic Press Ltd

which can occur at very low C_2H_4 concentrations, although this may not necessarily have an adverse effect on the plant. The failure of flowers to open can represent a serious economic loss, with carnations representing a particular problem: there have been reports of their flowers remaining in bud ('sleepy') in florists shops near traffic lights in the narrow streets of Dutch cities. Of course the traffic lights were not the direct cause, but rather the emissions from accelerating vehicles! Unlike symptoms on many other species, *Cattleya* orchid species show acute visible injury in the form of a shrivelling of the sepals ('dry sepal'), thereby rendering the bloom valueless. As vehicle emissions grew, severe problems arose in California (Figure 10.1), with many growers abandoning orchid production in the vicinity of large cities e.g. in the Bay Area in 1962 damage to over one million flowers resulted in an economic loss of \$100 000 (Abeles, 1985). In the UK there have been reports of similar damage close to a major airport, although there was uncertainty as to whether road traffic or aircraft emissions were responsible. It should be noted that *Cattleya* orchids can show dry sepal symptoms after exposure to only 2 ppb for 24 hours (Davidson, 1949) (Table 10.3), indicating that risk of damage occurs at widespread ambient levels.

Examples of clear demonstration of the effects of ambient levels of C_2H_4, other than in the well-documented case of orchids, are very limited. Much of our understanding of effects of ambient levels of pollutants comes from studies in which plants are exposed to ambient or filtered air in various types of chambers. Charcoal filters are normally employed for this purpose and they do not remove C_2H_4; even if they did, then the mix of other pollutants precludes clear causal attribution. Thus it is possible that C_2H_4 may be having adverse yet undetected impacts in the field. Indeed, Taylor (1984) writing in the first edition of this book noted that 'the chronic effects of ethylene in the field may be responsible for many of the symptoms of leaf senescence, chlorosis, defoliation and premature drop of fruits and blossoms, but its involvement in such problems is difficult to assess'. However, there are a limited number of investigations where the effects of ambient C_2H_4 have been studied by use of chambers equipped with 'Purafil' filters ($KMnO_4$ impregnated alumina), which can remove c. 75% of this pollutant. One of the few chamber studies in which C_2H_4 was filtered out of ambient air was conducted by Abeles and Heggestad (1973), and included fumigation treatments with 25, 50 and 100 ppb (Table 10.4). In this case cucumber, dwarf marigolds, soybean and *Phaseolus vulgaris* bean were grown in chambers located outside Washington DC, where the mean ambient C_2H_4 concentration was 39 ppb. Filtration increased flower and fruit numbers in soybean and *Phaseolus*, and only 25 ppb fumigation treatment reduced these almost to zero, with similar effects on numbers of cucumber and marigold flowers; at 100 ppb an interesting phenomenon was observed in the form of a switch from a predominance of male flowers in the controls to a predominance of females. A more recent study was conducted by Reid and Watson (1985) in Calgary, using Purafil filtered air as the control. In this case 10 ppb produced a small stimulation of seed number and weight, but at 150 ppb and above reductions occurred; in the case of sunflower there was a consistent pattern of C_2H_4 producing smaller leaves and fewer stomata, with a reduced shoot/root ratio.

A limited number of investigations have examined the influence of environmental conditions on C_2H_4 injury. Barden and Hanan (1972) fumigated *Dianthus caryophyllus*

Table 10.4 Effects of filtration and fumigation experiments with C_2H_4 at Beltsville, MD (Abeles and Heggestad, 1973)

	Concentration	Cucumber flower no.		Marigold flower no.	Soybean flower no.	Soybean fruit no.	Kidney bean fruit no.	
		♂	♀				Mature	Immature
Filtered air	8 ppb	18	1	5	22	8	8	18
Ambient air	39 ppb	14	2	5	16	2	5	5
Fumigation	25 ppb	3	1	2	2	0	1	4
	50 ppb	3	1	1	0	0	0	0
	100 ppb	0	8	0	0	0	0	0

(carnation) plants with 3–227 ppb C_2H_4 and showed that as the temperature increased from 2–21 °C, there was a progressive reduction in flower-keeping life. More recently Mortensen (1989) fumigated lettuce with 120 ppb at photon flux densities ranging from 50–150 μmol m^{-2} s^{-1} at temperatures between 16 and 24 °C, resulting in dry weight reductions of 25–50%: the largest effect occurred at the intermediate light level (100 μmol m^{-2} s^{-1}), but there was no interaction in this case with temperature. Thus there are clearly inconsistencies and massive gaps in knowledge concerning impacts of climatic conditions on C_2H_4 injury. Abeles (1973) reported that dry sepal on orchids was much more common at a number of locations during the winter months. However, this appeared to be associated with calm conditions and temperature inversions, resulting in C_2H_4 accumulating at night. A recent paper has examined the role of climate on C_2H_4-induced epinasty in potato near a petrochemical complex in The Netherlands (Tonneijck *et al.*, 1999), and showed that this response was positively and negatively related to temperature and vapour pressure deficit, respectively. The degree of epinasty in this study was considerably less than would be predicted from laboratory studies, probably reflecting differences in environmental conditions interacting with the C_2H_4 response. The mean C_2H_4 concentrations in the field were 62 ppb/30 ppb by night and day, respectively. This did not elicit any reduction in tubers' yield and contrasts markedly with the report of Hall *et al.* (1957) in which severe effects on cotton in the form of weakened stems, compacted internodes and earlier flowering occurred near a polyethylene plant in Texas where C_2H_4 concentrations were estimated as c. 3000 ppb; doubtless the last 40 years have seen major reductions in emissions from such sources in the developed world generally.

Tonneijck *et al.* (1999) explained the interactions with vapour pressure deficit and temperature on the basis of changes in stomatal conductance, the stomata being the route of C_2H_4 into the plant and thus to its site of action. There have been a number of studies in recent years which have examined the influence of C_2H_4 on stomatal physiology and photosynthesis processes. For example, Taylor and Gunderson (1988) demonstrated that in soybean reductions of both stomatal conductance and the chloroplasts' ability to assimilate CO_2 took place in the presence of C_2H_4, although the latter was greater. The same authors subsequently demonstrated that these effects are direct and not caused by C_2H_4-induced changes in leaf orientation to the light (Gunderson and Taylor, 1991). However, the relevance of this research to the ambient must be viewed with caution in view of the C_2H_4 concentrations employed being orders of magnitude higher than those prevailing in the field.

Ethylene is a pollutant which has been monitored in the field using various species of plant as biomonitors. One common technique employed is to measure the angle of epinasty observed, and tomato and African marigold have been employed for this purpose. A problem here is that epinasty may be induced by other factors, but recently a good relationship has been determined between this response and ambient C_2H_4 levels near a Dutch polyethylene complex (Tonneijck *et al.*, 1999). Also in The Netherlands *Petunia nyctaginiflora* cv. White Joy has been used extensively as a biomonitor, on the basis of flower bud abortion and production of smaller flowers (Posthumus, 1984). Pleijel *et al.* (1994) have employed *Petunia hybrida* as a biomonitor for ethylene at different distances from a Swedish motorway and showed that flowers appeared faster,

Table 10.5 Relative concentrations of several unsaturated hydrocarbons that produce biological response similar to that produced by ethylene (Abeles and Gahagan, 1968)

Compound	Abscission	Pea stem growth inhibition	Tobacco growth inhibition	Epinasty
Ethylene	1	1	1	1
Propylene	60	100	100	500
Acetylene	1250	2800	100	500
1-Butene	100 000+	270 000	200	500 000
1, 3-Butadiene	100 000+	5 000 000	–	–

but were smaller as the road was approached. It would seem that *Cattleya* orchids could make a good biomonitor in view of the extreme and specific sensitivity to C_2H_4 in the form of dry sepal, but so far as is known this has not been carried out, possibly in view of economic and unfavourable climatic considerations.

Table 10.5 shows the relative sensitivity of a number of plant responses to ethylene in relation to some other unsaturated hydrocarbons. It is apparent that the latter are orders of magnitude less toxic. Thus it would seem that they have no relevance in the field, except in the case of industrial accidents.

In conclusion it can be stated that the geographical distribution of C_2H_4 damage to plants in the field remains unknown. Falling concentrations from vehicles in the developed world seem to indicate that any problems are likely to be declining, although ambient levels in cities are still high enough to cause orchid dry sepal. Thus it would seem probable that other adverse impacts may essentially be restricted to point sources in the form of polyethylene plant. However, with growing motor vehicle emissions in the developing world (see Chapter 21) it would seem likely that C_2H_4 may well be having major impacts on vegetation in and near their urban areas, but this has not been investigated.

VOCs other than ethylene

As previously described the data relating to the phytotoxic effects of VOCs other than ethylene are limited. Considering that there are up to 13 000 VOCs in the environment there is obviously a limit on all the compounds that can be evaluated, however, there is surprisingly little information available. The majority of the studies have related to plant uptake of VOCs from the standpoint of using them as biomonitors (Wagrowski and Hites, 1997; Nakajima *et al.*, 1994), reservoirs for the cycling of these compounds (Simonich and Hites, 1994; Horstmann and McLachlan, 1998) or as potential pathways in the transmission of these often carcinogenic compounds to humans (Jones and Duarte Davidson, 1997; Thomas *et al.*, 1998). The uptake and the subsequent fate of semi-volatile organic compounds (SOCs) and persistent organic pollutants (POPs) is now relatively well understood (McLachlan, 1999), for the VOCs this is less certain. This is possibly because deposition to the cuticle of POP and SOC is easily related to the octanol:water partition coefficient (K_{ow}) of these compounds as a result of their wax-loving nature, whereas the VOC may additionally be under stomatal control and subject to further plant metabolism (McLachlan, 1999).

The levels of VOCs in the atmosphere are studied today mostly for their impact on photochemical oxidant formation and many countries have national networks for these measurements. Much of the recent research on VOCs and plants has centred on their possible role in new forest decline.

Trichloroacetic acid (TCA)

One of the key compounds investigated in relation to new forest decline is trichloroacetic acid (TCA). TCA is a known herbicide; in sensitive plants TCA inhibits the enzymes responsible for the conversion of ammonia to amides and phytotoxic levels of ammonia accumulate (Ashton and Crafts, 1973). TCA is formed in the atmosphere as a product of the breakdown of C1 and C2 halocarbons. Symptoms of TCA toxicity observed have been: a curling of the needles, increases in the number of plastoglobules, reduced number of thylakoids and inhibition of wax layer development (Kristen *et al.*, 1992; Heitefuss, 1989; Plumacher and Dombrowsky, 1997). The principal pathways for the formation of TCA are believed to be the reactions of 1,1,1-trichloroethane (CH_3CCl_3) and tetrachloroethene with hydroxyl radicals in the atmosphere (Frank and Frank, 1990). In addition to direct deposition TCA may enter the transpiration stream from soil following wash out, as a result of its high water solubility. TCA is also formed within plant cells as a result of the detoxification of the dry deposition of C_2 volatile chlorocarbons by monoxygenase P450 (Plumacher and Schroeder, 1994). There is also the possibility that TCA is released in soils (Hoekstra *et al.*, 1999), but the strength of this source is open to debate (Jordan *et al.*, 1999).

Atmospheric concentrations of TCA in the atmosphere are not frequently reported, but one study recorded levels between 6 and 500 pg m^{-3} (Jordan *et al.*, 1999), while rain concentrations in Germany were reported to be 0.14 $\mu g\ l^{-1}$ median, 0.86 $\mu g\ l^{-1}$ maximum (Stattelberger *et al.*, 1997). It should be noted that after year-on-year rises in atmospheric concentrations of the C1 and C2 halocarbons between 1965 and 1990, their atmospheric concentrations are now showing signs of decline as a result of the Montreal Protocol (Prinn *et al.*, 1995; Frank and Frank, 1990). There may also be local enhancement of TCA concentrations as a result of industrial activities such as pulp mills using chlorine bleaching agents (Juuti *et al.*, 1993, 1995).

TCA was found at concentrations between 10 and 130 ng g^{-1} in Scots pine needles in Finland (Frank *et al.*, 1992). Norokorpi and Frank (1995) reported a significant increase in the loss and chlorosis of pine needles in Finnish forests in relation to TCA with two distinct populations: a resistant group with a gradient of 0.32% defoliation per unit (ng g^{-1}) TCA and 0.78% per unit for the sensitive population. One possible reason for these population differences may be the enzyme activities of glutathione transferase, monoxgyenase P450. The enzymes are responsible for conjugation and cleavage of xenobiotics, respectively. As stated above, the latter may actually increase internal TCA concentrations. In a national study across Finland no relationship between TCA concentration in the foliage and the level of defoliation was observed, but a significant effect of TCA on lichen biomass was observed (Juuti *et al.*, 1996). Plumacher and Dombrowsky (1997) concluded that the total deposition to forests in the vicinity of Berlin was 30–200 fold lower than TCA doses from a single TCA application, so unless plants were in the vicinity of a local source there was unlikely to be a phytotoxic impact. In one of the few long-term exposure studies aerial deposition in fog resulted in disruption of epicuticular waxes which was not observed from purely root uptake, but both pathways resulted in a decrease in chloroplast size with increasing TCA concentrations in needles (Sutinen *et al.*, 1997). In the vicinity of the Caspian Sea, which is highly polluted with organic chemicals, needle concentrations of TCA were in the range 3 to 69 g kg^{-1}; this is one million fold higher than those in other studies and the conclusion that such concentrations may be having an effect on the local vegetation seems valid (Weissflog *et al.*, 1999).

Nitrophenols

Automobile exhaust is a primary source of nitrophenols (Nakajima *et al.*, 1993), and they are also formed in the atmosphere by the reaction of benzene and alkylated aromatics with hydroxyl radicals and NO_x (Heterich and Hermann, 1990). Lüttke *et al.* (1997) proposed that the mononitrophenols were principally from car exhaust, while the dinitrophenols were formed in the atmosphere in the liquid phase. Gaseous concentrations of all the compounds were always higher than in the liquid phase, with 4 nitrophenol the most abundant compound with a concentration range 14–70 ng m^{-3}, while the other compounds were in the range 1.2–30 ng m^{-3}. These values were at a background site, but at a site, influenced by urban discharges the cumulative mononitrophenols were 0.35 nmol m^{-3} and the cumulative dinitrophenols were 0.09 nmol m^{-3} (Heterich and Hermann, 1990). Values of 2,4 dinitrophenol in rain at an Austrian background site were 1.61 $\mu g\ l^{-1}$, but 2,4 dinitrophenol was often only 20% of the total nitrophenol burden (Weiss *et al.*, 1997).

The nitrophenols were suggested to be involved in forest decline by Rippen *et al.* (1987), while Shea *et al.* (1983) observed phytotoxic effects of 2,4 dinitrophenol at 20 to 200 $\mu g\ l^{-1}$. Pigment bleaching and ultrastructural changes in mitochondria and mesophyll chloroplasts were observed in leaves exposed in cuvettes to 6–10 mg m^{-3}: these are very high concentrations, but the authors argued that the exposure was only for a short duration and therefore at the constant prolonged exposure in the environment phytoxicity may well be observed (Kristen *et al.*, 1992). The herbicidal action of this group of compounds operates by the uncoupling of oxidative phosphorylation and at high concentrations photosynthetic phosphorylation (Kaufmann, 1976). Nitrophenols are readily taken up by plants from both the gaseous and liquid phase. Weiss *et al.* (1997) recorded a mean concentration of 60 $\mu g\ kg^{-1}$ (max. 160 $\mu g\ kg^{-1}$) for total nitrophenols in pine needles at background sites in Austria. Natangelo *et al.* (1999) found higher total nitrophenol concentrations in leaves of a more damaged forest site (313 ng g^{-1}) compared with 148 ng g^{-1} at a non-damaged site.

The phytotoxic effects of other compounds have only been sporadically investigated and the results are summarised in Table 10.6. Generally the conclusions of Smidt (1998)

Table 10.6 Concentrations of VOCs in the air of forested and congested areas compared to their effective concentrations (adapted from Smidt, 1993, 1998 and Schroeder, 1993)

Compound	Biochemical and physiological effects	Concentrations in ambient air	Effective concentrations
PAN	Formation of radicals and lipid peroxides, chlorosis, necrosis, bronzing	<2 ppb[C]	16 ppb (8 hrs)
Methane		1700 ppb[F]	1 000 000 ppb
Formaldehyde	Reduced viability of pollen	9–13 ppb (24 hrs)[C]	16 ppb (year)
Tetrachloromethane	Increased POD and GST activity, degradation of pigments, decline in photosynthesis, chlorosis, necrosis, leaf drop, change in wax structure	<0.2 ppb[F]	25 ppb (24 hrs)
1,1,1-trichloroethene		0.02–0.29 ppb[F]	25 ppb (24 hrs)
Tetrachloroethene		<3 ppb (24 hrs)[F]	6–20 ppb (+UV, 48 hrs–6 wks)
Formic acid in fog[1]	Increase in detoxification enzymes	1 ppm[F]	1 ppm
Halons	Increased GST, increase in xanthophylls, decrease in plant dry weight	2 ppt	10 ppb

F = forested, C = congested (urban). [1]Tezuka *et al.* (1998).

hold, that unless there is a localised elevation of the VOC concentration as a result of a significant source it seems unlikely that these VOCs cause direct phytotoxic effects individually. Schubert *et al.* (1995) when investigating the effects of VOCs on tobacco pollen germination, found that the 25% effective dose (ED_{25}) for 1,2 dichloroethane, trichloroethene, tetrachhloroethene, chloroform and benzene were 4170, 15 890, 4980, 6770 and 11 420 ppm, respectively. These values are very high compared to those previously observed when screening for physiological affects (Table 10.6). The halons have also been observed to have phytotoxic effects with a decline in transpiration and dry weight as a result of exposure (Debus and Schroeder, 1991).

Tolerance mechanisms

From Table 10.6, it can be seen that the detoxification enzymes are frequently enhanced when plants are exposed to VOCs. One particular path of enquiry has been the action of glutathione-S-transferases which conjugate xenobiotics with glutathione: these conjugates are non-toxic and are subsequently metabolised. This process has been widely studied in relation to herbicide tolerance. Wolf (1992) found that damaged trees on the Wank mountain in Bavaria had a different GST isozyme to that in healthy trees and proposed that the damaged trees may lack the 'right' isozyme to detoxify the xenobiotics they were exposed to. Glucosyl transferases have been shown to conjugate chloro- and nitrophenols to glucosides. The antioxidant enzymes superoxidase dismutase and ascorbate peroxidase are elevated following exposure to formic and acetic acid (Ogawa *et al.*, 1998). It could therefore be hypothesised that these VOC pollutants put the plants under stress as a result of a redirection of their normal metabolism to increased synthesis of detoxification enzymes.

Overall conclusions

A wide survey of the literature leads to the conclusion that under prevailing atmospheric levels of VOCs there is little evidence for direct damage by any one compound, other than C_2H_4. Where there are local sources this may not be the case. Undoubtedly there is evidence of VOCs causing plant stress, but the effects seem relatively benign in the wider environment. However, further research is required particularly investigating the effects of pollutant mixtures (see Chapter 12). Whether there is a cost to the plant by the induction of detoxification enzymes also needs investigation, again mixtures of VOCs may overcome these systems.

Finally, there are still many compounds whose phytotoxic effects have not been assessed. Considering the huge number of compounds in the environment rapid screening methods need to be developed. Possible approaches are systems such as the pollen tube test referred to above, although this seemed relatively insensitive, DNA adducts as proposed by Weber-Lofti *et al.* (1998) although these lack compound specificity, or the use of quantitative structure activity relationships. It is only when such systems are in place that a full evaluation of VOC effects can be undertaken and most importantly dose-response relationships determined to allow sensitive regions to be identified using approaches such as the critical loads concept.

References

Abeles, F. and Gahagan, III, H.B. (1968). Abscission: the role of ethylene, ethylene analogues, carbon and oxygen. *Plant Physiol.*, **43**, 1255–1258.

Abeles, F.B. and Heggestad, H.E. (1973). Ethylene: an urban air pollutant. *J. Air Pollut. Control Assoc.*, **23**, 517–521.

Abeles, F.B. (1973). Ethylene in plant biology. Chapter 11. *Air Pollution and Ethylene Cycle*. Academic Press, New York.

Abeles, F.B. (1985). Sources of ethylene of horticultural significance. In: Roberts, J.A. and Tuckers, G.A. (eds). *Ethylene and Plant Development*. Butterworths, London, pp. 287–296.

Ashton, F.M. and Crafts, A.S. (1973). *Mode of Action of Herbicides*. John Wiley & Sons, New York.

Barden, L.E. and Hanan, J.J. (1972). Effect of ethylene on carnation keeping life. *J. Amer. Soc. Hort. Sci.*, **97**, 785–788.

Davidson, O.W. (1949). Effects of ethylene on orchid flowers. *Proc. Am. Soc. Hort. Sci.*, **53**, 440–446.

Debus, R. and Schroeder, P. (1991). Responses of *Petunia hybrida* and *Phaseolus vulgaris* to fumigation with difluorochlorobromomethane (Halon 1211). *Chemosphere*, **21**, 1499–1505.

Frank, H. and Frank, W. (1986). Photochemical activation of chloroethenes leading to destruction of photosynthetic pigments. *Experientia* **42**, 1267–1269.

Frank, W. and Frank, H. (1990). Concentrations of airborne C_1 and C_2 halocarbons in forest area in west Germany: Results of three campaigns in 1986, 1987 and 1988. *Atmospheric Environment*, **24A**: 1735–1739.

Frank, H., Scholl, H., Sutinen, S. and Norokorpi, Y. (1992). Trichloroacetic acid in conifer needles in Finland. *Annals of Botany Fennici*, **29**, 263–267.

Gunderson, C.A. and Taylor, G.E. (1991). Ethylene directly inhibits foliar gas exchange in *Glycine max*. *Plant Physiol.*, **95**, 337–339.

Hall, W., Truchelut, G.B., Leinweber, C.L. and Herrero, F.A. (1957). Ethylene production by the cotton plant and its effects under experimental and field conditions. *Physiol. Plantarum*, **10**, 306–317.

Heck, W.W. and Pires, E.G. (1962). *Effect of ethylene on horticultural and agronomic plants*. Texas Agricultural Experiment Station Report No: MP 613.

Heitefuss, R. (1989). *Crop and plant protection – The practical foundations*. Wiley, Chichester.

Heterich, R. and Hermann, R. (1990). Comparing the distribution of nitrated phenols in the atmosphere of two German hill sites. *Environmental Technology*, **11**, 961–972.

Horstmann, M. and McLachlan, M.S. (1998). Atmospheric deposition of semivolatile organic compounds to two forest canopies. *Atmospheric Environment*, **32**, 1799–1809.

Hoekstra, E.J., deLeer, E.W.B. and Brinkman, U.A.T. (1999). Mass balance of trichloroacetic acid in the soil top layer. *Chemosphere*, **38**, 551–563.

Jones, K.C. and Duarte Davidson, R. (1997). Transfers of airborne PCDD/Fs to bulk deposition collectors and herbage. *Environmental Science and Technology*, **31**, 2937–2943.

Jordan, A., Frank, H., Hoekstra, E.J. and Juuti, S. (1999). New directions: exchange of comments on 'The origins and occurrence of trichloroacetic acid'. *Atmospheric Environment* **33**, 4525–4527.

Juuti, S., Hirvonen, A., Tarhanen, J., Holopainen, J.K. and Ruuskanen, J. (1993). Trichloroacetic acid in pine needles in the vicinity of a pulp-mill. *Chemosphere*, **26**, 1859–1868.

Juuti, S., Norokorpi, Y., Helle, T. and Ruuskanen, J. (1996). Trichloroacetic acid in conifer needles and arboreal lichens in forest environments. *Science of the Total Environment*, **180**, 117–124.

Juuti, S., Norokorpi, Y. and Ruuskanen, J. (1995). Trichloroacetic acid (TCA) in pine needles caused by atmospheric emissions of kraft pulp-mills. *Chemosphere*, **30**, 439–448.

Kaufmann, D.D. (1976). In *Phenols. Herbicides, Chemistry, Degradation and Mode of Action*. (eds P.C. Kearney and K. Kearney). Marcel Dekker, New York, pp. 665–707.

Kristen, U., Lockhausen, J., Petersen, W., Schult, B. and Strube, K. (1992). Veranderungen an Fichtennadeln nach Begasung mit 2,4-dintrophenol, Benzaldehyd, Fufural, Trichloroethan und Trichloressigsaure. Luftverunreinigungen und Waldschaden am Standort 'Postturm', Forstamt Farchau/Ratzeburg. W. a. B. Michaelis, J., GKSS-Forschungszentrum Geesthacht. GKSS 92/E/100: 341-352.

Lüttke, J., Scheer, V., Levsen, K., Wunsch, G., Cape, J.N., Hargreaves, K.J., Storeton-West, R.L., Acker, K., Wieprecht, W. and Jones, B. (1997). Occurrence and formation of nitrated phenols in and out of cloud. *Atmospheric Environment*, **31**(16), 2637–2648.

McLachlan, M.S. (1999). Framework for the interpretation of measurement of SOCs in plants. *Environmental Science and Technology*, **33**, 1799–1804.

Mortensen, L.M. (1989). Effect of ethylene on growth of greenhouse lettuce at different light and temperature levels. *Scientia Horticulturae*, **39**, 97–103.

Nakajima, D., Teshima, T., Ochiai, M., Tabata, M., Suzuki, J. and Suzuki, S. (1994). Determination of 1-nitropyrene retained in leaves in roadside trees. *Bulletin of Environmental Contamination and Toxicology*, **53**, 888–894.

Natangelo, M., Mangiapan, S., Bagnati, R., Benfenati, E. and Fanelli, R. (1999). Increased concentrations of nitrophenols in leaves from a damaged forestal site. *Chemosphere*, **38**, 1495–1503.

Norokorpi, Y. and Frank, H. (1995). Trichloroacetic acid as a phytotoxic air pollutant and the dose response relationship for Scots pine. *The Science of the Total Environment*, **160/161**: 249–463.

Ogawa, T., Takenaka, C. and Tezuka, T. (1998). Responses of anti-oxidant enzymes to mist containing sulfuric or organic acid in Hinoki Cypress (*Chamaecyparis obtusa*) seedlings. *Environmental Sciences*, **6**, 185–196.

Pleijel, H., Ahlfors, A., Skärby, L., Pihl, G., Sellden, G. and Sjödin, A. (1994). Effects of air pollutant emissions from a rural motorway on *Petunia* and *Trifolium*. *Sci. Tot. Environ.*, **146/147**, 117–123.

Plumacher, J. and Dombrowsky, P. (1997). *Volatile chlorinated hydrocarbons (VCHs) and trichloroacetic acid (TCA) in forest areas of Berlin and its surroundings. Organic xenobiotics and plants: Impact, metabolism, toxicology*, Federal Environment Agency, Vienna, Austria.

Plumacher, J. and Schroeder, P. (1994). Accumulation and fate of C_1/C_2-chlorocarbons and trichloroacetic acid in spruce needles from an Austrian mountain site. *Chemosphere* **29**, 2467–2476.

Posthumus, A.C. (1984). Monitoring levels and effects of air pollutants. In: Treshow, M. (ed.). *Air Pollution and Plant Life*. Wiley, Chichester, pp. 73–95.

Prinn, R.G., Weiss, R.F., Miller, B.R., Huang, J., Alyea, F.N., Cunnold, D.M., Fraser, P.J., Hartley, D.E. and Simmonds, P.G. (1995). Atmospheric trends and lifetime of CH_3Cl_3 and global OH concentrations. *Science* **269**(5221), 187–192.

Reid, D.M. and Watson, K. (1985). Ethylene as an air pollutant. In: Roberts, J.A. and Tuckers, G.A. (eds). *Ethylene and Plant Development*. Butterworths, London, pp. 277–286.

Rippen, N., Zietz, E., Frank, R., Knacker, T. and Klopffer, W. (1987). Do airborne nitrophenols contribute to forest decline? *Environmental Technology Letters*, **8**, 475–482.

Sawada, S. and Totsuka, T. (1986). Natural and anthropogenic sources and fate of atmospheric ethylene. *Atmos. Environ.*, **20**, 821–832.

Schubert, U. Wisanowski, L. and Kull, U. (1995). Determination of phytotoxicity of several volatile organic compounds by investigating the germination pattern of tobacco pollen. *Journal of Plant Physiology*, **145**, 514–518.

Schroeder, P. (1993). Detoxification and metabolism of xenobiotics in *Picea* and *Pinus*. (eds. P. Schroeder, H. Frank and B. Rether). *The proceedings of the 2nd IMTOX workshop on volatile organic pollutants: Levels, fate and ecotoxicological impact*. pp. 12–26. Wissenschafts Verlag.

Shea, P.J., Weber, J.B. and Overcash, M.R. (1983). Biological activities of 2,4-dinitrophenol in soil plant systems. *Residue Reviews*, **87**, 1–41.

Simonich, S.L. and Hites, R.A. (1994). Vegetation-atmosphere partitioning of polycyclic aromatic hydrocarbons. *Environmental Science and Technology*, **28**, 939–943.

Smidt, S. (1993). Emissions and inputs of VOCs in Austria. Their possible contribution to forest decline. (eds. P. Schroeder, H. Frank and B. Rether) *The proceedings of the 2nd IMTOX workshop on volatile organic pollutants: Levels, fate and ecotoxicological impact*. pp. 12–26. Wisfenschafts Verlag.

Smidt, S. (1998). Assessment of the relevance of VOCs to forest trees – general remarks. In: *Proceedings of the 4th IMTOX-Workshop*, Eds: Federal Environment Agency, Vienna, Conference Papers Vol. 24, ISBN 3-85457-439-8.

Squier, S.A., Taylor, G.E., Selvidge, W.J. and Gunderson, C.A. (1985). Effect of ethylene and related hydrocarbons on carbon assimilation and transpiration in herbaceous and woody species. *Environ. Sci. Technol.*, **19**, 432–437.

Stattelberger, R., Gotz, B., Riss, A., Lorbeer, G. and Steiner, E. (1997). *Nitrophenols and trichloroacetic acid in precipitation in Austria. Organic xenobiotics and plants: Impact, metabolism, toxicology*, Federal Environment Agency, Vienna, Austria.

Sutinen, S., Juuti, S. and Ryyppo, A. (1997). Long-term exposure of Scots pine seedlings to monochloroacetic and trichloroacetic acid: effects on the needles and growth. *Annales Botanici Fennici*, **34**, 265–273.

Taylor, O.C. (1984). Organismal responses of higher plants to atmospheric pollutants: photochemical oxidants and others. In: Treshow, M. (ed.). *Air Pollution and Plant Life*. Wiley, Chichester, pp. 215–238.

Taylor, H.J., Ashmore, M.R. and Bell, J.N.B. (1987). *Air Pollution Injury to Vegetation*. Institution of Environmental Health Officers, London.

Taylor, G.E. and Gunderson, C.A. (1988). Physiological site of ethylene effects on carbon dioxide assimilation in *Glycine max*. L. Merr. *Plant Physiol.*, **86**, 85–92.

Tezuka, T., Ogawa, T., Matsumoto, K., Katou, K., Ishizaka, Y. and Takenaka, C. (1998). Organic acids in acid fog are an important effector contributing to forest stress. *Environmental Sciences*, **6**, 99–100.

Thomas, G., Sweetman, A.J., Ockenden, W.A., Mackay, D. and Jones, K.C. (1998). Air-pasture transfer of PCBs. *Environmental Science and Technology*, **32**, 936–942.

Tonneijck, A.E.G., ten Berge, W.F., Jansen, B.P. and Bakker, C. (1999). Epinastic response of potato to atmospheric ethylene near polyethylene manufacturing plants. *Chemosphere*, **39**, 1617–1628.

Wagrowski, D.M. and Hites, R.A. (1997). Polycyclic aromatic hydrocarbon accumulation in urban, suburban and rural vegetation. *Environmental Science and Technology*, **31**, 279–282.

Weber-Lofti, F., Guillemaut, P., Rether, B. and Dietrich, A. (1998). Use of plant DNA adducts as molecular bio-indicators of air pollution: field studies in the north east of France. *Proceedings of the 4th IMTOX-Workshop*. Federal Environment Agency, Vienna, Conference Papers Vol. 24 ISBN 3-85457-439-8.

Weiss, P., Lorbeer, G., Stephan, C. and Svabenicky, F. (1997). *Short chain aliphatic halocarbons, trichloroacetic acid and nitrophenols in spruce needles of Austrian background forest sites. Organic xenobiotics and plants: Impact, metabolism, toxicology*. Federal Environment Agency, Vienna, Austria.

Weissflog, L., Manz, M., Popp, P., Elansky, N., Arabov, A., Putz, E. and Schuurmann, G. (1999). Airborne trichloroacetic acid and its deposition in the catchment area of the Caspian Sea. *Environmental Pollution*, **104**, 359–364.

Wolf, A. (1992). GST activity and isosymes in healthy and damaged spruce trees. *Proceedings of the 2nd IMTOX workshop on volatile organic pollutants; levels, fate and ecotoxicological impact*. Garmish Partenkirken, Germany, Wissenschafts – Verlag.

11 Effects of particulates

A. FARMER

Particulate pollution

Particulates exist in the atmosphere in many forms. These may range from sub-micron aerosols to visible dust particles. Some of these particulates are from natural sources, while others are derived from human activity. The relative impact that particulates have on vegetation depends on their individual size, cumulative quantity and their chemistry. There is currently significant concern over particulate pollution, particularly for small particulates, either of less than 10 μm diameter (PM_{10}) or less than 2.5 μm diameter ($PM_{2.5}$), QUARG (1996). However, the emphasis on these types of particulates (e.g. in developing air quality standards in the European Union or the United States) is due to concern that they may have an adverse effect on human health (Farmer, 1997). This is not a focus of this book and there will, therefore, be little discussion of these types of particulates here. However, vegetation may play a role in managing such particulates and a brief consideration of this issue will be given at the end of this chapter.

Particulate sources

Particulates may be derived from a number of sources. Primary particulates are derived directly from a source. These include wind-blown soils, quarries, domestic and industrial combustion and vehicle exhausts. Secondary particulates are those formed due to the interaction between other compounds in the atmosphere, e.g. nitrate particles formed from the oxidation of nitrogen oxides.

The main primary particulate sources that are known to cause significant impacts on vegetation are:

- Mineral extraction, particularly quarries.
- Mineral and other processing industries, e.g. cement and fertilizer works.
- Road use, especially for unpaved roads.
- Vehicle exhausts
- Natural sources, e.g. soil blow, dust storms and volcanoes.

Particulate size

While small particulates are of most concern for human health impacts, much larger particulates are important in affecting vegetation. Particulates can impact on plants in

Air Pollution and Plant Life, second edition. Edited by J.N.B. Bell and M. Treshow. ISBN 0 471 49090 3 (HB), 0 471 49091 1 (PB).

a number of ways and those impacts mediated via smothering of leaves or altering soil chemistry are not dependent on particle size (except in that size may affect transport and deposition rates and, therefore, the occurrence of the problem). Where particulates may block leaf stomata or enter cells, the size of stomata is important. These vary between species, but the diameter generally lies between 8 and 10 μm. It can be seen, therefore, that relatively large particulates can play important physiological roles in plant responses.

A number of studies of particulate impacts on vegetation have examined the particle sizes from sources causing impacts. Table 11.1 summarizes the results found.

It is evident, therefore, that most particulate sources generate particles of widely different sizes. However, the proportion of particles of different sizes also varies. Generally, particulate sources will show a greater abundance of smaller particles compared to larger. For example, while Ninomiya *et al.* (1971) found a very large range of particle sizes (Table 11.1), 40% of those derived from cars running on leaded fuel were less than 0.1 μm in diameter.

It is also important to note that particle size is important in determining the distance that particles might travel before deposition. A study on unpaved roads by Everett (1980), for example, showed that most particles of more than 50 μm diameter were deposited within 8 m of the edge of the road. At 30 m from the road, few particles larger than 20 μm diameter were found.

Particulate chemistry

The chemistry of particulates is also important in determining the potential effects that they may have on vegetation. Particulates which are inert will impact on plants in a purely physical manner. However, those with active chemistry might cause direct physiological (e.g. toxicological) responses on/in leaves or affect substrate chemistry.

Relatively few studies have been undertaken with inert particulates and field examples of these are usually limited to particulate sources from relatively slow weathering rocks, e.g. granites.

Many particulate sources known to cause extensive problems for vegetation are calcareous in origin. A number of these are derived from limestone quarries. However, many studies have also stressed the importance of dust derived from cement kiln processing. Limestone is calcium carbonate, which is alkaline. However, cement kiln dust may largely consist of calcium oxide. This is not only alkaline, but also it is highly reactive with water and is known to cause extensive damage to plants exposed

Table 11.1 Particulate sizes for a range of different particulate types

Particulate type/source	Particulate diameter	Reference
Cement kiln dust	80–90% <30 μm	Darley (1966)
Motor vehicles	Range 0.01–5000 μm	Ninomiya *et al.* (1971)
Urban road dust	Mostly 3–100 μm	Thompson *et al.* (1984)
Fly ash	1–2000 μm	Krajickova and Mejstrik (1984)
Coal dust	3–100 μm	Rao (1971)

to it. For example, Darley (1966) noted that solutions of cement kiln dust had a pH of 12.0 and also contained a number of heavy metals and bisulphite. The contaminants are potentially toxic, while sufficient quantities of the caustic dust could 'dissolve' leaf tissues.

Road dust may be derived from a variety of sources. Unpaved roads may be covered with limestone gravel and so generate alkaline particulates. However, in urban areas many of the particulates are derived from motor vehicles' exhausts. Such particulates may contain a variety of heavy metal contaminants (Santelmann and Gorham, 1988) and be potentially toxic.

Other particulate sources also have their own specific chemistry. Most notable is that derived from coal processing and handling. Coal dust often contains relatively high levels of sulphur and fluoride (Rao, 1971), which may result in direct toxic effects on vegetation.

Levels of deposition

A key factor in determining the level of impact that particulate pollution will have on vegetation is the quantity of particulates deposited. Many studies of particulate impacts on plants provide little or no assessment of the rates of deposition. However, this information is important particularly in management issues, e.g. in undertaking an Environmental Impact Assessment of a particulate producing process. It may be relatively simple to calculate predicted deposition rates, but the consequences for vegetation response may be unclear due to lack of published information on the consequences of these levels of deposition.

For roads the levels of deposition depend on the number of vehicles. Roberts *et al.* (1975) found that an unpaved dry gravel road with an average daily traffic (ADT) of 250 cars produced mean atmospheric particulate concentrations of 584 $\mu g\ m^{-3}$, while a paved road with an ADT of 18 000 produced mean atmospheric particulate concentrations of 463 $\mu g\ m^{-3}$. Everett (1980) found that deposition along an unpaved Alaskan road was around 10 $g\ m^{-2}d^{-1}$ in summer. This deposition rate declined logarithmically with distance from the road, but deposition was still measurable up to 1 km away.

Many factors influence particulate deposition, including wind strength and precipitation. Stronger winds will disperse larger particulates over longer distances, while rain will wash out particulates from the atmosphere. Rain may also remove particulates from plant surfaces, but rarely from soils.

Deposition of particulates may occur through a number of processes. As stated above larger particles will be deposited more quickly due to their larger settling velocities and they are, therefore, found more frequently closer to emission sources (Boubel *et al.*, 1994). Processes leading to deposition include:

- Sedimentation (simple settling of particles from the atmosphere).
- Diffusion.
- Turbulence (especially around vegetation surfaces).
- Removal by rainfall or occult deposition (mist and fog).

A key factor affecting deposition of particulates is surface roughness. Increasing roughness will increase deposition. Vegetation itself may significantly increase the roughness of the land surface. This may be on a large scale (e.g. the presence of trees) or a small scale (e.g. finely divided leaves or plant forms such as mosses). Wet surfaces also lead to increased deposition (Pye, 1987). Thus the nature of surrounding vegetation is important in assessing any likely dispersion of particulates from sources. It may, therefore, be used in managing particulate pollution (see later).

Routes for impacts on vegetation

It is possible to distinguish three principal routes/processes for the ways that particulate pollution can affect vegetation. These are:

- Direct deposition onto leaf surfaces.
- Blocking leaf stomata and/or uptake into leaf tissues.
- Deposition onto substrates (e.g. soil) and indirect effects via changes in substrate chemistry.

The degree of impact from these different processes depends on the different characteristics of particulates described earlier.

Deposition onto leaf surfaces

The levels of deposition will be influenced by the roughness of leaf surfaces and on whether such surfaces are wet. Rough surfaces will not only cause higher deposition rates, but may also reduce the potential for deposited particulates to be removed during rainfall. The levels of accumulated deposition will also depend on the lifespan of leaves, with longer-lived leaves having a greater time period over which to accumulate deposited pollutants. Increase in surface roughness might be caused by the shape of leaves or the presence of leaf hairs. Mosses, with their finely divided thalli, also present very rough surfaces and are efficient at trapping particulates. Manning and Feder (1980), for example, noted that forest canopies increase turbulent deposition considerably due to their increased surface roughness and, therefore, concluded that this vegetation type was the most efficient at trapping particulates. Roughness may also be increased at the very small scale. Thus the stomata of many leaves are slightly elevated above the epidermal cell layer and may cause local turbulence (Burkhardt *et al.*, 1995). This may aid in deposition to these structures (see next section).

Within the constraints of the physical processes controlling particulate deposition to any surfaces, a wide range of particle sizes may be deposited to leaf surfaces. Indeed, very large particles can be observed on leaves close to appropriate sources, such as quarries.

Impacts via the stomata

Given that the size of stomata is generally between 8 and 10 μm in diameter, particle size is important in affecting the likely impacts that any particulate source might have.

Particles that are of similar diameter to stomata are likely to become lodged into the stomatal openings. Smaller particles may pass through into the leaf tissues, while larger particles will be excluded.

Most plant leaves (though not some plant surfaces such as those of bryophytes and lichens) are protected by a cuticle, therefore potential chemical or toxicological effects are more likely from those particles which penetrate the leaf than those which remain on the surface. Thus the chemistry of the small particulate fraction within an overall sample is important in predicting impacts. An exception to this is cement dust which is so caustic as to be able to dissolve the cuticular layer.

The stomata on many leaves often predominate on the lower leaf surface. Thus simple deposition of particulates through settling onto leaf surfaces is unlikely to lead to stomatal impacts. Blocking of, or entry into, stomata will require deposition through turbulent motion around and below the leaf surface.

Deposition onto substrates

Like deposition onto leaf surfaces, levels of deposition onto substrates will depend on surface roughness and wetness. Indeed, particulate accumulation in soils in many plant communities does not occur directly. Particulates may initially be deposited into vegetation and then be transferred to the underlying soil by run off due to rainfall, or during leaf-fall. The effects of particulates via substrates are generally through changes in substrate chemistry. Therefore, particle size is of little importance compared to its chemical composition.

Responses of vegetation to particulates

Introduction

A wide range of studies have been undertaken examining the impacts of particulates on vegetation. This section will describe the most important of these, emphasizing the different types of impacts that may occur and focusing on the most sensitive species and communities. For a more complete list of references to research that has been undertaken, the reader is directed to Farmer (1993) and Beckett *et al.* (1998).

Physiological responses

A range of physiological responses have been noted for different plant species. Some are due to the direct physical action of the presence of particulates. These include the blocking of stomata, which alters transpiration rates. Particulates from road sources have also been noted to cause elevated temperatures in leaves, which will affect metabolic functions. Effects such as reduced photosynthesis can be the result of simple shading (thus also stimulating increased chlorophyll levels) or a decline in metabolic function due to cell structural damage or toxicity effects. A summary of some relevant studies is provided in Table 11.2.

Table 11.2 Examples of physiological responses demonstrated by plants subject to particulate pollution

Reference	Species	Particulate source	Physiological response
Darley (1966)	*Phaseolus vulgaris*	Cement kiln	Reduced photosynthesis
Singh and Rao (1981)	*Triticum aestivum*	Cement	Reduced transpiration
Oblisami *et al.* (1978)	*Gossypium hirsutum*	Cement factory	Increased chlorophyll levels
Borka (1980)	*Helianthus annuus*	Cement kiln	Reduced photosynthesis
Eveling (1969)	*Pisum sativum*	Inert dust	Increased water loss
Ricks and Williams (1974)	*Quercus petraea*	Smokeless fuel plant	Changes to gas exchange
Fluckiger *et al.* (1979)	*Populus tremula*	Inert silica gel	Reduced diffusive resistance

Table 11.3 Examples of injury caused to plants from the deposition of particulates

Reference	Species	Particulate source	Response
Czaja (1961)	*Beta vulgaris*	Cement factory	Cell plasmolysis
Czaja (1961)	Eight tree species	Cement	Cell destruction, bark peeling
Taylor *et al.* (1986)	*Avena sativa*, *Phaseolus vulgaris*	Cement kiln	Acute injury
Bleweiss *et al.* (1985)	*Cedrus libani*	Central-heating smoke	Dieback of branches and death of trees
Rao (1971)	*Mangifera indica*, *Citrus limon*	Coal dust	Leaf lesions

Injury

A number of studies have also demonstrated direct injury to plants exposed to particulate pollution. This is usually due to the active chemistry or toxic properties of specific particulates. Some of these might occur following entry into leaf tissue. However, highly reactive particulates such as cement dust are know to cause cell and leaf damage even when deposited onto leaf surfaces protected by a cuticle. Examples of responses found are given in Table 11.3.

Table 11.4 Examples of growth and reproductive responses of plants to particulate pollution

Reference	Species	Particulate source	Response
Singh and Rao (1968)	*Triticum aestivum*	Cement	Reduced vegetative and reproductive growth
Parsatharathy *et al.* (1975)	*Zea mays*	Cement	Reduced vegetative and reproductive growth
Borka (1980)	*Helianthus annuus*	Cement	Reduced growth
Brandt and Rhoades (1973)	*Acer rubrum*, *Quercus prinus*, *Q. rubra*	Limestone quarry	Reduced growth
Brandt and Rhoades (1973)	*Liriodendron tulipifera*	Limestone quarry	Increased growth
Braun and Fluckiger (1987)	*Abies alba*	Urban road	Reduced growth
Havas and Huttunen (1972)	*Pinus sylvestris*	Fertilizer factory	Young trees increased growth, old trees decreased growth

Growth and reproduction

Ultimately, recording physiological responses to particulate pollution may have little consequence to the plants being studied, unless this translates into a growth or reproductive response. These latter responses would indicate a probable change in the fitness of the individual plants in natural communities or impacts on crop yield in managed systems. Table 11.4 summarizes some examples of growth and reproductive responses to particulate pollution. Note that not all responses are negative.

Indirect responses

It is important to note that many air pollutants not only cause direct changes to plant function, but they may also lead to increases in indirect adverse effects from pests and pathogens. This has also been noted with particulate pollution (see Chapter 20). Thus Taylor *et al.* (1986) noted that the deposition of cement kiln dust caused an increase in leaf-spotting fungus in *Beta vulgaris* and an increase in aphid numbers in *Medicago sativa*. Manning (1971) found that limestone dust resulted in changes to the phylloplane flora of woodland species such as *Sassafras albidium* and *Vitis vulpina*. Leaves with particulate pollution showed an increase in the number of fungi and bacteria on their surfaces, although there was no change in the types of fungi or bacteria present. Very heavy particulate pollution, however, resulted in a decrease in phylloplane flora. In contrast the needles of *Tsuga canadensis* showed increased fungal, but decreased bacterial populations in areas with particulate pollution.

Changes to growth and reproductive rates may not only influence the behaviour of pests and pathogens, but they also have consequences for the general structure of plant communities. The differential impact of particulate pollution on the fitness of different species in a community will tend to cause a change in its species composition. Thus Brandt and Rhodes (1972) found community changes in deciduous woodland in Virginia due to deposition of limestone dust. In unpolluted woodlands, the dominant trees were *Quercus prinus*, *Quercus rubra* and *Acer rubrum*. However, in polluted areas, the dominant tree species were *Quercus alba*, *Quercus rubra* and *Liriodendron tulipifera*. Most species showed a negative growth response to the particulate pollution, except for *L. tulipifera*, which demonstrated a growth promotion.

Changes to substrate chemistry

Particulate deposition (in sufficient quantities) may change the chemistry of the substrates on which plants grow. This is most obviously seen with soils, but it may also affect tree bark or other substrates which support epiphytic communities. It is sometimes assumed that particulates generated from rocks underlying individual plant communities are unlikely to affect the chemistry of overlying soils as the soil chemistry is likely to be similar to that of the hard geology. However, this is often not the case. This is particularly so where soils have developed for many years over calcareous rocks (e.g. limestone). Such soils are subject to continual inputs of rain, which is mildly acidic and, therefore, are also subject to slow leaching of base cations. In southern England, for example, this may lead to the occurrence of acidic heathland communities where the underlying geology is limestone.

A number of examples have been studied where particulate deposition has altered soil chemistry and caused significant changes in the species composition of plant communities. These include:

Heathlands (Etherington, 1978; Warne in Farmer, 1993). Communities studied in Wales and Derbyshire showed that natural heathlands had low soil pHs (e.g. pH 3–4) and supported typical ericaceous shrubs species. However, particulates produced by neighbouring limestone quarries resulted in an elevated soil pH and decline of these species. Indeed these changes also had negative consequences for a range of invertebrate species dependent on the heathlands.

Grasslands (e.g. Grime, 1970). Particulates from limestone quarries in Derbyshire were found to cause changes to acidic grassland communities, with a shift from calcifuge to calcicole species.

Tundra communities (Spatt and Miller, 1981; Walker and Everett, 1987). Some of the most extensive research on the effects of particulate pollution on natural plant communities has been undertaken on arctic vegetation subject to calcareous dust derived from unpaved roads. Different vegetation types vary in their sensitivity to particulate deposition (Table 11.5). It is also clear that it may take a number of years for some impacts to be observed.

The most sensitive plant communities are those dominated by *Sphagnum* mosses. Spatt and Miller (1981) found a decline in *Sphagnum lense* with levels of deposition of 1.0–2.5 g $m^{-2}d^{-1}$ and could even detect some impacts where deposition was as low as 0.07 g $m^{-2}d^{-1}$. Most *Sphagnum* species are particularly sensitive as they are completely

Table 11.5 The effects of road dust on different northern Alaskan plant communities over different exposure periods. Redrafted from Klinger *et al.* (1983)

Vegetation type	Time period (years)	Effect of particulate pollution from the road
Dry prostrate shrub tundra	2	Little or no effect
	10	Mosses and lichens smothered near the road. Low herbs do less well than sedges and shrubs
Moist graminoid tundra	2	Possible increase in herbs
	10	Intolerant species decline (e.g. *Cassiope tertragona*, *Tomethypnum nitens* and *Catascopium nigritum*) Tolerant species include *Dryas integrifolia*, *Drepanocladus* spp., *Salix* spp., *Campyllum stellatum* and *Carex* spp.
Wet graminoid tundra	2	Increase in algae and possibly in standing crop
	10	Mosses, including *Catascopium nigritum* die near the road. Other mosses more tolerant
Aquatic graminoid tundra	10	No effects except close to the road, where some aquatic mosses may be smothered

intolerant of a combination of elevated pH and the presence of calcium ions – precisely the conditions which result from the deposition of calcareous particulates.

In some areas the impacts on the vegetation were so severe that overall plant cover was reduced, which resulted in an increase of thawing rates in the spring. The impacts on roadside vegetation were reduced where the road passed through vegetation communities growing on locally occurring limestone rocks.

Epiphytic communities. Lichens are well known as being sensitive to a wide range of air pollutants (Bates and Farmer, 1992), (see Chapter 17). This also includes particulates. Table 11.6 summarizes some key studies relating to different particulate types. Lichens can be sensitive to general smothering by particulates. However, lichen species are highly sensitive to the chemistry of the substrate on which they grow. Many tree species have acidic or mildly acidic bark and small changes to bark pH can be very important in determining species occurrence. It is not surprising that the deposition of calcareous particulate pollution can lead to significant changes in community composition. Interestingly, polluted trees often exhibit the presence of species typical of rock surfaces, including limestone rocks, thus demonstrating powerfully the role of chemistry rather than other factors in the presence of these species.

Using vegetation to manage particulate pollution

It can be seen from the above discussions of the effects of particulates on vegetation, that some plants are efficient at trapping airborne particulates. While previous discussion has focused on the adverse effects that this has for the plants concerned, this

Table 11.6 Examples of studies undertaken on different types of particulates which lead to a change in epiphytic lichen communities

Reference	Location	Results
Gilbert (1976)	Derbyshire, England. Deciduous woodland	Dust from limestone quarries resulted in heavy contamination of trees. The bark pH of *Fraxinus excelsior* was increased from 3.5 to 6.5. This changed the lichen flora to one typical of saxicolous communities or those of hypotrophicated bark. These included *Caloplaca decipiens*, *Catillara chlybeia*, *Lecanora calcarea*, *Lecanora campestris* and *Lecidella scabra*. Intolerant species included those typical of acidified bask, including *Hypogymnia physodes* and *Parmelia saxatilis*
Pihlstrom (1982, 1987)	Southern Finland. *Pinus sylvestris* forest	Calcareous dust was found to raise bark pH from 4.0 to 8.0. The saxicolous lichen *Xanthoria* spp. increased in abundance
Wittman and Turk (1988)	Austria. Alpine forest	Alkaline particulate pollution resulted in a complete change in lichen communities from the *Pseudovernietum furfuraceae* to the *Physcietum adscendentis*
Pocs (1990) and Veneklaas (1990)	Tropical rain forest in Tanzania and the Andes, respectively	Both authors noted that alkaline particulates derived from volcanic sources created locally abundant basophilic floras, which are usually less abundant in the tropics

property also offers some scope for those managing air pollution to use selected plants to reduce airborne particulate concentrations.

The simplest (and oldest) direct management tool has been to use densely foliated trees around particulate point sources, e.g. quarries. For example, Spitsyna and Skripal'shchikova (1991) noted that a 15 km wide stand of birch trees was able to intercept 50% of the particulates produced from an open-cast coal mine. Fast-growing evergreen species are most beneficial, providing rapid and year-round protection. Any adverse effects that the particulates may cause is usually compensated for by the vigorous growth rates of the species planted.

However, Beckett *et al.* (1998) have also emphasized the importance that trees play in improving urban air quality more generally. Trees may either act directly as particle traps, or, create locally altered climatic conditions (due to the action of transpiration), which reduce particulate concentrations.

In Chicago, McPherson *et al.* (1994) found that trees removed 234 tons of PM_{10} in 1991, improving average hourly air quality by 0.4%. In a similar study Nowak *et al.* (1997) found trees caused an improved air quality in Philadelphia of 0.72%. While these might not seem to be dramatic improvements, they are significant in comparison to the economic costs of other measures to improve air quality. Beckett *et al.* (1998), therefore, recommended that pollution tolerant tree species should be more widely planted in urban areas, particularly close to particulate pollution sources (e.g. roads) as an active management tool to improve urban air quality.

Conclusions

Particulate pollution is, therefore, a general term including a wide variety of different pollutants derived from many different sources. They may have many different impacts on vegetation through a variety of different pathways. For a few species the effects may be positive, but, for most species, particulates (in sufficient quantity) can lead to damage and growth reduction. This can result in a decline in crop yields or changes to natural communities.

However, there is still too little known about some types of particulates (e.g. from different rock types with contrasting chemistries) and how different levels of deposition affect plant communities over varying time periods. Such information is essential for making informed management decisions, such as granting planning permission for quarrying operations.

Finally, the properties of some plants to remove particulates can be used as an active management tool, to screen emission sources or help provide general improvements in urban air quality.

References

Bates, J.W. and Farmer, A.M. (1992). *Bryophytes and Lichens in a Changing Environment*. Clarendon Press, Oxford.

Beckett, K.P., Freer-Smith, P.H. and Taylor, G. (1998). Urban woodlands: their role in reducing the effects of particulate pollution. *Environmental Pollution*, **99**, 347–369.

Bleweiss, G., Werker, E. and Peleg, M. (1985). Air pollution effects on *Cedrus libani* trees – a case study. *Environmental Conservation*, **12**, 70–73.

Borka, G. (1980). The effect of cement dust pollution on growth and metabolism of *Helianthus annuus*. *Environmental Pollution (Series A)*, **22**, 75–79.

Boubel, R.W., Fox, D.L., Turner, D.B. and Stern, A.C. (1994). *Fundamentals of Air Pollution*. Academic Press, San Diego.

Brandt, C.J. and Rhoades, R.W. (1972). Effects of limestone dust accumulation on composition of a forest community. *Environmental Pollution*, **3**, 217–225.

Brandt, C.J. and Rhoades, R.W. (1973). Effects of limestone dust accumulation on lateral growth of forest trees. *Environmental Pollution*, **4**, 207–213.

Braun, S. and Fluckiger, W. (1987). Does exhaust from motorway tunnels affect the surrounding vegetation? In *Air Pollution and Ecosystems*, eds. P. Mathy and D. Reidel, Kluwer, Dordrecht, pp. 665–670.

Burkhardt, J., Peters, K. and Crossley, A. (1995). The presence of structural surface waxes on coniferous needles affects the pattern of dry deposition of fine particulates. *Journal of Experimental Botany*, **46**, 823–831.

Czaja, A.T. (1961). Uber das Problem der Zementstaubwirkung auf Pflanzen. *Staub*, **22**, 228–232.

Darley, E.F. (1966). Studies of the effect of cement-kiln dust on vegetation. *Journal of the Air Pollution Control Association*, **16**, 145–150.

Etherington, J.R. (1978). Eutrophication of limestone heath soil by limestone quarrying dust and implications for conservation. *Biological Conservation*, **13**, 309–319.

Eveling, D.W. (1969). Effects of spraying plants with suspensions of inert dusts. *Annals of Applied Biology*, **64**, 139–151.

Everett, K.R. (1980). Distribution and properties of road dust along the northern portion of the haul road. In: *Environmental Engineering and Ecological Baseline Investigations along the Yukon River-Purdhoe Bay Haul Road*. Eds J. Brown and R. Berg. US Army Cold Regions Research and Engineering Laboratory, CRREL Report 80-19. pp. 101–128.

Farmer, A.M. (1993). The effects of dust on vegetation – a review. *Environmental Pollution*, **79**, 63–75.

Farmer, A.M. (1997). *Managing Environmental Pollution*. Routledge, London.

Fluckiger, W., Oertli, J.J. and Fluckiger, H. (1979). Relationship between stomatal diffusive resistance and various applied particle sizes on leaf surfaces. *Zeitschrift Pflanzenphysiologie*, **91**, 173–175.

Gilbert, O.L. (1976). An alkaline dust effect on epiphytic lichens. *Lichenologist*, **8**, 173–178.

Grime, J.P. (1970). *People and Plants in Derbyshire*. Derbyshire Naturalists Trust, Matlock.

Havas, P. and Huttunen, S. (1972). The effect of air pollution on the radial growth of Scots pine (*Pinus sylvestris* L.). *Biological Conservation*, **4**, 361–368.

Krajickova, A. and Mejstrik, V. (1984). The effect of fly-ash particles on the plugging of stomata. *Environmental Pollution*, **36**, 83–93.

Manning, W.J. (1971). Effects of limestone dust on leaf condition, foliar disease incidence, and leaf surface microflora of native plants. *Environmental Pollution*, **2**, 69–76.

Manning, W.J. and Feder, W.A. (1980). *Biomonitoring Air Pollutants with Plants*. Applied Science Publishers, London.

McPherson, E.G., Nowak, D.J. and Rowntree, R.E. (1994). *Chicago's Urban Forest Ecosystem: Results of the Chicago Urban Forest Climate Project*. USDA General Technical Report NE-186.

Ninomiya, J.S., Bergman, W. and Simpson, B.H. (1971). Automotive particulate emissions. In: *Proceedings of the Second International Clean Air Congress*, eds H.M. Englund and W.T. Beery. Academic Press, New York, pp. 663–671.

Nowak, D.J., McHale, P.J., Ibarra, M., Crane, D., Stevens, J.C. and Luley, C.J. (1997). Modelling the effects of urban vegetation on air pollution. In: *22nd NATO/CCMS International Technical Meeting on Air Pollution Modelling and its Application*, pp. 276–282.

Oblisami, G., Pathmanabhan, G. and Pathmanabhan, C. (1978). Effect of particulate pollutants from cement-kilns on cotton plants. *Indian Journal of Air Pollution Control*, **1**, 91–94.

Parthasarathy, S., Arunuachalam, N., Natarajan, K., Oblisami, G. and Rangaswami, G. (1975). Effect of cement dust pollution on certain physical parameters of maize crop and soils. *Indian Journal of Environmental Health*, **17**, 114–120.

Pihlstrom, M. (1982). Kalkstensdammets inverkan pa epifytiska lavar i Kalstrand, Sibbo (Sydfinland). *Memoranda Socias Fauna Flora Fennica*, **58**, 102–112.

Pihlstrom, M. (1987). The influence of airborne limestone dust on epiphytic lichens in southern Finland. *International Botanical Congress Abstracts*, **14**, 420.

Pocs, T. (1990). The exploration of the East African bryoflora. *Tropical Bryology*, **2**, 177–191.

Pye, K. (1987). *Aeolian Dust and Dust Deposits*. Cambridge University Press, Cambridge.

QUARG (1996). *Airborne Particulate Matter in the United Kingdom*. Third Report of the Quality of Urban Air Review Group. Department of the Environment, London.

Rao, D.N. (1971). A study of the air pollution problem due to coal unloading in Varanasi, India. In: *Proceedings of the Second International Clean Air Congress*, eds H.M. Englund and W.T. Beery. Academic Press, New York, pp. 273–276.

Ricks, G.R. and Williams, R.J.H. (1974). Effects of atmospheric pollution on deciduous woodland part 2: effects of particulate matter upon stomatal diffusion resistance in leaves of *Quercus petraea* (Mattuschka) Leibl. *Environmental Pollution*, **6**, 87–109.

Santelmann, M.V. and Gorham, E. (1988). The influence of airborne road dust on the chemistry of *Sphagnum* mosses. *Journal of Ecology*, **76**, 1219–1231.

Singh, S.N. and Rao, D.N. (1968). Effect of cement dust pollution on soil properties and on wheat plants. *Indian Journal of Environmental Health*, **20**, 258–267.

Singh, S.N. and Rao, D.N. (1981). Certain responses of wheat plants to cement dust pollution. *Environmental Pollution (Series A)*, **24**, 75–81.

Spatt, P.D. and Miller, M.C. (1981). Growth conditions and vitality of *Sphagnum* in a tundra community along the Alaska pipeline haul road. *Arctic*, **34**, 48–54.

Spitsyna, N.T. and Skripal'shchikova, L.N. (1991). Phytomass and dust accumulation of birch forests near open-pit mines. *Soviet Journal of Ecology*, **22**, 345–359.

Taylor, H.J., Ashmore, M.R. and Bell, J.N.B. (1986). *Air Pollution Injury to Vegetation*. Institution of Environmental Health Officers.

Thompson, J.R., Mueller, P.W., Fluckiger, W. and Rutter, A.J. (1984). The effect of dust on photosynthesis and its significance for roadside plants. *Environmental Pollution (Series A)*, **34**, 171–190.

Veneklaas, E. (1990). Nutrient fluxes in bulk precipitation and throughfall in two montane tropical rain forests, Columbia. *Journal of Ecology*, **78**, 974–992.

Walker, D.A. and Everett, K.R. (1987). Road dust and its environmental impact on Alaskan taiga and tundra. *Arctic and Alpine Research*, **19**, 479–489.

Wittman, H. and Turk, R. (1988). Immissionsokologischer Untersuchungen uber den epiphytischen Flechtenbewuchs in der Umbebung des Magnesitwerks in Hochfilzen (Tirol/Osterreich). *Central Gesampt Fortwesen*, **105**, 35–45.

12 Effects of increased nitrogen deposition

R. BOBBINK AND L.P.M. LAMERS

Introduction

The emissions of ammonia (NH_3) and nitrogen oxides (NO_x) have strongly increased in the second half of the last century. Nitrogen oxides originate mainly from burning of fossil fuel by industry and traffic, whereas ammonia is volatized in great quantities from intensive agricultural systems, such as dairy farming and intensive animal husbandry. Because of short- and long-range transport of these nitrogenous compounds, atmospheric nitrogen (N) deposition clearly increased in many natural and semi-natural ecosystems. Atmospheric N deposition can be 20–60 kg N ha^{-1} yr^{-1} in non-forest ecosystems in Europe and the USA, instead of the estimated background inputs of 1–3 kg N ha^{-1} yr^{-1} in the early 1900s (e.g. Galloway, 1995; Asman *et al.*, 1998; see Chapter 4).

The availability of nutrients is one of the important factors which determine the species composition of ecosystems. N is the limiting nutrient for plant growth in many natural and semi-natural ecosystems, especially of oligotrophic and mesotrophic habitats. Most of the plant species from such conditions are adapted to nutrient-poor conditions, and can only survive or compete successfully on soils with low N availability (e.g. Tamm, 1991; Aerts and Chapin, 2000). In addition, the N cycle in ecosystems is complex and strongly regulated by (micro)biological processes and it is thus likely that many changes can occur in plant growth, interspecific relationships and soil-based processes as a result of increased deposition of air-borne N pollutants.

The sequence of events which occurs when N inputs have increased in a region with originally low deposition rates, is highly complicated. Many processes interact and can operate over different time scales. The severity of the effects of airborne N deposition depends on (1) the duration and total amount of the inputs, (2) the form of the N input, (3) the intrinsic sensitivity of the (plant) species present and (4) the abiotic conditions in the ecosystem. Acid neutralizing capacity (ANC), soil nutrient availability, and soil factors which influence the nitrification potential and N immobilization rate, are especially of importance. The last two items (3 and 4) can be influenced by both past and present land use and by management. As a consequence, high variation in sensitivity to N deposition has been observed between different ecosystems. Despite this highly diverse sequence of events, the following main effect 'categories' can be recognized.

Air Pollution and Plant Life, second edition. Edited by J.N.B. Bell and M. Treshow. ISBN 0 471 49090 3 (HB), 0 471 49091 1 (PB).

(1) *Direct toxicity of nitrogen gases and aerosols to individual species.* The direct toxicity of dry and wet deposition can be an important effect on the above-ground plant parts of individual plants. These effects have been mostly studied for crops and young trees, but studies with native herbaceous or dwarf-shrub species in open-top chambers (OTC) have clearly demonstrated changes in physiology and reductions in growth at high concentrations of air-borne N pollutants (e.g. Pearson and Stewart, 1993). Direct effects of toxicity are not treated in this chapter, but in Chapter 7.

(2) *Accumulation of nitrogen compounds, resulting in changes of species composition.* Enhanced N inputs result in a gradual increase in the availability of N in the soil. This leads to an increase in plant productivity in N-limited vegetation and thus higher litter production. Because of this, N mineralization will gradually increase, which may cause enhanced plant productivity and in the longer term competitive exclusion of characteristic species by relatively fast-growing nitrophilic species. The rate of N cycling in the ecosystem is clearly enhanced in this situation. When the natural N deficiencies in an ecosystem are fully fulfilled, plant growth becomes restricted by other resources, such as phosphorus (P), and productivity will not increase further. N concentrations in the plants will, however, mostly increase, which may affect the palatability of the vegetation. Finally, the ecosystem becomes 'N-saturated', which leads to (an increased risk of) N leaching from the soil to the deeper groundwater.

(3) *Soil-mediated effects of acidification.* This long-term process, caused by inputs of acidifying compounds, may lead to loss of ANC, lower pH, increased leaching of base cations, increased concentrations of potentially toxic metals (e.g. Al), decrease in nitrification and accumulation of litter (e.g. Van Breemen *et al.*, 1982; Ulrich 1983, 1991). In this situation acid-resistant plant species will become dominant, and several species typical of intermediate pH disappear. As a consequence of the reduced nitrification rate, ammonium becomes the dominant N species in the soil, which may affect the growth of sensitive species.

(4) *Increased susceptibility to secondary stress and disturbance factors such as drought, frost, pathogens or herbivores.* The sensitivity of plants to stress or disturbance factors may be significantly affected by deposition of N pollutants. The resistance to plant pathogens can be lowered because of lower vitality of the individuals as a consequence of the different impacts of pollutants, whereas increased (organic) N contents of plants can also result in increased herbivory. Furthermore, N-related changes in plant physiology, biomass allocation (root/shoot ratios) and mycorrhizal infection can also influence the susceptibility of plant species to drought or frost.

The increased inputs of airborne N compounds can lead to severe and complex changes in the functioning and structure of ecosystems. A overview of the possible sequence of events is given in Figure 12.1. Because of the overall complexity and the difficulties in generalizing about all effects, the impacts of increased atmospheric N deposition on non-forest ecosystems will be treated by showing the consequences over a range of well-studied ecosystems, mostly of high conservation values. The following

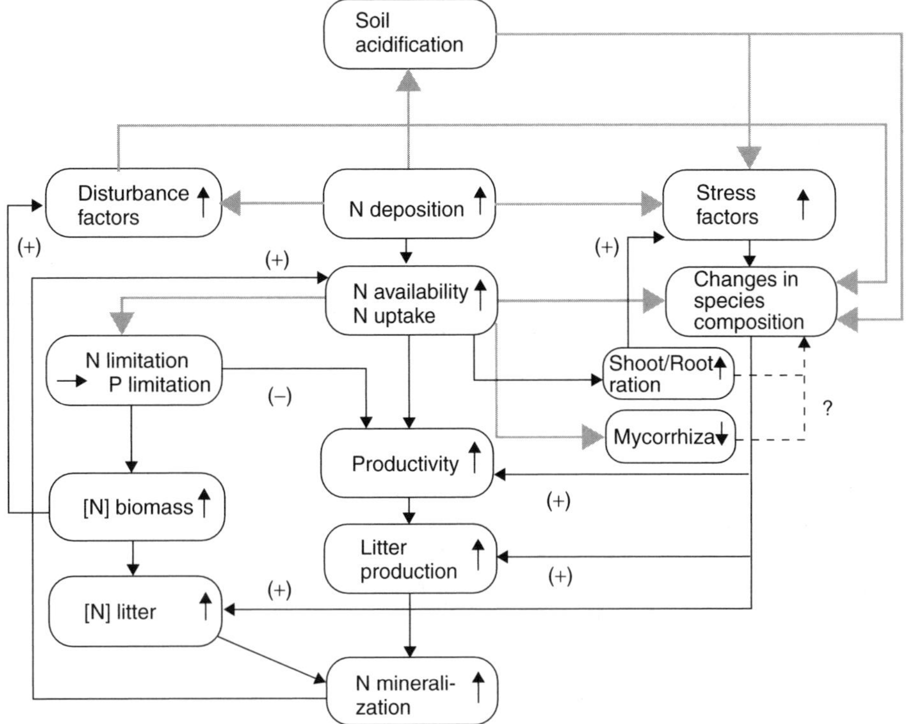

Figure 12.1 Scheme of the main impacts of increased N deposition on non-forest ecosystems. ↑ indicates increase; ↓ decrease; solid arrow: effect will occur in the short term (<5 yrs); tinted arrow: indicates long-term effect. (+): positive feedback; (−): negative feedback. Adapted and published with permission from Aerts and Bobbink (1999)

categories are discussed in detail: raised bogs, heathlands, and species-rich grasslands. In addition, the different effects of reduced or oxidized N deposition will be shown, if relevant. The effects of N deposition on tree growth and forests or woodland are not included in this chapter. For information on this topic, see Pearson and Stewart (1993) and Fangmeier *et al.* (1994). Finally, a synopsis of the most important effects and the sensitivity to N deposition is given.

Effects of N enrichment on raised bogs

Raised bogs are common in the boreal and Atlantic parts of Europe and America. They receive most of their nutrients from the atmosphere and it is to be expected that these natural ecosystems are particularly sensitive to N deposition. Raised bogs differ clearly from most other ecosystems in that they are strongly dominated by mosses, especially *Sphagnum* species (peat mosses). Dead *Sphagnum* material decomposes very slowly in these wet and acid conditions, favouring the development of thick peat layers. The peat mosses efficiently capture air-borne components, which are the principal nutrient input into these ombrotrophic ('rain-fed') peatlands (Malmer *et al.*, 1994). Few vascular plants are able to deal with this harsh environment and they are only present in low densities. Characteristic plant species include sedges (*Carex*, *Eriophorum*), heathers (*Andromeda*, *Calluna*, and *Erica*) and insectivorous plant genera (*Drosera* and *Sarracenia*). The latter species compensate for low nutrient concentrations (N) by trapping and digestion of insects (Ellenberg, 1988a). Without airborne N pollution, the input of N is very low (<5 kg N ha^{-1} yr^{-1}) and largely provided by background N deposition and N_2 fixation via cyanobacteria associated with the peat mosses and other plants (Chapman and Hemond, 1982).

When N input starts to increase in an unpolluted bog area, the following sequence of events can be generalized. N is rapidly taken up by *Sphagnum* at low levels of N deposition, because this element limits its growth in many raised bogs (e.g. Aerts *et al.*, 1992). N concentration of the peat mosses will not increase in this situation, because of the growth stimulation. At higher N inputs (ca. 12–18 kg N ha^{-1} yr^{-1}) N is no longer growth-limiting and starts to accumulate in the plant material of the mosses as free N and N-rich amino acids (e.g. Nordin and Gunnarsson 2000), as in many other species under N enrichment (Figure 12.2). The most characteristic raised bog *Sphagnum* species are replaced under these conditions by more nitrophilous *Sphagnum* species (like *Sphagnum recurvum*) and shallow rooting vascular plants. This shift in peat moss composition is clearly found in a raised bog near Kiel in Germany after two years of N additions (15 kg N ha^{-1} yr^{-1}) (Lütke Twenhöven, 1992). *Sphagnum fallax* (= *recurvum*), a species typical of hollows and the hummock-builder *S. magellanicum* responded in different ways. *S. fallax* increased significantly following the addition of N in bog hollows, but only slightly on the bog lawns and as a result it outcompeted *S. magellanicum* in the hollows and, if water supply was sufficient, also on the lawns. On the bog hummocks N addition, especially nitrate, reduced the growth of both bog mosses, probably by increased light competition with vascular plants (see later). High N inputs are clearly supraoptimal for the growth of characteristic hummock-forming *Sphagnum* species, as demonstrated by restricted development in growth experiments and transplantation studies between clean and polluted locations in the UK. In areas with high N loads, such as the Pennines, the growth of *Sphagnum* is, in general, lower than in unpolluted areas (Lee and Studholme, 1992). After transplantation of *Sphagnum* from an 'unpolluted' site to a bog in the southern Pennines, a rapid increase in N content from ca. 12 to 20 mg N g dry wt^{-1} was observed (Press *et al.*, 1986). Furthermore, a large increase in N-containing amino acids (arginine) in the shoots of these peat mosses was found after application of N, indicating a nutritional imbalance of the species (Tomassen *et al.*, 1999).

Thus, at intermediate N loads, hummock-forming *Sphagnum* species are replaced by more common peat mosses. P limits the growth of *Sphagnum* at higher N inputs, as shown by Malmer (1988) and Aerts *et al.* (1992) in field fertilization trials in Scandinavia. Still, at N loads above ca. 20 kg N ha^{-1} yr^{-1} the N concentrations in *Sphagnum* plant material do not increase any more, clearly suggesting that the extra atmospheric N load is no longer bound in the peat mosses (Figure 12.2). Because of this, it is likely that the availability of N in the soil will increase significantly above inputs of 15–20 kg N ha^{-1} yr^{-1}, thereby making the bog ecosystem sensitive to establishment

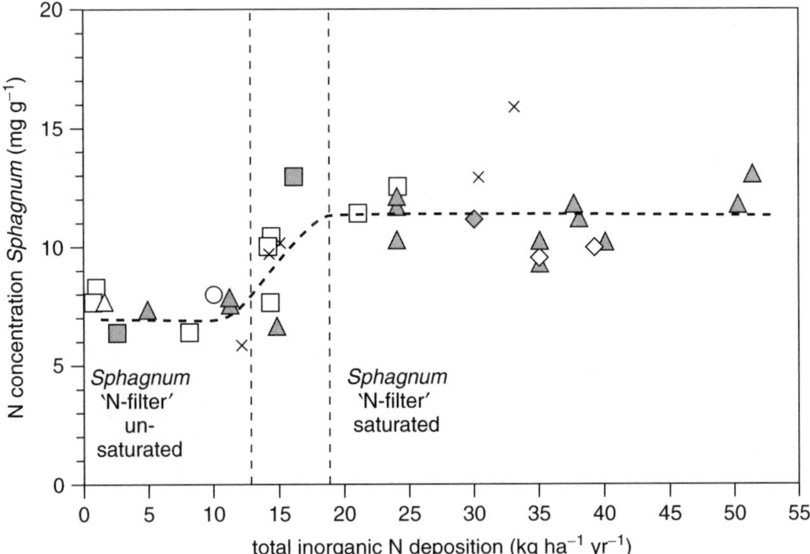

Figure 12.2 The N concentration of raised bog *Sphagnum* species (apical parts) in Europe and USA in relation to total atmospheric N inputs. Published with permission from Lamers *et al.* (2000)

of competitive vascular plants, which require high N supply, such as *Eriophorum* and *Molinia caerulea*. A comparison between low-input Irish and high-input Dutch raised bogs clearly demonstrated an increase in ammonium concentration in the top peat water from ca. 3 in Ireland to 105 μmol l^{-1} in The Netherlands. In line with this, *M. caerulea* and *Betula* strongly increased in the high-N sites, but not in the low-N sites (Lamers *et al.*, 2000). A N-induced shift in competitive strength between *Sphagnum* spp. and vascular plants has also been demonstrated in mesocosm studies (Heijmans *et al.*, 2001).

A decline of the original bog vegetation, together with an increase in more N-dependant species (e.g. *M. caerulea*, *Deschampsia flexuosa*) and trees (*Betula pubescens*) has also been observed in other areas with high ammonium deposition, above ca. 10–15 kg N ha^{-1} yr^{-1} (Aaby, 1990, 1994; Risager, 1998) (Figure 12.3). Hogg *et al.* (1995) have studied the effects of both N additions and cutting of the dominant grass species *Molinia caerulea*, on the growth of *Sphagnum* species in the ombrotrophic part of a small valley mire near York (UK). The growth of *S. palustre* and *S. fimbriatum* was reduced by 50% by the addition of 12 kg N ha^{-1} yr^{-1} during two years. Where *M. caerulea* was abundant and *Sphagnum* growth was poor, enhanced N inputs had no effect, but cutting *M. caerulea* in summer was beneficial to these bog mosses, re-invigorating their growth. This clearly correlates with the general trend in continental ombrotrophic bogs in areas with high N loads; the hollows and wet lawns have become dominated by *Sphagnum* species which do not form hummocks, and the hummocks by vascular plants such as *M. caerulea* and *Eriophorum* species.

(a)

(b)

Figure 12.3 Vegetation changes in a Danish bog (Draved Mose) from 1955 (a) via 1961 (b) to 1987 (c). In this period the atmospheric N input increased from 7 kg N ha^{-1} yr^{-1} to at least 15 kg N ha^{-1} yr^{-1} in the 1980s. In 1955 the hollows were well defined and the bog expanse was treeless. In 1961 two *Eriophorum* species had increased, but the bog was still treeless. In 1987, 26 years later, the hummock–hollow pattern was dramatically less distinctive and the bog was extensively invaded by *Betula pubescens* and *Molinia caerulea*. No changes in hydrology occurred in the observation period. Photographs kindly provided by Prof. D. Aaby and Dr M. Risager; Aaby, 1990; 1994; Risager, 1998

(c)

Figure 12.3 *(Continued)*

In addition to its effects on the development and competitive interactions of the bog-forming *Sphagnum* species, enhanced N input can also influence the competitive relationships between vascular plants in nutrient-deficient vegetation. The effects of the supply of extra N on the population ecology of *Drosera rotundifolia* has been studied in a four-year fertilization experiment in Swedish ombrotrophic bogs (Redbo-Torstensson, 1994). It was demonstrated that experimental applications above 10 kg N ha^{-1} yr^{-1} (as NH_4NO_3 at ambient deposition of 5 kg N ha^{-1} yr^{-1}) caused a significant decrease in the establishment of new individuals and the survival of the plants. This decrease in the total density of this characteristic insectivorous bog species was not caused by toxic effects of N, but by enhanced competition for light with tall species such as *Eriophorum* and *Andromeda*, which responded positively to the increased N inputs.

A further detrimental effect of the increased N content of the bog mosses could be an increased decay rate of the peat, as N content and C : N ratios strongly influence decomposition rates (e.g. Swift *et al.*, 1979). Almost all of the deposited N will accumulate in ombrotrophic bogs, because of the very low rate of denitrification and the high retention capacity of the vegetation. The effects of N addition upon moorland peats have recently been studied in the UK (Yesmin *et al.*, 1996a). The results of the first 1–1.5 years indicate that (1) nitrification is relatively unimportant in these peaty soils (see also Proctor, 1995), and (2) C : N ratios of the peat decrease significantly with higher, but realistic N loads. Several studies have suggested that decomposition in bogs may be rate-limited by N (Coulson and Butterfield, 1978; Luken and Billings, 1985), although others failed to find this relationship (Rochefort *et al.*, 1990). The comparison between a high-N and low-N site in Swedish peatlands showed that although potential *Sphagnum* decomposition rates were higher at the first location, N addition was unable to stimulate these rates for both locations (Aerts *et al.*, 2001). In another study, the decomposition of *Sphagnum* peat in Swedish ombrotrophic bogs was studied along a gradient of N deposition (Hogg *et al.*, 1994). The results of this decay experiment indicated that the short-term decomposition rate of *Sphagnum* peat was more influenced by the P content of the material than by N, although some relation with N supply was observed. The differences between the studies seem to indicate that decomposition rates are controlled through chemical changes in the living plants (e.g. concentrations of phenolic compounds) and subsequent changes in litter quality (Aerts *et al.*, 1999, 2001). Apart from the effects on the decay of *Sphagnum* peat, decomposition rates are expected to be influenced by changes in litter composition. The increased production of vascular plants mentioned above provides more easily degradable and N-enriched litter (both above- and below-ground), leading to stimulation of decomposition and mineralization (Lamers *et al.*, 2000). Further evidence is, however, necessary to clarify the long-term effects of enhanced N supply on the decay of peat.

Based upon these field studies, it has become clear that increased N loads affect raised and other ombrotrophic bog ecosystems, especially because of the high retention capacity of the mosses and the closed N cycling. The growth of several, especially hummock-building, mosses is negatively affected by increased N inputs. N availability in the rooting zone of the vascular plants therefore increases resulting in competition between the prostrate dominants (e.g. *Eriophorum*, *M. caerulea*) and the subordinate plant species and thus, in reduced typical plant diversity. An overview of the cascade of effects in raised bogs is given in Figure 12.4. Additional long-term studies with enhanced N (both NO_y and NH_x) are necessary, also to clarify the effects of decreased

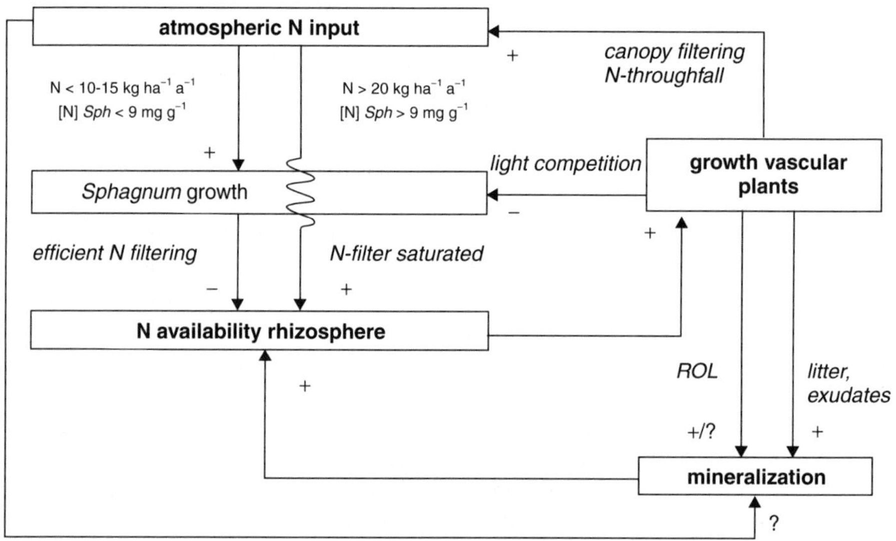

Figure 12.4 The increased N availability in the rhizosphere produces a number of feedbacks, intensifying the unfavourable changes in the functioning of raised bogs. [N] *Sph*, N concentration in *Sphagnum*; ROL, radial oxygen loss by roots. Published with permission from Lamers *et al*. (2000)

C/N ratios on the decay rate of the peat and on the invasion rate of trees on the hummocks.

Consequences of increased nitrogen inputs in heathlands

Heathlands have for a long time played a prominent role in the west European landscape. Various types of ecosystems have been described as heath, but the term is here used only when the dominant life form is that of small-leaved evergreen dwarf shrubs. Trees and tall shrubs are absent or very sparse, and the canopy is 1 m or less above soil surface. Forbs and grasses may form discontinuous strata, and frequently there is a ground layer of lichens and mosses (De Smidt, 1979; Gimingham *et al.*, 1979). Dwarf-shrub heaths occur in various parts of the world, especially in montane habitats, but are nowhere so extensive as in the Atlantic and sub-Atlantic parts of temperate Europe. Inland lowland heathlands are artificially made ('semi-natural') although they have existed for a number of centuries. In these heathlands the succession towards woodland has been prevented by mowing, burning, sheep grazing and sod removal. As in other regions, European heaths can be divided into four categories according to broad differences in habitat: (1) dry lowland heathlands; (2) wet lowland heathlands; (3) upland *Calluna* heath/moorland; and (4) alpine/arctic heathlands. For a more detailed floristic and structural description see Gimingham *et al.* (1979) and Ellenberg (1988a). In this section, the complex consequences of increased atmospheric N inputs are illustrated with the well-studied effects in dry lowland heaths and upland *C. vulgaris* heaths.

Lowland dry heaths

Lowland heaths belong to the lowland, temperate parts of Europe and are widely dominated by some Ericaceae, especially *Calluna vulgaris* under dry soil conditions and *Erica tetralix* at wet locations. These communities are found on nutrient-poor mineral soils with a low pH (3.5–4.5). Originally, the balance of nutrient input and output was in equilibrium in these lowland heaths. The original land use involved a regular, periodic removal of nutrients from the ecosystem via grazing and sod removal (e.g. Heil and Aerts, 1993). Sod removal was practised less systematically in many Scandinavian and British heathlands and *C. vulgaris* was renewed here by burning at regular intervals (Gimingham *et al.*, 1979). The traditional land use ceased in the early 1900s and the area occupied by this ecosystem decreased strongly all over its distribution area (Ellenberg, 1988a). Several lowland heath areas have become nature reserves in the last 60 years, because of their conservation value. Despite this, many lowland heaths have become dominated by grass species in western Europe. An evaluation, using aerial photographs, has for instance demonstrated that more than 35% of Dutch heaths have been altered into grassland in the 1980s (Van Kootwijk and Van der Voet, 1989). It has been suggested that the strong increase in atmospheric N deposition might be a significant factor in the observed transition to grasslands. However, N addition only clearly affected the competitive interactions between the heather and the grasses in young heaths and it has become clear that the observed changes have been caused by a complicated and interacting sequence of events at different time scales. In this section an overview is given on these strongly interrelated changes.

Increased plant productivity and N accumulation

Despite their low stature, the variable *C. vulgaris* canopies have a strong filtering effect on air pollutants. Bulk deposition accounted for only ca. 35–40% of total atmospheric input N and total N deposition was between 30–45 kg N ha^{-1} yr^{-1} in heathlands in the eastern part of The Netherlands in the late 1980s. Clearly more than 70% of the total N input was deposited as ammonium or ammonia (Bobbink *et al.*, 1992; Bobbink and Heil, 1993). It is thus likely that N loads have also gradually increased in many other heathland sites in European agricultural regions in the last decades of the twentieth century (Buijsman *et al.*, 1987; Sutton *et al.*, 1993; see Chapter 4).

As in many terrestrial ecosystems, it is to be expected that increased N inputs will cause higher biomass production of the vegetation. Many studies indeed showed increased plant productivity of the dwarf shrubs after experimental N enrichment in heathlands in several NW European countries (e.g. Heil and Diemont, 1983; Van der Eerden *et al.*, 1991; Aerts and Heil, 1993; Power *et al.*, 1995; Cawley *et al.*, 1998). This clearly indicates that most of these lowland dry heath ecosystems are primarily limited by N, although some inland dry heaths in Denmark seem to be limited by P or K (Riis-Nielsen, 2001). An illustrative example of the growth stimulation of *C. vulgaris* was found in a field experiment in the UK. It was set up in 1989 to assess the long-term impacts of realistic N loads on a lowland dry heathland in southern Britain (Uren, 1992; Uren *et al.*, 1997; Power *et al.*, 1995, 1998a). After 7 years, application of ammonium sulphate (7.7 and 15.4 kg N ha^{-1} yr^{-1}, ambient load 10–15 kg N ha^{-1} yr^{-1}) has not resulted in any negative effects upon *C. vulgaris*. Indeed, a significant stimulation of flower production, shoot density, and litter production occurred and the canopy in the highest N treatment was 50% taller than in the control plots after 6 years. The increased shoot growth in the N-treated vegetation is not reflected in root growth, resulting in an increased shoot to root ratio. N shoot concentrations increased somewhat, too, but only significantly at the start of the experiment (years 1 and 3). The significantly enhanced litter production and the observed stimulation of decomposer activity could have long-term effects on the N cycling and accumulation in the system.

It is already well known that the amount of organic material and N increases during secondary heathland succession (Chapman *et al.*, 1975; Gimingham *et al.*, 1979). The accumulation of organic matter and of N was quantified after sod removal in dry heaths in The Netherlands by Berendse (1990). He found a large increase in plant biomass, soil organic matter and total N storage in the first 20–30 years of succession. N mineralization was low in the first 10 years (ca. 10 kg N ha^{-1} yr^{-1}), but strongly increased in the next 20 years to 50–110 kg N ha^{-1} yr^{-1} (Figure 12.5). Regression analysis suggested an annual increase of N in the system of ca. 33 kg N ha^{-1} yr^{-1}. These values are in good agreement with the measured N inputs in Dutch heathlands (Bobbink *et al.*, 1992). Organic matter in the soil increased rapidly during secondary succession after sod removal, which removed almost all of the soil surface organic matter. This process is likely to be accelerated by the enhanced biomass and litter production of the dwarf-shrubs caused by the extra N inputs. N accumulation in the

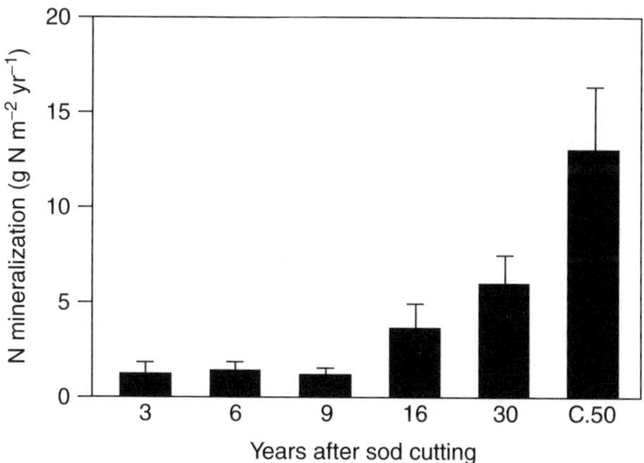

Figure 12.5 Net annual N mineralization (g m^{-2} yr^{-1} ± sd) in heathlands plots differing in age (years) after sod cutting in a Dutch heathland (Stabrecht); published with permission from Berendse (1990)

system increased too, and hardly any N was lost because there was high ammonium immobilization in the soils and hardly any N leaching to deeper layers in Dutch, British or Danish heathlands (De Boer, 1989; Van Der Maas, 1990; Power *et al.*, 1998a; Kristensen and McCarty, 1999)). This phenomenon is also obvious in British upland *C. vulgaris* heath (see next section).

It is thus clear that the productivity of *C. vulgaris* increased in most lowland heaths with N enrichment and that these heaths strongly retain and accumulate N in the system. Gradually, N mineralization will be enhanced, and the availability of N will thus become higher. Because of this, the risk of increased growth or invasion of nitrophilous species (e.g. grasses) is certainly enhanced.

Changes in competitive interactions

Competition experiments in containers and in the field have clearly demonstrated an important effect of increased N availability on the competitive interactions between *C. vulgaris* and grasses in the early phase of secondary succession in dry lowland heath. After experimental N additions grasses strongly outcompeted heather, if the total cover of the vegetation was still low (Heil and Diemont, 1983; Roelofs, 1986; Heil and Bruggink, 1987; Aerts *et al.*, 1990). However, *C. vulgaris* clearly is a better competitor in mature heath vegetation at (very) high N loads than the grasses if its canopy remains closed (Aerts *et al.*, 1990; Aerts, 1993). It is thus evident that increased concentrations of N in the soil are not enough to cause the shift from dwarf-shrub to grass dominance in these dry lowland heaths. To alter the competitive interactions between *C. vulgaris* and the grasses, opening of the closed *C. vulgaris* canopy is a prerequisite, otherwise the grasses cannot profit from the increased availability of N due to light deprivation. Heather beetle outbreaks and N-induced secondary stresses (especially winter injury and drought) are probably the main processes which can open the canopy, and are thus crucial for the dramatic shift in species composition observed in lowland heath in high N-load regions.

Heather beetle outbreaks

Outbreaks of heather beetle (*Lochmaea suturalis)*, a chrysomelid beetle, can occur frequently in dry lowland heaths. It forages exclusively on the green parts of *C. vulgaris*. Outbreaks of the beetle can lead to the opening of closed *C. vulgaris* canopy over large areas, greatly reducing light interception (Berdowski, 1987, 1993) and leading to enhanced growth of understorey grasses, such as *Deschampsia flexuosa* or *Molinia caerulea* (Figure 12.6). Insect herbivory is generally affected by the nutritive value of the plant material, with N concentration especially important (e.g. Crawley, 1983). Correlative field studies in low and high N input areas and experimental applications of N to heathlands have shown increased concentrations of this element in the (green) parts of *C. vulgaris* at high N fluxes (e.g. Heil and Bruggink, 1987; Bobbink and Heil, 1993; Pitcairn *et al.*, 1995). Brunsting and Heil (1985) found that the growth of the larvae increased after foraging on the leaves of *C. vulgaris* with higher N concentrations in a rearing experiment. After field additions of ammonium sulphate in a roofed heath experiment, total number or biomass of the first-stage larvae of the beetle was not affected by the treatments, but the development of subsequent larval stages was significantly accelerated (Van der Eerden *et al.*, 1990). Larval growth rates and adult weight of heather beetles were found to be significantly higher when these insects were reared on *C. vulgaris* plants which were collected in the British lowland heath experiment, after 7 years of relatively low N addition (Power *et al.*, 1998b).

Heather beetle larvae were also cultivated on shoots of *C. vulgaris* taken from plants which had been fumigated with ammonia in open-top chambers (12 months; 4–105 $\mu g\ m^{-3}$) (Van der Eerden *et al.*, 1991). After 7 days, both the mass and development rate of the larvae were clearly increased at higher ammonia concentrations. In addition, the growth of heather beetles instars was also significantly stimulated on *C. vulgaris* from UK heaths after ammonia fumigation, with relatively high concentrations, probably caused by the enhanced N concentrations in the plant material (Uren, 1992). Also for larvae of winter moth (*Operophtera brumata*) it has been demonstrated that increased N fluxes on *C. vulgaris* stands stimulated larval development and growth rate (Kerslake *et al.*, 1998). It is thus likely that the frequency and intensity of insect outbreaks can be stimulated by increased atmospheric N loads. This is supported by the observations of Blankwaardt (1977), who reported that from 1915 onwards heather beetle outbreaks occurred at ca. 20-year intervals in The Netherlands, whereas in the last 15 years of the observation period, the interval has been less than 8 years. In addition, it was observed that *C. vulgaris* plants were more severely damaged in N-fertilized vegetation during a heather beetle outbreak, both in The Netherlands (Heil and Diemont, 1983) and in Denmark (Tybirk *et al.*, 1995; Riis-Nielsen, 2001). It is thus likely that N enrichment influences the frequency and severity of beetle outbreaks, although the controlling processes need further research.

(a)

(b)

Figure 12.6 The heathland area Oud Reemsterveld (The Netherlands) in 1979 (dominated by *Calluna vulgaris*) just before a heather beetle outbreak (a) and 3 years later (1982) (b) completely dominated by *Deschampsia flexuosa* (photographs kindly made available by G.W. Heil)

Secondary stresses

It has been shown that frost sensitivity increased in some tree species with increasing concentrations of air pollutants (e.g. Aronsson, 1980; Dueck *et al.*, 1991). This increased susceptibility is sometimes correlated with enhanced N concentrations in the leaves or needles. Impacts of N deposition on the frost sensitivity of *C. vulgaris* could be possible, because of its evergreen growth form and the observed increased foliar N contents, associated with higher N loads (e.g. Heil and Bruggink, 1987; Pitcairn *et al.*, 1995). Furthermore, it is suggested that the observed die-back of the *C. vulgaris* shoots in the successive severe winters of the mid-1980s in The Netherlands was, at least partly, caused by increased winter injury.

Van der Eerden *et al.* (1990) studied the effects of ammonium sulphate and ammonia upon frost sensitivity in *C. vulgaris.* Fumigation with ammonia of *C. vulgaris* plants in open-top chambers over 4–7-month periods (100 $\mu g\ m^{-3}$) revealed that frost sensitivity was not affected in autumn (September or November), but in February, just before growth started, frost injury increased significantly at $-12\,^{\circ}C$ (Van der Eerden *et al.*, 1991). A study in greenhouse chambers during the winter period in England also demonstrated increased frost sensitivity of heather plants, but the applied concentrations of ammonia were very high (140–280 $\mu g\ m^{-3}$) (Uren, 1992). Van der Eerden *et al.* (1991) studied the frost sensitivity of *C. vulgaris* vegetation artificially sprayed with different levels of ammonium sulphate (3–91 kg N $ha^{-1}\ yr^{-1}$). After 5 months the frost sensitivity of *C. vulgaris* increased slightly, although significantly, compared with the control in vegetation treated with the highest level of ammonium sulphate (400 $\mu mol\ l^{-1}$; 91 kg N $ha^{-1}\ yr^{-1}$). The sensitivity decreased, however, again 2 months later and no significant effects of the ammonium sulphate application upon frost hardiness were measured at that time. The effects of 7 years of low levels of N addition (7.7 and 15.4 kg N $ha^{-1}\ yr^{-1}$) on frost sensitivity of *C. vulgaris* were, however, hardly present (Power *et al.*, 1998b).

In conclusion, although (very) high levels of ammonia or ammonium sulphate seem to increase the frost sensitivity of *C. vulgaris* plants, the significance of this phenomenon at ambient N loads is very uncertain and application of ammonium nitrate may even reduce frost sensitivity (see next section). Currently, the relationship between frost sensitivity and N inputs is insufficiently quantified to be a main component in triggering the opening of the lowland dry heath canopy under high N deposition.

Summer ‘browning’ of the *C. vulgaris* canopies was frequently seen in dry summers in the 1980s in The Netherlands – the decade with the highest N loads. This suggested that N enrichment stimulated the sensitivity of *C. vulgaris* to periods of drought, probably by reduced root growth with respect to the development of the shoot or by a decrease in mycorrhizal infection. These effects might have major implications for the capacity of plants to deal with water (or nutrient) stress. The partitioning of biomass is very plastic and mostly determined by nutrient availability and light intensity (e.g. Brouwer, 1962, 1983). It has been shown that most plant species studied allocated more biomass to the shoots than to the roots at higher nutrient concentrations (e.g. Poorter and Nagel, 2000). This also held for *C. vulgaris* and other heathland species in a pot experiment over 2 years: root weight ratio (RWR) significantly decreased with increasing N additions (Aerts *et al.*, 1991). However, field validation of these lower root to shoot ratios as a response to N enrichment is, surprisingly, scarce. In hardly any of the numerous field experiments in dry lowland heaths, was this effect studied or found. An indication of its importance was observed after long-term N applications in a British dry lowland heathland (Power *et al.*, 1998a). They found a small reduction in root to shoot ratio after 7 years of N addition (15.4 kg N ha^{-1} yr^{-1}), compared with the untreated *C. vulgaris* vegetation. These data demonstrated that enhanced N inputs might lead to a decreased root to shoot ratio, which implies a higher transpiring surface compared to the water uptake surface. At ample water supply, this will not cause any problem to the *C. vulgaris* plants. They also found higher water losses from *C. vulgaris* plants from the N-treated vegetation, compared with the control situation, but no differences in water potential of the shoots (Power *et al.*, 1998b). However, it might lead to severe growth reduction or even local 'browning' and die-back of this species in cases of severe drought episodes, but this is still mostly speculative. Recently, experimental summer drought on N-enriched lowland heath plots significantly increased the vulnerability of *C. vulgaris* to this implied stress (Cawley *et al.*, 1998).

Besides the changes in root to shoot ratios, ericoid mycorrhizal infection of the roots of the heathers could also be influenced by an increase in N load. Until now, the outcome of measurements on this has scarcely been studied and the outcomes are highly variable, as for AM mycorrhizal infection, too (Aerts and Bobbink, 1999). Some studies on the effects of increased N availability on ericoid mycorrhizal infection of *C. vulgaris* root showed no effects (Johansson, 2000), in others N load stimulated the infection of mycorrhizas (Caporn *et al.*, 1995), whereas other studies showed restricted infection after N treatment (Yesmin *et al.*, 1996b). At this moment the importance of this phenomenon in the decline of *C. vulgaris* and the shift to grass dominance is not at all clear.

It is obvious that the sensitivity of *C. vulgaris* to drought stress might be increased by the shift in root to shoot ratio, but the importance of this process has to be clarified in field conditions under long-term N applications.

Conclusion

The impacts of increased N inputs to dry inland heaths are complex and occur at different time scales. Firstly, increased N availability stimulates biomass and litter production of *C. vulgaris* in most situations. N is strongly retained in the system, gradually leading to higher N mineralization rates in the soil. However, the species remains the stronger competitor with respect to grasses, even at very high N availability, if the canopy is not opened. The shift from dwarf shrub to grass dominance is clearly triggered by opening of the canopy caused by heather beetle attacks, winter injury or drought. After die-back of the *C. vulgaris* shoots, grasses quickly profit from the increased light intensity, together with the high N availability because of N accumulation. Within a few years, this may (locally) lead to a drastic increase in grass cover. Because of the stochastic behaviour of several processes (e.g. heather beetle outbreaks, winter injury) and the many long-term processes which interact with them, it is very difficult to clarify experimentally these relationships without long-term (10–20 years) and large-scale experiments.

Upland C. *vulgaris* heaths

It has been suggested that the *C. vulgaris* heaths present in the upland areas of Britain and other mountainous parts of Europe are also likely to be sensitive to N deposition (Hornung *et al.*, 1995). These communities are characterized by a dominance of dwarf shrubs (*C. vulgaris*), a high abundance of bryophyte species and acidic, peaty soils. The abiotic conditions are colder and wetter than in lowland heathlands. Current N deposition in the UK is thought to be partly responsible for the decline of certain mosses and lichen species in these heathlands in recent decades (Thompson and Baddeley, 1991; Woodin and Farmer, 1993; Baddeley *et al.*, 1994).

The effects of N inputs on upland *C. vulgaris* heaths have been recently studied experimentally and in field surveys in the UK. In many of the studied sites the N contents of *C. vulgaris* and of bryophytes was significantly higher in areas with high N inputs. N contents in current heather and moss samples were also higher than in historical plant material (Pitcairn *et al.*, 1995). Altitudinal transect studies in upland *C. vulgaris* heaths have shown that shoot N concentrations of upland vegetation increase with altitude, whilst P concentrations show no definite trends. This increase can be linked to enhanced N deposition at high altitude due to the orographic enhancement of deposition (Metcalfe *et al.*, 1999). Reciprocal monolith transplants of *C. vulgaris* and *Nardus stricta* between polluted and unpolluted hills have, furthermore, shown that the patterns of N deposition are reflected in the plant shoots (Hicks *et al.*, 1995). Results from experimental application of N mists (ammonium nitrate; 3–60 kg N ha^{-1} yr^{-1}) to a number of upland plants over 1–2 years caused increased growth of above- and below-ground plant parts, especially of *N. stricta* and *Vaccinium vitis-idaea* (Leith *et al.*, 1999).

The impacts of ammonium nitrate applications (0–40–80–120 kg N ha^{-1} yr^{-1}; ambient load ca. 20 kg N ha^{-1} yr^{-1}) upon upland *C. vulgaris* heath been examined in North Wales (UK) from summer 1989 up to the present (Caporn *et al.*, 1994, 1995; Lee and Caporn, 1998; Carroll *et al.*, 1999). In the first 3 years *C. vulgaris* benefited from N inputs; shoot growth, flowering, and litter production were significantly stimulated even at low N addition, this being very comparable with the consequences of N inputs in lowland heath. Frost sensitivity of *C. vulgaris* decreased in most years with increasing N additions and a small increase in N concentrations of the shoots and litter was observed. Mycorrhizal infection of *C. vulgaris* was not reduced after 3 years. Within the first experimental period no significant changes in species composition were found (Caporn *et al.*, 1994, 1995). In the subsequent period, large N accumulation was measured in the canopy and in the litter of the N-treated vegetation, and soil inorganic N concentrations and soil microbial biomass also increased. However, even after 10 years of the highest N application (140 kg N ha^{-1} yr^{-1}) N leaching from the soil was still extremely low (J.A. Lee, pers. com.). Mosses and lichens disappeared from below the *C. vulgaris* canopy in all N treatments, probably because of changes in canopy architecture (light deprivation) and enhanced litter accumulation. In the winter of 1993/94 acute 'winter browning' of *C. vulgaris* shoots was observed, most notably in the two highest N treatments. There was, in addition, a significant increase in the proportion of winter injured shoots in all N treatments in later years, as shown for spring 1996 in Figure 12.7. The observed damage resulted, most likely, from desiccation, rather

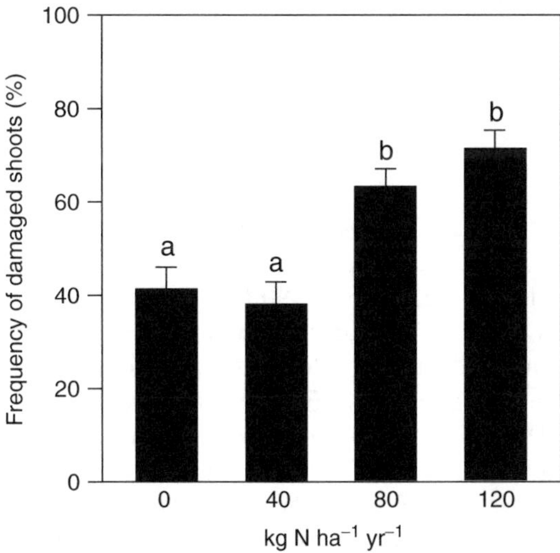

Figure 12.7 Effects of 7 years of ammonium nitrate additions (in kg N ha^{-1} yr^{-1}) on the frequency (means $\pm$ SE) of winter damaged *C. vulgaris* shoots in upland *C. vulgaris* heath in N. Wales (published with permission from Carroll *et al.*, 1999). Columns sharing a letter are not significantly different

than frost injury (Lee and Caporn, 1998; Carroll *et al.*, 1999). Clear adverse effects of increased N inputs have become obvious in these upland *C. vulgaris* heaths, especially because of the length of the latter study, which is still running after 12 years. This type of long-term study, which has often been underestimated or even neglected, is very important when interpreting the effects of increased N loads.

Most of these British studies in upland heaths indicate effects on ecosystem function at rather low N loads, together with changes in growth and physiology of the dominant plants. Changes in vascular plant diversity have at present not been found, but it has been shown that lichen and bryophyte species are very sensitive.

Effects of atmospheric nitrogen inputs on species-rich grasslands

Semi-natural grasslands with traditional agricultural use have been an important part of the landscape in Europe. These ecosystems contain many rare and endangered plant and animal species and a number of them have been set aside as nature reserves because of their conservation importance (e.g. Ellenberg, 1988a; Woodin and Farmer, 1993). They are generally nutrient poor, because of long agricultural use with low inputs of manure combined with nutrient removal by grazing or hay making. They are characterized by many species of low stature and by nutrient-poor soil status (Ellenberg Jr., 1988b). To maintain high species diversity, fertilization has to be avoided and it is thus likely that these species-rich grasslands will be affected by increased N load. Numerous types of semi-natural grasslands occur around Europe (e.g. Ellenberg, 1988a). In this section we will only treat the effects of N enrichment in two well-studied grassland types, i.e. dry well-buffered calcareous grasslands and acidic, weakly buffered grasslands on poor soils. These systems may serve as an indication for the (possible) effects of N inputs in other grassland types.

Calcareous grasslands

Calcareous grasslands (FESTUCO-BROMETEA) are present on well-buffered shallow soils ($pH_{topsoil}$: 7–8; $CaCO_3$ ca. 10%), often rendzinas, low in P and N with subsoils of different kinds of limestone ($CaCO_3$ >90%). A considerable part of the European calcareous grasslands are MESOBROMION communities; a temperate grassland type (Willems, 1982). Plant productivity is low but calcareous grasslands are among the most species-rich plant communities in Europe and contain a large number of rare and endangered species. They have decreased markedly in area during the second half of the last century (e.g. Wolkinger and Plank, 1981; Ratcliffe, 1984) and the remnants have became nature reserves in several European countries. Specific management is needed to maintain the characteristic calcareous vegetation and prevent natural succession towards woodland (e.g. Wells, 1974; Dierschke and Engels, 1991).

A gradual increase of one grass species (*Brachypodium pinnatum*) was observed on many sites in Dutch calcareous grassland nature reserves in the late 1970s/early 1980s, although the management (hay making in autumn) had not changed since the mid-1950s. It was suggested that increased atmospheric N inputs (from 10–15 kg N in the 1950s to 30–40 kg N ha^{-1} yr^{-1} in the 1980s) caused this drastic change in vegetation composition of these calcareous grasslands on Cretaceous limestone (pH 7–8) (Bobbink and Willems, 1987). The effects of N enrichment have, therefore, been investigated in two field experiments in The Netherlands (Bobbink *et al.*, 1988; Bobbink, 1991). Application of ammonium nitrate (50–100 kg N ha^{-1} yr^{-1}) resulted in a drastic increase of *B. pinnatum* and a strong reduction in species diversity (including several Dutch Red List species), caused by the change in vertical structure of the grassland vegetation (Figure 12.8). In addition to vascular plants, many characteristic lichens and mosses have also disappeared from calcareous grasslands in The Netherlands between 1950 and 1985 (During and Willems, 1986). Experiments by Van Tooren *et al.* (1990) have shown this to be partly caused by the (indirect) effects of extra N input. The effects of N supply on the massive expansion of *B. pinnatum* and the reduction of plant species number have also been observed in a 5-year field study using a factorial design (Willems *et al.*, 1993).

B. pinnatum has proved to have a very efficient nitrogen acquisition and subsequent reallocation of the nutrient from its senescent shoots into its extensive rhizome system. The extra N is redistributed below ground and leads to increased growth in the following years, despite very high N:P ratios (30–35; Bobbink, 1991). As a result, the fast-growing species progressively monopolized (>75%) the nitrogen storage in both the above-ground and below-ground compartments of the vegetation as N availability increased (Bobbink *et al.*, 1989; De Kroon and Bobbink, 1997). It is striking that *B. pinnatum* can increase its biomass despite the well-known low P concentrations in the calcareous grassland soil. This can be explained by its capacity to form plant material very low in P, to recycle P between the rhizomes and the shoot, and its ability

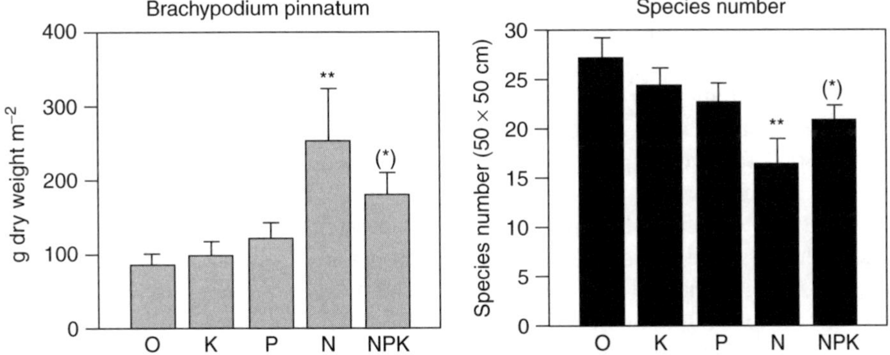

Figure 12.8 Above-ground biomass of *Brachypodium pinnatum* (g m^{-2}) and phanerogamic species number (per 50 × 50 cm) after three years of N addition (as ammonium nitrate; 100 kg N ha^{-1} yr^{-1}) in Dutch calcareous grasslands (adapted with permission from Bobbink, 1991)

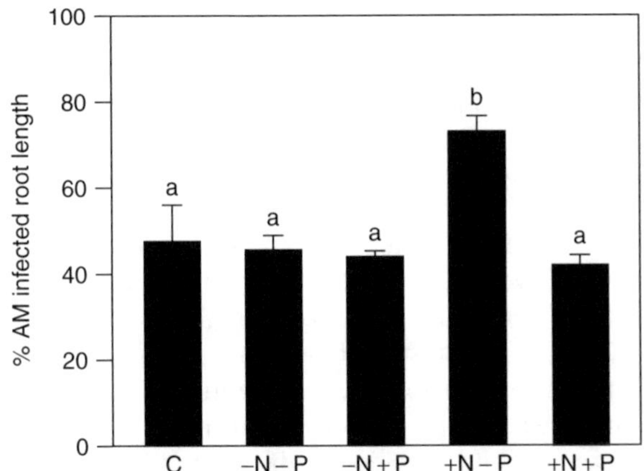

Figure 12.9 AM infection (% of root length; mean ± SE) in *B. pinnatum* after 7 years of differential nutrient addition at a Dutch calcareous grassland. The complete fertilization treatment (+**N** + **P**) added all potentially limiting elements including N, P, K, Mg, Zn, Cu, Mn, Mo and B. The nutrients were annually applied in April and in September; complete fertilization, except for N and P (−**N** − **P**); complete fertilization minus N (−**N** + **P**); complete fertilization minus N (+**N** − **P**); unfertilized plots, which received only water (**C**). The +N − P treatment was significantly different from all other treatments. Based upon unpublished results of Bobbink *et al.*

to acquire extra P from the soil when N inputs are high. In this respect, it is remarkable that arbuscular mycorrhizal (AM) infection of the roots of this grass species strongly increased after 7 years of N addition on a Dutch calcareous grassland site (Figure 12.9).

Nitrogen cycling and accumulation in calcareous grassland can be significantly influenced by two major outputs: (i) leaching from the soil, and (ii) removal with management regimes. N losses by denitrification in dry calcareous grasslands are negligible (<1 kg N ha^{-1} yr^{-1}) (e.g. Mosier *et al.*, 1981). N leaching to deeper soil layers appeared to be very low (0.7 kg N ha^{-1} yr^{-1}) in untreated Dutch calcareous grassland, equalling only 2% of the total N deposition measured (Van Dam *et al.*, 1989; Van Dam, 1990). After two-weekly spraying of ammonium sulphate (50 kg N ha^{-1} yr^{-1}) for two years, nitrate leaching increased to 3.5 kg N ha^{-1} yr^{-1}, but was still only a small proportion (4%) of the total N load (Van Dam *et al.*, 1989). It is thus evident that this calcareous grassland ecosystem retained almost all N in the system, due to a combination of enhanced plant uptake (Bobbink, 1991) and increased immobilization in the soil (Van Dam, 1990). The most important output of N from the grasslands is by exploitation or management. From the 1950s onwards almost all of the calcareous grasslands in The Netherlands were mown in autumn with removal of the hay. The annual nitrogen removal through hay varies slightly between years and sites, but is in general between 15–20 kg N ha^{-1} (Bobbink, 1991; Bobbink and Willems, 1991). Legume species (*Leguminosae*), also occurring in calcareous vegetation, form an additional N input for the vegetation through N fixation (ca. 5 kg N ha^{-1} yr^{-1}). Using a steady state N mass balance model (De Vries, 1994) it has become clear that all input above 15–20 kg N ha^{-1} yr^{-1} will be strongly accumulated in this well-buffered calcareous grassland system.

Outside The Netherlands, the effects of enhanced N inputs on calcareous grassland have hardly been investigated in continental Europe. Neitzke (2001) investigated a calcareous grassland zone adjacent to fertilized agricultural lands in the Eifel (Germany). She found a strong dominance of *B. pinnatum* and some other meadow grasses in these borders. This observation could be explained by an increased net N mineralization of ca. 30–35 kg N ha^{-1} yr^{-1}, a clear indication that these calcareous grasslands are also sensitive to increased N loads. However, data from a fertilization experiment in the 'alvar' grassland, a thin-soiled vegetation over flat limestone, on the Swedish island of Öland, suggest that this vegetation hardly responded to application of N or P (Sykes and Van der Maarel, 1991). Only irrigation in combination with these nutrients caused an increase in production and in the grass cover. This is probably explained by the low annual precipitation in this area (400–500 mm).

In calcareous grasslands in Britain, addition of 50–100 kg N ha^{-1} yr^{-1} in combination with P sometimes stimulated dominance of grasses (Smith *et al.*, 1971; Jeffrey and Pigott, 1973), particularly of *Festuca rubra*, *F. ovina* or *Agrostis stolonifera*. However, *B. pinnatum* or *Bromus erectus*, the most frequent species in continental calcareous grassland, were absent from these sites. Following a survey of data from a number of conservation sites in southern England, Pitcairn *et al.* (1991) concluded that *B. pinnatum* had expanded in the UK during the twentieth century. They considered that much of the early spread could be attributed to a decline in grazing pressure but that more recent increases had, in some cases, taken place despite grazing or mowing, and might be related to increased N input. This was partly confirmed by Hewins and Ling (1998), who observed lower plant diversity and more aggressive grass species in calcareous grasslands in parts of the Cotswolds (UK), receiving higher ammonia concentrations. From ^{15}N measurements and N additions, Unkovich *et al.* (1998) concluded that low N mineralization was limiting plant growth in the British calcareous grassland studied, and that the soil acts as a strong N sink. Application of N clearly increased plant production in this unimproved calcareous grassland on Jurassic limestone. The results of pot or mesocosm experiments are, however, ambiguous. Treatment with N (20–40–80 kg N ha^{-1} yr^{-1}) for 2 years to eight forbs and one grass (*B. pinnatum*), planted together in pots, did not result in *B. pinnatum* becoming dominant (Wilson *et al.*, 1995). In another study with chalk grassland mesocoms (SE England), however, the biomass of *B. pinnatum* significantly increased after 2 years' N addition (13–140 kg N ha^{-1} yr^{-1}), even under a 8-weekly summer defoliation regime (Bryant and Taylor, 1998). It is thus to be expected that plant productivity and expansion of aggressive grasses in British calcareous grasslands is in several situations not limited by N, but probably by P.

This is supported by long-time field studies in a calcareous grassland on carboniferous limestone (pH = 6.5) in Derbyshire in the UK (Morecroft *et al.*, 1994; Carroll *et al.*, 1997; Lee and Caporn, 1998). The effects of N additions (30–70–140 kg N ha^{-1} yr^{-1} as ammonium nitrate and 140 kg N ha^{-1} yr^{-1} as ammonium sulphate) have been studied since 1989 up to the present. In the first 3-year period no changes resulting from differential growth stimulation in different species were found, but plant N concentrations, nitrate reductase activities and soil nitrogen mineralization rates clearly increased with enhanced inputs of N, even at the lowest application level. Morecroft *et al.* (1994) concluded that plant growth appeared to be limited by P, but functional processes of the nitrogen cycling were certainly affected by extra N inputs. In 1996, after 6 full years of treatment, there was no evidence at all that any of the N additions caused an increase in plant cover. On the contrary, a significant reduction in the cover of the dicotyledons and sedges (including typical plant species) was found, especially in the ammonium sulphate treatment. This can be exemplified by the development of *Thymus praecox* in the N treated vegetation. Its cover was significantly reduced in the highest N treatments in the years 1994 to 96. In addition, bryophyte cover was clearly reduced by the N treatments. It is evident that soil acidification, as a consequence of N addition, plays a major role in the observed reduction of some species. Soil pH decreased to around 6.0–6.2 in the vegetation treated with ammonium nitrate (>35 kg N ha^{-1} yr^{-1}), compared with 6.7 in the control vegetation. After 6 years of ammonium sulphate treatment it declined even to ca. 5.1, probably due to the full nitrification of the ammonium applied, leading to extra proton production (Figure 12.10). Soil N mineralization rates were still higher in the N-treated vegetation in 1996, probably caused by an increased activity of the micro-organisms or an increase in their population sizes (Carroll *et al.*, 1997; Lee and Caporn, 1998). Recent soil-microbial studies demonstrated that 7 years of N addition significantly stimulated root surface phosphomonoesterase activity in this calcareous grassland, although this was much less clear than in an upland *C. vulgaris* heath (Johnson *et al.*, 1999). It indicates that the cycling of nutrients other than N, such as P, can have an important role in ecosystems that have received long-term chronic N inputs.

In conclusion, strong differences between Dutch/German and British studies may exist, because the impacts of increased N loads can be influenced by P limitation. However, increased N availability from atmospheric deposition is probably of major importance in a number of N-limited European calcareous grasslands. Its effects are

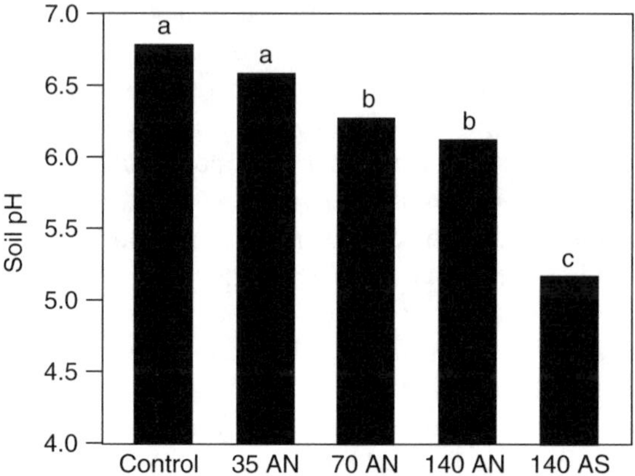

Figure 12.10 Top soil $pH_{(H2O)}$ (0–5 cm) values in a British calcareous grassland which received 3.5–14 g N m_{-2} yr_{-1} as ammonium nitrate (AN) or 14 g N as ammonium sulphate (AS) during 7 years; redrawn with permission from Carroll *et al.* (1997). Groups sharing a letter are not significantly different

indicated by enhanced growth of some 'tall' grasses, especially of stress-tolerant competitors (Grime, 1979) which have a slightly higher potential growth rate and show efficient nitrogen utilization. In general, typical calcareous plant species of low stature will decline in this situation. Nitrogen retention in this system is high with hardly any leaching, and N mineralization can be enhanced by N inputs. In calcareous grasslands showing P limitation, plant productivity and species composition will not respond to extra N load. However, N cycling, as indicated by N mineralization and soil microbial processes, increases considerably in these P-limited ecosystems. This may strongly enhance the risk of nitrate leaching to the groundwater. In addition, the consequences of soil acidification, especially by ammonium inputs, can still have major effects on the species composition of calcareous grasslands relatively low in soil $CaCO_3$ (see also next section). More long-term (ca. 10–20 years) and comparative studies are needed to verify conclusions on the impacts of excess N inputs over a range of calcareous grassland ecosystems.

Acidic grasslands

As described before, a remarkable transition from dwarf-shrub to grass-dominated vegetation has been observed in lowland heaths in recent decades. Species diversity of (slightly) acidic grasslands (NARDETEA) seems also to be affected and many characteristic herbaceous species of these acidic grasslands (pH 4.5–6.0) (e.g. *Arnica montana*, *Antennaria dioica*, *Pedicularis sylvatica*, *Polygala serpyllifolia*, *Succisa pratensis* and *Thymus serpyllum)* have declined or become locally extinct in The Netherlands. The distribution of these species is related to small-scale, spatial variability of the soils in the heathland landscape on Pleistocene sandy deposits. It has been suggested that acidifying deposition has caused such drastic abiotic changes that these species can no longer survive (Van Dam *et al.*, 1986). Grasses or ericoid species now dominate species-poor stands on formerly biodiverse sites. Enhanced N fluxes to nutrient-poor heathland soils lead to increased N availability and, because most of the deposited N in north-west Europe originates from ammonia/ammonium deposition, it may also cause soil acidification. Whether eutrophication or acidification, or a combination of both processes, is important depends on soil pH, ANC and nitrification rate. Roelofs *et al.* (1985) found that in dwarf-shrub dominated heathland soils nitrification was inhibited at pH 4.0–4.2, and ammonium accumulated while nitrate concentrations decreased to almost zero at these or lower pH values. In contrast, nitrification rates were much higher in soils supporting endangered species, due to the slightly higher pH and ANC. In soils within the pH range of 4.5–6.0, extra acidity is buffered by cation exchange processes (Ulrich, 1983, 1991). If, however, base saturation (mainly calcium) has been depleted, pH falls and this may cause the decline of species generally associated with somewhat higher soil pH values. Furthermore, any pH decrease may indirectly result in increased leaching of base cations, increased aluminium mobilization and thus enhanced Al/Ca ratios of the soil (Van Breemen *et al.*, 1982; Ulrich, 1991). In addition, the decrease of the soil pH may inhibit nitrification, resulting in ammonium accumulation and increased NH_4/NO_3 ratios.

Hydroculture experiments have been used to study the effects of pH change on root growth and survival rate of several of the endangered species (*A. montana*, *A. dioica*, *Viola canina*, *Hieracium pilosella* and *Gentiana pneumonanthe*) and the dominant species (*Molinia caerulea* and *Deschampsia flexuosa*) (Van Dobben, 1991). The dominant plants showed a lower pH optimum (3.5 and 4.0, respectively) than the endangered species (4.2–6.0) but the latter survived very low pH without visible injuries during the short experimental period. In field surveys, soil characteristics of several threatened acidic grassland species have been compared with those of the dominant species (Houdijk *et al.*, 1993; Roelofs *et al.*, 1996; De Graaf, 2000; Roem and Berendse, 2000). Endangered species were generally present on soils with higher pH, lower N concentration, and lower Al/Ca ratio than the dominant grasses or heathers. In addition, NH_4/NO_3 ratios were lower than those in the dwarf-shrub dominated soils. Fennema (1990, 1992) demonstrated that soils from locations where *A. montana* is still present had higher pH and lower Al/Ca ratios than soils of former *A. montana* stands, but he found no differences in total soil N concentrations and NH_4/NO_3 ratios. Both studies indicate that the decline of these endangered species of acidic grasslands is associated with high Al/Ca ratios, and in many cases also increased NH_4/NO_3 ratios. In hydroculture experiments, however, no significant effects of Al and Al/Ca on root growth or survival rates have been observed at high nutrient levels (Pegtel, 1987; Kroeze *et al.*, 1989; Van Dobben, 1991). In contrast, De Graaf *et al.* (1997) showed *A. montana* to be very sensitive to enhanced Al/Ca ratios at intermediate or low nutrient levels. Another Red-list species (*Cirsium dissectum*) also appeared to be very sensitive to high ammonium concentrations and NH_4/NO_3 ratios (De Graaf *et al.*, 1998) (Figure 12.11). Pot experiments with acidic grassland soil have indicated that increased NH_4/NO_3 ratios (due to ammonium accumulation) impaired the uptake of essential cations and caused decreased vitality of *T. serpyllum*. This species only remained vital in artificially buffered soils where nitrification rates were high enough to establish a balance between nitrate and ammonium, despite high N input (Houdijk, 1993; Houdijk *et al.*, 1993).

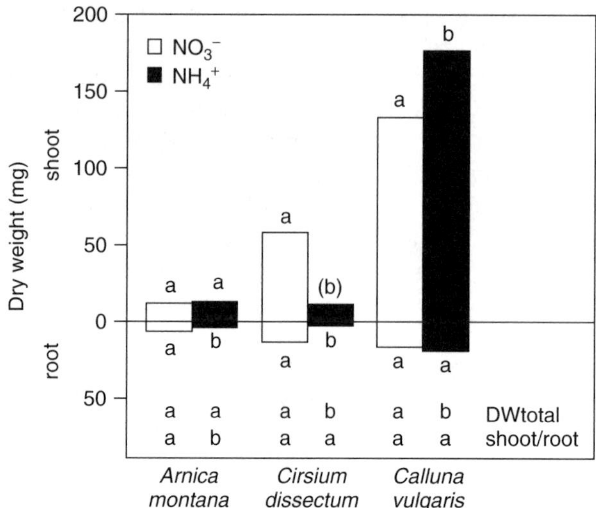

Figure 12.11 Shoot and root dry weight (mg) of *Arnica, Cirsium* and *Calluna* grown on nitrate (100 μmol 1^{-1}) or ammonium (100 μmol 1^{-1}) as sole N source. For each species, significant differences between treatments are indicated by different letters; published with permission from De Graaf *et al.* (1998)

In addition to the above effects, it has been hypothesized that the mutualistic relation between AM infection and these endangered species is adversely affected by the enhanced inputs of N. Field and greenhouse studies have revealed hardly any adverse effects of low pH or ammonia treatment on AM infection of *A. montana* and *A. dioica*, both highly threatened species in lowland acidic grasslands. Application of ammonium, however, sometimes caused a small, but significant decrease in AM infection. Treatments with aluminium, however, reduced AM infection much more distinctly, especially of *A. montana* (Heijne, 1995).

The long-term effects of additions to an acidic grassland (*Festuca-Agrostis-Galium* community) have been followed from 1989 to date in Derbyshire in the UK (Morecroft *et al.*, 1994; Carrol *et al.*, 1997; Lee and Caporn, 1998; Carroll *et al.*, 2000). After the first 3 years, no significant change in species composition was observed, with only the moss *Rhytidiadelphus squarrosus* declining after ammonium sulphate additions. Shoot N concentrations, nitrate reductase activities and soil nitrogen mineralization rates were significantly higher at nitrogen inputs >35 kg N ha^{-1} yr^{-1}, as in the calcareous grassland site of this study (see previous section). Six years of N applications have resulted in marked changes in the abundance of the bryophyte species. A dose-related reduction in bryophyte cover, significant at all levels of N applications, was found in the second half of the experimental period (Carroll *et al.*, 2000). No increase in plant productivity was found after 6 years of N additions, which clearly demonstrated (at least present) P limitation of this acidic grassland site. Like in the nearby calcareous site, total plant cover decreased somewhat, especially in the vegetation treated with ammonium sulphate. Even in 1996 the N mineralization rate was still higher in the N-treated vegetation. The addition of ammonium sulphate caused greater botanical change, especially in bryophyte cover, than the addition of other nitrogenous compounds. This could most probably be attributed to its acidifying effects; the pH decreased to 4.1 in this situation, compared with 4.4 in the control treatment.

In conclusion, increases in grasses and decreases in plant diversity have been observed at intermediate levels of N inputs in several acidic grasslands. Interactions with P limitation can, however, modify the outcome. The effects of enhanced ammonium (sulphate) inputs or ammonium accumulation in the soil are more severe than the effects of nitrate addition in these weakly to moderately buffered grasslands (pH 4.5–6.5) because of the negative effects of soil acidification. It has been shown that several endangered plant species of slightly acidic habitats are especially sensitive to secondary acidification effects such as aluminium and/or ammonium toxicity.

Concluding remarks

In this chapter it has been shown that increased atmospheric inputs of N (both in reduced and oxidized form) have a wide variety of effects on the structure and functioning of terrestrial non-forest ecosystems. Because of the huge variation in biotic and abiotic conditions, the impacts have been presented for a number of typical and illustrative categories of natural and semi-natural ecosystems, mostly of high conservation value. The understanding of the ecological consequences of increased atmospheric N inputs (including NH_x) on terrestrial ecosystems has increased steadily since the early 1980s due to the growing amount of research carried out in response to the drastic increase in N inputs in NW Europe in the second half of the twentieth century. As a result, it is nowadays possible to make reliable estimates of the sensitivities of a large number of ecosystems to increased N inputs (Table 12.1).

The sequence of events which occurs when N loads increase in a region with originally low deposition rates is highly complicated. Many processes interact and they may operate over different time scales. One of the most obvious effects is the accumulation of N compounds, resulting in enhanced N cycling, N availability and increased plant productivity, because many natural terrestrial ecosystems are N limited. This may lead to competitive exclusion of characteristic species by more nitrophilous plants, especially under nutrient-poor conditions. On very nutrient-poor soils, plant diversity sometimes increases as a result of the invasion of species which were not tolerant to the original harsh conditions. In this case, however, the native flora of these environments disappears. Long-term increased atmospheric N deposition can finally lead to a transition from N-limited to P-limited plant production. This transition has consequences for ecosystem productivity and internal nutrient cycling, as is also the case in originally P-limited ecosystems under N enrichment.

High inputs of acidifying N compounds can also cause soil acidification (with reduced ANC and increased concentrations of toxic metals such as aluminium), especially after nitrification of ammonium in weakly or moderately buffered environments. In this situation, acid-resistant plant species have become dominant, in parallel with the disappearance of several endangered plants typical of intermediate pH values. In addition, the related change in the balance between ammonium and nitrate may also affect

Table 12.1 Overview of the impacts of air-borne N inputs on semi-natural and natural ecosystems, including estimated effect thresholds. Overall sensitivity is given as: very sensitive: first effects observed at 5–10 kg N ha^{-1} yr^{-1}; sensitive: between 10–20; moderately sensitive: 20–40; not sensitive: > 40 kg N ha^{-1} yr^{-1}. + = high; ± = moderately; 0 = not important; ? = unknown (published with permission from Bobbink, 1998)

	Overall sensitivity	Important effects		Secondary Stresses	Indicators
		Eutrophication	Acidification		
Wetlands					
Soft-water bodies	very sensitive	+	+	?	Decline isoetid water plants
Ombrotrophic bogs	very sensitive	+	0	?	Decrease typical mosses, increase graminoids, N accumulation
Mesotrophic fens	mod. sensitive	+	±	?	Increase tall graminoids, decline diversity and typical mosses
Marshes	not sensitive	0	0	0	–
Species-rich grasslands					
Calcareous grasslands	(mod.) sensitive	+	0	?	Increase grasses, decline diversity, increased leaching.
Neutral-acidic grassl.	mod. sensitive	+	+	?	Increase grasses, decline diversity, leaching increased
Mountainous grassl.	sensitive?	?	?	?	Changes in species composition

Heathlands					
Dry lowland heath	sensitive	+	±?	+	Increase grasses, decline heather and mosses; N accumulation
Wet lowland heath	sensitive	+	±?	?	Increase grasses, decline diversity, N accumulation
Species-rich heaths/ weakly buffered grassl.	sensitive	±	+	?	Decline endangered species
Upland *Calluna* heath	mod. sensitive	+	0	+?	N accumulation, sec. stresses
Arctic/alpine heath	sensitive?	?	?	?	?
Forest understorey					
Acidic coniferous forest	sensitive	+	±	?	Increase nitrophilic species; N accumulation; leaching
Acidic deciduous forest	sensitive	+	±	?	Increase nitrophilic species; N accumulation; leaching

the performance of several plant species. In several of the ecosystems discussed, effects of both increased N availability and soil acidification have occurred simultaneously and are thus difficult to unravel.

Finally, it has been demonstrated in some ecosystems (e.g. dry heaths, upland *C. vulgaris* heaths) that the susceptibility of the vegetation to secondary stresses or disturbance factors (pathogens and herbivores; winter injury and drought) has been affected by atmospheric N pollutants, although data are lacking for most other ecosystems. Increased shoot to root ratios and decreased mycorrhizal infection can, in addition, increase the sensitivity to stresses, such as summer drought periods. At present, however, many questions remain on the generality of these observations and on their ecological significance.

In order to conserve global biodiversity in natural and semi-natural ecosystems it is crucial to control N emissions to the atmosphere. Only in this way, will it be possible to reduce or prevent the obvious negative effects on ecosystem diversity and functioning in the future. Although our knowledge of the effects of airborne N inputs clearly increased in the last 20 years, only a small proportion of all possible impacts have been elucidated. It will therefore be a main challenge in future to unravel N impacts in all their complexity for a world-wide selection of (sensitive) ecosystems, especially in interaction with other foreseen threats, such as global warming and increased CO_2 concentrations. These efforts are crucial to develop necessary, but realistic, reduction programmes for N pollutants and to verify robust dynamic ecosystem models which are reliable enough to calculate long-term effect thresholds for atmospheric N deposition.

References

Aaby, B. (1990). Overvågning af hojmoser 1989. *Skov-og Naturstyrelsen*, Miljoministeriet, pp. 1–89.

Aaby, B. (1994). Monitoring Danish raised bogs. *Mires and Man. Mire conservation in a densely populated country – the Swiss experience* (ed A. Grünig), Kosmos, Birmensdorf, pp. 284–300.

Aerts, R. (1993). Competition between dominant plant species in heathlands. *Heathlands: Patterns and processes in a changing environment* (eds R. Aerts and G.W. Heil), pp. 125–151. Kluwer, Dordrecht.

Aerts, R., Berendse, F., De Caluwe, H. and Schmitz, M. (1990). Competition in heathland along an experimental gradient of nutrient availability. *Oikos*, **57**, 310–318.

Aerts, R. and Bobbink, R. (1999). The impacts of atmospheric nitrogen deposition on vegetation processes in non-forest ecosystems. In: S.J. Langan (ed.) *The impact of nitrogen deposition on natural and semi-natural ecosystems*. Kluwer, Dordrecht, pp. 85–122.

Aerts, R., Boot, R.G.A. and Van der Aart, P.J.M. (1991). The relation between above- and belowground biomass allocation patterns and competitive ability. *Oecologia*, **87**, 551–559.

Aerts, R. and Chapin, F.S. (2000). The mineral nutrition of wild plants revisited: a re-evaluation of processes and patterns. *Advances in Ecological Research*, **30**, 1–67.

Aerts, R. and Heil, G.W. (eds) (1993). *Heathlands: Patterns and processes in a changing environment*. Kluwer, Dordrecht.

Aerts, R., Verhoeven, J.T.A. and Whigham, D.F. (1999). Plant-mediated controls on nutrient cycling in temperate fens and bogs. *Ecology*, **80**, 2170–2181.

Aerts, R., Wallén, B. and Malmer, N. (1992). Growth-limiting nutrients in *Sphagnum*-dominated bogs subject to low and high atmospheric nitrogen supply. *Journal of Ecology*, **80**, 131–140.

Aerts, R., Wallen, B., Malmer, N. and De Caluwe, H. (2001). Nutritional constraints on *Sphagnum*-growth and potential decay in northern peatlands. *Journal of Ecology*, **89**, 292–299.

Aronsson, A. (1980). Frost hardiness in Scots pine. II. Hardiness during winter and spring in young trees of different mineral status. *Studia Forestalia Suecica*, **155**, 1–27.

Asman, W.A.H., Sutton, M.A. and Schjorring, J.K. (1998). Ammonia: emission, atmospheric transport and deposition. *New Phytologist*, **139**, 27–48.

Baddeley, J.A., Thompson, D.B.A. and Lee, J.A. (1994). Regional and historical variation in the nitrogen content of *Racomitrium lanuginosum* in Britain in relation to atmospheric nitrogen deposition. *Environmental Pollution*, **84**, 189–196.

Berdowski, J.J.M. (1987) *The catastrophic death of Calluna vulgaris in Dutch heathlands*. PhD thesis, University of Utrecht, Utrecht.

Berdowski, J.J.M. (1993). The effect of external stress and disturbance factors on *Calluna*-dominated heathland vegetation. *Heathland: Patterns and processes in a changing environment* (eds R. Aerts and G.W. Heil), pp. 85–124. Geobotany 20, Kluwer, Dordrecht.

Berendse, F. (1990). Organic matter accumulation and nitrogen mineralization during secondary succession in heathland ecosystems. *Journal of Ecology*, **78**, 413–427.

Blankwaardt, H.F.H. (1977). Het optreden van de heidekever (*Lochmaea suturalis* Thomson) in Nederland sedert 1915. *Entomologische Berichten*, **37**, 34–40.

Bobbink, R. (1991). Effects of nutrient enrichment in Dutch chalk grassland. *Journal of Applied Ecology*, **28**, 28–41.

Bobbink, R. (1998). Impacts of tropospheric ozone and airborne nitrogenous pollutants on natural and semi-natural ecosystems: a commentary. *New Phytologist*, **139**, 161–168.

Bobbink, R., Bik, L. and Willems, J.H. (1988). Effects of nitrogen fertilization on vegetation structure and dominance of *Brachypodium pinnatum* (L.) Beauv. in chalk grassland. *Acta Botanica Neerlandica*, **37**, 231–242.

Bobbink, R., Den Dubbelden, K.C. and Willems, J.H. (1989). Seasonal dynamics of phytomass and nutrients in chalk grassland. *Oikos*, **55**, 216–224.

Bobbink, R. and Heil, G.W. (1993). Atmospheric deposition of sulphur and nitrogen in heathland ecosystems. *Heathland: Patterns and processes in a changing environment* (eds R. Aerts and G.W. Heil), pp. 25–50. Geobotany 20, Kluwer, Dordrecht.

Bobbink, R., Heil, G.W. and Raessen, M.B.A.G. (1992). Atmospheric deposition and canopy exchange processes in heathland ecosystems. *Environmental Pollution*, **75**: 29–37.

Bobbink, R. and Willems, J.H. (1987). Increasing dominance of *Brachypodium pinnatum* (L.) Beauv. in chalk grasslands: a threat to a species-rich ecosystem. *Biological Conservation*, **40**, 301–314.

Bobbink, R. and Willems, J.H. (1991). Impact of different cutting regimes on the performance of *Brachypodium pinnatum* in Dutch chalk grassland. *Biological Conservation*, **56**, 1–21.

Brouwer, R. (1962). Distribution of dry matter in the plant. *Netherlands Journal of Agricultural Science*, **10**, 399–408.

Brouwer, R. (1983). Functional equilibrium: sense or nonsense. *Netherlands Journal of Agricultural Science*, **31**, 335–348.

Brunsting, A.M.H. and Heil G.W. (1985). The role of nutrients in the interaction between a herbivorous beetle and some competing plant species in heathlands. *Oikos*, **44**, 23–26.

Bryant, J.B. and Taylor, G. (1998). Simulated atmospheric nitrogen deposition affects growth, physiology and competitive interactions in model chalk grassland swards. *Book of abstracts*, NERC Air Pollution Effects Research, Centre for Ecology and Hydrology, Edinburgh.

Buijsman, E., Maas, J.M. and Asman, W.A.H. (1987). Anthropogenic NH_3 emissions in Europe. *Atmospheric Environment*, **21**, 1009–1022.

Caporn, S.J.M., Risager, M. and Lee, J.A. (1994). Effects of nitrogen supply on frost hardiness in *Calluna vulgaris* (L.) Hull. *New Phytologist*, **128**, 461–468.

Caporn, S.J.M., Song, W., Read, D.J. and Lee, J.A. (1995). The effect of repeated nitrogen fertilization on mycorrhizal infection in heather (*Calluna vulgaris* (L.) Hull). *New Phytologist*, **129**, 605–609.

Carroll, J.A., Campbell, C.D., Caporn, S.J.M., Cawley, L., Johnson, D., Leake, J.R., Lee, J.A., Lei, Y., Morecroft, M., Read, D.J. and Taylor, A. 1997. *Natural vegetation responses to atmospheric nitrogen deposition: Critical levels and loads for nitrogen for vegetation growing on contrasting soils*. Final report May 1994-1997. Department of the Environment Research Contract Reference EPG 1/3/11; University of Sheffield, Sheffield.

Carroll, J.A., Caporn, S.J.M., Cawley, L., Read, D.J. and Lee, J.A. (1999). The effect of increased deposition of atmospheric nitrogen on *Calluna vulgaris* in Upland Britain. *New Phytologist*, **141**, 423–431.

Carroll, J.A., Johnson, D., Morecroft, M.D., Taylor, A., Caporn, S.J.M. and Lee, J.A. (2000). The effect of long-term nitrogen additions on the bryophyte cover of upland acidic grasslands. *Journal of Bryology*, **22**, 83–89.

Cawley, L.E., Caporn, S.J.M., Carroll, J.A., Cresswell, N. and Stronach, I.M. (1998). Influence of elevated nitrogen on drought tolerance in lowland heath. *Book of abstracts*, NERC Air Pollution Effects Research, Centre for Ecology and Hydrology, Edinburgh.

Chapman, R.R. and Hemond, H.F. (1982). Dinitrogen fixation by surface peat and *Sphagnum* in an ombrotrophic bog. *Canadian Journal of Botany*, **60**, 538–543.

Chapman, S.B., Hibble, J. and Rafael, C.R. (1975). Net aerial production by *Calluna vulgaris* on lowland heath in Britain. *Journal of Ecology*, **63**, 233–258.

Coulson, J.C. and Butterfield, J. (1978). An investigation of the biotic factors determining the rates of plant decomposition on blanket bog. *Journal of Ecology*, **66**, 631–650.

Crawley, M.J. (1983). *Herbivory–the dynamics of animal/plant interactions*. Blackwell, Oxford.

De Boer, W. (1989). *Nitrification in Dutch heathland soils*. PhD thesis, Agricultural University of Wageningen, Wageningen.

De Graaf, M.C.C. (2000). *Exploring the calcicole–calcifuge gradient in heathlands*. PhD thesis, University of Nijmegen, Nijmegen.

De Graaf, M.C.C., Bobbink, R., Roelofs, J.G.M. and Verbeek, P.J.M. (1998). Differential effects of ammonium and nitrate on three heathland species. *Plant Ecology*, **135**, 185–196.

De Graaf, M.C.C., Bobbink, R., Verbeek, P.J.M. and Roelofs, J.G.M. (1997). Aluminium toxicity and tolerance in three heathland species. *Water, Air and Soil Pollution*, **98**, 229–239.

De Kroon, H. and Bobbink, R. (1997). Clonal plant performance under elevated nitrogen deposition, with special reference to *Brachypodium pinnatum* in chalk grassland. *The ecology and evolution of clonal plants* (eds H. De Kroon and J. Van Groenendael), pp. 359–379, Backhuys Publishers, Leiden.

De Smidt, J.T. (1979). Origin and destruction of Northwest European heath vegetation. *Werden und Vergehen von Pflanzengesellschaften* (eds O. Wilmanns and R. Tüxen), pp. 411–435. J. Cramer, Vaduz.

De Vries, W. (1994). *Soil response to acid deposition at different regional scales*. PhD thesis, Agricultural University of Wageningen, Wageningen.

Dierschke, H. and Engels, M. (1991). Response of a *Bromus erectus* grassland (Mesobromion) to abandonment and different cutting regimes. *Modern ecology: Basic and applied aspects* (eds G. Esser and D. Overdieck), pp. 376–397. Elsevier, Amsterdam.

Dueck, Th.A., Dorel, F.G., Ter Horst, R. and Van Der Eerden, L.J. (1991). Effects of ammonia and sulphur dioxide on frost sensitivity of *Pinus sylvestris*. *Water, Air and Soil Pollution*, **54**, 35–49.

During, H.J. and Willems, J.H. (1986). The impoverishment of the bryophyte and lichen flora of the Dutch chalk grasslands in the thirty years 1953–1983. *Biological Conservation*, **36**, 143–158.

Ellenberg, H. (1988a). *Vegetation ecology of Central Europe*. Cambridge University Press, Cambridge.

Ellenberg, H. Jr. (1988b). Floristic changes due to nitrogen deposition in central Europe. *Nord (Miljörapport)*, **15**, 375–383.

Fangmeier, A., Hadwiger-Fangmeier, A., van der Eerden, L.J. and Jäger, H. (1994). Effects of atmospheric ammonia on vegetation–a review. *Environmental Pollution*, **86**, 43–82.

Fennema, F. (1990). Effects of exposure to atmospheric SO_2, NH_3 and $(NH_4)_2SO_4$ on survival and extinction of *Arnica montana* L. and *Viola canina* L. Report no. 90/14 RIN, Arnhem.

Fennema, F. (1992). SO_2 and NH_3 deposition as possible causes for the extinction of *Arnica montana* L. *Water, Air and Soil Pollution*, **62**, 325–336.

Galloway, J.N. (1995). Acid deposition: perspectives in time and space. *Water, Air and Soil Pollution*, **85**, 15–24.

Gimingham, C.H., Chapman, S.B. and Webb, N.R. (1979). European heathlands. *Ecosystems of the world, 9A* (ed R.L. Specht), pp. 365–386, Elsevier, Amsterdam.

Grime, J.P. (1979). *Plant strategies and vegetation processes*. John Wiley & Sons, Chichester, UK.

Heijmans, M.M.P.D., Berendse, F., Arp, W.J., Masselink, A.K., Klees, H., De Visser, W. and Van Breemen, N. (2001). Effects of elevated carbon dioxide and increased nitrogen deposition on bog vegetation in The Netherlands. *Journal of Ecology*, **89**, 268–279.

Heijne, B. (1995). *Effects of acid rain on vesicular-arbuscular mycorrhiza of herbaceous plants in dry heathland*. PhD thesis, Utrecht University, Utrecht.

Heil, G.W. and Aerts, R. (1993). General introduction. *Heathland: Patterns and processes in a changing environment* (eds R. Aerts and G.W. Heil), pp. 1–24. Geobotany 20, Kluwer, Dordrecht.

Heil, G.W. and Bruggink, M. (1987). Competition for nutrients between *Calluna vulgaris* (L.) Hull and *Molinia caerulea* (L.) Moench. *Oecologia*, **73**, 105–108.

Heil, G.W. and Diemont, W.H. (1983). Raised nutrient levels change heathland into grassland. *Vegetatio*, **53**, 113–120.

Hewins, E.J. and Ling, K.A. (1998). The impacts of management and atmospheric ammonia deposition on plant communities of calcareous grasslands. *Book of abstracts*, NERC Air Pollution Effects Research, Centre for Ecology and Hydrology, Edinburgh.

Hicks, W.K., Leith, I.D., Woodin, S.J. and Fowler, D. (1995). Upland vegetation and enhanced nitrogen deposition. *Abstract book 'Acid Reign '95'*, p. 229, Kluwer, Dordrecht.

Hogg, E.H., Malmer, N. and Wallén, B. (1994). Regional and microsite variation in potential decay rate of *Sphagnum magellanicum* in south Swedish raised bogs. *Ecography*, **17**, 50–59.

Hogg, P., Squires, P. and Fitter, A.H. (1995). Acidification, nitrogen deposition and rapid vegetational change in a small valley mire in Yorkshire. *Biological Conservation*, **71**, 143–153.

Hornung, M., Sutton, M.A. and Wilson, R.B. (eds) (1995). *Mapping and modelling of critical loads for nitrogen–a workshop report*. Institute of Terrestrial Ecology, Penicuik.

Houdijk, A.L.F.M. (1993). *Atmospheric ammonium deposition and the nutritional balance of terrestrial ecosystems*. PhD Thesis, University of Nijmegen, Nijmegen.

Houdijk, A.L.F.M., Verbeek, P.J.M., Van Dijk, H.F.G. and Roelofs, J.G.M. (1993). Distribution and decline of endangered herbaceous heathland species in relation to the chemical composition of the soil. *Plant and Soil*, **148**, 137–143.

Jeffrey, D.W. and Pigott, C.D. (1973). The response of grasslands on sugar-limestone in Teesdale to application of phosphorus and nitrogen. *Journal of Ecology*, **61**, 85–92.

Johansson, M. (2000). The influence of ammonium nitrate on the root growth and ericoid mycorrhizal colonization of *Calluna vulgaris* (L.) Hull from a Danish heathland. *Oecologia* **123**, 418–424.

Johnson, D., Leake, J.R. and Lee, J.A. (1999). The effects of quantity and duration of simulated pollutant nitrogen deposition on root surface phosphatase activities in calcareous and acid grasslands: a bioassay approach. *New Phytologist*, **141**, 433–442.

Kerslake, J.E., Woodin, S.J. and Hartley, S.E. (1998). Effects of carbon dioxide and nitrogen enrichment on a plant–insect interaction: the quality of *Calluna vulgaris* as a host for *Operophtera brumata. New Phytologist*, **140**, 43–53.

Kristensen, H.L. and McCarty, G.W. (1999). Mineralization and immobilization of nitrogen in heath soil under intact *Calluna*, after heather beetle infestation and nitrogen fertilization. *Applied Soil Ecology* **13**, 187–198.

Kroeze, C., Pegtel, D.M. and Blom, C.J.C. (1989). An experimental comparison of aluminium and manganese susceptibility in *Antennaria dioica*, *Viola canina, Filago minima* and *Deschampsia flexuosa. Acta Botanica Neerlandica*, **38**, 165–172.

Lamers, L.P.M., Bobbink, R. and Roelofs, J.G.M. (2000). Natural nitrogen filter fails in polluted raised bogs. *Global Change Biology*, **6**, 583–586.

Lee, J.A., Baddeley, J.A. and Woodin, S.R. (1989). Effects of acidic deposition on semi-natural vegetation. *Acidification in Scotland*, pp. 94–111, Scottish Development Department, Edinburgh.

Lee, J.A. and Caporn, S.J.M. (1998). Ecological effects of atmospheric reactive nitrogen deposition on semi-natural terrestrial ecosystems. *New Phytologist*, **139**, 127–134.

Lee, J.A. and Studholme, C.J. (1992). Responses of *Sphagnum* species to polluted environments. *Bryophytes and lichens in a changing environment* (eds J.W. Bates and A.M. Farmer), pp. 314–332, Clarendon Press, Oxford.

Leith, I.D., Hicks, W.K., Fowler, D. and Woodin, S.J. (1999). Differential responses of UK upland plants to nitrogen deposition. *New Phytologist*, **141**, 277–289.

Luken, J.O. and Billings, W.D. (1985). The influence of microtopographic heterogeneity on carbon dioxide efflux from a subarctic bog. *Holarctic Ecology*, **8**, 306–312.

Lütke Twenhöven, F. (1992). Untersuchungen zur Wirkung stickstoffhaltiger Niedersläge auf die Vegetation von Hochmooren. *Mitteilungen der Arbeitsgemeinschaft Geobotanik in Schleswig-Holstein und Hamburg*, **44**, 1–172.

Malmer, N. (1988). Patterns in the growth and the accumulation of inorganic constituents in the *Sphagnum* cover on ombrotrophic bogs in Scandinavia. *Oikos*, **53**, 105–120.

Malmer, N., Svensson, B.M. and Wallén, B. (1994). Interactions between *Sphagnum* moss and field layer vascular plants in the development of peat-forming systems. *Folia Geobotanica et Phytotaxonomica*, **29**, 483–496.

Metcalfe, S.E., Fowler, D., Derwent, R.G., Sutton, M.A., Smith, R.I. and Whyatt, J.D. (1999). Spatial and temporal aspects of nitrogen deposition. *The impact of nitrogen deposition on natural and semi-natural ecosystems* (ed S.J. Langan), pp. 15–50. Kluwer, Dordrecht.

Morecroft, M.D., Sellers, E.K. and Lee, J.A. (1994). An experimental investigation into the effects of atmospheric nitrogen deposition on two semi-natural grasslands. *Journal of Ecology*, **82**, 475–483.

Mosier, A.R., Stillwel, M., Paton, W.J. and Woodmansee, R.G. (1981). Nitrous oxide emissions from a native shortgrass prairie. *Journal of the Soil Science Society of America*, **45**, 617–619.

Neitzke, M. (2001). Analysis of vegetation and nutrient supply in calcareous grassland border zones to determine critical loads for nitrogen. *Flora*, **196**, 292–303.

Nordin, A. and Gunnarsson, U. (2000). Amino acid accumulation and growth of *Sphagnum* under different levels of N deposition. *Ecoscience*, **7**, 474–480.

Pearson, J. and Stewart, G.R. (1993). The deposition of atmospheric ammonia and its effects on plants. *New Phytologist*, **125**, 283–305.

Pegtel, D.M. (1987). Effects of ionic Al in culture solutions on the growth of *Arnica montana* L. and *Deschampsia flexuosa (L.) Trin. Plant and Soil*, **102**, 85–92.

Pitcairn, C.E.R., Fowler, D. and Grace, J. (1991). *Changes in species composition of semi-natural vegetation associated with the increase in atmospheric inputs of nitrogen*. Report Nature Conservancy Council, Institute of Terrestrial Ecology, Edinburgh, UK.

Pitcairn, C.E.R., Fowler, D. and Grace, J. (1995). Deposition of fixed atmospheric nitrogen and foliar nitrogen content of bryophytes and *Calluna vulgaris* (L.) Hull. *Environmental Pollution*, **88**, 193–205.

Poorter, H. and Nagel, O. (2000). The role of biomass allocation in the growth response of plants to different levels of light, CO_2, nutrients and water: A quantative review. *Australian Journal of Plant Physiology*, **27**, 595–607.

Power, S.A., Ashmore, M.R., Cousins, D.A. and Ainsworth, N. (1995). Long-term effects of enhanced nitrogen deposition on a lowland dry heath in southern Britain. *Water, Air and Soil Pollution*, **85**, 1701–1706.

Power, S.A., Ashmore, M.R. and Cousins, D.A. (1998a). Impacts and fate of experimentally enhanced nitrogen deposition on a British lowland heath. *Environmental Pollution*, **120**, 27–34.

Power, S.A., Ashmore, M.R., Cousins, D.A. and Sheppard, L.J. (1998b). Effects of nitrogen addition on the stress sensitivity of *Calluna vulgaris. New Phytologist*, **138**, 663–673.

Press, M.C., Woodin, S.J. and Lee, J.A. (1986). The potential importance of an increased atmospheric nitrogen supply to the growth of ombrotrophic *Sphagnum* species. *New Phytologist*, **103**, 45–55.

Proctor, M.C.F. (1995). The ombrogenous bog environment. *Restoration of temperate wetlands* (eds B.D. Wheeler, S.C. Shaw, W. Fojt and R.A. Robertson) pp. 287–303, John Wiley & Sons, Chichester, UK.

Ratcliffe, D.A. (1984). Post-medieval and recent changes in British vegetation: the culmination of human influence. *New Phytologist*, **98**, 73–100.

Redbo-Torstensson, P. (1994). The demographic consequences of nitrogen fertilization of a population of sundew, *Drosera rotundifolia. Acta Botanica Neerlandica*, **43**, 175–188.

Riis-Nielsen, T. (2001). *Effects of nitrogen on the stability and dynamics of Danish heathland vegetation*. PhD-thesis, University of Copenhagen, Copenhagen.

Risager, M. (1998). *Impacts of nitrogen on* Sphagnum *dominated bogs with emphasis on critical load assessment*. PhD thesis, University of Copenhagen, Copenhagen.

Rochefort, L., Vitt, D.H. and Bayley, S.E. (1990). Growth, production, and decomposition dynamics of *Sphagnum* under natural and experimentally acidified conditions. *Ecology*, **71**, 1986–2000.

Roelofs, J.G.M. (1986). The effect of airborne sulphur and nitrogen deposition on aquatic and terrestrial heathland vegetation. *Experientia*, **42**, 372–377.

Roelofs, J.G.M., Bobbink, R., Brouwer, E. and De Graaf, M.C.C. (1996). Restoration ecology of aquatic and terrestrial vegetation on non-calcareous sandy soils in The Netherlands. *Acta Botanica Neerlandica*, **45**, 517–542.

Roelofs, J.G.M., Kempers, A.J., Houdijk, A.L.F.M. and Jansen, J. (1985). The effect of airborne ammonium on *Pinus nigra* var. *maritima* in The Netherlands. *Plant and Soil*, **84**, 45–56.

Roem, W.J. and Berendse, F. (2000). Soil acidity and nutrient supply ratio as possible factors determining changes in plant species diversity in grassland and heathland communities. *Biological Conservation*, **92**, 151–161.

Smith, C.T., Elston, J. and Bunting, A.H. (1971). The effects of cutting and fertilizer treatment on the yield and botanical composition of chalk turfs. *Journal of the British Grassland Society*, **26**, 213–219.

Sutton, M.A., Pitcairn, C.E.R. and Fowler, D. (1993). The exchange of ammonia between the atmosphere and plant communities. *Advances in Ecological Research*, **24**, 301–393.

Swift, M.J., Heal, O.W. and Anderson, J.M. (1979). *Decomposition in terrestrial ecosystems*. Blackwell, Oxford.

Sykes, M.T. and Van Der Maarel, E. (1991). Spatial and temporal patterns in species turnover in the limestone grasslands of Öland, Sweden. *Abstracts 34th IAVS Symposium on mechanisms in vegetation dynamics*, p. 36, Eger, Hungary.

Tamm, C.O. (1991). *Nitrogen in terrestrial ecosystems. Questions of productivity, vegetational changes, and ecosystem stability*. Springer Verlag, Berlin.

Thompson, D.B.A. and Baddeley, J.A. (1991). Some effects of acidic deposition on montane *Racomitrium lanuginosum* heaths. *The effects of acid deposition on nature conservation in Great Britain* (eds S.J. Woodin and A.M. Farmer), pp. 17–28. NCC, Peterborough.

Tomassen, H.B.M., Bobbink, R., Peters, C.H.J., Van der Ven, P.J.M. and Roelofs, J.G.M. (1999). *N critical loads for acidic grasslands, dry dune grasslands and ombrotrophic bogs*. Report University of Nijmegen/Utrecht University. Nijmegen, 52 pp (in Dutch).

Tybirk, K., Bak, J. and Henriksen, L.N. (1995). Basis for mapping of critical loads in Nordic sensitive terrestrial ecosystems. *TemaNord*, **610**, 7–69.

Ulrich, B. (1983). Interaction of forest canopies with atmospheric constituents: SO_2, alkali and earth alkali cations and chloride. *Effects of accumulation of air pollutants in forest ecosystems* (eds B. Ulrich and J. Pankrath), pp. 33–45, D. Reidel, Dordrecht.

Ulrich, B. (1991). An ecosystem approach to soil acidification. *Soil acidity* (eds B. Ulrich and M.E. Summer), pp. 28–79, Springer Verlag, Berlin.

Unkovich, M., Jamieson, J., Monaghan, R. and Barraclough, D. (1998). Nitrogen mineralisation and plant nitrogen acquisition in a nitrogen-limited calcareous grassland. *Environmental and Experimental Botany*, **40**, 209–219.

Uren, S.C. (1992). *The effects of wet and dry deposited ammonia on Calluna vulgaris*. PhD thesis, Imperial College, University of London.

Uren, S.C., Ainsworth, N., Power, S.A., Cousins, D.A., Huxedurp, L.M. and Ashmore, M.R. (1997). Long-term effects of ammonium sulphate on *Calluna vulgaris. Journal of Applied Ecology*, **34**, 207–216.

Van Breemen, N., Burrough, P.A., Velthorst, E.J., Dobben, H.F. van, Wit, T. de, Ridder, T.B. and Reijnders H.F.R. (1982). Soil acidification from atmospheric ammonium sulphate in forest canopy throughfall. *Nature*, **299**, 548–550.

Van Dam, D. (1990). *Atmospheric deposition and nutrient cycling in chalk grassland*. PhD thesis, University of Utrecht, Utrecht.

Van Dam, D., Bobbink, R., Heil, G.W. and Heijne, B. (1989). Nitrogen and sulphur cycling in chalk grassland; the influence of acid rain. *Man and his ecosystem, Volume 2* (eds L.J. Brasser and W.C. Mulder), pp. 201–206, Elsevier, Amsterdam.

Van Dam, D., Van Dobben, H.F., Ter Braak, C.F.J. and De Wit, T. (1986). Air pollution as a possible cause for the decline of some phanerogamic species in The Netherlands. *Vegetatio*, **65**, 47–52.

Van Der Eerden L.J., Dueck, Th.A., Elderson, J., Van Dobben, H.F., Berdowski, J.J.M. and Latuhihin M. (1990). *Effects of NH_3 and $(NH_4)_2SO_4$ deposition on terrestrial semi-natural vegetation on nutrient-poor soils*. Report IPO/RIN, Wageningen.

Van Der Eerden, L.J., Dueck, Th.A., Berdowski, J.J.M., Greven, H. and Van Dobben, H.F. (1991). Influence of NH_3 and $(NH_4)_2SO_4$ on heathland vegetation. *Acta Botanica Neerlandica*, **40**, 281–297.

Van Der Maas, M.P. (1990). *Hydrochemistry of two Douglas fir ecosystems and a heather ecosystem in the Veluwe*, The Netherlands. Report, Agricultural University of Wageningen.

Van Dobben, H.F. (1991). Integrated effects (Low vegetation). *Acidification research in The Netherlands. Final Report of the Dutch Priority Programme on Acidification* (eds G.J. Heij and T. Schneider), pp. 464–524, Elsevier, Amsterdam.

Van Kootwijk, E.J. and Van der Voet, H. (1989). *De kartering van heidevergrassing in Nederland met de Landsat Thematic Mapper sattelietbeelden*. Report RIN 89/2, Arnhem.

Van Tooren, B.F., Odé, B., During, H.J. and Bobbink, R. (1990). Regeneration of species richness in the bryophyte layer of Dutch chalk grasslands. *Lindbergia*, **16**, 153–160.

Wells, T.C.E. (1974). Some concepts of grassland management. *Grassland ecology and wildlife management* (ed E. Duffey), pp. 163–174, Chapman and Hall, London.

Willems, J.H. (1982). Phytosociological and geographical survey of Mesobromion communities in Western Europe. *Vegetatio*, **48**, 227–240.

Willems, J.H., Peet, R.K. and Bik, L. (1993). Changes in chalk grassland structure and species richness resulting from selective nutrient additions. *Journal of Vegetation Science*, **4**, 203–212.

Wilson, E.J., Wells, T.C.E. and Sparks, T.H. (1995). Are calcareous grasslands in the UK under threat from nitrogen deposition? An experimental determination of a critical load. *Journal of Ecology*, **83**, 823–832.

Wolkinger, F. and Plank, S. (1981). *Dry grasslands of Europe*. Council of Europe, Strasbourg.

Woodin, S.J. and Farmer, A.M. (1993). Impacts of sulphur and nitrogen deposition on sites and species of nature conservation importance in Great Britain. *Biological Conservation*, **63**, 23–30.

Yesmin, L., Gammack, S.M. and Cresser, M.S. (1996a). Changes in N concentrations of peat and its associated vegetation over 12 months in response to increased deposition of ammonium sulphate or nitric acid. *Science of the Total Environment*, **177**, 281–290.

Yesmin, L., Gammack, S.M. and Cresser, M.S. (1996b). Effects of atmospheric nitrogen deposition on ericoid mycorrhizal infection of *Calluna vulgaris* growing in peat soils. *Applied Soil Ecology*, **4**, 49–60.

13 Effects of wet deposited acidity

T.W. ASHENDEN

Introduction

The concept that 'acid rain' could damage vegetation was first raised in 1872 by Robert Angus Smith, the Chief Alkali Inspector of the UK in his reports on the rainfall around Manchester (Smith, 1872). However, it was not until 100 years later that any serious attention was paid to the subject. Then, several Scandinavian countries raised the possibility that long-distance transport of acid pollutants from the industrial countries of northern Europe was occurring and that the pollutants were being deposited in rainfall and causing long-term acidification of lakes and possible damage to forests. A major Norwegian project was initiated in 1972 to study 'Acid Rain Effects on Forests and Fish' and this was followed by several international conferences (USDA, 1976; NATO, 1978). It is now accepted that wet deposition is a major route for the transfer of potentially damaging quantities of sulphur, nitrogen and hydrogen ions from the atmosphere to ecosystems. Impacts on vegetation may be direct, by contact with plant foliage, or indirect, via longer-term soil mediated responses. In this chapter, we will consider only the direct effects of wet acid deposition on vegetation.

Nature of acid deposition

Rain, hail and snow

Pollutant gases and particulates are scavenged by the water droplets which form clouds, and may be transferred large distances before being deposited in precipitation. Without additions of these pollutants, rainwater would be slightly acidic because it contains dissolved atmospheric carbon dioxide in the form of carbonic acid. Therefore, 'acid rain' is taken to be snow, hail or rain which is more acid (i.e. lower pH) than pH 5.6. Large networks of rainfall monitors exist and, over the past 20 years, the presence of pollutants has been found to increase substantially the acidity of rainfall. In the UK, annual means down to pH 4.1 and monthly mean acidities as low as pH 3.4 have been reported in rural areas (RGAR, 1990).

In urban areas, rainfall may be slightly more acidic because of the 'washout' of high levels of gaseous pollutants during rainfall events (RGAR, 1990). None the less rain, hail and snow events at pH values below pH 3.4 are considered extremely rare and

Air Pollution and Plant Life, second edition. Edited by J.N.B. Bell and M. Treshow. ISBN 0 471 49090 3 (HB), 0 471 49091 1 (PB).

wet deposition at ambient acidities is widely believed to have little scope for direct impacts on vegetation.

Mist, fog and cloud

Mist, fog and cloud droplets are generally more acid than rainfall collected in the same region. The relative surface area of the finer droplets is large and allows increased uptake of acid gases. Also, pollutant aerosols may act as condensation nuclei with resultant higher ionic concentrations than in rain drops where condensation processes have already resulted in growth and dilution. Concentrations of pollutants in fog and cloud may be typically 10 times those found in rain and acidities between pHs 2 and 3 have frequently been reported (see review by Cape, 1993).

In coastal areas which are subject to advective fog and in upland areas which are cloaked for prolonged periods in low cloud or mist, wet deposition to plant canopies through direct interception of fog, mist and cloud droplets is considered to be of major importance. Indeed, it has been estimated that these processes, referred to as 'occult deposition' may account for over 25% of the pollutant inputs even in high rainfall areas (RGAR, 1990). The higher concentrations of pollutants in mist, fog and cloud events put vegetation in these regions at risk of direct adverse impacts of acid deposition.

Post-deposition concentration of water droplets

Once deposited on to vegetation, the acidity of water droplets does not remain constant. At the end of a rain, mist or cloud event, evaporation may occur and lead to marked increases in solute concentrations. Such increases have been observed to result in more than a 10-fold increase in hydrogen ion concentrations (that is a reduction of over 1 pH unit) (Frevert and Klemm, 1984). Hence, exposure of plant surfaces to highly acidic droplets could occur after any rain, mist, fog or cloud event. This potential post-deposition concentration of precipitation droplets needs to be taken into consideration when evaluating the direct impacts of wet deposition on vegetation.

Visible leaf injury

Visible injury has been observed in a range of species exposed to acid precipitation below pH 3.4. Injury has been shown to take the form of leaf lesions, chlorosis, necrosis, wilting of leaf tips, accelerated senescence and premature abscission of leaves (Jacobson, 1984). The degree of injury may be related to factors which influence the capture and retention of droplets and thus the effective exposure period. These include leaf morphology, surface wettability, temperature, humidity and air turbulence. The frequency and intensity of precipitation events are also important.

Sensitivity to foliar injury varies greatly between species. Simulated rain at pH 3.4 has been shown to cause visible lesions on leaves of sensitive crops such as radishes, soybeans and beets (Evans, 1984). However, exposures to mists at pH 2.5 have been found to cause no visible leaf damage to a range of leguminous crop species (Ashenden and Bell, 1989). Similarly, for trees, visible injury symptoms have been reported after exposure to rain at pH 3.4 for hybrid poplar (Evans *et al.*, 1978) while Cape (1993) lists a range of experiments in which no visible injury has been found at pH 3 or greater. In general, however, conifers are considered to be less susceptible to injury than broad leaved species (Percy, 1991). Little research has been conducted on the effects of wet deposition on native plant species. However, several studies have failed to induce symptoms of visible injury in exposures at pH 2.5 (Ashenden *et al.*, 1991; Edge *et al.*, 1994).

There is some evidence that, on an individual plant, susceptibility to visual injury may vary according to leaf age. Exposure to pH 3.4 rain was found to cause visible lesions on young leaves of bracken (*Pteridium aquilinum*) but not on older ones (Evans and Curry, 1979). Similarly, Caporn and Hutchinson (1987) found visible injury to cotyledons of young cabbage seedlings after just one exposure to pH 3.0 rain. Whilst successive treatments induced substantial further damage to the cotyledons, injury to true leaves on the same plants was slight.

Effects on growth and productivity

Crops

Acid precipitation may result in both depression of plant growth because of the toxic effects of acidity or in growth stimulation because of foliar fertilisation with sulphate and/or nitrate. For crop species there have been numerous studies but results have been conflicting. Several authors have reported that simulated rain at high levels of acidity (below pH 3.0) may reduce primary production (Harcourt and Farrar, 1980; Amthor, 1984). However, responses differ markedly between species and cultivars. In a study of 28 crops, Lee *et al.* (1981) found rain at pH 3.0–3.5 to reduce growth in five but increase growth in another six species. No consistent effects were found for the other crops under study. With less acid rain, most researchers have found little effect on crop growth and several reviews (Lee *et al.*, 1981; Irving, 1983; Jacobson, 1991) have concluded that effects below pH 3 are likely to be minimal. However, Evans *et al.* (1982, 1983) found rain at pH 4 to reduce yields of field grown soybeans compared with exposures at pH 5.6. Similarly Ashenden and Bell (1987) calculated growth reductions of 9–34% in different dry weight fractions of winter barley grown on a range of soil types, in response to the critical pH range of rainfall of 3.5–4.5 (Table 13.1). At an even lower acidity of rainfall, growth reductions were reported for seedlings of broad bean (*Vicia faba*) exposed to simulated rain at pH 4.5 compared with ones supplied with pH 5.6 rain (Ashenden and Bell, 1989).

Trees

Evaluation of the effects of wet acid deposition on the growth of trees is more difficult. Experiments cannot be conducted over the whole life period of the trees and thus interpretations are based on relatively short-term exposures, usually of seedlings or saplings,

Table 13.1 Predicted yield reductions for different dry weight fractions of winter barley (*Hordeum vulgare* L.) in response to rainfall of pHs 3.5, 4.0 and 4.5. Weight reductions are calculated as an average for a range of soil types and expressed as a percentage of yields obtained at pH 5.6. Reproduced by permission of Kluwer Academic Publishers, *Plant and Soil*, vol 98 (1997), Ashenden and Bell, 'Yield reduction in winter barley growth on a range of soils', pp. 433–437, Table 2

Dry weight fraction	PH of rain		
	4.5	4.0	3.5
Stems	12.0	16.8	21.5
Ears	19.4	26.9	33.7
Total shoot	10.1	14.3	18.4
Total plant	9.2	13.1	16.8

After Ashenden and Bell (1987).

and extrapolations made to predict impacts on mature trees growing over many years. The majority of these studies have indicated that a pH below 3 is required for adverse effects on tree growth (see reviews by Binns, 1984; Cape, 1993). However, Percy (1986) reported reduced shoot apex height in Scots pine (*Pinus sylvestris*) exposed to pH 4.6 rain over only a five-week period. In contrast to this observation, Abrahamsen (1980) found increased shoot height after two years and no effect on growth after five years for the same species in response to rainfall at pH 3.0. Increased height of seedlings of birch (*Betula pendula*) was found after exposure to even more acid rainfall applications of pH 2.5 and despite visible leaf injury symptoms (Ashenden and Bell, 1988). This observation clearly indicates that, at least in the short term, visible leaf injury may not be indicative of the impacts of acid deposition on growth.

Some studies have attempted to make longer term evaluations of the impacts of wet acid deposition on forests by using tree ring analysis. One such study in Sweden showed reduced growth of conifer trees in those areas which experienced the greatest rates of acid deposition (Jonsson, 1977). However, a similar study in Norway failed to confirm such a correlation (Abrahamsen *et al.*, 1976). A major difficulty in the interpretation of these long-term studies is the likely complication that there are interactions between the impacts of acid deposition with other climatic variables and soil-mediated influences on the growth of the trees. Some of these issues will be considered in Chapter 15 with respect to forest decline where acid deposition is claimed to be one of the contributing factors.

Native vegetation

There have been comparatively fewer experimental studies on the impacts of acid rain on native herbaceous species. A summary of studies which have shown effects on plants with exposures to realistic acidities of rainfall is shown in Table 13.2. Once again, there are no clear trends in plant growth responses. Irving (1983) found reductions in root growth for tall fescue (*Festuca arundinacea*) and cocksfoot (*Dactylis glomerata*) in response to rain at pH 4.0 but not at a more acidic pH 3.5. Meanwhile, in studies of six upland plant species, five showed increased growth in response to pH 2.5 fogs whilst the remaining species, birdsfoot trefoil (*Lotus corniculatus*) showed reduced growth at pH 2.5 and pH 3.5 but increased growth at pH 4.5 compared with a control pH 5.6 fog treatment (Ashenden *et al.*, 1991; Edge *et al.*, 1994). Such differential responses of species to acidic deposition may have important implications under competitive situations in semi-natural ecosystems. Reduced growth in a species as a response to pollutants will make it less able to survive in competition with more resistant species. Similarly, if a species does not show growth stimulation comparable to competing species in response to pollutants, it will decline.

Lower plant species, such as lichens and bryophytes, are likely to be much more susceptible to acid deposition because of a lack of a protective cuticle (see Chapter 17). Each rainfall or mist event will bring plant cells directly into contact with the dissolved pollutants. Applications of simulated acid rain have been shown to result in growth reductions and visible injury at pH 3.5 for field plots of feather mosses and reindeer lichens, while exposures to pH 4 and above have little effect (Hutchinson *et al.*, 1986; Scott *et al.*, 1989). However, the main bulk of evidence for impacts of wet acidity on these plant groups comes from longer term investigations. Gilbert (1986) reported on the gradual decline of the large foliose lichen *Lobaria pulmonaria* in Monks Wood, Northumberland. This species was found thriving on 20 oak trees in 1965 but progressively declined until completely lost by 1984. During this period, the average pH of rainfall was about 4.2 and the pH of the bark on the oak trees decreased from 5.2 in the 1960s to 4.7 in 1984. Similar long-term degradation of ecosystems because of continuous inputs of wet acid deposition have been suggested for upland heathlands and ombrotrophic mires but in these more emphasis has been placed on the damaging effects of sulphur and, in more recent years, nitrogen influxes rather than acidity *per se* (see reviews by Lee *et al.*, 1988 and 1992).

Effects on assimilate partitioning

Evidence from experimental exposures to acid rain and fog treatments suggests that changes in overall plant growth may be accompanied by an alteration in the partitioning of assimilates. Several studies on crops have revealed biomass reductions at high acidities to be primarily because of a decrease in root rather than shoot growth (Ferenbaugh, 1976; Lee *et al.*, 1981; Jacobson *et al.*, 1986). Similarly, growth stimulation in grasses at pH 2.5 has been shown to be accompanied by a shift in favour of shoot growth. Such alteration in partitioning of assimilates is often found in response to gaseous forms of pollutants. Mansfield (1988) has suggested that the reductions in root mass may be a result of decreased carbohydrate translocation by a possible inhibition of sieve-tube loading. Whatever the mechanism, it is clearly a potentially deleterious growth effect since water loss:uptake ratios may increase and plants may be expected to become more susceptible to drought at later stages of development.

Table 13.2 Effects of wet acid deposition on native herbaceous species

Species	Treatment	Effect
Festuca arundinacea	30 mm × 20 weeks pH 4.0 or pH 3.5 rain	Reduced root growth with pH 4.0; no effect with pH 3.5
Festuca arundinacea	30 mm × 9 weeks pH 4.0 rain	Increased total plant weight
Dactylis glomerata	30 mm × 12 weeks pH 4.0 or pH 3.5 rain	Reduced root growth with pH 4.0; no effect with pH 3.5
Lolium perenne	30 mm × 20 weeks pH 4.0 or pH 3.5 rain	Reduced root growth with both pH 3.5 and pH 4.5
Lolium perenne	24 mm rain at pH 4.5 plus 6 mm mist at pH 2.5, 3.5 or 4.5 for 17 weeks	No effect with pH 3.5 mist. Increased root and total plant weight with pH 2.5 and 4.5 mists
Holcus lanatus	24 mm rain at pH 4.5 plus 6 mm mist at pH 2.5, 3.5 or 4.5 for 13 weeks	Increased shoot and total plant weight with pH 2.5 mist
Lotus corniculatus	24 mm rain at pH 4.5 plus 6 mm mist at pH 2.5, 3.5 or 4.5 for 18 weeks	Increased total plant weight with pH 4.5 mist. Decreased shoot and total plant weights with pH 3.5 and pH 2.5 mist
Anthoxanthum odoratum	24 mm rain at pH 4.5 plus 6 mm mist at pH 2.5, 3.5 or 4.5 for 28 weeks	Increased root, shoot and total plant weights with pH 2.5 mist
Poa alpina	24 mm rain at pH 4.5 plus 6 mm mist at pH 2.5, 3.5 or 4.5 for 63 weeks	Increased root, shoot and total plant weights with pH 2.5 mist
Epilobium brunnescens	24 mm rain at pH 4.5 plus 6 mm mist at pH 2.5, 3.5 or 4.5 for 21 weeks	Increased root, shoot and total plant weights with pH 2.5 mist

After Irving (1983), Edge *et al.* (1994), Ashenden *et al.* (1991).

Effects on reproduction

All stages of the reproductive processes of plants have been found to be affected by rainfall acidity but responses once again vary substantially between species and few consistent effects have been observed. Jacobson *et al.* (1987) found reduced numbers of female flowers, dry mass of flowers and immature fruit in cucumber plants exposed to rainfall acidities below pH 3.4. Similarly, reductions in flower production were found at pH 4.5 and below compared with a pH 5.6 fog exposure for birdsfoot trefoil

(*Lotus corniculatus*) but, in contrast, an induction of production of flowering stems in perennial ryegrass (*Lolium perenne*) and increased flower production for New Zealand willow herb (*Epilobium brunnescens*) was reported in response to a pH 2.5 fog (Edge *et al.*, 1994; Ashenden *et al.*, 1991).

Pollen germination and pollen tube growth are well known to be sensitive indicators of atmospheric pollutants. In a series of studies reviewed by Cox (1987) the LD_{50} dosage (i.e. pH for a 50% failure of germination) was found to be between 3.95 and 3.63 for a range of broad-leaved tree species. Understory vegetation was found to be less sensitive (LD_{50} between pH 3.58 and 3.14) and conifers were the least sensitive of the plant groups tested (LD_{50} between pH 3.19 and 2.94).

Studies on several crop species have shown reduced seed production with exposure to acid rain. Simulated rainfall at pH 3.1 reduced both the numbers of pods per plant and seeds per pod for pinto beans (Evans and Lewin, 1980). In studies on soybeans, numbers of pods were reduced by exposures to pH 2.3–4.0 rain but seed numbers per pod and individual seed weights were unaffected (Evans *et al.*, 1981b). Similarly, for winter barley, rainfall acidities of pH 2.5 and 3.5 reduced production of ears but without effects on grains per ear or thousand grain weights, compared with plants grown in exposures to ambient pH 5.6 rains.

Germinative capacity in response to rainfall acidity again varies greatly between species. Percy (1986) found pH 2.6 rain to inhibit germination in four, but have no effect on seven other tree species. One of the most sensitive species found to date is red maple (*Acer rubrum*) which showed inhibition of germination capacity at pH 4.0 (Raynal *et al.*, 1982). Similarly, in studies on three fern species, Lawrence and Ashenden (1993) found little or no germination of spores at pH 2.5 and greatly reduced germination at pH 3.5 compared with spores exposed to less acid (>pH 4.5) mists. In contrast, however, several other studies have shown increased seed germination with increased acidities of rainfall (Lee and Weber, 1979; Percy, 1986).

Alterations in gas and water vapour exchange

There have been comparatively few studies on the physiological changes caused by wet acid deposition and many of the studies which have been conducted are confounded by interactions with soil conditions. However, reductions in net photosynthesis have been found for plants of white clover exposed to pH 2.5 and 3.5 acid mists compared with plants exposed to a pH 5.6 mist (Ashenden *et al.*, 1995). An earlier study by Ferenbaugh (1976) failed to find effects of pH 2.0–2.5 rains on net photosynthesis in kidney bean (*Phaseolus vulgaris*) but did observe increased respiration and reduced starch and sugar contents of leaves compared with plants exposed to pH 5.7 rainfall. For red spruce, exposures to acid mist and rain at pH 3 and 3.8 produced a large increase in the ratio of photosynthesis to respiration (McLaughlin and Tjoelker, 1992). Failure to demonstrate reductions in photosynthesis at lower rainfall acidities where reduced growth has been found for sensitive crops may be related to the fact that most measurements are taken over short time scales and not during misting events, because of technical difficulties.

Generally, where reported, exposure to wet acid deposition results in increased rates of transpiration. Evans *et al.* (1981a) found diffusion resistance much lower in leaves of *Phaseolus vulgaris* exposed to pH 2.7–3.4 rain compared with foliage exposed to pH 5.7 rainfall. Similarly, for trees, Mengel *et al.* (1989) observed increased transpiration for Norway spruce exposed to pH 3 mist and Leonardi and Flückiger (1989) found lower stomatal diffusion resistance at night and rapid water loss from detached leaves of beech exposed to pH 3 mist. In Norway spruce, Barnes *et al.* (1990) found reduction in stomatal control for plants exposed to mist at pH 3.6. These results suggest that foliage which has been exposed to acidic precipitation may subsequently be more vulnerable to drought conditions.

Invisible changes to leaf surfaces

There have been numerous reports of changes in the structure and properties of leaf surfaces, without visible injury symptoms, in response to acid mists and rain. These changes may be indicative of a long-term weathering and degradation of leaf cuticles in response to repeated rain/mist events and may occur at lower acidities than those required for visible injury.

Wax structure

Reports of abraded waxy cuticles and changes in the microscopic organisation of leaf surfaces have been made for several species. Rinallo *et al.* (1986) found fusion of fibrils, cracking of stomatal plugs and extensive erosion of epicuticular waxes for needles of Norway spruce exposed to mist at pH 3.5. Similarly, structural changes in surface waxes together with changes in chemical composition of waxes have been detected for seedlings of red spruce at pH 3 and Sitka spruce at pH 3.4 (Percy *et al.*, 1990; Percy and Baker, 1990). At a lower acidity of pH 4, simulated rain was found to cause changes to wax structure but without any reduction in the total amount of surface wax for Scots pine (Turunen and Huttunen, 1991).

Wettability

The wettability of a leaf is described in terms of the angle of contact between a liquid droplet and the leaf surface. A smaller contact angle denotes a greater attraction of the droplet to the leaf surface and thus a greater wettability. Such an increase in wettability could be expected to increase the effective dose of droplet acidity to the leaf and thus increase the potential for injury. Several studies have confirmed leaf wettability to be correlated with the degree of visible injury caused by acid rain (Haines *et al.*, 1985; Caporn and Hutchinson, 1986).

Exposure to acid rain and mist treatments at pHs of only 4.2–4.6 has been demonstrated to increase wettability of leaves for several species without visible injury symptoms (Percy and Baker, 1988 and 1989). These increases in wettability will reflect subtle changes in leaf surface properties which in the longer term may be expected to increase progressively effective pollutant dose.

Cuticular transpiration

Leaf cuticles are at the interface of the plant and its atmospheric environment. They form a protective barrier around internal plant tissues and occupy all parts of the leaf surface other than stomatal openings. Thus, changes in cuticular permeability may be important in affecting uptake of pollutants and rates of water loss. While cuticular transpiration accounts for only 2–5% of total plant transpiration, these losses cannot be regulated like stomatal transpiration.

Several studies have shown cuticular permeability to be affected by wet acid deposition. Barker and Ashenden (1992) found a significant correlation between water permeability of detached cuticles and pH of fog to which plants had been exposed. Similarly, Mengel *et al.* (1989) reported increased cuticular transpiration for needles of *Picea abies* exposed to acid fogs and pointed out that, in the long term, such increases in uncontrolled water loss could increase the susceptibility of plants to drought.

Spectral reflectance characteristics

Non-photographic spectroradiometry has been shown to be potentially a very sensitive indicator of subtle changes to leaf characteristics caused by acid rain. By measuring radiance in four wave bands (green, 0.5–0.6 μm; red, 0.6–0.7 μm; near infrared, 0.76–1.1 μm; and near-middle infrared, 1.35–1.75 μm) for canopies of birch seedlings, it has proved possible to distinguish between plants exposed to pH 5.6 rain and ones exposed to higher acidities of rainfall ranging from only pH 4.5 (Ashenden and Williams, 1988). The actual minor changes to the leaf surfaces, or structures within the leaves, which caused these changes in light reflective characteristics have not been identified. However, there is clearly a longer term potential for using this technique to identify areas of vegetation affected by wet acid deposition.

Foliar fertilisation and leaching

It is well known that the sulphur and nitrogen found in acid rain and mist may wash over plants, pass into the soil, be taken up as plant nutrients and then result in growth stimulation (Taylor *et al.*, 1986). However, an alternative route for uptake is directly via the foliage. Evans *et al.* (1981a) demonstrated direct uptake of tritiated water and sulphate into leaves of *Phaseolus vulgaris* over a range of rainfall pHs. While sulphate entered leaves faster at pH 2.7 than at pH 5.7, tritiated water entered foliage at similar rates for all pHs indicating that absorption of materials by leaf surfaces is a selective process. Subsequently, in a study where mist solutions were prevented from entering the rooting medium, a growth stimulation was found in red spruce seedlings exposed to pH 3 mist (Jacobson *et al.*, 1990).

A much more frequent observation is the loss of substances from leaves. Ions, particularly Ca^{2+}, Mg^{2+} and K^+, are readily lost from the surfaces of all wet leaves by cation exchange processes. However, the rate of leaching has been found to be greatly enhanced by increasing acidities of rainfall (Wood and Bormann, 1975; Adams and Hutchinson, 1984; Leonardi and Flückiger, 1989; Barker and Ashenden, 1992). The leaching rates of Ca^{2+} tend to be largest followed by Mg^{2+} with the leaching behaviour of K^+ being less consistent (Adams and Hutchinson, 1984; Barker and Ashenden, 1992). Some researchers have suggested that these effects of foliar leaching may result in nutrient deficiency within leaves and decreases of calcium and magnesium have been found in needles of red spruce seedlings after long-term exposures to acid mist (Jacobson *et al.*, 1989). However, other researchers have argued that losses of ions from leaching do not occur at biologically significant rates and point out that leached ions will wash into the soil and be reabsorbed by plant roots (Mengel *et al.*, 1987). It seems likely that any nutrient deficiencies from this process would develop over years rather than in the short term and be confined to plants growing on nutrient-poor soils.

Relative importance of sulphate and nitrate in wet deposition

The acidity of wet deposition is mainly attributed to a combination of sulphuric and nitric acids, the proportions of which may vary with location and precipitation event. While most research has considered the effects of acid rain/mist/fog/cloud events in terms of hydrogen ion concentrations (i.e. pH), a number of studies have shown that plant responses may be affected by the anions present. In general, where differences occur, precipitation containing predominantly sulphate has been found to be more toxic than precipitation at the same pH but predominantly containing the nitrate anion. Cape *et al.* (1991), for example, found mist made with sulphuric acid to defoliate seedlings of red spruce after a few weeks while a mist at the same pH of 2.5, made with nitric acid, had no apparent effect. Similarly, Rinallo *et al.* (1986) showed much greater damage to surface structures of leaves exposed to sulphuric acid alone than to a mixture of sulphuric acid and nitric acids or nitric acid alone at the same pH.

References

Abrahamsen, G. (1980). Impact of atmospheric sulphur deposition on forest ecosystems. In: *Atmospheric Sulphur Deposition* (Eds D.S. Shriner, C.R. Richmond and S.E. Lindberg) pp. 397–415. Ann Arbor Science Publishers, Ann Arbor.

Abrahamsen, G., Bjor, K., Hornvedt, R. and Tveite, B. (1976). Effects of acid precipitation on coniferous forests. In: *Impact of Acidic Precipitation on Forests and Freshwater Ecosystems in Norway* (Ed F.H. Brakke) pp. 36–63. SNSF Report LFR6/76.

Adams, C.M. and Hutchinson, T.C. (1984). A comparison of the ability of leaf surfaces to neutralize acid rain drops. *New Phytol.*, **97**, 463–478.

Amthor, J.S. (1984). Does acid rain directly influence plant growth? Some comments and observations. *Environ. Pollut.*, Ser A. **36**, 1–6.

Ashenden, T.W. and Bell, S.A. (1987). Yield reductions in winter barley grown on a range of soils and exposed to simulated acid rain. *Plant and Soil*, **98**, 433–437.

Ashenden, T.W. and Bell, S.A. (1988). Growth responses of birch and Sitka spruce exposed to acidified rain. *Environ. Pollut.*, **51**, 153–162.

Ashenden, T.W. and Bell, S.A. (1989). Growth responses of three legume species exposed to simulated acid rain. *Environ. Pollut.*, **62**, 21–29.

Ashenden, T.W., Bell, S.A. and Rafarel, C.R. (1995). Responses of white clover to gaseous pollutants and acid mist: implications for setting critical levels and loads. *New Phytol.*, **130**, 89–96.

Ashenden, T.W., Rafarel, C.R. and Bell, S.A. (1991). Exposures of two upland plant species to acidic fogs. *Environ. Pollut.*, **74**, 217–225.

Ashenden, T.W. and Williams, J.H. (1988). Differences in spectral characteristics of birch canopies exposed to simulated acid rain. *New Phytol.*, **109**, 79–84.

Barker, M.G. and Ashenden, T.W. (1992). Effects of acid fog on cuticular permeability and cation leaching in holly (*Ilex aquifolium*). *Agriculture, Ecosystems and Environment*, **42**, 291–306.

Barnes, J.D., Eamus, D. and Brown, K.A. (1990). The influence of ozone, acid mist and soil nutrient status on Norway spruce (*Picea abies* (L.) Karst). I. Plant water relations. *New Phytol.*, **114**, 713–720.

Binns, W.O. (1984). *Acid Rain and Forestry*. Forestry Commission Research and Development Paper 134. Forestry Commission, Edinburgh.

Cape, J.N. (1993). Direct damage to vegetation caused by acid rain and polluted cloud: definition of critical levels for forest trees. *Environ. Pollut.*, **82**, 167–180.

Caporn, S.J.M. and Hutchinson, T.C. (1986). The contrasting response to simulated acid rain of leaves and cotyledons of cabbage (*Brassica oleracea* L.). *New Phytol.*, **103**, 311–324.

Caporn, S.J.M. and Hutchinson, T.C. (1987). The influence of temperature, water and nutrient conditions during growth on the response of *Brassica oleracea* L. to a single, short treatment with simulated acid rain. *New Phytol.*, **106**, 251–259.

Cox, R.M. (1987). The response of plant reproductive processes to acidic rain and other air pollutants. In: *Effects of Atmospheric Pollutants on Forests, Wetlands and Agricultural Ecosystems* (Eds T.C. Hutchinson and K.M. Meema) pp. 155–170. Springer-Verlag, Berlin and Heidelburg.

Edge, C.P., Bell, S.A. and Ashenden, T.W. (1994). Contrasting growth responses of herbaceous species to acidic fogs. *Agriculture, Ecosystems and Environment*, **51**, 293–299.

Evans, L.S. (1984). Acid precipitation effects on terrestrial vegetation. *Ann. Rev. Phytopathol.*, **22**, 397–420.

Evans, L.S. and Curry, T.M. (1979). Differential responses of plant foliage to simulated acid rain. *Am. J. Bot.*, **66**, 953–962.

Evans, L.S., Curry, T.M. and Lewin, K.F. (1981a). Responses of leaves of *Phaseolus vulgaris* to simulated acid rain. *New Phytol*., **88**, 403–420.

Evans, L.S., Gmur, N.F. and Da Costa, F. (1978). Foliar response of six clones of hybrid poplar to simulated acid rain. *Phytopathology*, **68**, 847–856.

Evans, L.S. and Lewin, K.F. (1980). Growth, development and yield responses of pinto beans and soybeans to hydrogen ion concentrations of simulated acid rain. *Environ. Exp. Bot.*, **21**, 103–113.

Evans, L.S., Lewin, K.F., Conway, C.A. and Patti, M.L. (1981b). Seed yields (quantity and quality) of field-grown soybeans exposed to simulated acid rain. *New Phytol.*, **89**, 459–470.

Evans, L.S., Lewin, K.F., Cunningham, E.A. and Patti, M.J. (1982). Effects of simulated acid rain on yields of field-grown crops. *New Phytol.*, **91**, 429–441.

Evans, L.S., Lewin, K.F., Patti, M.J. and Cunningham, E.A. (1983). Productivity of field-grown soybeans exposed to simulated acid rain. *New Phytol.*, **93**, 377–388.

Ferenbaugh, R.W. (1976). Effects of simulated acid rain on *Phaseolus vulgaris* L. (Fabaceae). *Am. J. Bot.*, **63**, 283–288.

Frevert, T. and Klemm, O. (1984). Wie ändern sich pH-Werte im Regen-und Nebelwasser beim Abtrocknen auf Pflanzenoberfl ächen? *Archiv für Meteorologie, Geophysik und Bioklimatologie Ser.*, **B34**, 75–81.

Gilbert, O.L. (1986). Field evidence for an acid rain effect on lichens. *Environ. Pollut.*, Ser A, **40**, 227–231.

Haines, B.L., Jernstedt, J.A. and Neufeld, H.S. (1985). Direct foliar effects of simulated acid rain. II Leaf surface characteristics. *New Phytol.*, **99**, 407–416.

Harcourt, S.A. and Farrar, J.F. (1980). Some effects of simulated acid rain on the growth of barley and radish. *Environ. Pollut.*, **22**, 69–73.

Hutchinson, T.C., Dixon, M. and Scott, M. (1986). The effect of simulated acid rain on feather mosses and lichens of the boreal forest. *Water Air Soil Pollut.*, **31**, 409–416.

Irving, P.M. (1983). Acid precipitation effects on crops: a review and analysis of research. *J. Environ. Qual.*, **12**, 442–453.

Jacobson, J.S. (1984). Effects of acidic aerosol, fog, mist and rain on crops and trees. *Phil. Trans. Royal Soc. Lond.*, **B305**, 327–338.

Jacobson, J.S. (1991). The effects of acid precipitation on crops. In: *Acid Deposition in Europe* (Eds M.J. Chadwick and M. Hutton) pp. 81–98. Stockholm Environment Institute, York.

Jacobson, J.S., Bethard, T., Heller, L.I. and Lassoie, J.P. (1990). Response of *Picea rubens* seedlings to intermittent mist varying in acidity, and in concentrations of sulfur-, and nitrogen-containing pollutants. *Physiol. Plant.*, **78**, 595–601.

Jacobson, J.S., Lassoie, J.P., Osmeloski, J. and Yamada, K. (1989). Changes in foliar elements in red spruce seedlings after exposure to sulfuric and nitric acid mist. *Water, Air and Soil Pollut.*, **48**, 141–159.

Jacobson, J.S., Osmeloski, J., Yamada, K. and Heller, L. (1987). The influence of simulated acidic rain on vegetative and reproductive tissues of cucumber (*Cucumis sativus* L.). *New Phytol.*, **105**, 139–147.

Jacobson, J.S., Troiano, J.J., Heller, L.I. and Osmeloski, J. (1986). Influence of sulfate, nitrate and chloride in simulated acidic rain on radish plants. *J. Environ. Qual.*, **15**, 301–304.

Jonsson, B. (1977). Soil acidification by atmospheric pollution and forest growth. *Water Air Soil Pollut.*, **7**, 497–501.

Lawrence, P.A. and Ashenden, T.W. (1993). Effects of acidic gases and mists on the reproductive capability of three fern species. *Environ. Pollut.*, **79**, 268–270.

Lee, J.A., Caporn, S.J.M. and Read, D.J. (1992). Effects of increasing nitrogen deposition and acidification on heathlands. In: *Acidification Research, Evaluation and Policy Applications* (Ed T. Schneider). Elsevier Science, Amsterdam.

Lee, J.A., Press, M.C., Studholme, C. and Woodin, S.J. (1988). Effects of acidic deposition on wetlands. In: *Acid Rain and Britain's Natural Ecosystems* (Eds M.R. Ashmore, J.N.B. Bell and C. Garretty), pp. 27–37. Imperial College Centre for Environmental Technology, London.

Lee, J.J., Neely, G.E., Perrigan, S.C. and Grothaus, L.C. (1981). Effect of simulated sulfuric acid rain on yield, growth and foliar injury of several crops. *Environ. Exper. Bot.*, **21**, 171–185.

Lee, J.J. and Weber, D.E. (1979). The effects of simulated acid rain on seedling emergence and growth of eleven woody species. *Forest Sci.*, **25**, 393–398.

Leonardi, S. and Flückiger, W. (1989). Effects of cation leaching in mineral cycling and transpiration: investigation with beech seedlings, *Fagus sylvatica L. New Phytol.*, **111**, 173–179.

McLaughlin, S.B. and Tjoelker, M.J. (1992). Growth and physiological changes in red spruce saplings associated with acid deposition levels at high elevation sites in the Southern Appalachians, USA. *Forest Ecol. Manage.*, **51**, 43–51.

Mansfield, T.A. (1988). Establishing the critical physiological effects of air pollutants on plants. *Aspects Appl. Biol.*, **17**, 1–7.

Mengel, K., Hogrebe, A.M.R. and Esch, A. (1989). Effect of acidic fog on needle surface and water relations of *Picea abies*. *Physiologia Plantarum*, **75**, 201–207.

Mengel, K., Lutz, H.J. and Breininger, M. Th. (1987). Aswaschung von Nährstoffen durch sauren Nebel aus jungen untakten Fichten (*Picea abies*). *Z. Pflanzenernaehr. Bodenkd.*, **150**, 61–68.

NATO (1978). *Effects of Acid Precipitation on Vegetation and Soils*. NATO, Toronto.

Percy, K. (1986). The effects of simulated acid rain on germinative capacity, growth and morphology of forest tree seedlings. *New Phytol.*, **104**, 473–484.

Percy, K. (1991). Effects of acid rain on forest vegetation: morphology and non-mensurational growth effects. In: *Effects of Acid Rain on Forest Resources*. Proceedings of a conference held in Ste. Foy. Quebec, Forestry Canada, Ottawa, pp. 97–110.

Percy, K.E. and Baker, E.A. (1988). Effects of simulated acid rain on leaf wettability, rain retention and uptake of some inorganic ions. *New Phytol.*, **108**, 75–82.

Percy, K.E. and Baker, E.A. (1989). Effect of simulated acid rain on foliar uptake of RB^+ and $SO_4{}^{2-}$ by two clones of Sitka Spruce (*Picea sitchensis* (Bong) Carr.). In: *Air Pollution and Forest Decline* (Eds J.B. Butcher and I. Bucher-Wallin), pp. 493–495. IUFRO, Birmensdorf, Switzerland.

Percy, K.E. and Baker, E.A. (1990). Effects of simulated acid rain on epicuticular wax production, morphology, chemical composition and on cuticular membrane thickness in two clones of Sitka spruce (*Picea sitchensis* (Bong) Carr.). *New Phytol.*, **116**, 79–87.

Percy, K.E., Krause, C.R. and Jensen, K.F. (1990). Effects of ozone and acidic fog on red spruce needle epicuticular wax ultrastructure. *Can. J. Forest Res.*, **20**, 117–120.

Raynal, D.J., Roman, J.R. and Eichenlaub, W.M. (1982). Response of tree seedlings to acid precipitation. I Effect of substrate acidity on seed germination. *Environ. Exp. Bot.*, **22**, 377–383.

RGAR (1990). United Kingdom Review Group on Acid rain. *Acid Deposition in the United Kingdom 1981–1985*. Warren Spring Laboratory, Stevenage.

Rinallo, C., Raddi, P., Gellini, R. and Di Lonardo, V. (1986). Effects of simulated acid deposition on the surface structure of Norway spruce and silver fir needles. *Eur. J. Forest Pathol.*, **16**, 440–446.

Scott, M.G., Hutchinson, T.C. and Feth, M.J. (1989). A comparison of the effects on Canadian boreal forest lichens of nitric and sulphuric acids as sources of rain acidity. *New Phytol.*, **111**, 663–671.

Smith, R.A. (1872). *Air and Rain–The Beginnings of Chemical Climatology*. Longman, London.

Taylor, Jr., G.E., Norby, R.J., McLaughlin, S.B., Johnson, A.H. and Turner, R.S. (1986). Carbon dioxide assimilation and growth of red spruce (*Picea rubens* Sarg.) seedlings in response to ozone, precipitation chemistry and soil type. *Oceologia* (Berlin), **70**, 163–171.

Turunen, M. and Huttunen, S. (1991). Effect of simulated acid rain on epicuticular wax of Scots pine needles under northerly conditions. *Can. J. Bot.*, **69**, 412–419.

USDA (1976). *Proceedings of First International Conference on Acid Rain*. (Eds. L.S. Dochinger and T.A. Seliga). Technical report NE-23. United States Department of Agriculture Forest Service, Upper Darby, Pennsylvania.

Wood, T. and Bormann, F.H. (1975). Increases in foliar leaching caused by acidification of an artificial mist. *Ambio*, **4**, 169–171.

14 Effects of pollutant mixtures

A. FANGMEIER, J. BENDER, H.-J. WEIGEL AND H.-J. JÄGER

Introduction

Plants in their natural environment are important mediators in the exchange of a wide variety of different gaseous and particulate compounds between the atmosphere and the biosphere (Table 14.1), i.e. the vegetation may either act as a source or a sink for these compounds. For example, as part of the overall biogeochemical cycles of the major elements several trace gases (CO, CO_2, CH_4, NMHC, or VOC, N_2, NO, NO_2, N_2O, SO_2, H_2S etc.) are emitted from terrestrial ecosystems via the plant canopy into the atmosphere where they act as greenhouse gases or where they may contribute to the causes of regional or local air pollution problems.

Once airborne compounds are deposited to the biosphere they may affect plant performance and ecosystem properties. Referring to their impact on vegetation they can be broadly divided (see also Table 14.1) into

- essential nutrient compounds which act as macro- or micronutrients (e.g. the gases CO_2, SO_2, NO, NO_2, NH_3 and particulate NH_4, NO_3-N, SO_4-S, P, Ca, Fe, Mg) and
- compounds which may cause adverse or toxic effects (e.g. the gaseous pollutants O_3, SO_2, NO_2, HF, PAN, NMHC or VOC, metals like Pb, Cd, Hg) or excess nutrient substances (e.g. N, S, Zn, Al) which alter normal patterns of growth and development in ecosystems (Dämmgen and Weigel, 1998).

Plants as well as other organisms in ecosystems are seldom exposed to single airborne chemicals or pollutants, respectively, but almost always to a number of compounds together. The pattern of these pollutant mixtures varies spatially and temporally, e.g. as co-occurrences can be simultaneous and/or sequential. Due to the occurrence of such mixtures any assessment of air pollutant effects on plants and ecosystems should consider possible interactions of pollutants with regard to their effects. Past and current air pollution impact research has been dominated by single pollutant studies and relatively few attempts have been made to address the issue of effects of pollutant mixtures. This chapter will review our current knowledge of effects of mixtures of air pollutants on higher plants (crops, wild plants and forest trees). Emphasis will be on gaseous pollutants but examples of interactions of gases with other types of pollutants will also be described. Whenever possible only studies relevant to ambient conditions with

Air Pollution and Plant Life, second edition. Edited by J.N.B. Bell and M. Treshow. ISBN 0 471 49090 3 (HB), 0 471 49091 1 (PB).

Table 14.1 Atmospheric compounds involved in element flux between vegetation and atmosphere (after Dämmgen and Weigel 1998)

• H_2O-vapor, CO_2, CH_4, N_2O, NO_2, O_3	→ trapping of infrared radiation, contribution to the greenhouse effect
• NH_3, CO, HC	→ effects on reactivity of the atmosphere
• CH_4, CO_2, SO_2, NO_2, NO, NH_3 (gases), NH_4-/NO_3-N, SO_4-S, P, Ca, K, Fe, Mg (particles)	→ involved in nutrient cycling, act as macro- and micronutrients
• O_3, SO_2, NO_2, HF, H_2O_2, PAN, NMHC/VOC (gases), heavy metals (e.g. Pb, Cd, Hg), surplus nutrients (bioavailable forms of N, S, Zn, Al)	→ potentially toxic, affecting 'normal' growth and performance of organisms, populations and ecosystems

Abbreviations: Al: aluminum; Ca: calcium; Cd: cadmium; CH_4: methane; CO: carbon monoxide; CO_2: carbon dioxide; Fe: iron; HC: hydrocarbons; HF: fluoride; Hg: mercury; H_2O: water vapor; H_2O_2: hydrogen peroxide; H_2S: hydrogen sulfide; H_2SO_4: sulfuric acid; HNO_2: salpetric acid; HNO_3: nitric acid; K: potassium; Mg: magnesium; N: nitrogen; N_2O: laughing gas; N_2O_5: dinitrogenpentoxide; NH_3: ammonia; NH_4^+: ammonium; NH_4NO_3: ammonium nitrate; NO: nitrogen monoxide; NO_2: nitrogen dioxide; NO_3^-: nitrate; NO_X: NO + NO_2; NMHC: non-methane hydrocarbons; O_3: ozone; P: phosphorus; PAN: peroxyacetylnitrate; Pb: lead; S: sulfur; SO_2: sulfuric acid; SO_4^{2-}: sulfate; VOC: volatile organic compounds; Zn: zinc.

respect to the pollutant exposure concentration and the plant growth conditions are considered.

Relevant pollutant scenarios

Studies of effects of air pollutant mixtures have been confined to a limited number of pollutants and are mostly restricted to gaseous pollutants. Most of the available literature on pollutant combinations is older than 10 years and has primarily dealt with mixtures of the gases SO_2, NO_2 and O_3 (reviewed by Ormrod, 1982; Runeckles, 1984; Reinert, 1984; Kohut, 1985; Mansfield and McCune, 1988). This is because the air pollution climate up to the 1990s was characterized by the widespread occurrence of SO_2 and NO_2 at phytotoxic concentrations in addition to high O_3 levels, i.e. these pollutants were recognized as being a major part of the pollution problems in Europe and North America. However, most of these early studies with pollutant interactions involved short-term exposures at acute concentrations.

The pollution climate in many parts of the developed world has changed during the last decades (Dämmgen and Weigel 1998). For example, concentrations of SO_2 have declined considerably and higher levels are usually restricted to wintertime when the energy demand is high, while ground level O_3 concentrations remain the most significant threat to vegetation during the growing season (Stockwell *et al.*, 1997; Kley *et al.*, 1999). A summertime situation for the daily variation of O_3 and SO_2 concentrations in a rural area of Germany is shown in Figure 14.1(a). More recent information on plant responses to pollutant combinations is scarce and research has primarily focused on interactions of O_3 with one other pollutant (Wolfenden *et al.*,

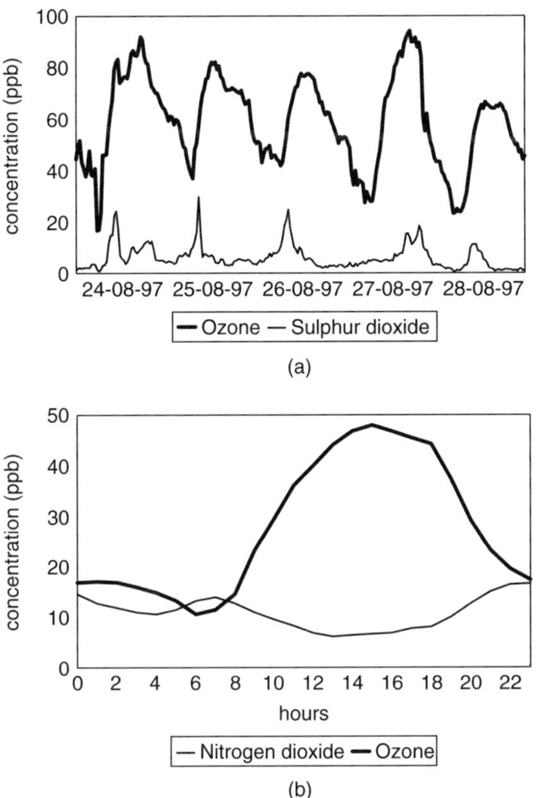

Figure 14.1 (a): Example of a period of co-occurrences of O_3 and SO_2 at Hasenholz (Brandenburg, Germany) in August 1997 (Dämmgen, unpublished results), and (b): mean diurnal profile of O_3 and NO_2 concentrations in ambient air at the FAL Braunschweig (Bender *et al.*, 1991). Reproduced by permission of Bender *et al.* Figures 1 and 2 in 'Response of nitrogen metabolism in beans' from *The New Phytologist* (1971) issue **119**, pp. 261–7, Blackwell Science

1992; Bender and Weigel, 1993, 1994; Barnes and Wellburn, 1998). Only a few studies have considered multiple pollutant interactions.

Currently, a simultaneous occurrence of air pollutants like SO_2, NO, NO_2, O_3 or NH_3 at phytotoxic levels is unusual. Co-occurrences of e.g. SO_2, NO_2 and O_3 are only of short duration and far less frequent than sequential or combined sequential/concurrent exposures (USEPA, 1996), although pollutants which are emitted from similar sources (e.g. SO_2 and NO_2 produced by fossil fuel combustion) tend to show similar spatial and temporal patterns. In particular, concentrations of NO_2 and O_3 (both constituents of photochemical smog) vary during the day in patterns that normally result in sequential exposures (Figure 14.1(b)).

Pollutants like O_3 and NO_2 are all pervasive and of widespread concern. On more local scales or in the vicinity of point sources, mixtures of these prevalent gases with the other pollutants listed in Table 14.1 (e.g. NH_3, HF, VOC, heavy metals) are also of concern with respect to their effects on the local vegetation.

Classification of interactions and analytical requirements

Similar to the situation in toxicology upon exposure of an organism to a mixture of chemicals the (sequential or concurrent) exposure of plants to mixtures of air pollutants has repeatedly been shown to modify the magnitude and nature of the response to individual pollutants. However, attempts to classify the mode of action of air pollutant mixtures on plants remain scarce and the terminology used to describe the interactions has not been standardized (Feron *et al.*, 1998). One of the most frequently used concepts describing possible interactions of two air pollutants in a mixture (pollutant A and B) makes a distinction between *no joint action* when one of the pollutants induces no plant responses, and *joint action* when both pollutants result in some plant responses (USEPA, 1984; Figure 14.2). The concept of joint action includes the sub-categories (i) *additive* effects, when effect_{AB} is equal to $\text{effect}_A + \text{effect}_B$ (which is not an interaction) and (ii) *interactive* effects, when effect_{AB} is not equal to $\text{effect}_A + \text{effect}_B$. If the sum of the two effects_{AB} is less than $\text{effect}_A + \text{effect}_B$, the interaction is termed *antagonistic* ('less-than-additive'). Alternatively, if the sum of the two effects_{AB} is greater than $\text{effect}_A + \text{effect}_B$, the interaction is called *synergistic* ('more-than-additive').

It remains open if this concept is also valid where more than two pollutants in a mixture have to be considered. Moreover, the type of interaction may change with the concentrations of the individual components in a mixture. Therefore, for more complex pollutant mixtures the use of the terms *more* or *less than additive* instead of synergism and antagonism is preferable as they seem biologically more meaningful and can be handled more easily with statistical procedures.

Experimental assessments of interactive effects of pollutants in mixtures require complete factorial designs that consider all the pollutants of concern and all their possible combinations (Oshima and Bennett, 1979). Numerous studies exist which allow only individual comparison of treatments because there were not enough experimental units (e.g. exposure chambers, field plots) available to assess the main and

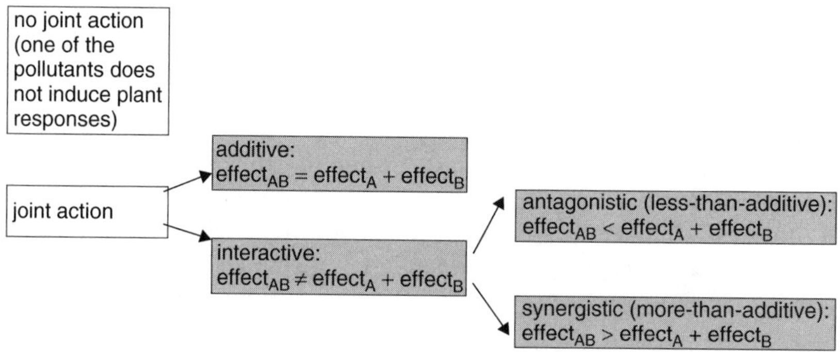

Figure 14.2 Scheme of possible actions of two pollutants (A and B) in combination

interactive effects of each pollutant involved (USEPA, 1984). For example, a full examination of the effects of a three-pollutant mixture requires eight separate treatments, not considering any replication. Thus, for a statistically sound number of replications sufficient experimental units must be available to run all treatments at the same time. Furthermore, because of the large number of exposure chambers required, very few studies have been able to incorporate a range of concentrations of the individual pollutants in mixtures. With respect to the number of species listed in Table 14.1 it is obvious that not all possible combinations of pollutants occurring in the field have been tested.

Assessments of effects of pollutant mixtures and especially of gaseous pollutants have also been carried out by comparing the response of plants grown in ambient air with those grown in air from which the relevant pollutants were fully or partially removed by e.g. filtration (Heagle, 1989; Manning and Krupa, 1992; Ashmore and Bell, 1994; see later). Similarly, treatments consisting of filtered air serving as controls or references, respectively, were compared to treatments where filtered air was 'spiked' with a set of the most prevailing pollutants at the particular site of the study (Küppers *et al.*, 1994; Fangmeier, 1989). These kinds of approaches only allow a site-specific assessment of the potential adverse effects of the respective pollution climate on the vegetation but may not be suitable for attributing effects to a specific pollutant because of the difficulty in partitioning the effects to the individual pollutants and their interaction.

Mechanisms of single pollutant effects

Air pollutants have two basic modes of action on vegetation: they may either be potentially phytotoxic, or they may serve as plant nutrients at low deposition rates and turn out to be toxic at higher deposition rates.

Pollutants that are solely phytotoxic have no function in the regular metabolism of a plant organism. The most prominent example of such an air pollutant which does not turn up in the metabolic pathways of organisms is represented by tropospheric ozone (see Chapters 5 and 6). The uptake of such pollutants may cause chronic effects (i.e. a reduction of vitality, such as reduced growth and productivity, without producing visible injuries) or acute effects (such as visible leaf lesions). The occurrence of such effects depends on the ability of the organ or of the organism to detoxify the pollutant.

Other pollutants or their products formed during atmospheric transport, however, occur in the regular metabolism of organisms. By this means, they can contribute to the nutrition of plants. This holds for most of the nitrogen and sulfur containing compounds and also for essential heavy metals. The uptake of such a pollutant will cause adverse effects for a plant if the uptake rate exceeds the assimilation capacity for that particular compound. Assimilation of the compound may then be regarded as an efficient detoxification process at the individual level. However, even the successful assimilation of such compounds may turn out to be adverse in ecological terms by affecting the ecological properties of a species or the stability of an ecosystem. Such adverse effects connected with the successful assimilation of pollutants have been observed frequently in ecosystems exposed to nitrogen deposition (see Chapter 12). Since nitrogen is a limiting nutrient in nearly all natural terrestrial ecosystems, deposition of nitrogen at low levels leads to enhanced growth and productivity. This N deposition makes the systems more susceptible to other stresses, e.g. it has been demonstrated for conifer trees and heathland shrubs such as *Calluna vulgaris* that plants are less tolerant to drought and frost after moderate exposure to NH_3 (see Fangmeier *et al.*, 1994) which may be attributed to reduced root growth and decreased uptake of soil water and soil nutrients such as K^+, Mg^{2+} and Ca^{2+}. Furthermore, attack by pests appears to be more severe under low to moderate nitrogen deposition which has been shown for the interplay between heather (*Calluna vulgaris*) and the heather beetle (*Lochmaea suturalis*) (Berdowski, 1993).

An important mechanism which is frequently involved in the occurrence of acute effects is mediated by reactive oxygen species. These may form during uptake of oxidants such as ozone, but also due to SO_2 uptake. There are many other environmental stresses involving reactive oxygen species (Bartosz, 1997), such as chilling or drought. Oxidative stress may also be brought about by intensive irradiation under conditions where stomatal closure prevails ('photooxidation'). In this case, the chloroplasts are the locations within the cell that suffer most from oxidative stress. During evolution, plants have 'learned' to cope with such stress in various ways. Concerning photooxidative stress, antioxidant systems, consisting of a cycle of various enzymes and metabolites and involving photorespiration, have evolved in chloroplasts. Unless the capacity of these systems is not exceeded, no bleaching will occur. However, oxidative stress brought about by pollutants is more similar to stress caused by pests such as phytopathogenic fungi in incompatible host–pathogen interactions (Kangasjärvi *et al.*, 1994), as the location of first occurrence of oxidants is represented by the apoplast and not by the cell interior. Plants can defend themselves against oxidants in the apoplast by building up some antioxidative potential in the apoplast. Ascorbic acid appears to be one of the most important antioxidants acting as a radical scavenger in this respect. A high flux of free radicals in the apoplast eventually initiates the hypersensitive response (Dixon *et al.*, 1996; Delledonne *et al.*, 1998) leading to rapid, localized cell death which may be seen as necrotic lesions of the leaf (Schraudner *et al.*, 1998).

Another basic mode of action of pollutants that may lead to cell death is a disturbance of the cellular acid-base regulation. Any pollutant that reacts acidic (such as SO_2, NO_x) or alkaline (such as NH_3) is capable of disturbing the acid-base regulation. Plant cells possess a pH-stat system which keeps the cellular pH at the desired level (Raven, 1988; Pfanz, 1995). The capacity of this cellular pH-stat system depends, among others, on the mode of nitrogen nutrition of a particular species which in turn is associated with its basic functional type (Pearson and Stewart, 1993; Fangmeier *et al.*, 1994). Fast-growing ruderal species, including most crops and fast-growing pioneer tree species, usually prefer nitrate (NO_3^-) as a nitrogen source, whereas slow-growing perennials and climax species usually prefer ammonium (NH_4^+). Since approximately 0.78 mol excess OH^- are yielded in a cell when 1 mol NO_3^- is assimilated, whereas 1.22 mol H^+ are formed during assimilation of 1 mol NH_4^+ (Raven, 1988), plant species preferring nitrate uptake and assimilation are much less prone to cellular acidification. Correspondingly, the activity of nitrate reductase (the first enzyme involved in the NO_3^- assimilation pathway), and the cellular buffering capacity appear to be suitable parameters to predict a species' sensitivity to acidifying air pollutants (Soares *et al.*, 1995). As soon as the capacity of the pH-stat system of a cell is exceeded, the cell will inevitably die.

Heavy metals involve another basic mechanism of action. They are mainly taken up from soil, whereas atmospheric deposition onto plant surfaces represents only a minor source of uptake into the tissues (though high amounts can accumulate on plant surfaces). Mobility within higher plants differs significantly between heavy metal species. For example, cadmium is highly mobile within plants, whereas lead is rather immobile. Once taken up into the plant cell, many heavy metals (e.g. arsenic, lead, cadmium and mercury) may bind to thiol groups of proteins and may thereby disturb their functioning. Detoxification of heavy metals within plant tissues involves glutathione (a thiol compound consisting of three amino acids: glutamate, cysteine and glycine, also belonging to the antioxidant cycle to detoxify reactive oxygen species) phytochelatins, i.e. polypetides with many SH-containing amino acids, and many other complex builders (for a review see di Toppi and Gabbrielli, 1999). These scavenge heavy metal ions and prevent their binding to structural or enzyme proteins.

In the field, one pollutant never acts alone, rather, organisms and ecosystems are exposed to hundreds of different compounds which are interacting in numerous ways. For a mechanistic understanding of the mode of interaction of different pollutants it would be necessary to detect how one pollutant affects the susceptibility of an organism to another pollutant. This can work by several mechanisms: one pollutant may modify the capacity of an organism to detoxify or to assimilate another compound, or one pollutant may modify the rate of uptake of another pollutant via modifications of stomatal behavior. Such a conceptual model of pollutant interactions has been provided by Barnes and Wellburn (1998). As air pollutants represent only one aspect of the chemical climatic conditions which vary strongly in space and time, we need also to understand how the interactions between pollutants vary according to any other environmental condition. In this respect, we are only at the very beginning of a possible understanding.

Interactions of two pollutants

Ozone and sulfur dioxide

The combination of these two gases has been studied quite intensively. In most of the older studies, more-than-additive effects on growth and yield have been reported (reviewed by Reinert *et al.*, 1975; Ormrod, 1982; Runeckles, 1984). More recent work carried out in the framework of the European Crop Loss Assessment Network, which was conducted between 1985 and 1991, supports the finding of synergistic effects of $O_3 + SO_2$ (Bender and Weigel, 1993). However, no significant interactions were detected concerning the yield response of eight different crop species exposed to O_3 and SO_2 at realistic concentrations in open-top field chamber systems within the U.S. National Crop Loss Assessment Program (Heagle *et al.*, 1988). There are also reports of less-than-additive effects of the two pollutants. e.g., Ashmore and Önal (1984) found leaf injury in barley after exposure to O_3 alone, but less injury after exposure to $O_3 + SO_2$.

In contrast, field surveys for symptoms of foliar injury on Saskatoon serviceberry (*Amelanchier alnifolia*) indicated that symptoms could clearly be attributed to chronic SO_2-exposures, but were exacerbated by the presence of O_3 (Krupa and Legge 1999). Based on an overall examination of both field survey and literature data Krupa and Legge (2001) concluded that Saskatoon serviceberry can be used as a biological indicator of chronic sulfur dioxide exposures, even in the presence of phytotoxic O_3 levels.

To understand how these two gases may interact in different manners, one must keep in mind that exposure to ozone at low concentrations will activate stress response pathways which successfully detoxify the reactive oxygen species built after O_3 entry into the leaf interior. This may also protect a plant from SO_2 as long as the threshold at which the acid-base regulation is disturbed by SO_2 is not exceeded. In this case, the two gases may act in a less-than-additive manner. If both pollutants are present at higher concentrations and taken up at respective amounts, more-than-additive interactions may occur. The complex response to combinations of O_3 and SO_2 and the dependence of the interaction on the exposure concentrations is illustrated in Figure 14.3 for spring rape (based on data obtained by Adaros *et al.*, 1991a).

Ozone and gaseous ammonia

Relatively few experiments have been carried out on the interactive effects of tropospheric ozone and ammonia (NH_3). The basic modes of action of these two pollutants differ considerably (compare earlier section). The few experimental studies that are available have either dealt with forest trees, such as Douglas fir (*Pesudotsuga menziesii*) or Scots pine (*Pinus sylvestris*), or with bush bean (*Phaseolus vulgaris*). In Douglas fir, antagonistic effects of O_3 and NH_3 on net photosynthesis and stomatal function were observed (Van Hove and Bossen, 1994), as at least in the later stages of a five-month fumigation period NH_3 appeared to delay O_3 effects. In *Pinus sylvestris*, Pérez-Soba *et al.* (1995) found less effects on mycorrhizal infection and on the enzyme glutamate dehydrogenase under $O_3 + NH_3$ exposure than for each pollutant alone. Correspondingly, Dueck *et al.* (1998) stated that O_3 ameliorated the effect of NH_3 on drought sensitivity of Scots pine. However, additive effects on leaf injury and pod yield were reported for bean (*Phaseolus vulgaris* L. cv. Pros) in an open-top chamber study conducted by Tonneijck and van Dijk (1998). Thus, no clear conclusions can be drawn on the interactions between O_3 and NH_3. However, in three of the four studies cited above, antagonistic effects were observed. This may be attributed to the fact that in all of these studies NH_3 was applied at concentrations not causing acute effects. Rather, the plants could successfully assimilate gaseous ammonia and improve their nitrogen status. By this means, their resistance to O_3 effects might have been increased. It may not be excluded from these data that O_3 and NH_3 might act synergistically if applied at higher concentrations.

Ozone and nitrogen dioxide

Both O_3 and NO_2 are constituents of photochemical 'smog', i.e. this combination must be regarded as one of the most common under field conditions. Surprisingly, however,

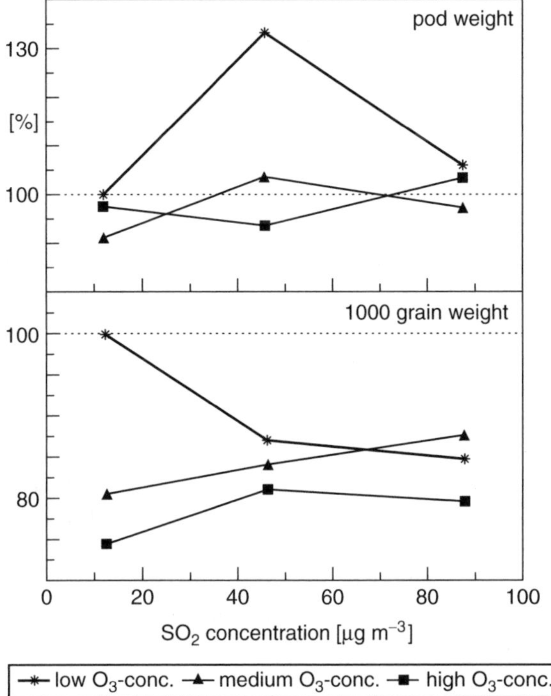

Figure 14.3 Interactive effects of O_3 and SO_2 on yield parameters of spring rape (*Brassica napus* cv. Callypso). Results are expressed as percentage of controls (charcoal-filtered air). Based upon data presented in Adaros *et al.* (1991a). Reprinted from *Environmental Pollution*. vol. 72, Adaros *et al.*, 'Single and interactive effects of low levels of O_3, SO_2 and NO_2 on the growth and yield of spring rape', Figure 1, pp. 269–296, with permission from Elsevier Science

the combined action of O_3 and NO_2 on plants has received only little attention in pollutant mixture research (Barnes and Wellburn 1998). In many regions concentrations of these pollutants vary during the day in patterns that result in sequential rather than in simultaneous exposures (Lefohn *et al.*, 1987; Bender *et al.*, 1991; Figure 14.1(b)).

In most of the previous studies dealing with mixtures of O_3 and NO_2 a simultaneous exposure regime has been selected rather than sequential exposure. Simultaneous combinations of O_3 and NO_2 can cause reductions in growth and productivity in a number of species (Kress and Skelly, 1982; Reinert, 1984; USEPA, 1984; Ito *et al.*, 1985). However, the majority of these studies have employed concentrations which are considerably higher than those observed in polluted environments. Guderian and Tingey (1987) therefore concluded that short-term exposures to relatively high concentrations of both pollutants may result in synergistic (more-than-additive) effects on plants.

More realistic studies using sequential exposures of near-ambient concentrations of O_3 and NO_2 were conducted by Runeckles and Palmer (1987), Goodyear and Ormrod (1988), Adaros *et al.* (1991a, b) and Bender *et al.* (1991) and included a number of crop species (wheat, barley, bush bean, radish, tomato, oil-seed rape). In general, these experiments provided only little evidence of statistically significant interactions (additive effects predominate). When interactions were observed, the mode of interaction on several growth and yield parameters was mostly antagonistic (Bender and Weigel, 1994) i.e. the presence of NO_2 counteracted the adverse effects of O_3. Nevertheless, a high variability in the responses was found, e.g. from year to year or between different response variables. The sensitivity of plants to this pollutant combination seems also to be influenced by the stage of plant development. For example, a shift in nitrogen metabolism from stimulation to depression during the plants' development was observed in a detailed mechanistic study on the response of *Phaseolus vulgaris* to the sequence O_3-NO_2 (Bender *et al.*, 1991). Both NO_2 and the O_3-NO_2 sequence stimulated the *in vitro* activity of enzymes linked with nitrogen assimilation (nitrate and nitrite reductase, glutamine synthetase, glutamate dehydrogenase), which were consistent with the antagonistic effect of the pollutant mixture on plant growth (Figure 14.4). However, during anthesis, after prolonged exposure to the pollutants, the stimulatory effect of NO_2 on nitrogen metabolism was no longer observed, and the stimulation of plant growth by NO_2 decreased with time (Bender *et al.*, 1991). As a general conclusion one can state that antagonistic effects predominate when plants are exposed to a realistic sequential combination of O_3 and NO_2, whereas synergistic interactions are more likely if the two gases are applied simultaneously at higher concentrations (Bender and Weigel, 1994; Barnes and Wellburn, 1998).

Ozone and acid deposition

Concerns over the possible role of acid rain or acid mist and O_3 in the forest decline syndrome has led to several studies with forest tree species. However, considerably

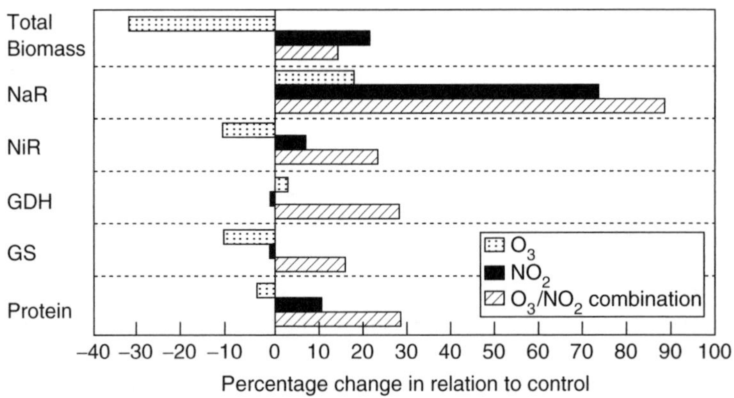

Figure 14.4 Effects of O_3 and NO_2, alone or in sequential combination, on total plant biomass and parameters associated with N metabolism in leaves of bush bean (*Phaseolus vulgaris* L. cv. Rintintin). Measurements were made during the vegetative growth stage of the plants (Bender *et al.*, 1991). NaR: nitrate reductase; NiR: nitrite reductase; GDH: glutamate dehydrogenase; GS: glutamine synthetase

less information exists about interactions on crops and other herbaceous plant species. Although plant effects of acid deposition generally involve both direct effects on foliage (e.g. visible damage, foliar leaching, cuticular weathering) and indirect effects through the soil (e.g. change in pH or deposition of sulfate or nitrate), it appears that direct effects of acidic wet deposition on plant leaves are less important as they have usually been observed at relatively unrealistic levels of pH (<3). Furthermore, there is evidence that the composition of the simulated acid rain and not the overall acidity determines plant responses (Wellburn, 1994). The impacts of wet acidic deposition alone (usually simulated acid rain) have been studied extensively in laboratory or greenhouse experiments. There are also univariate studies using large-scale facilities in the field in mature forest stands (e.g. the Solling roof project, Bredemeier *et al.*, 1998), so that a sound database has emerged on the single effects of acid deposition, even for field conditions.

In contrast, nearly all studies on the combined effects of acid deposition and O_3 have been conducted in small or medium-scale exposure systems (such as glasshouse fumigation chambers or open-top chambers), mostly with seedlings/saplings rather than with larger trees. This apparent scaling problem in the experimental approaches adopted makes it difficult to derive conclusions on the responses of forest trees to combined exposures to acidic deposition and O_3. A comprehensive literature review indicated that the majority of studies on O_3/acid deposition interactions have not demonstrated the existence of statistically significant interactions (USEPA, 1996). In more than 75% of recent reports of studies on over 30 species there was either no effect of one or other of the pollutants or the effects of both pollutant stresses were simply additive. However, in other studies where statistically significant interactions between O_3 and acid rain/mist have been reported, the interactions were mostly antagonistic. For example, a few studies with conifer species such as Norway spruce, red spruce, loblolly pine or slash pine showed that the adverse effects of O_3 on tree growth or physiology were reversed by application of simulated acid rain (e.g. Barnes *et al.*, 1990; Kohut *et al.*, 1990; Qiu *et al.*, 1992; Dean and Johnson, 1992). The results were partly explained by a beneficial fertilizer effect due to the nitrate and sulfate present in the acid rain solution applied. However, it must be noted that these conclusions were derived from short-term experiments that usually have been performed over one or two seasons, so that a fertilizer effect of the extra input of sulfur and nitrogen could be an initial effect but, in the longer term, acid deposition may reduce soil buffering capacity and increase leaching of nutrients from the soil.

Sulfur dioxide and nitrogen dioxide

Interactive effects of SO_2 and NO_2 have been studied rather intensively between c. 1980 and 1990 when high atmospheric concentrations of both SO_2 and NO_2 prevailed over large regions across Europe. In the meantime, SO_2 concentrations have dropped considerably, while NO_x concentrations have remained largely unchanged. Therefore, less attention is presently paid to these single pollutants though they still represent a major problem elsewhere in the world (e.g. in nearly any industrialized region of the developing countries). See Chapter 21.

Several studies on the interactive effects of SO_2 and NO_2, many of them conducted in the United Kingdom, have demonstrated that perennial grass species and trees are very susceptible to the combination of these two pollutants (Ashenden and Mansfield, 1978; Wellburn, 1982; Freer-Smith, 1984) and that the effects are more than additive. However, it must be taken into account that rather high concentrations were applied in most of these studies. It is difficult to draw conclusions on the interaction of SO_2 and NO_2 at lower concentrations only causing subtle effects. Studies under conditions more related to ambient air carried out in Australia by Murray *et al.* (1992, 1994) revealed that cereal grain yields were stimulated by the mixture of SO_2 and NO_2, but clover growth was retarded.

Nevertheless, two important findings can be extracted from these studies. One is that exposure during wintertime was always more effective than during summer. This must be attributed to the fact that the assimilation capacity, and therefore the degree of detoxification, for both SO_2 and NO_2 is higher at higher growth rates, i.e. in summer. Another finding is related to a biochemical explanation of more-than-additive effects. NO_2 alone is toxic only at very high concentrations (Segschneider, 1995) never occurring in ambient air even under worst case conditions. At lower concentrations, it can readily be metabolized and utilized in the normal nitrogen metabolism as the activities of nitrate- and nitrite reductase are co-regulated in order to prevent the accumulation of nitrite (NO_2^-) which acts as a free radical. However, in the presence of SO_2 the up-regulation of nitrite reductase is prevented (Wellburn *et al.*, 1981). By this means nitrite may accumulate and cause toxic effects.

Sulfur dioxide and ammonia

Several investigations on the interactive effects of SO_2 and NH_3 have been conducted in The Netherlands late in the 1980s when it became clear that ammonia represented probably the most important air pollutant in this country and that SO_2 pollution, which was still occurring at rather high levels at that time, could significantly interact with NH_3. However, antagonistic as well as synergistic effects have been observed in fumigation experiments. In all these experiments, the pollutants were applied at reasonably low concentrations.

Dueck (1990) found more-than-additive effects on the survival rate of heather (*Calluna vulgaris*) seedlings when they were exposed for eight months either to NH_3 alone (reduction of survival by 20%), to SO_2 alone (survival reduced by 50%) or to SO_2 plus NH_3 (survival reduced by 87%). Synergistic effects were also observed when frost hardiness of Scots pine was tested after exposure to SO_2 and NH_3, alone or in combination, for five months during wintertime (Dueck *et al.*, 1990). However, the combination of SO_2 plus NH_3 caused antagonistic effects on leaf gas exchange of poplar (*Populus euramericana*). When ammonia was applied alone, net photosynthesis (P_{max}) and stomatal conductance (g_s) were increased, whereas exposure to SO_2 alone reduced both P_{max} and g_s. Positive effects of NH_3 were no longer existent when both pollutants were applied together (van Hove *et al.*, 1991). Antagonistic effects were also detected in Scots pine, where nitrogen metabolism (activity of nitrogen assimilation enzymes) was enhanced by NH_3, but decreased by SO_2 (Pérez-Soba *et al.*, 1994).

From the experiments conducted so far no clear conclusions can be drawn on the mode of interaction between SO_2 and NH_3. Both gases are capable of disturbing the acid-base regulation within plant tissues, and both gases may alter the nutritional status of a plant. The mode of interaction will depend on plant functional type, on nutritional status and on many other environmental conditions.

Sulfur dioxide and hydrogen fluoride

Hydrogen fluoride represents the most acidic chemical found in the environment. Acute injury on vegetation has often been observed around glass factories and other industries from where fluoride is emitted. Only few experimental investigations have been conducted to test the interactive effects of SO_2 and HF. Bennett and Hill, (1974) used acute concentrations of these two air pollutants and found additive effects on the inhibition of apparent photosynthesis. More recently, the response of wheat (Davieson *et al.*, 1990) and of soybean, maize, peanut and bean (Murray and Wilson, 1990) to both pollutants was studied at more realistic concentrations. In these studies conducted in open-top field chambers, less-than-additive effects were observed for wheat (on shoot surface area), and for soybean, peanut and bean (on yield, respectively).

Acid mist/acid rain and sulfur dioxide

Only little work has been published on the combined effects of acid rain and SO_2 from controlled exposure experiments. In a study with broad bean (*Vicia faba*) conducted in open-top field chambers, Adaros *et al.* (1988) did not detect any significant interactions between SO_2 exposure and acid rain treatment on yield or biomass at final harvest. However, at intermediate harvests both less-than-additive interactions (leaf dry weight, leaf area) and more-than-additive interactions (number of pods) were observed.

Interaction of more than two pollutants

There are some experimental studies which involved more than two different pollutants in mixture. These studies which were carried out in controlled environment conditions and in indoor and outdoor chambers may roughly be divided into experiments dealing either with more than two gaseous pollutants or with combinations of two or more gaseous pollutants and heavy metals or wet acidic deposition. Hardly any generalized conclusions or rules describing modes of interaction can be drawn from these few experiments which in addition were too diverse in their designs. Most of these studies were carried out more than 10 years ago and there has not been much further insight into this problem since then (compare Barnes and Wellburn, 1998).

Interaction of ozone, sulfur dioxide and nitrogen dioxide

Among those studies dealing with more than two gaseous pollutants, most of the work has involved sulfur dioxide, nitrogen dioxide and ozone, as these gases have been regarded as the most important pollutants affecting terrestrial ecosystems for the past decades.

Some work has been conducted on the response of nitrate reductase (NR) activity, peroxidase (POD) activity, and superoxide dismutase (SOD) activity in needles of spruce saplings (*Picea abies*) to SO_2 plus NO_2, O_3 plus NO_2, or O_3 plus SO_2 plus NO_2 (Klumpp *et al.*, 1989a, 1989b). These air pollutant treatments were combined with different supplies of magnesium and calcium to mimic soil nutrient depletion due to acid deposition. All of the enzyme activities responded in a different manner both to the pollutants and to soil nutrient supply when different needle ages were considered, and no consistent effects on pollutant interactions were found. In a field study, the forest floor vegetation of a melick-beech forest was exposed either to SO_2, to SO_2 plus NO_2, or to SO_2 plus NO_2 plus O_3 (Steubing *et al.*, 1989). It was shown in this study that spring geophytes with hygromorphic leaves represented the most susceptible functional type of plant species of this vegetation and that community structure was disturbed by the pollutants. The effects were most severe when ozone was added as a pollutant. However, as ozone was not applied as a single pollutant in this study, no conclusions on interactions can be drawn.

Adaros *et al.* (1991a, 1991b) investigated growth and yield responses of some crop species to SO_2, NO_2 and O_3, singly and in combination. For spring rape (*Brassica napus*) the authors report positive effects on yield of SO_2 or NO_2 alone, but antagonistic effects of the combined exposure to SO_2 plus NO_2. When O_3 exposure was combined with either SO_2 or NO_2, again antagonistic effects were observed. In spring barley and spring wheat, Adaros *et al.* (1991b) found mainly antagonistic effects for any of the combinations of the three gaseous pollutants, SO_2 or NO_2 or O_3. Only in one of the spring wheat cultivars tested (*Triticum aestivum* cv. Turbo), synergistic effects on yield were caused by the combination of O_3 with SO_2 and/or NO_2.

Heavy metals in combination with gaseous air pollutants

Little research has been done on possible interactive effects of air pollutants with pollutants that elicit their effects not solely through foliar uptake. Heavy metals, which conveniently comprise a group of trace metals with a metallic density of >5 g m^{-3} (e.g. Zn, Cd, Pb, Cu, Cr, Ni) may adversely affect plant and ecosystem health above particular threshold concentrations in the plant tissue or in the soil (Prasad and Hagemeyer, 1999). Their main route of entry into plants is uptake from the soil, although atmospheric deposition can also contribute to the metal input. Therefore, particularly plants growing on metal contaminated soils may be exposed to a combination of soil and air pollutants. As heavy metals have been shown to affect stomatal conductance (Poschenrieder and Barcelo, 1999), the possibility of interactions with gaseous air pollutants does exist.

A number of studies were carried out to assess interactive effects of SO_2 and heavy metals (Cu, Cd, Pb, Zn) on growth and yield parameters as well as on transpiration of plant species (Krause and Kaiser, 1977; Lamoreaux and Chaney, 1978; Toivonen and Hofstra, 1979; Grünhage and Jäger, 1981; Dueck, 1986). For example, Toivonen and Hofstra (1979) found that the uptake of SO_2 into plants was reduced by a concomitant exposure to Cu, while Dueck (1986) did not find interactive effects of SO_2 and Cu on ryegrass and clover after *in vivo* root uptake of the metal. According to Lamoreaux and Chaney (1978) a concurrent exposure of excised silver maple leaves to SO_2 and Cd resulted in a greater reduction of photosynthesis and transpiration than exposure to each of the pollutants alone.

Exposure of peas, cress and lettuce to O_3 and Ni and Cd via soil contamination at metal levels that were not phytotoxic *per se*, resulted in an increased sensitivity of the plants to O_3, while at metal levels that resulted in adverse effects the sensitivity of the plants to O_3 decreased (Czuba and Ormrod, 1974; Ormrod, 1977). Harkov *et al.* (1979) describe a synergistic type of interaction of O_3 and Cd, as tomato plants were more sensitive towards O_3 when they were simultaneously exposed to Cd *in vivo*.

Overall, from the very few studies that have been done with mixtures of gaseous air pollutants and heavy metals no consistent picture about possible interactive effects has emerged.

Acid rain in combination with more than one gaseous pollutant

Ashenden and coworkers tested the interactive effects of SO_2 plus NO_2, O_3, or O_3 plus SO_2 plus NO_2, in combination with acid mist applications at pH of 2.5, 3.5, 4.5, and 5.6. For white clover (*Trifolium repens*), all of the gaseous pollutant treatments caused reductions in biomass. This effect was also observed for acid mist applications at pH 2.5 (Ashenden *et al.*, 1995). The results suggested an antagonistic interaction between O_3 alone or O_3 plus SO_2 plus NO_2 and acid mist. In a similar experiment with the two grasses, *Lolium perenne* and *Agrostis capillaris*, Ashenden *et al.* (1996) found less-than-additive effects on biomass of *L. perenne* for the mixture O_3 plus SO_2 plus NO_2 than for O_3 alone or SO_2 plus NO_2. Surprisingly, acid mist application at pH 2.5 stimulated growth of *L. perenne* when grown in filtered air or with ozone alone. Most probably, *L. perenne* could utilize the N and S-compounds contained in the acid mist as nutrients without suffering from the proton load, as long as no additional acidic gases were applied. The responses of *Agrostis capillaris* to the gaseous pollutants were less pronounced than for *L. perenne*. Again, a growth stimulation under acid mist application at pH 2.5 was found, but no significant interactions between acid mist treatments and gaseous air pollutants occurred in *A. capillaris*.

Comparison of plant growth between filtered and non-filtered air

Comparison of responses of plants grown in environments with either filtered or non-filtered air can provide information on effects of the impact of current levels of all pollutants in ambient air and on the range of concentrations of the pollutants of immediate interest. In such comparisons the pattern of the air pollutants (e.g. the occurrence of peak values of all species in the mixture) is the same as in ambient air.

In a season-long exposure experiment in open-top chambers Weigel *et al.* (1987) compared the growth responses of a winter and a spring barley cultivar to filtered (CF) and non-filtered (NF) ambient air. Monthly mean winter- and summertime gas concentrations [$\mu g\ m^{-3}$] in ambient air during the experiment (Nov.–Aug.) were in

Table 14.2 Summary of effects of pollutant combinations

Pollutant mixture	Combined effect	Mode of combined action	Species most frequently investigated
$O_3 + SO_2$	Both synergistic and antagonistic	Radical attack (O_3, SO_2) + acidification (SO_2)	Crops, forest trees
$O_3 + NH_3$	Mostly antagonistic (at low [NH_3]) or additive	Radical attack (O_3) + better nitrogen nutrition (at low [NH_3]) **or** acidification (at high [NH_3])	Forest trees
$O_3 + NO_2$	Synergistic at high conc., antagonistic at low conc.	Radical attack (O_3) + better nitrogen nutrition (NO_2)	Crops and vegetables
O_3 + acid deposition	Mostly additive, sometimes antagonistic	Radical attack (O_3) + acidification **or** improved nutrient status for the short term due to nitrate and sulfate contained in acid deposition	Forest trees
$SO_2 + NO_2$	Mostly synergistic	Acidification and radical attack (SO_2) + nitrite toxicity (after NO_2 uptake) due to blocking of nitrite-reductase by SO_2	Perennial grasses, forest trees
$SO_2 + NH_3$	Both synergistic and antagonistic	Acidification (both gases), improved nitrogen nutrition in the short term (at low [NH_3])	Forest trees, heather
$SO_2 + HF$	Additive or antagonistic	Acidification (both gases)	Cereals and other crops
Acid mist/acid rain + SO_2	Too few studies, no consistent response	Acidification (any of the pollutants) + improved nutrient status (acid deposition)	Broad bean
$O_3 + NO_2 + SO_2$	No consistent response across different studies	Radical attack (O_3) + acidification (SO_2) + improved nitrogen nutrition or radical attack by nitrite (NO_2)	Forest trees, crops, natural vegetation
Heavy metals + SO_2	No consistent responses	Acidification (SO_2) + disturbance of structure and functioning of proteins (heavy metals)	Grasses, forest trees
Heavy metals + O_3	Often synergistic	Radical attack (O_3) + disturbance of structure and functioning of proteins (heavy metals)	Vegetable species
Acid rain + two or more gaseous pollutants	Only one study with diverse results depending on pollutant combination and plant species		White clover and two perennial grass species

the range of 34–127 for SO_2, 34–52 for NO_2 and 42–52 for O_3. Air filtration resulted in higher numbers of winter barley plants surviving frost damage as compared to the NF treatment. With spring barley there were no effects of the air filtration on grain yield but changes in vegetative growth parameters. Earlier year-round studies on the effect of CF and NF urban air in the UK on the growth of ryegrass resulted either in increases in shoot yield (Roberts *et al.*, 1983) or showed no yield differences between the treatments (Colvill *et al.*, 1983). Results of such experiments are difficult to compare, unless detailed information on the pollutant concentrations during the exposure and on the filtration efficiencies of the enclosure systems are given.

Conclusions

The major results of experiments with pollutant combinations are summarized in Table 14.2. Obviously, our understanding of effects of pollutant mixtures is far from satisfactory. This is for several reasons. As pointed out above, hardly any existing exposure system is suitable to run multifactorial experiments with a sufficient number of treatments and of treatment replications. Moreover, as pollutant effects do interact with climatic and plant culture conditions, results found at one particular site do not necessarily represent the general response of a species to pollutant mixtures. The situation is further complicated by the fact that the mode of interaction between pollutants depends on their concentration (or better: on their flux into, or their uptake by, the plant). Thus, there is no realistic chance for a substantial improvement of the current knowledge by the sole use of experimental investigations.

One might ask for modeling approaches to gain more insight on the effects of pollutant combinations. However, recent European-wide studies which involved both experimental and modeling approaches to test the interactive effects of climatic conditions, atmospheric CO_2 enrichment and tropospheric O_3 on crops such as wheat (ESPACE-wheat, Jäger *et al.*, 1999) or potato (CHIP, see http://www.var.fgov.be/chip.php) clearly demonstrated that the modeling tools are not yet sophisticated enough to predict accurately the effects of such combinations as observed in field experiments.

Most probably, we will eventually come to the conclusion that any pollutant concentration above background concentrations will have adverse effects if we do not only regard short-term effects but long-term effects like alterations of species composition, possible loss of biodiversity, or effects on ecosystem functioning. Even for one single pollutant, critical levels of pollutants (i.e. the pollution concentration which should not be exceeded to prevent adverse effects) had to be lowered as knowledge on the long-term effects increased. For example, critical levels for the annual average concentrations of NH_3 were estimated to be 300 $\mu g\ m^{-3}$ in 1979, 75 $\mu g\ m^{-3}$ in 1982, and only 8 $\mu g\ m^{-3}$ in 1991 (Fangmeier *et al.*, 1994). The latter value is still valid and is close to the background concentration in unpolluted areas, and it does not yet take into account possible interactions with other pollutants. The importance of pollutant combinations in determining plant responses and their importance to calculate reliable critical levels has been clearly recognized. However, with the existing experimental methodology it will be difficult to define reliable critical levels for pollutant combinations.

References

Adaros, G., Weigel, H.J. and Jäger, H.-J. (1988). Effects of sulphur dioxide and acid rain alone or in combination on growth and yield of broad bean plants. *New Phytol.* **108**, 67–74.

Adaros, G., Weigel, H.J. and Jäger, H.-J. (1991a). Single and interactive effects of low levels of O_3, SO_2 and NO_2 on the growth and yield of spring rape. *Environ. Pollut.* **72**, 269–286.

Adaros, G., Weigel, H.J. and Jäger, H.-J. (1991b). Concurrent exposure to SO_2 and/or NO_2 alters growth and yield responses of wheat and barley to low concentrations of O_3. *New Phytol.* **118**, 581–591.

Ashenden, T.W. and Mansfield, T.A. (1978). Extreme pollution sensitivity of grasses when SO_2 and NO_2 are present in the atmosphere together. *Nature* **273**, 142–143.

Ashenden, T.W., Bell, S.A. and Rafarel, C.R. (1995). Responses of white clover to gaseous pollutants and acid mist: implications for setting critical levels and loads. *New Phytol.* **130**, 89–96.

Ashenden, T.W., Bell, S.A. and Rafarel, C.R. (1996). Interactive effects of gaseous air pollutants and acid mist on two major pasture grasses. *Agric. Ecosyst. Environ.* **57**, 1–8.

Ashmore, M.R. and Bell, J.N.B. (1994). Clearing the air – How filtration experiments have demonstrated the hidden effects of air pollution on vegetation. In *Immissionsökologische Forschung im Wandel der Zeit*. Kuttler, W. and Jochimsen, M., eds., Essener Ökologische Schriften Bd. 4, pp. 53–60.

Ashmore, M.R. and Önal, M. (1984). Modification by sulfur dioxide of the responses of *Hordeum vulgare* to ozone. *Environ. Pollut.* **36**, 31–43.

Barnes, J.D. and Wellburn, A.R. (1998). Air pollutant combinations. In *Responses of Plant Metabolism to Air Pollution and Global Change*. De Kok, L.J. and Stulen, I., eds. (Leiden: Backhuys), pp. 147–164.

Barnes, J.D., Eamus, D. and Brown, K.A. (1990). The influence of ozone, acid mist and soil nutrient status on Norway spruce [*Picea abies* (L.) Karst.] I. Plant-water relations. *New Phytol.* **114**, 713–720.

Bartosz, G. (1997). Oxidative stress in plants. *Acta Physiol. Plant.* **19**, 47–64.

Bender, J. and Weigel, H.J. (1993). Crop responses to mixtures of air pollutants. In *Effects of Air Pollution on Agricultural Crops in Europe*. Jäger, H.-J., Unsworth, M.H., De Temmerman, L. and Mathy, P., eds. Air Pollution Research Report 46, pp. 445–453.

Bender, J. and Weigel, H.J. (1994).The role of other pollutants in modifying plant responses to ozone. In *Critical Levels for Ozone*. Fuhrer, L. and Achermann, B., eds., Schriftenreihe der FAC Liebefeld 16, pp. 240–247.

Bender, J., Weigel, H.J. and Jäger, H.-J. (1991). Response of nitrogen metabolism in beans (*Phaseolus vulgaris* L.) after exposure to ozone and nitrogen dioxide, alone and in sequence. *New Phytol.* **119**, 261–267.

Bennett, J.H. and Hill, A.C. (1974). Acute inhibition of apparent photosynthesis by phytotoxic air pollutants. In *Air Pollution Effects on Plant Growth*. Dugger, M., ed. (Amer.Chem.Soc.Symp.Ser.3), pp. 115–127.

Berdowski, J.J.M. (1993). The effect of external stress and disturbance factors on *Calluna*-dominated heathland vegetation. In *Heathland: Patterns and Processes in a Changing Environment. Geobotany 20*. Aerts, R. and Heil, G.W., eds. (Dordrecht: Kluwer), pp. 85–124.

Bredemeier, M., Blanck, K., Dohrenbusch, A., Lamersdorf, N., Meyer, A.C., Murach, D., Parth, A. and Xu, Y.J. (1998). The Solling roof project–site characteristics, experiments and results. *For. Ecol. Manage.* **101**, 281–293.

Colvill, K.E., Bell, R.M., Roberts, T.M. and Bradshaw, A.D. (1983). The use of open-top chambers to study the effects of air pollutants, in particular sulphur dioxide, on the growth of ryegrass *Lolium perenne* L. Part II –The long-term effect of filtering polluted urban air or adding SO_2 to rural air. *Environ. Pollut.* **31**, Ser. A, 35–55.

Czuba, M. and Ormrod, D.P. (1974). Effects of cadmium and zinc on ozone-induced phytotoxicity in cress and lettuce. *Can. J. Bot.* **52**, 645–649.

Dämmgen, U. and Weigel, H.J. (1998). Trends in atmospheric composition (nutrients and pollutants) and their interaction with agroecosystems. In *Sustainable Agriculture for Food, Energy and Industry: Strategies towards Achievement*. El Bassam, N., Behl, R. and Prochnow, B., eds. (James and James), pp. 85–93.

Davieson, G., Murray, F. and Wilson, S. (1990). Effects of sulphur dioxide and hydrogen fluoride, singly and in combination, on growth and yield of wheat in open-top chambers. *Agr. Ecosyst. Environ.* **30**, 317–325.

Dean, T.J. and Johnson, J.D. (1992).Growth response of young slash pine trees to simulated acid rain and ozone stress. *Can. J. For. Res.* **22**, 839–848.

Delledonne, M., Xia, Y.J., Dixon, R.A. and Lamb, C. (1998). Nitric oxide functions as a signal in plant disease resistance. *Nature* **394**, 585–588.

di Toppi, L.S. and Gabrielli, R. (1999). Response to cadmium in higher plants. *Environ. Exp. Bot.* **41**, 105–130.

Dixon, R.A., Lamb, C.J., Paiva, N.L. and Masoud, S. (1996). Improvement of natural defense responses. In *Engineering Plants for Commercial Products and Applications*. Collins, G.B. and Shepherd, R.J., eds. (New York: New York Academy of Science), pp. 126–139.

Dueck, T.A., Zuin, A. and Elderson, J. (1998). Influence of ammonia and ozone on growth and drought sensitivity of *Pinus sylvestris*. *Atmos. Environ.* **32**, 545–550.

Dueck, T.A. (1986). *Impact of Heavy Metals and Air Pollutants on Plants*. Academisch Proefschrift, Free University Press, Amsterdam.

Dueck, T.A. (1990). Effects of ammonia and sulphur dioxide on the survival and growth of *Calluna vulgaris* (L.) Hull seedings. *Funct. Ecol.* **4**, 109–116.

Dueck, T.A., Dorèl, F.G., Ter Horst, R. and Van der Eerden, L.J.M. (1990). Effects of ammonia, ammonium sulphate and sulphur dioxide on the frost sensitivity of Scots pine (*Pinus sylvestris* L.). *Water Air Soil Pollut.* **54**, 35–49.

Fangmeier, A. (1989). Effects of open-top fumigations with SO_2, NO_2, and ozone on the native herb layer of a beech forest. *Env. Exp. Bot.* **29**, 199–213.

Fangmeier, A., Hadwiger-Fangmeier, A., Van der Eerden, L.J.M. and Jäger, H.-J. (1994). Effects of atmospheric ammonia on vegetation–a review. *Environ. Pollut.* **86**, 43–82.

Feron, V.J., Cassee, F.R. and Groten, J.P. (1998). Toxicology of chemical mixtures: International perspectives. *Environ. Health Perspect.* **106**, 1281–1289.

Freer-Smith, P.H. (1984). The responses of six broadleaved trees during long-term exposure to SO_2 and NO_2. *New Phytol.* **97**, 49–61.

Goodyear, S.N. and Ormrod, D.P. (1988). Tomato response to concurrent and sequential NO_2 and O_3 exposures. *Environ. Pollut.* **51**, 315–326.

Guderian, R. and Tingey, D.T. (1987). *Notwendigkeit und Ableitung von Grenzwerten für Stickoxide*. Umweltbundesamt, Berichte 1/87 (Berlin: Erich Schmidt Verlag).

Grünhage, L. and Jäger, H.-J. (1981). Kombinationswirkungen von SO_2 und Cadmium auf *Pisum sativum* L.1. Ertrag, Schadstoffgehalte und Ionenhaushalt. *Angew. Bot.* **55**, 345–359.

Harkov, R., Clarke, B. and Brennan, E. (1979). Cadmium contamination may modify response of tomato to atmospheric ozone. *JAPCA* **29**, 1247–1249.

Heagle, A.S. (1989). Ozone and crop yield. *Annu. Rev. Phytopathol.* **27**, 397–423.

Heagle, A.S., Kress, L.W., Temple, P.J., Kohut, R.J., Miller, J.E. and Heggestad, H.E. (1988). Factors influencing ozone dose-yield response relationships in open-top field studies. In

Assessment of Crop Loss from Air Pollutants. Heck, W.W., Taylor, O.C. and Tingey, D.T., eds. (London: Elsevier Applied Science), pp. 141–180.

Ito, O., Okano, K., Kuroiwa, M. and Totsuka, T. (1985). Effects of NO_2 and O_3 alone and in combination on kidney bean plants: growth, partitioning of assimilates and root activities. *J. Exp. Bot.* **36**, 652–662.

Jäger, H.-J., Hertstein, U. and Fangmeier, A., Guest Editors (1999). The European Stress Physiology and Climate Experiment –Project 1: Wheat. *Eur. J. Agron.* **10**, 153–264.

Kangasjärvi, J., Talvinen, J., Utriainen, M. and Karjalainen, R. (1994). Plant defence systems induced by ozone. *Plant Cell Environ.* **17**, 783–794.

Kley, D., Kleinmann, M., Sandermann, H. and Krupa, S. (1999). Photochemical oxidants: state of the science. *Environ. Pollut.* **100**, 19–42.

Klumpp, A., Küppers, K. and Guderian, R. (1989a). Nitrate reductase activity of needles of Norway spruce fumigated with different mixtures of ozone, sulfur dioxide, and nitrogen dioxide. *Environ. Pollut.* **58**, 261–271.

Klumpp, G., Guderian, R. and Küppers, K. (1989b). Peroxidase- und Superoxiddismutase-Aktivität sowie Prolingehalte von Fichtennadeln nach Belastung mit O_3, SO_2 und NO_2. *Eur. J. For. Pathol.* **19**, 84–97.

Kohut, R.J. (1985). The effects of SO_2 and O_3 on plants. In *Sulfur Dioxide and Vegetation*. Winner, W.E., Mooney, H.A. and Goldstein, R.A., eds. (Stanford: Stanford University Press), pp. 296–312.

Kohut, R.J., Laurence, J.A., Amundson, R.G., Raba, R.M. and Melkonian, J.J. (1990). Effects of ozone and acidic precipitation on the growth and photosynthesis of red spruce after two years of exposure. *Water Air Soil Pollut.* **51**, 277–286.

Krause, G.H.M. and Kaiser, H. (1977). Plant response to heavy metals and sulphur dioxide. *Environ. Pollut.* **12**, 63–71.

Kress, L.W. and Skelly, J.M. (1982). Response of several eastern forest tree species to chronic doses of ozone and nitrogen dioxide. *Plant Dis.* **66**, 1149–1152.

Krupa, S.V. and Legge, A.H. (1999). Foliar injury symptoms of Saskatoon serviceberry (*Amelanchier alnifolia* Nutt.) as a biological indicator of ambient sulfur dioxide exposures. *Environ. Pollut.* **106**, 449–454.

Krupa, S.V. and Legge, A.H. (2001). Saskatoon serviceberry and ambient sulfur dioxide exposures: study sites re-visited, 1999. *Environ. Pollut.* **111**, 363–365.

Küppers, K., Boomers, J., Hestermann, C., Hanstein, S. and Guderian, K. (1994). Reaction of forest trees to different exposure profiles of ozone-dominated air pollutant mixtures. In *Critical Levels for Ozone*. Fuhrer, J. and Achermann, B., eds., Schriftenreihe der FAC Liebefeld 16, pp. 98–110.

Lamoreaux, R.J. and Chaney, W.R. (1978). Photosynthesis and transpiration of excised silver maple leaves exposed to cadmium and sulphur dioxide. *Environ. Pollut.* **17**, 259–268.

Lefohn, A.S., Davis, C.E., Kones, C.K., Tingey, D.T. and Hogsett, W.E. (1987). Co-occurrence patterns of gaseous air pollutant pairs at different minimum concentrations in the United States. *Atmos. Environ.* **18**, 2521–2526.

Manning, W.J. and Krupa, S.V. (1992). Experimental methodology for studying the effects of ozone on crops and trees. In *Surface Level Ozone Exposure and their Effects on Vegetation*. and Lefohn, A.S., ed. (Chelsea: Lewis Publishers), pp. 93–156.

Mansfield, T.A. and McCune, D.C. (1988). Problems of crop loss assessment when there is exposure to two or more gaseous pollutants. In *Assessment of Crop Loss from Air Pollutants*. Heck, W.W., Taylor, O.C. and Tingey, D.T., eds. (London: Elsevier Applied Science), pp. 317–344.

Murray, F. and Wilson, S. (1990). Yield responses of soybean, maize, peanut and navy bean exposed to SO_2, HF and their combination. *Env. Exp. Bot.* **30**, 215–223.

Murray, F., Wilson, S. and Monk, R. (1992). NO_2 and SO_2 mixtures stimulate barley grain production but depress clover growth. *Env. Exp. Bot.* **32**, 185–192.

Murray, F., Wilson, S. and Samaraweera, S. (1994). NO_2 increases wheat grain yield even in the presence of SO_2. *Agr. Ecosyst. Environ.* **48**, 115–123.

Ormrod, D.P. (1977). Cadmium and nickel effects on growth and ozone sensitivity of pea. *Water Air Soil Pollut.* **8**, 263–270.

Ormrod, D.P. (1982). Air pollutant interactions in mixtures. In *Effects of Gaseous Air Pollutants in Agriculture and Horticulture*. Unsworth, M.H. and Ormrod, D.P., eds. (Butterworths: London), pp. 307–311.

Oshima, R.J. and Bennett, J.P. (1979). Experimental design and analysis. In *Handbook of Methodology for the Assessment of Air Pollution Effects on Vegetation*. Heck, W.W., Krupa, S.V. and Linzon, S.N., eds. (Pittsburg: Air Poll. Control Assoc.), pp. 4-1-22.

Pearson, J. and Stewart, G.R. (1993). The deposition of atmospheric ammonia and its effects on plants. *New Phytol.* **125**, 283–305.

Pérez-Soba, M., Dueck, T.A., Puppi, G. and Kuiper, P.J.C. (1995). Interactions of elevated CO_2, NH_3 and O_3 on mycorrhizal infection, gas exchange and N metabolism in saplings of Scots pine. *Plant Soil* **176**, 107–116.

Pérez-Soba, M., Van der Eerden, L.J.M. and Stulen, I. (1994). Combined effects of gaseous ammonia and sulphur dioxide on the nitrogen metabolism of the needles of Scots pine trees. *Plant Physiol. Biochem.* **32**, 539–546.

Pfanz, H. (1995). Apoplastic and symplastic proton concentrations and their significance for metabolism. In *Ecophysiology of Photosynthesis*. Schulze, E.D. and Caldwell, M.M., eds. (Berlin: Springer), pp. 103–122.

Poschenrieder, C.H. and Barcelo, J. (1999): Water relations in heavy metal stressed plants. In *Heavy Metal Stress in Plants. From Molecules to Ecosystems*. Prasad, M.N.V. and Hagemeyer, J., eds. (Berlin-Heidelberg: Springer), pp. 207–229.

Prasad, M.N.V. and Hagemeyer, J., eds. (1999): *Heavy Metal Stress in Plants. From Molecules to Ecosystems* (Berlin-Heidelberg: Springer).

Qiu, Z., Chappelka, A.H., Somers, G.L., Lockaby, B.G. and Meldahl, R.S. (1992). Effect of ozone and simulated acidic precipitation on above and below-ground growth of loblolly pine (*Pinus taeda*). *Can. J. For. Res.* **22**, 582–587.

Raven, J.A. (1988). Acquisition of nitrogen by the shoots of land plants: its occurrence and implications for acid-base regulation. *New Phytol.* **109**, 1–20.

Reinert, R.A. (1984). Plant response to air pollutant mixtures. *Ann. Rev. Phytopath.* **22**, 421–442.

Reinert, R.A., Heagle, A.S. and Heck, W.W. (1975). Plant responses to pollutant combinations. In *Responses of Plants to Air Pollution*. Mudd, J.B. and Kozlowski, T., eds. (New York: Academic Press), pp. 157–177.

Roberts, T.M., Bell, R.M., Horsman, D.C. and Colvill, K.E. (1983). The use of open-top chambers to study the effects of air pollutants, in particular sulfur dioxide, on the growth of ryegrass *Lolium perenne* L. Part I – Characteristics of modified open-top chambers used for both air-filtration and SO_2 fumigation experiments. *Environ. Pollut.* **31**, Ser. A, 9–33.

Runeckles, V.C. (1984). Impact of air pollutant combinations on plants. In *Air Pollution and Plant Life*. Treshow, M., ed. (New York: John Wiley & Sons), pp. 239–285.

Runeckles, V.C. and Palmer, K. (1987). Pretreatment with nitrogen dioxide modifies plant response to ozone. *Atmos. Environ.* **21**, 717–719.

Schraudner, M., Moeder, W., Wiese, C., Van Camp, W., Inzé, D., Langebartels, C. and Sandermann, H. (1998). Ozone-induced oxidative burst in the ozone biomonitor plant, tobacco Bel W3. *Plant Journal* **16**, 235–245.

Segschneider, H.J. (1995). Auswirkungen atmosphärischer Stickoxide (NOx) auf den pflanzlichen Stoffwechsel: Eine Literaturübersicht. *Angew. Bot.* **69**, 60–85.

Soares, A., Ming, J.Y. and Pearson, J. (1995). Physiological indicators and susceptibility of plants to acidifying atmospheric pollution: a multivariate approach. *Environ. Pollut.* **87**, 159–166.

Steubing, L., Fangmeier, A., Both, R. and Frankenfeld, M. (1989). Effects of SO_2, NO_2, and O_3 on population development and morphological and physiological parameters of native herb layer species in a beech forest. *Environ. Pollut.* **58**, 281–302.

Stockwell, W.R., Kramm, G., Scheel, H.E., Mohnen, V.A. and Seiler, W. (1997). Ozone formation, destruction and exposure in Europe and the United States. In *Forest Decline and Ozone*. Sandermann, H., Wellburn, A.R. and Heath, R.L., eds. (Springer: New York), pp. 1–38.

Toivonen, P.M. and Hofstra, G. (1979). The interaction of copper and sulphur dioxide in plant injury. *Can. J. Plant Sci.* **59**, 475–479.

Tonneijck, A.E.G. and Van Dijk, C.J. (1998). Responses of bean (*Phaseolus vulgaris* L. cv. Pros) to chronic ozone exposure at two levels of atmospheric ammonia. *Environ. Pollut.* **99**, 45–51.

USEPA (U.S. Environmental Protection Agency) (1984). *A Review and Assessment of the Effects of Air Pollutant Mixtures on Vegetation–Research Recommendations*. EPA-600/3-84. U.S. Environmental Protection Agency, Corvallis, Oregon.

USEPA (U.S. Environmental Protection Agency) (1996). *Air Quality Criteria for Ozone and Related Photochemical Oxidants. Vol. II*. EPA/600/P-93/004bF. U.S. Environmental Protection Agency, Office of Research and Development, Washington, DC.

van Hove, L.W.A. and Bossen, M.E. (1994). Physiological effects of five months exposure to low concentrations of O_3 and NH_3 on Douglas fir (*Pseudotsuga menziesii*). *Physiol. Plant.* **92**, 140–148.

van Hove, L.W.A., van Kooten, O., van Wijk, K.J., Vredenberg, W.J., Adema, E.H. and Pieters, G.A. (1991). Physiological effects of long term exposure to low concentrations of SO_2 and NH_3 on poplar leaves. *Physiol. Plant.* **82**, 32–40.

Weigel, H.J., Adaros, G. and Jäger, H.J. (1987). An open-top chamber study with filtered and non-filtered air to evaluate the effects of air pollutants on crops. *Environ. Pollut.* **47**, 213–244.

Wellburn, A.R. (1982). Effects of SO_2 and NO_2 on metabolic function. In *Effects of Gaseous Air Pollution in Agriculture and Horticulture*. Unsworth, M.H. and Ormrod, D.P., eds. (London: Butterworth), pp. 169–187.

Wellburn, A.R. (1994). *Air Pollution and Climate Change. The Biological Impact* (Essex: Longman Scientific & Technical).

Wellburn, A.R., Higginson, C., Robinson, D. and Walmsley, C. (1981). Biochemical explanations of more than additive inhibitory effects of low atmospheric levels of sulphur dioxide plus nitrogen dioxide upon plants. *New Phytol.* **88**, 223–237.

Wolfenden, J., Wookey, P.A., Lucas, P.W. and Mansfield, T.A. (1992). Action of pollutants individually and in combination. In *Air Pollution Effects on Biodiversity*. Barker, J.R. and Tingey, D.T., eds. (New York: Van Nostrand Reinhold), pp. 72–92.

15 Forest decline and air pollution: an assessment of 'forest health' in the forests of Europe, the Northeastern United States, and Southeastern Canada

J.L. INNES AND J.M. SKELLY

Introduction

At the outset of this chapter it is important to note that both authors believe air pollutants to be of considerable importance to the long-term health and productivity of the forests of Europe and eastern and western North America. There is abundant evidence of direct and indirect air pollutant induced effects to forest ecosystems on both continents (Chappelka and Samuelson, 1998; Innes, 1993; Innes and Oleksyn, 2000; Skelly, 2000). The nature and extent of the impacts varies among different components of the forest ecosystem and with different types of air pollutants. Acidic soils, for example, can be very sensitive to acidic deposition. Lichens are widely recognized as being sensitive to sulphur dioxide. Many vascular plant species are sensitive to nitrogen deposition. Forest streams can be susceptible to acidification, with adverse effects on stream fauna. Ozone and other pollutants are known to have direct effects on animals, although the amount of available information is less than that for plants (with the exception of effects on human health). Ozone has been directly implicated in changes in the composition of tree species in some Californian forests (Miller and McBride, 1999); this is one of the few cases where air pollution at a regional scale has been documented as leading to forest decline, although even in this case it is a change in forest composition that has occurred rather than the destruction of the forest.

The variety of impacts that air pollution can have on forest ecosystems brings into focus the concept of 'forest health'. While it is often possible to state whether or not an individual tree is healthy, it is much more difficult to state whether a forest is healthy. This is because a forest consists of more than trees. In most cases the health of an individual tree is equated with its vigour. Vigour is a non-measurable quantity that combines the growth rate of the tree with its ability to withstand stress. Vigorous trees tend to be fast-growing, but not all fast-growing trees are vigorous. An individual tree is usually considered as healthy if its growth is not being impaired

Air Pollution and Plant Life, second edition. Edited by J.N.B. Bell and M. Treshow. ISBN 0 471 49090 3 (HB), 0 471 49091 1 (PB).

by any factor. Thus, site conditions can reduce the health of a tree by limiting its access to water or nutrients. Alternatively, a pathogen can reduce the health of a tree by impairing the biological functions of a tree in such a way as to reduce its growth. With increasing age following maturity, the vigour of a tree, expressed by both its growth rate and its ability to withstand stress, declines. There is generally a certain amount of latitude in determining the health of a tree. The vast majority of trees have some foliage affected by insects or disease, with the proportion increasing during the lifetime of the foliage. Similarly, patches of wood in the stem or branches of a tree may be infected by a pathogen without adversely affecting the general vigour of the tree. These endemic levels of stress are not generally considered as affecting the health of the tree. However, if the levels increase to a point whereby the normal biological processes of the tree are affected, then the tree is considered as unhealthy.

While it is possible to consider stand health in relation to the health of the trees comprising the stand, under the new ecosystem-based paradigm that has to a lesser or greater extent been accepted in forestry, it is not possible to extend the concept of tree health to forests. This is because the identification of forest health is based on a wide range of other parameters, with these parameters differing between individuals (e.g., Kolb *et al.*, 1994). For example, a production forester would consider a healthy forest to be one in which all the trees were healthy and growing well. However, a forest ecologist would expect to find some unhealthy and dead trees within a forest, as these would indicate the normal functioning of ecological processes within the forest. The health (from the viewpoint of someone interested in increasing timber production) of a stand may be increased by nitrogen fertilization, but this could adversely affect the forest ecosystem as viewed by a forest ecologist. Consequently, the attainment of tree, stand and forest health can potentially be mutually exclusive.

To some extent, the differing concepts can be merged, but this is essentially a theoretical exercise. For example, it is possible to speak of forest health within a certain 'normal' range. Dead, dying and unhealthy trees would be expected in a healthy forest, as long as these did not exceed 'normal' levels. The problem with such an approach is that the normal levels of mortality are often unknown in forests, particularly in those forests which are managed and which may therefore have weaker or older trees selectively removed. In Switzerland, for example, annual mortality rates of trees during the 1990s were about 0.5%, which has been considered as within the normal range of variation (Dobbertin, 1998). This figure represents the actual mortality of trees in the annual inventory, and makes no allowance for in-growth, which more than compensates for mortality within the plots (Dobbertin, 1998). However, this mortality rate is lower than that encountered in natural and semi-natural forests with gap-phase dynamics (about 2% a year), as the trees that are most likely to die are not selectively removed. In another example of the difficulties associated with defining normal conditions, a large area of forest in British Columbia was affected in 2001 by mountain pine beetles (*Dendroctonus ponderosae* [Hopkins]). Upwards of 25 000 ha. of forest had >90% tree mortality and over 7 million ha. were affected by lower rates of mortality. Given that a major limitation for such outbreaks is the occurrence of very cold winter temperatures, does the recent absence of such temperatures, possibly associated with globally increasing temperatures, represent a 'normal' situation?

One response to these difficulties has been to substitute the term 'forest health' with 'forest condition'. However, this term is equally difficult to define, although there is a tendency to recognize forest condition as a descriptive term that has no relationship to causal mechanisms. Thus, in Europe, defoliation is reported as an indicator of forest condition. In fact, defoliation is probably better seen as an indicator of tree health, as the term when strictly used clearly implies a loss of foliage. However, in the European surveys, most countries actually assess crown transparency rather than defoliation (although the latter term is used in the annual reports), and crown transparency can be classed as an indicator of crown condition as it does not necessarily imply a loss of foliage.

As a result of these problems, most assessments of 'forest health' are in practice restricted to the health of individual trees within plots. In Europe, such studies are often based on the visual assessment, from the ground, of 24 trees at a site. The most commonly used sampling design is based on a $16 \times 16\ km^2$ grid, so the 24 trees are assumed to be representative of a much larger area (25 600 ha.). Very few studies have attempted to look at other aspects of the forest ecosystem and no attempts are generally made to look at ecological indicators of ecosystem health, such as ecosystem resilience. Surveys of forest soils have been undertaken at a large scale in Europe (Anon, 1997). This is because soil chemistry may provide an alternative measure of forest health to tree condition. With increasing soil acidity, the risk of aluminium toxicity increases (Ulrich, 1983), although there are many complications to this broad generalization (Innes, 1993). Acidic deposition is believed to be responsible for the loss of base cations from the soil, and there is evidence from Solling (Germany) and other sites for changes in soil chemistry that correlate with changes in acidic deposition over time (Wesselink *et al.*, 1995). However, whether or not soil acidity, cation exchange capacity, base saturation or some other measure of soil chemistry can really be used as a measure of forest health remains disputed.

In contrast to the statement at the beginning of this chapter that air pollution impacts are well known, the identification of the direct and elusive indirect effects of regional-scale air pollutants in forests downwind of mesoscale transported pollutants is both complex and non-conclusive. As part of efforts expended by forestry agencies over the past two decades in determining the influence of regional-scale air pollutants on forests, numerous surveys were designed to assess forest health and/or forest condition across much of the North American and European continents (Anon, 1985a, 1985b; Bengtsson and Lindroth, 1985; Binns *et al.*, 1985; Duplat, 1985; Liedeker, 1986; Lonsdale, 1986; Millers and Miller-Weeks, 1986; Neumann and Stowasser, 1986; Magasi, 1988; Landmann and Bouhot-Delduc, 1995). Many of these surveys emerged as part of a much larger research effort specifically designed to determine the effects of acidic deposition on forest trees (e.g., De Vries and Leeters, 1996). It is interesting to note that few of the initial surveys collected foliage from the observed trees within the survey plots for diagnosis of causes. Instead, ground observers recorded yellowing and transparency (and, in some cases, defoliation) of crowns (if observed); causes of these innocuous symptoms were often left undetermined at the time of field observations.

It is important to understand that various forms of chlorosis are commonly observed in many biotic and abiotic disease- and insect-caused maladies, and defoliations at any time of the year are likewise due to a myriad of possible causes. In some cases, canopy transparency (e.g., in Switzerland) was only recorded if the causes were unknown (i.e., only unexplained transparency was recorded), eliminating any possibility of identifying a pollution effect if secondary pathogens were active. In other cases (e.g., with the European Community survey), defoliation was sometimes ascribed to one of a few classes (e.g., insect damage), although in the majority of cases, the causes of defoliation were not recorded. As a result of an error in the design of the forms, it is not known whether these 'unknown causes' actually reflect an unknown cause or whether the observer simply failed to enter the cause. In both Europe and North America, even the classic diseases, insect activities, and environmental stressors such as simple droughts or nutritional disturbances were left as non-determined and therefore as unrecorded; for the most part, critical information on stand histories has been similarly ignored in these surveys.

As a result of these limitations in survey design and execution, the concepts of a general forest decline, i.e., 'Waldsterben' or 'neuartige Waldschäden' as taking place within many European and North American forests with unknown causes was easily fostered; the maladies with 'unknown causes' were then ascribed to have their foundations in air pollution (Schütt and Cowling, 1985; Schütt, 1988). Several reviews have been critical of the basic concepts of a general Waldsterben having taken place in the past several decades and/or as predicted for the future welfare of our forests (Kandler, 1990, 1993; Skelly, 1989, 1990; Innes, 1992, 1993; Skelly and Innes, 1994; Kandler and Innes, 1995). These have drawn attention to some of the problems associated with the design and analysis and interpretation of the surveys. Although research has largely stopped (it is difficult to undertake cause–effect research on an effect that is so transitory), many of the surveys have continued and have even been intensified.

More recent, well-designed, national-level surveys, e.g., Canada's Forest Insect and Disease Survey (FIDS) as the foundation for the Acid Rain National Early Warning System, ARNEWS (Magasi, 1988) and the North American Maple Project (Millers *et al.*, 1992) have produced some very important results concerning the actual states of tree health within areas of high pollution deposition. These surveys have recorded a number of different tree crown variables while also recording the presence of symptoms incited by disease pathogens, insects, and environmental stresses. In this manner the symptom etiologies are instantly ascribed to the most likely causes and the potential role of acidic depositions in directly causing the symptoms has become better defined.

It is important to also note that the plausible predispositional role of acidic deposition and other forms of pollution leading to either increased or decreased biological interactions has not been addressed in this chapter (see Chapter 20). Such an interactive role of 'acid rain' cannot be properly evaluated during the tree health surveys as currently in existence; this is an entirely different area seriously lacking in cause–effect research efforts to date. However, we must remember that simply stating and restating that such interactions are possible does not bring them into reality.

Reports of forest health – eastern North America

Canadian surveys – ARNEWS

The Acid Rain Early Warning System (ARNEWS) of Forestry Canada was established in 1984 to detect early indications of air pollution and climate change damage to Canada's forests (Forestry Canada, 1992). Specific objectives were established: (1) to detect possible damage to forest trees, vegetation and soils caused by air pollution; (2) to identify forest damage that is not attributable to management practices or natural causes; and (3) to monitor forest vegetation and soils for changes in forest health.

The ARNEWS survey was initiated during a very emotionally driven period which strongly encouraged the finding of acid rain 'effects' throughout the forests of eastern Canada; concerns were expressed at all levels of forest science and were fostered by the more popular presentation of the supposed threat of acid rain to Canada's forests (Environment Canada, 1982; Edwards, 1987, Woodman and Cowling, 1987). This led to the survey being aimed at determining that acidic deposition was the only plausible explanation for the presence of dead and dying trees in the forests of Canada, with attention focusing particularly on the condition of sugar maple (*Acer saccharum* Marsh.).

The ARNEWS plots were established to be representative of the tree species and forest types of Canada but with a major concentration of plots near the major centres of population along Canada's southern border; survey plots were established within known zones of high, medium, and low sulphate and nitrate atmospheric deposition. Preferred plot selection sites were also selected on soils thought to be sensitive to acidic deposition. Plots (trees) were observed annually with assessments made for mortality, tree crown condition, visible foliar symptoms, regeneration, ground vegetation, and foliage retention. Tree health assessments were conducted by the staff of Canada Forestry's Forest Insect and Disease Survey (FIDS) (Canadian Forest Service, 1993); FIDS was composed of individuals trained in matters of forest protection. When damage was detected on survey trees, a search was made to determine the causes involved. Known causal agents were identified and recorded in the field with specimens needing further diagnosis collected for laboratory identification by trained diagnosticians. Thus the survey and detection procedures as carefully developed within the ARNEWS programme offered an excellent methodology for the proper evaluation of forest tree health.

Results through to 1997 clearly did not suggest a major role of acidic precipitation in causing major problems for the health of Canada's forests. Indeed the following conclusions about sugar maple have been taken directly from the 1992 ARNEWS report (Hall, 1993):

> No mortality was found in 1992; the extent of mortality to date has been normal and was attributable to natural thinning.

> The assessment of the ARNEWS plots in 1992 indicates that there is no large-scale decline in the health of Canadian forests that can be attributed directly to atmospheric pollution. This conclusion is similar to that reached from previous assessments. Evidence of the

classic symptoms of air pollution were sought but few indications of pollution damage were found.

It is possible, of course, that trees have been weakened or stressed by external factors such as air pollution and that this stress is not apparent.

The effects of insects, diseases, droughts, and storms were observed frequently.

There has been no indication of a large-scale decline in the health of our forests.

With the cancellation of ARNEWS and the disbanding of staff responsible for the Forest Insect and Disease Survey, there has been a marked change in the nature of the reporting. More recent statements associated with the reporting on indicators by the Canadian Council of Forest Ministers (CCFM, 2000) have moved away from citing direct damage caused by air pollution, as observations are no longer being made. Instead, it reports that critical loads (defined by Nilsson and Grennfelt (1988) as 'quantitative estimates of an exposure to one or more pollutants below which significant harmful effects on specified sensitive elements of the environment do not occur according to present knowledge') for acidic deposition are frequently exceeded in eastern Canada, and that critical levels for ozone are exceeded in the southern Maritimes, the Windsor–Quebec corridor and the lower Fraser Valley of British Columbia. Instead of conducting field investigations to determine the extent of any health problems, the CCFM report cites an unpublished study that suggests that an annual productivity loss of 10% occurs when there is an exceedance amounting to 500 equivalents of acid deposition per hectare per year.

United States–Canadian surveys: NAMP

The North American Maple Project (NAMP) was initiated in 1987 via a Memorandum of Understanding and Special Project Agreement involving representative agencies of the US Department of Agriculture–Forest Service and Forestry Canada (Millers *et al.*, 1992). The stated purpose of the special international project was to monitor and evaluate sugar maple condition, particularly in relation to pollution and stand management history. Specific objectives were to determine: (1) the rate of change in sugar maple tree condition ratings from 1980 to 1990; (2) if the rate of change in sugar maple tree condition ratings were different among various levels of sulphate and nitrate deposition and between sugarbush stands (managed for sugar/syrup operations) versus undisturbed and more natural sugar maple forests; varying levels of initial stand decline conditions were also selected; and (3) the possible causes of the hypothesized sugar maple decline and the geographical relationships between potential causes and the extent of any noted decline.

Within NAMP, a total of 171 plot clusters (5 plots each) were established from Wisconsin (US) to Ontario (CN) and from New Hampshire (US) to Nova Scotia (CN). Plots were evenly distributed across managed and undisturbed stands and spread across sulphate wet deposition zones ranging from 10 to 35 kg^{-1} ha^{-1} yr^{-1} and nitrate wet deposition zones of 8 to 15 kg^{-1} ha^{-1} yr^{-1}. Among many other observations each tree was evaluated for branch dieback, foliage transparency and discoloration, dwarfed foliage and insect defoliation; overall 15 000 trees were observed including 11 000 sugar maples with ca. 7000 of these in the dominant and co-dominant crown classes.

The report concluded (Millers *et al.*, 1992):

> More than 90% of the sugar maples examined were considered healthy; of the 10% found to be somewhat less than healthy, approximately 86% of the maples with more than 50% dieback had major bole and/or root damage.
>
> The condition of sugar maple stands managed for sap production was essentially no different from stands classified as unmanaged, though statistically some minor differences were found.
>
> Likewise no important differences were observed between sugar maples growing in high vs. light sulfate wet deposition zones.
>
> Most of the crown condition improvements seem to be related to decreased insect damage from pear thrips (*Taeniothrips inconsequens* Uzel) in Massachusetts and Vermont and forest tent caterpillar (*Malacasoma disstria* Hübner), in northern Ontario. Wisconsin sugar maples had higher dieback and thinner crowns, probably as the result of the drought that began in 1988.

Based on our current knowledge of the health of the Canadian and United States sugar maple forests, it is obvious that the suggestion of a cataclysmic loss, i.e. '... whole forests of sugar maple succumbing to a mysterious plague' (Edwards, 1987), has not been confirmed by either the ARNEWS or NAMP surveys. Likewise the declaration of Woodman and Cowling (1987) that 'Visible injury and mortality among sugar maple and other hardwood trees have been observed in Quebec, Ontario, New England, New York, and Pennsylvania' as followed later in the same section of their article by... 'Air pollution is believed to be a primary or contributing causal factor because no other more plausible explanation has been suggested' should no longer be considered as tenable.

The findings of investigations of site-specific forest species declines such as the etiological studies recently completed by Kolb and McCormick (1993) within four sugar maple stands in Pennsylvania should be reviewed and incorporated into future survey efforts. They concluded that an observed reduction in basal area increment between declining and healthy trees was first evident following a series of defoliations and droughts of the 1960s and early 1970s followed by repeated severe damages by the pear thrips into the 1980s. Major reductions in basal area increment were recorded following the 1988 growing season, which has been noted as one of the warmest and driest and with the highest ozone exposures on record (Comrie, 1994). Long-term records of stand conditions and individual tree health are manifestly important in diagnosing the more immediately present symptoms of decline.

USA – Forest health monitoring

USA – the Pennsylvania surveys

A well-defined gradient of total atmospheric deposition has been described along the Allegheny Plateau in north-central Pennsylvania. The decreasing gradient of pollutant deposition exists from a west to east fashion due to prevailing winds and various source strengths inclusive of the heavily industrialized and urban Ohio Valley situated upwind and to the west of Pennsylvania. Wet sulphate deposition drops from 40 to 25 kg^{-1} ha^{-1} yr^{-1} across a distance of 176 km (west to east); H^+, nitrate and ammonium ions follow similar trends of decrease across the gradient. The area receives precipitation that is among the most acidic in the United States (Lynch, 1990). The purpose of the survey was to determine if anomalies in overall tree condition occurred in the forests along the gradient and to determine their possible relationships to the gradient parameters (Nash *et al.*, 1992).

The forest condition survey was conducted during the growing seasons of 1988 – 1989 with northern red oak (*Quercus rubra* (L.) Wang.), white oak (*Q. alba* L.), red maple (*Acer rubrum* L.) and black cherry (*Prunus serotina* Ehrh.) being the most intensively studied tree species. Observations were made by professional forest pathologists and entomologists within stands across four designated core areas along the decreasing deposition gradient; five trees of each species within two plots of each core area were evaluated for general tree health characteristics and foliage health via collected samples for field and/or laboratory diagnostics.

Results of the two-year intensive survey demonstrated (Nash *et al.*, 1992):

> No observable major trends in tree crown, branch, or main stem maladies were associated in an obvious cause and effect pattern with the deposition gradient.
>
> The four examined canopy species appeared in a reasonably good state of health.
>
> Several symptoms were attributable to known diseases and insects.
>
> Because many forest insects and diseases are cyclic, symptoms with distribution patterns suggesting 'gradient effects' (for example insect borers on white oak) merit further long-term studies.
>
> Our surveys of forest condition along this gradient for 1988 and 1989 offer no evidence of anomalies in the health of forest species along the described gradient.
>
> The finding of a 'no effect' should be considered as significant at this time because, often, only findings that report deleterious effects due to acidic depositions within forests are well received.

Although there were no dramatic shifts in forest tree health observed across the Pennsylvania deposition gradient other direct and indirect effects to forest vegetation have been shown; i.e. observable differences in lichen species diversity and heavy metal contamination have occurred along the gradient. As part of the overall air pollution deposition pattern across north-central Pennsylvania, tropospheric ozone exposures follow a similar decreasing trend of exposures in a west to east fashion (Simini *et al.*, 1992; Comrie, 1994). Foliar symptoms and seedling height growth of ozone-sensitive black cherry and yellow-poplar (*Liriodendron tulipifera* L.) were demonstrated to be associated with the ozone exposures using open-top chambers at three sites along the gradient; less ozone-sensitive red oak and red maple seedlings showed no responses within the chamber study (Simini *et al.*, 1992). These results serve as a clear reminder of the potential for significant effects due to air pollutants within the forests of central Pennsylvania; further studies are being conducted in scaling ozone effects to larger canopy size trees under natural forest conditions of exposure (Fredericksen *et al.*, 1995). Although controlled experiments with seedlings have been valuable for revealing the 'principles' of ozone action on woody plants, these findings lack ecological significance as long as validation at forest sites is missing (Matyssek and Innes, 1999). In addition, physiological, phenological and developmental differences related to tree size make inferences of ozone-induced effects from seedlings when scaled up to mature trees very uncertain (Fredericksen *et al.*, 1996; Samuelson *et al.*, 1996; Kolb *et al.*, 1997).

USA – Virginia: Shenandoah National Park surveys

A survey seeking to determine the incidence and severity of ozone-induced foliar symptoms on three sensitive eastern hardwood tree species was conducted during the late summer seasons of 1991 to 1993 (Hildebrand *et al.*, 1996). The objectives of the study were also to determine relationships between the observed symptoms and cumulative ambient ozone exposures experienced during the growing season from budbreak until sample collections. The study was unique in that three Trend Plots were established immediately adjacent to certified air quality monitoring stations positioned at strategic places across the length of the Shenandoah National Park; elevational gradients were also involved as well as microsite influences within and between the Trend Plots. In addition, mature canopy trees were selected for study with each tree climbed for sample collections of upper and mid-crown outside branchlets. Species investigated included black cherry, yellow-poplar, and white ash (*Fraxinus americana* L.). In addition to scoring the typical upper surface stipple induced by ozone exposures, crown transparency, density, percent live crown, and defoliation were recorded for each of the 30 trees per species per Trend Plot. A similar study was conducted concurrently in the Great Smoky Mountains National Park of North Carolina and Tennessee (Chappelka *et al.*, 1992).

Results of the three-year Shenandoah Trend Plot investigations included the following observations:

> All three species at all three sites exhibited foliar symptoms of ozone induced injury.

> Black cherry appeared as most consistent in foliar injuries as related to ozone exposures. Regression analysis indicated that the best model to describe injury increase with cumulative ozone concentration was exponential, particularly when foliar injury was related to the ozone exposure statistics of >SUM 60 and W126. (Lefohn *et al.*, 1988)

> Symptom recurrence inclusive of both black cherry and yellow-poplar at all sites sampled from 1991 to 1993 was 87%. (White ash was not consistently sampled at all sites in all years.)

Showman (1991), Hildebrand *et al.* (1996) and Schaub *et al.* (1999) also point to the importance of fully understanding environmental factors as they may directly influence foliar symptom expression within and between survey plots. As observed within the Shenandoah Trend Plot studies, yellow-poplar exhibited the most symptoms at the plot having the lowest ozone exposures over the three-year period of the study; injury was reported on a constant 67% of the trees. Their explanations centred on the location of the yellow-poplar stand being situated along a streambed with resultant constantly favourable ozone uptake and perhaps larger internal dose being experienced at this particular microsite within the Shenandoah National Park.

Reports of forest health – Europe

European surveys (United Nations Economic Commission for Europe, UNECE, and European Union, EU)

By far the biggest programme of forest health monitoring is conducted in Europe under the joint auspices of the UNECE and the EU. This international survey is based on a 16 × 16 km grid of plots that extends across almost the whole of the European land area. The survey involves the assessment of crown defoliation (transparency) and discoloration. In some countries, soil chemistry and foliage chemistry have been assessed in the same plots. The programme began in 1985, and has gradually extended across the whole of the UNECE region, such that in 1998, defoliation data were obtained from 127 455 trees on 5695 plots. This represents a major increase since 1987, when data from 27 250 trees on 1099 plots were obtained.

There is considerable controversy over the value of defoliation and discoloration as indicators of stress induced by air pollution (e.g. Innes, 1998), although there is some consensus that defoliation (when it is correctly assessed) can be interpreted as an indication of stress. However, in practice, defoliation is rarely determined and the reported figures more often refer to transparency, which cannot be readily equated with stress (e.g. Ferretti, 1997). Discoloration is generally extremely rare (it is normally associated with drought or with nutrient deficiencies), although its extent shows substantial year-to-year fluctuations. Despite the early recognition of the problems associated with surveys of defoliation and discoloration (e.g. Innes, 1988a, 1988b), they have continued to be used as they are considered to be the only suitable indicators for use in large-scale, 'representative' inventories.

A clear increase in the proportions of trees with more than 25% defoliation (transparency) assessed in the basic 16 × 16 km grid network of plots has been identified in Europe for the period 1989–1998 (Anon, 1999). This increase conceals many regional and species-specific variations, such that individual regions may not show an overall increase in defoliation during the period. Considerable care needs to be taken when comparing regional trends, as assessment standards between countries are variable, and have a marked effect on the reported distribution of defoliation in Europe (Ghosh *et al.*, 1997; Klap *et al.*, 1997).

The explanation of the increases in defoliation (transparency) have been difficult, and it is now generally recognized that the 16 × 16 km grid inventories will not provide information on cause–effect processes. Attempts have been made to establish whether there are any correlations between the distribution of defoliated trees (or the spatial pattern of trends in defoliation) and environmental variables (e.g. Klap *et al.*, 1997), but these have mainly served to reinforce the conclusion that surveys of defoliation do not provide a particularly useful contribution to cause–effect studies. Some reports have stressed the failure of the epidemiological approach in assessing air pollution impacts because the studies have not demonstrated the expected link between defoliation and pollution; the scientific basis for such arguments should be seriously questioned. Other studies (e.g. Webster *et al.*, 1996) have been more successful, and have demonstrated that the distribution of defoliated trees can be related (albeit weakly) to environmental variables such as soil water availability. Many of these difficulties probably arise from the inappropriateness of crown transparency as a response variable and the problems associated with determining site-specific values for the forcing factors.

It is difficult to draw any conclusions from the international UNECE survey. The increase in defoliation that has been recorded seems clear, but there are many potential sources of error that are not openly discussed. These include factors such as changes in the reference trees used to determine crown transparency. For example, in the 1999 report (reporting on the condition of trees in 1998), a graph is shown plotting the progressive increase in transparency of trees sampled in all years between 1989 and 1998 (Anon, 1999). However, in the country report of France, it is stated that:

> The general increase in defoliation as recorded between 1995 and 1997 is very uncertain due to changes in methodology introduced in France: a more precise definition of the 'reference tree' has led to the unexpected result that defoliation is now more 'severe' than before. This was confirmed by independent data, i.e. litter fall data available on Level II[1] plots. Therefore it is difficult to carry out analyses of the relationships between crown condition and causes and effects with data of most recent years.

As a result, considerable care should be taken over the reports published by the UNECE. These should not be viewed as scientifically valid, and they are not subject to any form of independent, scientific, peer-review process. They essentially represent the viewpoint of the lead agency, namely the German Ministry for Food, Agriculture and Forestry. The reports ignore the results of scientific programmes designed to investigate the impacts of air pollutants on forests, and consistently misuse the basic scientific principles used in establishing cause and effect relationships. Instead, as indicated above for the case of France, it is the national or provincial reports that frequently provide the most objective and scientifically valid information.

[1] Level II refers to the more detailed sampling programme undertaken under the auspices of the UNECE and EC, where a variety of environmental variables are recorded in addition to tree condition.

European national studies

Most countries conduct an inventory at a more detailed scale than 16 × 16 km. The density of plots varies from country to country, but the end result is a plethora of data on tree defoliation and discoloration. Some of these data have been analysed by individual countries. In most cases, the studies have attempted to identify correlations between the spatial pattern of defoliation and the spatial patterns of pollutants. Such an approach suffers from the problems associated with correlations between pollutants and other environmental variables (i.e. confounding factors). In addition, the precise environmental factors determining the transparencies of the trees are extremely difficult to quantify, making the identification of the independent variables difficult. For example, the occurrence of short-term moisture stress at a critical time (e.g. bud set) may not be well represented in climatic data based on monthly or even annual averages. In addition, there are major uncertainties in relating environmental data to specific forest stands, as data specific to the stands are rarely available. Instead, models have to be used, and these may introduce substantial errors.

France

Some of the most careful analyses of purported forest decline have been conducted in France, particularly in the Vosges Mountains in the east of the country. Landmann (1995) acknowledges that forest health may have been poor in the 1970s and 1980s, but emphasizes that a lack of long-term data and misinterpretation of the available data have created problems. For instance, he states that the assumption of an unprecedented forest decline has been widely accepted to the extent that anyone questioning it has been perceived as denying that air pollution has an impact on forests, and that:

> The belief that the 'novel forest decline' is unique in its extent seems to stem from a careless interpretation of the results of forest health surveys and the scant attention that has been paid to the historical records of forest declines.
>
> ... available evidence gives indirect and a posteriori support to (1) the seriousness of the 1970s/1980s crisis, in that a substantial decrease in defoliation has occurred in fir and spruce in the eastern mountains, and (2) the plausibility of the suggestion that the impression of increasing damage in the early 1980s in France might have been created by the extension of surveys and the step-by-step awareness among foresters, rather than to a real progression of decline.
>
> Large-scale forest inventories do not give any clear indication of a substantial deterioration of the forest condition in France.
>
> ... the weight of evidence for a general long term increase of forest growth now seems overwhelming. This growth increase is difficult to reconcile with the idea of a large-scale, unprecedented forest decline. The threat of a general imbalance within ecosystems might be conjured up, but such imbalances are unlikely to occur similarly under varied ecological conditions.

> The initial emphasis laid on crown thinning as the main criterion for the definition of forest decline and on the evaluation of air pollution effects stimulated research in a first phase, as an increasing number of reports suggested increasing damage, but proved counter-productive later on as it became progressively evident that there was no large-scale decline in overall forest condition.

As with the former ARNEWS programme in Canada, the French studies of forest health tend to be undertaken by specialists in forest pathology. There is an effective reporting system for the whole country and this has enabled the issue of forest decline to be put into the appropriate context.

Finland

A major research programme ('HAPRO') investigated the effects of acidification in Finland (Kauppi *et al.*, 1990; Kauppi, 1992). Acidification was found to be present, and a variety of environmental impacts were identified. However, the results for forest trees did not indicate that a major problem exists, although more sensitive components of forest ecosystems, such as lichens, were affected. Kauppi (1992) reports:

> The growing stock and the growth in Finnish forests have increased substantially in recent decades. Changes in the structure and age distribution of forests have been the primary reasons for this. Acidifying deposition or other forms of air pollutants have not had a significant negative effect on wood production.
>
> A link was established between diameter growth and defoliation of individual trees. However, no spatial correlation was observed between defoliation and pollution load on a regional scale.

While there was no indication of regional-scale impacts, it is pointed out that the available time series is very short and that more information would be useful. In addition, pollution from local sources (e.g. fur farms) is identified as a cause of localized forest damage.

Norway

Aamlid *et al.* (2000) have described the condition of forests in Norway. As with many European countries they do not discount the possible impacts of air pollution. Their results are typical of many:

> There is no abnormal tree death in Norwegian forests. The forests ecosystems investigated, which are representative for much of the Norwegian forest, have a satisfactory health condition.

Many of the identified problems in Norway have been associated with climatic stress, as might be expected from a mountainous country that extends north of the Arctic Circle.

Switzerland

Switzerland was one of the countries actively involved in the promotion of the 'Waldsterben' concept in the early 1980s. It was widely believed that the forests of Switzerland were dying, and forest scientists did little, if anything, to alleviate such fears (see, for example, the extremely pessimistic view put forward by Schweingruber (1985) and Schweingruber *et al.* (1983)). However, at the start of the 1990s, there was increasing recognition that this view was incorrect. A National Research Programme funded by the Swiss National Science Foundation concluded (Haemmerli *et al.*, 1992) that:

> Assessment of crown condition alone is insufficient for the estimation of stand vitality. The condition of the mountain forest in Davos and in the Alptal is considered normal based on several indices (e.g. growth, nutrient supply), even though the actual needle loss level is high.
>
> The investigations indicate neither a temporal nor a spatial dependence of crown condition on air pollution. It can be concluded for the study sites that air pollutants have hardly affected the crown condition of spruce within the observation period. In a spatial point of view, the pollution on the alpine site Davos is generally low, while the needle loss level is high. On the other hand, the pollution on the Laegeren, in the densely populated lowland, is relatively high whereas the extent of crown transparency is small.

Despite the indications that air pollution had little involvement with observed patterns of crown condition (Innes *et al.*, 1997), concern continued through the 1990s. Surveys of crown condition continued, although at a steadily reducing intensity, with a $4 \times 4\ km^2$ sampling intensity in the early 1990s, $8 \times 8\ km^2$ intensity in the mid-1990s and a $16 \times 16\ km^2$ inventory in 1995–96 and 1998–99. Analyses of the data have revealed that defoliation in some species can be related to soil moisture conditions, with drier sites tending to have more defoliation (Webster *et al.*, 1996). The conclusion reached after 16 years of inventory and research (Brang, 1998) was that air pollution presented a risk to forests, but that there was insufficient evidence to quantify the extent of this risk. However, a large-scale decline in forest condition, threatening the future existence of forests, was excluded as a likely scenario.

European Forestry Institute Growth Trends Study

For a number of years, there have been reports of increasing rates of forest growth (Innes, 1991). These have not just been restricted to high-latitude or high-altitude forests; they have also been reported from more optimal growth areas such as the Black Forest of Germany. These reports led to the development of a Europe-wide study on growth rates which had the specific aim of determining whether growth rates were higher today than in the past (Spiecker *et al.*, 1996). The results of this study showed that over much of Europe, growth rates of trees are increasing. Only one area (the Kola Peninsula of northern Russia) had decreasing growth rates. As this area is very heavily impacted by pollution from the smelters at Nikel, Monchegorsk and Apatity, such a reduction in growth could be more or less expected.

There are a number of possible causes for the changes in growth rates, but these were not assessed in the study; they are being assessed in a major international project sponsored by the European Union. Potential causes include changes in management practices, increased CO_2 concentrations, increased nitrogen deposition and changed climatic conditions. It is likely that a combination of these is occurring, with the relative importance of each changing from site to site. For example, changed management can clearly be excluded as a cause of growth increases in unmanaged areas, yet many such areas are showing growth increases (Innes, 1991). Similarly, nitrogen deposition can be excluded as a cause of documented increases in tree growth rates in the pristine environment of southern Chile (Innes *et al.*, 2000), but may well be a factor in many parts of Europe and eastern North America.

The Global Forest Resources Assessment 2000

The Global Forest Resources Assessment of the United Nations Food and Agriculture Organization (FAO) provides fundamental information on the condition of forests globally. The part dealing with temperate and boreal forests was published in 2000 (Anon, 2000). While the report includes information on forest health, the European dataset is the same as used by the UNECE and has the same problems. The report indicates that pollution from local sources affected less than 200 000 ha of forest throughout the boreal and temperate zones (an area amounting to 2477 million ha.) in the mid-to late-1990s. To place this in perspective, moderate to severe defoliation caused by one species of insect, the forest tent caterpillar (*Malacosoma disstria* Hübner) amounted to over 19 million ha in 1991 in Canada alone. Also for relative comparison of economic importance, during late December 1999, storm 'Lothar' swept across southern Europe with excessively high winds; the equivalent of 280% of the allowable annual timber cut was blown down in Switzerland, 300% in southern France, and 250% within Baden-Württemberg in southern Germany (WSL, 2001).

Further information can be gained from the data on increment and felling (including salvage felling) that is provided in the FAO report. No region had a rate of felling that exceeded the rate of replacement by growth. The ratio of fellings to annual increment ranged from 13.5% (Russian Federation) to 67.4% (the Nordic countries of Europe). These figures paint a very positive picture that reinforces many of the statements made above.

Conclusions

The term forest decline is a misnomer; and forest tree (species) decline is proposed as a better descriptive term for the phenomenon we occasionally observe within a given species and then usually on a site-to-site and sometimes regional basis. The phenomenon is not new to forests of North America or Europe (Kandler, 1992; Millers *et al.*, 1988). In contrast to the suggestions of a series of forest declines taking place within our mutual forests at a level never before seen and at such intensities as to be considered a new disease phenomenon, i.e. 'Waldsterben' (Schütt and Cowling, 1985), the most recent surveys of North American forests have shown the forests to be in a generally acceptable state of health (Hall, 1993; Millers *et al.*, 1993; Nash *et al.*, 1992; Miller-Weeks *et al.*, 1994).

The report of Miller-Weeks *et al.* (1994) reviewed the findings of the Forest Health Reports of 1991 and 1992 (USDA Forest Service 1993, 1994) and concluded by clearly stating 'In general, the forests of the Northeastern Area (Missouri to Minnesota and West Virginia to Maine) are healthy; although forest health concerns arise for particular species or in localized areas.' Numerous pests and pathogens of several major tree species were reported with considerable importance given to those recently introduced to the forests of the northeast. Ozone was noted as a pollutant of concern to sensitive species such as black cherry, white ash, and white pine (*Pinus strobus* L.) but impacts were not presented; no mention was made of acid rain effects being detected.

As based upon the most recent surveys by qualified forest protection trained specialists, we should conclude for the moment that our forests are acceptably healthy and that diseases, insects, and natural environmental stressors are all continuing to influence the observations of notable changes on a species and stand (sometimes regional) level basis. We should also conclude that the importance of air pollutant effects such as those manifested by visible foliar injuries and less easily monitored physiological changes are going to require much more longer term and therefore costly research programmes. It is with great concern that we note a decline in field-based observation of forest health. Instead, increasing reliance is placed on models of predicted impacts. Where the models predict that critical loads or levels of pollutants are exceeded, it is assumed that effects are present. Along with those who originally designed the concept of critical levels, we would argue vigorously that such models indicate a level of risk but in no way indicate that effects are present.

Despite these findings, there continue to be claims about the deterioration in the health of trees in Europe and North America. Such claims are being made in the more popular literature (e.g. Little, 1995; Ayers *et al.*, 1998), on the Internet and in the scientific literature (Sandermann *et al.*, 1997). We believe that forest health in many areas is being adversely affected by air pollution. However, we would seriously question the extent to which tree health is being adversely impacted by air pollutants outside of specific, well-delimited regions with important point sources or concentrations of diffuse sources. Our conclusion is based on purposely designed surveys of tree condition as noted within this chapter that either indicate that trees and their collective forests are in good health or provide no indication that they are in poor health.

Finally, the concept of a general forest decline 'Waldsterben' occurring in our forests as proposed due to acid rain or other forms of regional scale air pollution (in the past or for the foreseeable future) should be discarded (Skelly and Innes, 1994; Kandler and Innes, 1995).

References

Aamlid, D., Solberg, S., Hylen, G., Tørseth, K. and Clarke, N. (2000). Forest damage and forest monitoring in Norway – annual report of the Norwegian Monitoring Programme for forest damage 1999. (English Internet Version, 10.11.2000). http://www.skogforsk.no/forskning/skogpatologi/ops/E_statusText99.html (accessed 27 July, 2000).

Anon (1985a). *Waldschadenserhebung* 1985. Bundesminister für Ernährung, Landwirtschaft und Forsten, Bonn.

Anon (1985b). *Résultats de l'inventaire Sanasilva des dégâts aux forêts, 1985*. Office fédéral des forêts et de la protection du paysage, Berne; Institut federal de recherches forestières, Birmensdorf.

Anon (1997). *Forest Soil in Europe. Results of a Large-Scale Soil Survey*. United Nations Economic Commission for Europe, Geneva, and European Commission, Brussels.

Anon (1999). *Forest Condition in Europe. Results of the 1998 Crown Condition Survey*. United Nations Economic Commission for Europe, Geneva, and European Commission, Brussels.

Anon (2000). Forest resources of Europe, CIS, North America, Australia, Japan and New Zealand. Main Report. Geneva Timber and Forest Study Papers 17. United Nations, New York and Geneva.

Ayers, H., Hager, J. and Little, C.E. (eds.) (1998). *Appalachian Tragedy. Air pollution and Tree Death in the Eastern Forests of North America*. Sierra Club Books, San Francisco.

Bengtsson, G. and Lindroth, S. (1985). *Första resultaten fran statens skogsskadeinventeringen 1985. Institut for skogstaxering*. Sveriges Lantbruksuniversitet, Umeå.

Binns, W.O., Redfern, D.B., Rennolls, K. and Betts, A.J.A. (1985). *Forest Health and Air Pollution: 1984 Survey*. Forestry Commission Research and Development Paper 142. Forestry Commission, Edinburgh.

Brang, P. (ed.) (1998). *Sanasilva-Bericht 1997. Gesundheit und Gefährdung des Schweizer Waldes – eine Zwischenbilanz nach 15 Jahren Waldschadenforschung*. Berichte der Eidgenössische Forschungsanstalt für Wald, Schnee und Landschaft 345.

Canadian Council of Forest Ministers (CCFM) (2000). *Criteria and Indicators of Sustainable Forest Management in Canada*. Canadian Council of Forest Ministers, Ottawa.

Canadian Forest Service (1993). *Helping Forests Stay Healthy: The Forest Insect and Disease Survey*. Cat. No. F042-201/1993E, Hull, Quebec.

Chappelka, A.H., Hildebrand, E., Skelly, J.M., Mangis, D. and Renfro, J.R. (1992). Effects of ambient ozone concentrations on mature eastern hardwood trees growing in Great Smoky Mountains National Park and Shenandoah National Park. *Presented to the 85th Annual Meeting of the Air and Waste Management Association, June, 1992. Kansas City, Kansas*. Paper No. 920150.04. Pittsburgh, PA.

Chappelka, A.H. and Samuelson, L.J. (1998). Ambient ozone effects on forest trees of the eastern United States: a review. *New Phytologist* **139**, 91–108.

Comrie, A.C. (1994). A synoptic climatology of rural ozone pollution at three forest sites in Pennsylvania. *Atmospheric Environment* **28**, 1601–1614.

De Vries, W. and Leeters, E.E.J.M. (1996). *Effects of Acid Deposition on 150 Forest Stands in The Netherlands. I. Chemical Composition of Humus Layer, Mineral Soil and Soil Solution*. DLO Winand Staring Centre for Integrated Land, Soil and Water Research. Report 69.1 Wageningen, The Netherlands.

Dobbertin, M. (1998). Ergebnisse der Sanasilva-Inventur: Sterberate, Nutzungsrate und Einwuchsrate. In: P. Brang (ed.) *Sanasilva-Bericht 1997. Gesundheit und Gefährdung des Schweizer Waldes – eine Zwischenbilanz nach 15 Jahren Waldschadenforschung*. Berichte der Eidgenössische Forschungsanstalt für Wald, Schnee und Landschaft, Birmensdorf 345, 24–27.

Duplat, P. (1985). Quelques questions à propos du réseau de surveillance du dépérissement des forêts en France. In: Schmid-Haas, P. (ed.) *IUFRO Conference 'Inventorying and Monitoring Endangered Forests'*. Zürich, Switzerland, 1985/08/19-24, pp. 155–159.

Edwards, J.W. (1987). Sounding taps for the sugar maple. *National Wildlife*, Oct–Nov.

Environment Canada (1982). *Downwind: The Acid Rain Story*. Cat. No. En 56-56/1982E. Ottawa, Ontario.

Ferretti, M. (1997). Forest health assessment and monitoring: issues for consideration. *Environmental Monitoring and Assessment* **48**, 45–72.

Forestry Canada (1992). *The Acid Rain National Early Warning System (ARNEWS): Canada's National Forest Health Monitoring Network*. Cat. No. F042-179/1992E, Ottawa, Ontario.

Fredericksen, T.S., Joyce, B.J., Skelly, J.M., Steiner, K.C., Kolb, T.E., Kouterick, K.B., Savage, J.E. and Snyder, K.R. (1995). Physiology, morphology, and ozone uptake of leaves of black cherry seedlings, saplings, and canopy trees. *Environmental Pollution* **89**, 273–283.

Fredericksen, T.S., Skelly, J.M., Steiner, K.C., Kolb, T.E. and Kouterick, K.B. (1996). Size-mediated foliar response to ozone in black cherry trees. *Environmental Pollution* **91**, 53–63.

Ghosh, S., Landmann, G., Pierrat, J.C. and Müller-Edzards, C. (1997). Spatio-temporal variation in defoliation. In: C. Müller-Edzards, W. De Vries and J.W. Erisman (eds), *Ten years of monitoring forest condition in Europe. Studies on temporal development, spatial distribution and impacts of natural and anthropogenic stress factors. Technical background report*. Geneva and Brussels, United Nations Economic Commission for Europe/European Commission, pp. 35–50.

Haemmerli, F., Kräuchi, N. and Stark, M. (1992). The Swiss National Research Program Forest Damage and Air Pollution: (NFP 14+). In: Schneider, T. (ed.) *Acidification Research, Evaluation and Policy Applications*. Elsevier, Amsterdam, pp. 449–459.

Hall, J.P. (1993). *ARNEWS Annual Report* 1992. Can. For. Serv. Inf. Report ST-X-7, Ottawa, Ontario.

Hildebrand, E.S., Skelly, J.M. and Fredericksen, T.S. (1996). Foliar response of ozone sensitive hardwood tree species from 1991–1993 in the Shenandoah National Park, VA. *Canadian Journal of Forest Research* **26**, 658–669.

Innes, J.L. (1988a) Forest health surveys – a critique. *Environmental Pollution* **54**, 1–15.

Innes, J.L. (1988b) Forest health surveys: problems in assessing observer objectivity. *Canadian Journal of Forest Research* **18**, 560–565.

Innes, J.L. (1991). High-altitude and high-latitude tree growth in relation to past, present and future climate change. *The Holocene* **1**, 168–173.

Innes, J.L. (1992). Forest decline. *Progress in Physical Geography* **16**, 1–64.

Innes, J.L. (1993). *Forest Health: Its Assessment and Status*. CAB International. Wallingford, UK.

Innes, J.L. (1998). Role of diagnostic studies in forest monitoring programmes. *Chemosphere* **36**, 1025–1030.

Innes, J.L., Ghosh, S., Dobbertin, M., Rebetez, M. and Zimmermann, S. (1997). Kritische Belastungen und die Sanasilva-Inventur. In: Eidgenössische Forschungstalt für Wald, Schnee und Landschaft (WSL) (ed.), *Forum für Wissen 1997: Säure- und Stickstoffbelastungen – Ein Risiko für den Schweizer Wald?* Birmensdorf, Eidgenössische Forschungsanstalt für Wald, Schnee und Landschaft, pp. 73–83.

Innes, J.L. and Oleksyn, J. (2000). *Forest Dynamics in Heavily Polluted Regions*. CABI Publishing, Wallingford, UK.

Innes, J.L., Schneiter, G., Waldner, P. and Bräker, O.U. (2000). Latitudinal variations in tree growth in southern Chile. In Roig, F.A. (comp.) *Dendrocronologia en América Latina*, EDIUNC, Mendoza, Argentina, pp. 177–192.

Kandler, O. (1990). Epidemiological evaluation of the development of Waldsterben in Germany. *Plant Disease* **74**, 4–12.

Kandler, O. (1992). Historical declines and diebacks of Central European forests and present conditions. *Environmental Toxicology and Chemistry* **11**, 1077–1093.

Kandler, O. (1993). Development of the recent episode of Tannensterben (fir decline) in Eastern Bavaria and the Bavarian Alps. In: R.F. Huettl and D. Mueller-Dombois (eds), *Forest Decline in the Atlantic and Pacific Region*. Springer, Berlin, pp. 216–228.

Kandler, O. and Innes, J.L. (1995). Air pollution and forest decline in Central Europe. *Environmental Pollution* **90**, 171–180.

Kauppi, P. (1992). Finnish research programme on acidification (HAPRO) 1985–1990. In: Schneider, T. (ed.) *Acidification Research, Evaluation and Policy Applications*. Elsevier, Amsterdam, pp. 431–442.

Kauppi, P., Anttila, P. and Kenttämies, K. (eds.) (1990). *Acidification in Finland*. Springer Verlag, Berlin.

Klap, J., Voshaar, J.O., de Vries, W. and Erisman, J.W. (1997). Relationships between crown condition and stress factors. In: C. Müller-Edzards, W. De Vries and J.W. Erisman (eds), *Ten years of monitoring forest condition in Europe. Studies on temporal development, spatial distribution and impacts of natural and anthropogenic stress factors. Technical background report*. Geneva and Brussels, United Nations Economic Commission for Europe/European Commission, pp. 277–302.

Kolb, T.E., Fredericksen, T.S., Steiner, K.C. and Skelly, J.M. (1997). Issues in scaling tree size and age responses to ozone: A review. *Environmental Pollution* **98**, 195–208.

Kolb, T.E. and McCormick, L.H. (1993). Etiology of sugar maple decline in four Pennsylvania stands. *Canadian Journal of Forest Research* **23**, 2395–2402.

Kolb, T.E., Wagner, M.R. and Covington, W.W. (1994). Concepts of forest health. *Journal of Forestry* **92**, 6–8.

Landmann, G. (1995). Forest decline and air pollution effects in the French mountains: A synthesis. In: Landmann, G. and Bonneau, M. (eds.) *Forest Decline and Atmospheric Deposition Effects in the French Mountains*. Springer Verlag, Berlin, pp. 407–452.

Landmann, G. and Bouhot-Delduc, L. (1995). Ground monitoring of crown condition of forest trees in the French mountains. In: Landmann, G. and Bonneau, M. (eds.) *Forest Decline and Atmospheric Deposition Effects in the French Mountains*. Springer Verlag, Berlin, pp. 3–40.

Lefohn, A.S., Lawrence, J.A. and Kohut, R.J. 1988. A comparison of indices that describe the relationship between exposure to ozone and reduction in the yield of agricultural crops. *Atmospheric Environment* **22**, 1229–1240.

Liedeker, H. 1986. *Evaluation of Forest-Dieback Symptoms on Norway Spruce* (Picea abies *Karst.*) *and Red Spruce* Picea rubens *Sarg.*) *in Europe and North America.* Diplomarbeit, Lehrstuhl fur Forstbotanik, Ludwig-Maximilians-Universitat, Munich.

Little, C.E. (1995). *The Dying of the Trees*. Penguin, New York.

Lonsdale, D. (1986). *Beech Health Study* 1985. Forestry Commission Research and Development Paper 146. Forestry Commission, Edinburgh.

Lynch, J.A. 1990. Spatial and temporal variability in atmospheric deposition overview: A Pennsylvania prospectus. In: Lynch *et al.*, (eds.). *Proceedings, Atmospheric Deposition in Pennsylvania: A Critical Assessment*. Environmental Resources Research Institute, Pennsylvania State University, University Park, Pennsylvania, pp. 50–62.

Magasi, L.P. (1988). *Acid Rain National Early Warning System: Manual on Plot Establishment and Monitoring*. Canadian Forest Service Internal Report DPC-X-25, Ottawa, Ontario.

Matyssek, R. and Innes, J.L. (1999). Ozone – a risk factor for trees and forests in Europe? *Water, Air, and Soil Pollution* **116**, 199–226.

Miller, P.R. and McBride, J.R. (eds.) (1999). *Oxidant Air Pollution Impacts in the Montane Forests of Southern California. A Case Study of the San Bernardino Mountains*. Springer-Verlag, New York.

Miller-Weeks, M., Burkman, W.G., Twardus, D. and Mielke, M. (1994). Forest health in the Northeastern United States. *Journal of Forestry* **92**, 30–33.

Millers, I., Shriner, D.S. and Rizzo, D. (1988). *History of Hardwood Decline in the Eastern United States*. General Technical Report NE 126, Radnor, PA, USDA Forest Service, Northeastern Forest Experiment Station.

Millers, I., Allen, D.C. and Lachance, D. (1992). *Sugar Maple Crown Conditions Improve Between 1988 and 1990*. USDA Forest Service NA-TP-03-92.

Millers, I., Allen, D.C., Lachance, D. and Cymbala, R. (1993). *Sugar Maple Crown Conditions Improve between 1988 and 1992*. NA-TP 03-93, USDA Forest Service, Northeastern Area State and Private Forestry; Forestry Canada.

Millers, R. and Miller-Weeks, M. (1986). *Field Manual: Cooperative Survey of Red Spruce and Balsam Fir Decline and Mortality in New York, Vermont, and New Hampshire*. USDA-Forest Service. N.E. Area, Durham, NH.

Nash, B.L., Davis, D.D. and Skelly, J.M. (1992). Forest health along a wet sulfate/pH deposition gradient in northcentral Pennsylvania. *Journal of Environmental Toxicology and Chemistry* **11**, 1095–1104.

Neumann, M. and Stowasser, S. (1986). *Waldzustandsinventur: zur Objektivität von Kronenklassifizierungen. Jahresbericht 1986*. Forstliche Bundesversuchsanstalt Wien, pp. 101–108.

Nilsson, J. and Grennfelt, P. (eds.) (1988). *Critical Loads for Sulphur and Nitrogen*. Nord 1988:97, Nordic Council of Ministers, Copenhagen, Denmark.

Samuelson, L.J., Kelly, J.M., Mays, P.A. and Edwards, G.S. (1996). Growth and nutrition of *Quercus rubra* L. seedlings and mature trees after three seasons of ozone exposure. *Environmental Pollution* **91**(3), 317–323.

Sandermann, H., Wellburn, A.L. and Heath, R.L. (eds.) *Forest Decline and Ozone. A Comparison of Controlled Chamber and Field Experiments*. Springer Verlag, Berlin.

Schaub, M., Zhang, J., Skelly, J.M., Steiner, K.C. and Davis, D.D. (1999). Influence of varying soil moisture on gas exchange and ozone injury in three hardwood species. In: *Critical Levels for Ozone – Level II*, Preliminary Background Papers prepared for a Workshop under the Convention on Long-Range Transboundary Air Pollution of the United Nations Economic Commission for Europe (UN/ECE), Gerzensee, Switzerland, April 11–15, 1999.

Schütt, P. (1988). Forest decline in Germany. In: *Proceedings of the US FRG Research Symposium: Effects of Atmospheric Pollutants on the Spruce-fir Forests of the Eastern United States and the Federal Republic of Germany*. General Technical Report NE-120. Broomall: US Department of Agriculture, Forest Service, Northeastern Forest Experiment Station, pp. 87–88.

Schütt, P. and Cowling, E.B. (1985). 'Waldsterben', a general decline of forests in Central Europe: Symptoms, development, and possible causes. *Plant Disease* **69**, 548–558.

Schweingruber, F.H. (1985). Abrupt changes in growth reflected in tree ring sequences as an expression of biotic and abiotic influences. In: P. Schmid-Haas (ed.) *Inventorying and Monitoring Endangered Forests*. IUFRO Conference, Zurich (Switzerland), 1985/08/19-24. Eidgenössische Anstalt für das forstliche Versuchswesen, Birmensdorf, Switzerland, 291–295.

Schweingruber, F.H., Kontic, R. and Winkler-Seifert, A. (1983). *Eine jahrringanalytische Studie zum Nadelbaumsterben in der Schweiz*. EAFV Bericht Nr. 253. Eidgenössische Anstalt für das forstliche Versuchswesen, Birmensdorf, Switzerland.

Showman, R.E. (1991). A comparison of ozone injury to vegetation during moist and drought years. *Journal of the Air and Waste Management Association* **41**(1), 63–64.

Simini, M., Skelly, J.M., Davis, D.D. and Savage, J.E. (1992). Sensitivity of four hardwood species to ambient ozone in North Central Pennsylvania. *Canadian Journal of Forest Research* **22**, 1789–1799.

Skelly, J.M. (1989). Forest decline versus tree decline: The pathological considerations. *Environmental Monitoring and Assessment* **12**, 23–27.

Skelly, J.M. (1990). On the importance of etiological accuracy during surveys to determine forest condition. *World Resources Review* **2**, 250–277.

Skelly, J.M. (2000). Tropospheric ozone and its importance to forests and natural plant communities of the northeastern United States. *NE Naturalist* **7**, 221–236.

Skelly, J.M. and Innes, J.L. (1994). Waldsterben in the forests of Europe and eastern North America: Fantasy versus reality? *Plant Disease* **78**, 1021–1032.

Spiecker, H., Mielikäinen, K., Köhl, M. and Skovsgaard, J.P. (eds.) *Growth Trends in European Forests*. Springer Verlag, Berlin.

Ulrich, B. (1983). A concept of forest ecosystem stability and of acid deposition as driving force for destabilization. In: Ulrich, B. and Pankrath, J. (eds.) *Effects of Accumulation of Air Pollutants on Forest Ecosystems*. D. Riedel, Dordrecht, pp. 1–32.

USDA-Forest Service (1993). *Northeastern Area Forest Health Report*. NA-TP-03–93. USDA Forest Service, Washington DC.

USDA-Forest Service (1994). *Northeastern Area Forest Health Report*, 1992. NA-TP-01-94. USDA Forest Service, Washington DC.

Webster, R., Rigling, A. and Walthert, L. (1996). An analysis of crown condition of *Picea, Fagus* and *Abies* in relation to environment in Switzerland. *Forestry* **69**, 347–355.

Wesselink, L.G., Meiwes, K.J., Matzner, E. and Stein, A. (1995). Long-term changes in water and soil chemistry in spruce and beech forests, Solling, Germany. *Environmental Science and Technology* **29**, 51–58.

Woodman, J.N. and Cowling, E.B. (1987). Airborne chemicals and forest health. *Environmental Science and Technology* **21**, 120–126.

WSL [Swiss Federal Research Institute WSL] (2001). *Der Orkan Lothar – Ereignisanalyse des Wintersturms vom* 26.12.1999. Eidgenössische Forschungsanstalt WSL, Birmensdorf.

16 Effects of acidic deposition on aquatic ecosystems

R. HARRIMAN, R.W. BATTARBEE AND D.T. MONTEITH

Background

Historical perspective

Amongst the many issues associated with freshwater pollution, the acidification of fresh waters is considered to be of major political and scientific importance. Evidence of damage to freshwater ecosystems has been reported over extensive areas of Europe and eastern regions of North America while current investigations in Asian countries are revealing similar problems. To date there is little evidence of widespread acidification of fresh waters in the southern hemisphere.

Although links between acid waters and fish mortalities were suggested as far back as the 1920s, the full extent of the effects of acidic deposition were not appreciated until the 1960s. During the 1970s the scientific debate intensified as the mechanisms and processes of acidification were slowly unravelled. By the late 1980s the weight of scientific evidence was such that virtually all European and North American countries had accepted the link between acidic emissions, long-range transport/deposition and surface water acidification. Initially the key issues revolved around the contribution of natural and man-made acidity to fresh waters and almost exclusively concentrated on the role of SO_2 as the key acidifying agent. Later, in the 1990s, the emphasis switched to the deposition of N compounds as national and international legislation ensured a significant decline in emissions of sulphur compounds.

Evidence of long-term acidification

Acidified surface waters are generally defined as those waters which are more acid than they would have been in natural circumstances (i.e. not subject to acidic deposition).

Acid waters (i.e. those with pH <5) are therefore not necessarily acidified but could be naturally acid due to high concentrations of dissolved organic matter, however, in most countries acidified waters reveal common chemical properties such as elevated non-marine sulphate, acid neutralising capacity (ANC) <0 and significant concentrations of ionic forms of aluminium. Common biological properties include the decline and elimination of fish and loss of sensitive invertebrate species.

Air Pollution and Plant Life, second edition. Edited by J.N.B. Bell and M. Treshow. ISBN 0 471 49090 3 (HB), 0 471 49091 1 (PB).

Unfortunately, validated long-term (>50 yr) chemical and biological records are very rare with the exception of biological material accumulated in lake sediments. The development of sophisticated palaeolimnological techniques, which ultimately provided a means of predicting past pH records from the succession of diatom species in dated sediment cores, was the key evidence in demonstrating the rapid acceleration of surface water acidification during the past 150 years.

In the UK, Battarbee *et al.* (1988) estimated declines in pH of 0.5–1.0 units in the most acidified regions of north Wales, northern England and south-west Scotland during the past 100 years compared with insignificant changes during the centuries prior to 1850.

Emission control policy

The first attempts at controlling acidic emissions commenced in 1979 when the United Nations Economic Commission for Europe (UNECE) drew up the Convention on Long Range Transboundary Air Pollution (LRTAP) which was signed by most European and North American countries. These protocols set realistic, across the board targets of 30% reductions in emissions of SO_2 by 1993 and, for oxides of nitrogen, a general reduction to 1987 levels.

A second sulphur protocol was agreed in Oslo in 1994. This 'effects-based' protocol used a more scientific approach, based on Critical Loads methodology, which attempted to provide 'best value for money' by targeting emission reductions to areas with the most sensitive ecosystems. Using this technique critical loads for sulphur have been calculated for UK fresh waters (CLAG, 1995) and the most sensitive systems have been identified. Having ratified the Oslo protocol in 1996 the UK is committed to an 80% reduction in sulphur emissions by 2010, based on 1990 levels. Discussions are currently underway within the LRTAP convention to produce a 'multi-pollutant: multi-effect' protocol which would set national limits on emissions of sulphur dioxide, nitrogen oxides, ammonia and volatile organic compounds.

Despite the general similarities in abatement strategies between North America and Europe the methods of emission control differ somewhat. For example, in the USA, an emission trading programme is used to keep costs to a minimum while in Europe the protection of the most sensitive ecosystems is given priority (NAPAP, 1998).

Effects on water quality

Chemical responses

Early in the acidification debate one of the most difficult issues was to distinguish between 'natural' and 'anthropogenic' acidification. It appeared possible that land-use changes (Rosenqvist, 1978) and the replacement of organic anions by strong acid anions (Krug and Frink, 1983) may explain recent changes in freshwater acidity. In the latter case, the authors proposed that a natural substitution of organic anions with strong acid anions would occur without any significant change in pH.

However, palaeoecological studies quickly revealed that rates of historical pH change were not only an order of magnitude greater in the past century for both clear and coloured lakes, but were associated with increased deposition of carbonaceous particles and elevated levels of catchment and atmospherically derived trace metals. Regional (Harriman *et al.*, 1987) and national (Henriksen *et al.*, 1998) surveys of lakes and streams revealed a common pattern of acidification linked to the loading of acidic pollutants. Sites where no significant change in pH was evident (from palaeoecological studies) were usually located in remote areas with low S deposition, e.g., northern Norway and Sweden, north west Scotland and northern Canada. Lakes and streams in these areas were characterised by low concentrations of non-marine sulphate and ANC >0. In addition most clearwater lakes and streams had a pH >5.5 while those with pH <5.5 were invariably coloured, indicating the important role of natural organic acidity.

In areas where sediment core analysis revealed a significant decline in pH (e.g., southern Norway and Sweden, southwest Scotland, northern England and eastern states of Canada and USA) the water quality of the most sensitive lakes and streams was characterised by pH <5.5, high non-marine sulphate concentrations, ANC <0 and significant concentrations of ionic aluminium. Both coloured and clearwater systems are affected by increased S deposition but response rates and subsequent recovery appear to be influenced by organic production in catchments (see 'recovery' section).

Land-use and climate change

Another major issue is the role of land-use change in the acidification process (Harriman *et al.*, 1994). Although forest growth can change the properties of soils in terms of exchangeable acidity, base cations and nutrient status there is little evidence of accelerated stream acidification in natural forest ecosystems in pristine areas. However, the onset of single-age plantation forestry has resulted in changes in streamflow, sediment load and nutrient leaching at various stages of the forest cycle (Hudson *et al.*, 1997). By far the greatest influence on surface water acidification is the cumulative effect of interception of acidic pollutants by the forest canopy. Streams draining completely forested catchments can be 30–40% more acid than those draining heathland catchments in areas subject to high deposition of acidic pollutants (Harriman and Morrison 1982).

Perhaps the most difficult and complex issue is that of climate change/water quality interactions. The potential links between climate change and freshwater acidification are numerous, ranging from direct effects on the catchment pools of C, N and S to indirect effects caused by changes in forest production, vegetation type and rainfall patterns (Schindler, 1998; Murdoch *et al.*, 1998; Skjelkvale and Wright, 1998). The net effect of these potentially conflicting issues is uncertain although preliminary studies suggest that, in the short-term, acidification of surface waters may increase in currently polluted regions due to increasing nitrate and sulphate leaching combined with increasing in-lake degradation of dissolved organic matter and subsequent increased penetration of UV radiation. In some areas of western Europe increased frequency of Atlantic gales may increase the loading of marine salts onto catchments with associated effects on acidic episodes and base cation leaching from catchments (Harriman *et al.*, 2000).

Long-term trends and models

In the early 1980s the nearly 30% decline in UK industrial emissions of SO_2 and associated reduction in sulphate deposition provided the first genuine opportunity to evaluate the response rates of aquatic ecosystems to significant deposition reductions. This assessment was based on either water quality data from established reference catchments or dynamic models because biological indicators (e.g., diatoms) were considered inappropriate over such a short time scale. Between 1985 and 1995 a significant decline (20–40%) in non-marine sulphate was recorded in many lakes and streams in western Europe (Cerny, 1995; Skjelkvale and Henriksen, 1995; Harriman *et al.*, 1995) and North America (Jeffries *et al.*, 1995; Driscoll *et al.*, 1995). However, pH, ANC and base cation responses varied considerably from site to site and region to region which inevitably confused the recovery signal. Between 1990 and 1999 sulphate deposition showed minimal change and any further decline in sulphate levels in surface waters tended to be masked by large year-to-year variability, possibly caused by climatic impacts.

During the past decade many hydrochemical models have been developed to deal with specific acidification issues. Probably the most widely and currently used is MAGIC (Model of Acidification of Groundwater in Catchments) (Cosby *et al.*, 1985) which was first used to predict chemical change in response to changing S deposition scenarios, was then developed further to incorporate land-use/afforestation impacts and is currently being used on a European scale to predict responses to the second S Protocol. A summary of the current development and usage of MAGIC is given in Jenkins *et al.* (1997). Future models will almost certainly need to incorporate nitrogen compounds in critical load predictions especially now that a multi-pollutant/multi-effect strategy is being developed.

Effects on aquatic biota

Fish

Because of their ubiquitous distribution, fish, particularly trout and salmon, have been used as the key indicator of biological damage resulting from freshwater acidification. Decline and loss of fish populations have been reported in sensitive lakes and streams of most countries subject to high deposition of sulphur compounds. In some countries (e.g., Norway and Canada) up to 25% of all lakes have been reported to have damaged fisheries (Hesthagen *et al.*, 1999; Jeffries, 1997).

Aluminium, pH and calcium have long been recognised as the most important elements in determining the toxicity of acidic water to fish (Rosseland *et al.*, 1990) while the additional protection of dissolved organic matter has only recently been recognised (Roy and Campbell, 1997). In many countries, salmonid species are considered to be the most valuable fish resource but studies have revealed that they are the most sensitive to acidic conditions. For early life stages (egg development, hatching) where the skin controls the ion regulation processes, pH appears to be the key toxic agent but once the gill takes over the respiratory functions aluminium, in its ionic form, and calcium assume greater importance. Waters with low calcium concentrations (<50 μeq l^{-1}) are more sensitive to acid conditions as gill permeability is greater, while ionic aluminium concentrations >50 μg l^{-1} are considered to be toxic. Salmon smolts are believed to be the most sensitive life stage, especially in terms of their seawater tolerance, and appear to be susceptible to acidic episodes of pH 5.5 and ionic aluminium concentrations as low as 20–30 μg l^{-1} (Staurnes *et al.*, 1995).

Many physiological mechanisms have been identified which affect different life stages during critical environmental conditions and these have been reviewed by Rosseland and Staurnes (1994), however, the general consensus is that gradual reproductive failure and consequent decline in recruitment, is the main cause of extinction from streams and lakes.

Invertebrates

Invertebrates in streams and lakes affected by acidification show similar physiological responses to fish. However, unlike fish, invertebrate biomass production does not correlate well with acidification indicators such as pH and aluminium because some invertebrate species are not acid-sensitive within the normal pH range of acidified waters. Therefore certain invertebrates have been used as 'indicator' species to determine the extent of acidification.

Responses to increasing acidification are usually measured in terms of species composition, species diversity, the relative tolerances of different taxonomic groups or presence/absence of key indicator species. This information can be converted to an appropriate acidification index and used as an indicator of biological recovery (Raddum *et al.*, 1988; Raddum and Fjellheim, 1994).

Algae and macrophytes

Broad relationships between freshwater acidity and aquatic plant community composition have been recognised since the early part of the twentieth century (e.g., Pearsall, 1921, Misra, 1938). However, there are few instances where algal and macrophyte changes associated with lake acidification have been directly recorded. Instead such changes have to be inferred by comparison with non-acidified lakes (e.g. Almer *et al.*, 1974; Hendrey *et al.*, 1976; Flower *et al.*, 1987), from whole lake, enclosure, or laboratory experiments (e.g. Yan and Stokes, 1978) or from sediment core records (Battarbee *et al.*, 1990).

One key concern is whether lake acidification leads to a reduction in primary production ('oligotrophication') as suggested by Grahn *et al.* (1974) in response to a decrease in the availability of phosphorus. Despite a number of studies, principally in Canada (e.g. Yan and Stokes, 1978; Nalewajko and Paul, 1985), this issue has not been resolved. The impact of acidification on the composition of algal communities is clearer than the impact on productivity. For example, Almer *et al.* (1974) surveyed the phytoplankton of 115 lakes in south-west Sweden and observed that there were fewer taxa in lakes with pH values below 5 and that these were mainly Dinophyta (*Peridinium inconspicuum, Gymnodium* spp.) Chlorophyta *(Ankisfrodesmus convolutus v. minutulus)* and Chrysophyta *(Dinobryon crenulatum, D. sertularia)*. Diatoms contributed to the phytoplankton biomass only when the pH was greater than 5.0, and Chrysophyta and Chlorophyta were much more numerous at levels above pH 5.8. These data agree with observations from Canada confirming the importance *of Peridinium* and *Gymnodium* spp. in very acid lakes (Yan and Stokes, 1979; Yan, 1979). The general lack of planktonic diatoms below about pH 5.5 noted from algological surveys is well documented from surface sediment surveys (Davis *et al.*, 1983; Charles, 1985; Flower, 1986) and from sediment core data (Renberg *et al.*, 1985). Studies such as the Surface Waters Acidification Project (SWAP) (Stevenson *et al.*, 1991) demonstrate that freshwater diatom species have pH preferences which can be very tightly defined in terms of pH optima (the pH at which the species is most abundant, relative to other species) and pH tolerance (the range of pH values in which the species occurs). This information has been applied, using a weighted averaging approach, to infer the pH of lakes through time from the species composition of sediment core samples, as illustrated for the Round Loch of Glenhead (Figure 16.1).

Spatial studies of aquatic macrophyte species distribution often cover large acidity gradients, ranging from very acid to alkaline (e.g., Palmer *et al.*, 1992). At this scale, pH co-varies with many other chemical parameters, including alkalinity, conductivity, base cation and phosphorus concentration and it is therefore not surprising that primary axes of species ordination plots usually correlate with all these variables. However, lakes

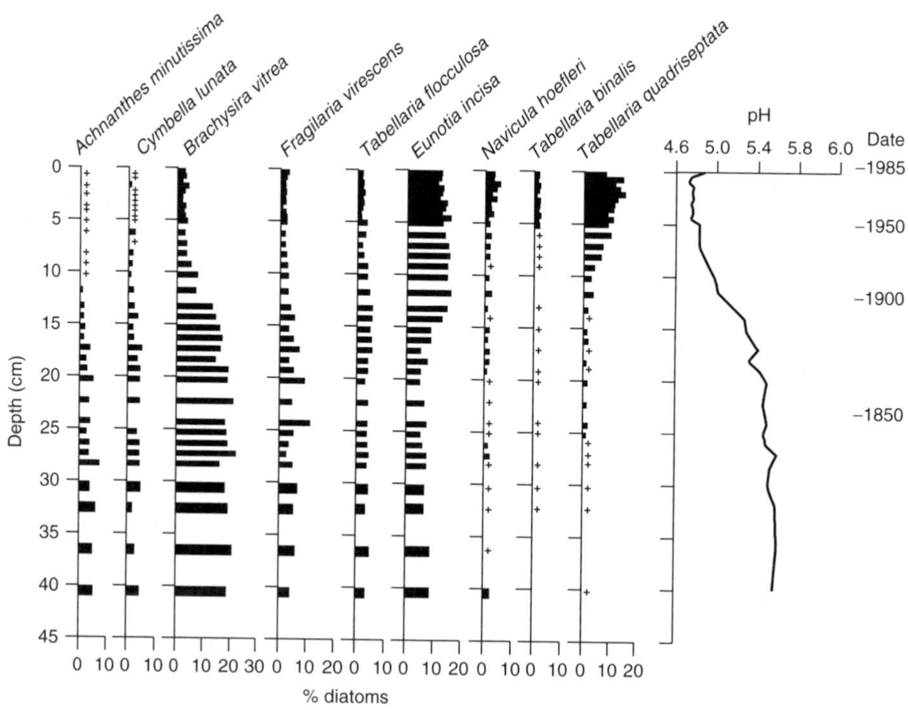

Figure 16.1 Historical changes in diatom species and inferred pH change in Round Loch of Glenhead, Galloway, south-west Scotland

which are prone to acidification are usually naturally slightly acid and oligotrophic, i.e. with low nutrient concentrations. The chemical transition undergone at individual acidifying lakes is therefore considerably more subtle than that represented by the larger acidity gradient in these spatial studies. Recent work on species changes in acidified waters points to changes in the availability of inorganic carbon and nitrogen as the driving factors.

The availability of carbon, required by freshwater plants for photosynthesis, is largely determined by the pH dependent equilibria of carbon dioxide (CO_2) bicarbonate (HCO_3^-) and carbonate (CO_3^{--}) (see Stumm and Morgan, 1970). CO_3^{--} is the dominant form of carbon in very alkaline waters, but, as pH falls, dominance shifts rapidly to HCO_3^- and then to CO_2. However, acidification also leads to an overall fall in the total inorganic carbon content of water so that CO_2 concentrations in acidified waters tend to be very low. Compounding this, the high viscosity of water results in very poor CO_2 diffusivity (10 000 times more slowly than in air). On the other hand, CO_2 may be available at high concentrations in the interstitial poor water of the sediments in which plants are rooted (e.g., Roelofs, 1983). Non-acidified but poorly buffered waters are usually poor in nitrogen, and sediment nitrate (NO_3) is often the dominant N source. However, nitrogen concentrations will increase when N deposition is a major contributor to total acidity, while reduced rates of nitrification in sediments at lower pH may lead to increases in ammonium relative to NO_3.

Aquatic vascular plants show various adaptations to aqueous carbon sequestration and these can be broadly summarised by a comparison of three main morphological groups: isoetids, elodeids and plants with floating leaves. The isoetid group contains a diverse range of plants, including the pteridophytes *Isoetes* spp., and angiosperms such as *Littorella dortmanna* and *uniflora*. In an apparent example of convergent evolution, isoetids show similar physiological mechanisms for surviving in a low carbon medium, and this is clearly an advantage in relatively acid water. Isoetids have rosette growth forms, and leaves are connected to roots by longitudinal air channels, or lacunae, which allow the transport of gases between the sediment and leaf chloroplasts. Sand-Jensen and Prahl (1982) found that virtually all CO_2 and O_2 exchange for submerged *L. dormanna* occurred via the roots, while Nielsen *et al.* (1991) found that, even when growing out of water, *L. uniflora* acquired 80% of its CO_2 supply from the exposed sediment. Isoetids also exhibit crassulacean acid metabolism which allows the fixation of carbon during hours of darkness, and have particularly low compensation points and low relative growth rates.

In contrast, elodeids often form dense stands in the water column which can provide important predation refuges for zooplankton. Since aerial tissues may be at considerable distance from the rooting zone, internal CO_2 diffusion resistance may be too high for a substrate source of carbon to be exploited. Carbon tends therefore to be obtained directly from the surrounding water either as CO_2, or as has been demonstrated for several alkaline preferring species, HCO_3^- (Maberly and Spence, 1983). Most elodeids have deeply lobed or segmented leaves which minimise boundary layer resistance to CO_2 diffusion. Perhaps as a result of limited carbon availability, elodeids are relatively poorly represented in acidic waters, and in the UK this component of the flora is mostly limited to *Myriophyllum alterniflorum*, *Callitriche hamulata*, *Potamogeton berchtoldii*, *Utricularia* spp. and *Juncus bulbosus*, while the macro-algal charophyte, *Nitella* spp., which also has an elodeid growth form, is common in mildly acid water. *Nitella* sp. is rarely found in waters with a pH of less than 6.0, suggesting that a HCO_3^- source of carbon may be important for growth, as has been demonstrated for other charophyte species.

Species with floating leaves, such as *Nuphar* spp., *Nymphaea* spp. and *Sparganium* (e.g., *S. angustifolium*), have the physiological advantage of an atmospheric source of CO_2. Their distribution in oligotrophic waters appears largely independent of acidity, and primarily limited by the availability of stable, fine-grained organic substrates which develop in relatively sheltered habitats.

Macrophyte diversity tends to fall with increasing acidity (e.g., Roberts *et al.*, 1985; Catling *et al.*, 1986). In general, and with the exception of *Juncus bulbosus*, elodeids tend to be lost first, and this is likely to result from changes in carbon and nitrogen availability. Consequently, Scandinavian and UK lakes below pH 5 are dominated by isoetid species, *Juncus bulbosus* and the acid-loving moss *Sphagnum auriculatum*. A number of studies have reported shifts in dominance from isoetids to *J. bulbosus* and *Sphagnum* as acidification continues. This has been supported by laboratory studies; for example in a comparison of physiological responses of *Littorella uniflora* and *J. bulbosus*, Roelofs *et al.* (1984) found that the more acid lakes had higher water column CO_2 concentrations. They suggested that carbon was mobilised from the lake sediments and released into the water column as acidity increased, and proposed this process favoured *J. bulbosus*, which can take up carbon through its leaves, over *L. uniflora* at lower pH. They also proposed that *J. bulbosus* was advantaged by an increase in ammonium relative to NO_3 as acidification reduced rates of nitrification. Laake (1976) found what appears to be a physiological influence of pH on *L. dortmanna*, recording a 75% reduction of photosynthesis when pH was lowered from 5.5 to 4.0. Suggestions that *Sphagnum* may have outcompeted isoetid species at a regional level (e.g., Grahn, 1985) has been contested by Borslett and Brettum, (1989) who could find no quantitative evidence that this has indeed taken place.

The replacement of isoetids by *J. bulbosus* and *Sphagnum flexuosum* in acidic pools in The Netherlands, where ammonium sulphate deposition is a principal contributor to acidification, has been attributed to a preference by the latter species for an ammonium form of nitrogen. *L. dortmanna, L. uniflora* and other angiosperms were found to be primarily dependent on sedimentary NO_3–N source (Schuurkes *et al.*, 1986).

Acidification may also result in a shift in the balance of higher plants and algae. In Sweden, acid lakes are often characterised by the development of extensive blue-green algal mats (Hendrey *et al.*, 1976; Lazarek, 1982, 1985) which are claimed to displace submerged macrophytes, especially *L. dortmanna* (Grahn, 1977, 1985). Filamentous green algae, especially *Mougeotia* spp., which develops dense epiphytic growths on macrophyte stems and leaves in relatively shallow water, are also invasive at low pH levels (Grahn *et al.*, 1974; Hendrey *et al.*, 1976).

Experimental liming can reverse the species balance, with declines observed in *J. bulbosus* and *Sphagnum* and increases in isoetids (e.g., Eriksson *et al.*, 1983; Raven,

1989). However, Roelofs *et al.* (1998) found *J. bulbosus* responded positively to liming and this has received considerable attention in the recent literature. Experiments by Lucassen *et al.* (1999) suggest that *J. bulbosus* may thrive when sediment ammonium and water-layer CO_2 concentrations are raised, and this can occur in limed lakes as a result of a shift in the interstitial HCO_3^-–CO_2 equilibrium following partial re-acidification.

Acidification in the future

Extent of chemical and biological recovery

Acidified waters in the most sensitive regions of Europe are expected to show further ecological recovery during the coming decade as the Oslo Sulphur Protocol takes effect. After the significant chemical changes in the 1980s and the relative stability in the early and mid-1990s there is recent evidence that sulphate concentrations in the most acidified waters are declining once again. An example of the historical pattern of sulphate decline, and associated increase in pH, is given for Loch Enoch, a key index site in south west Scotland (Figure 16.2) Comparisons of national surveys show a similar pattern of sulphate decline in most countries although the signal obtained is greatly dependent on the time scale used in the analysis (Stoddard *et al.*, 1999). Leaching of previously accumulated sulphate, especially from peaty soils, can also delay recovery and many catchments are currently leaching more sulphate than current deposition loadings would predict. Of great concern is the increasing importance of nitrogen as sulphate concentrations decline, particularly where episodic acidification may be critical. The lack of any clear relationship between N deposition and nitrate

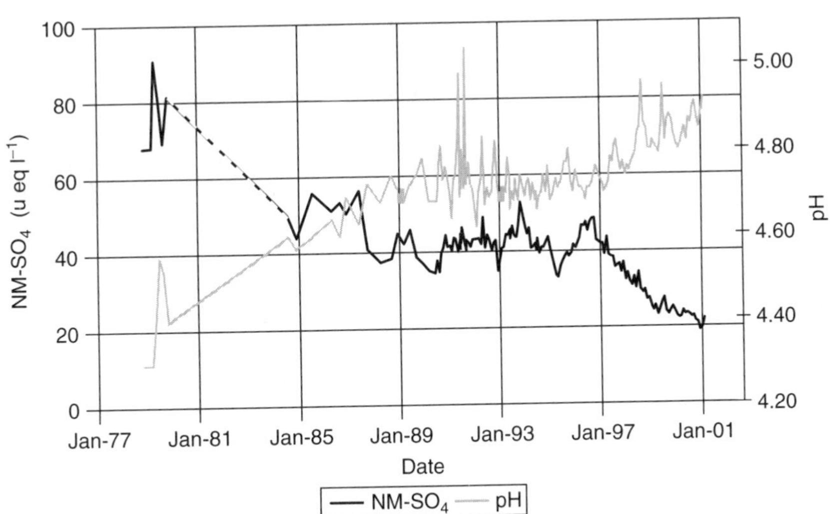

Figure 16.2 Long-term trends in non-marine sulphate and associated pH increase in Loch Enoch, Galloway, south-west Scotland

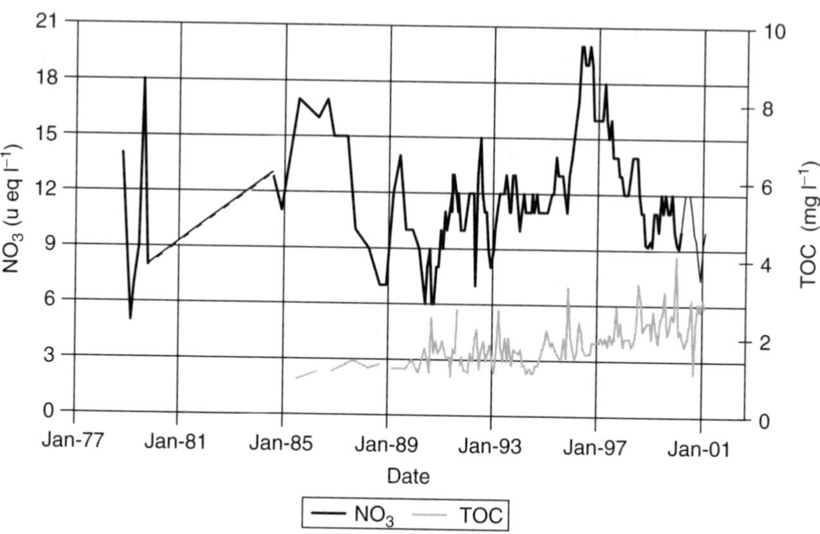

Figure 16.3 A comparison of nitrate and Total Organic Carbon (TOC) concentrations in Loch Enoch, Galloway, south-west Scotland

runoff to streams and lakes has complicated the development of acidity-based models incorporating both nitrogen and sulphur. Long- and short-term changes in climatic conditions may also produce spikes of both nitrate and organic carbon leaching, further compounding the difficulties in predicting long-term trends (Figure 16.3).

The MAGIC model provides a generally pessimistic view of chemical responses to the Oslo Sulphur Protocol, predicting only a small increase in pH at UK monitoring sites in the next 30–50 years, which will not be sustainable in the long term (Jenkins *et al.*, 1997).

Predictions of future biological recovery are probably no more than speculative as response times are likely to be considerably longer than those for chemical variables. Nevertheless, the general ecological improvements throughout the food chain, in what were the most seriously damaged areas (e.g., the lakes around Sudbury, Ontario), give some degree of optimism that recovery is a long-term possibility.

Unfortunately, the additional influence of climate change and ozone depletion is likely to modify catchment responses to emission reductions. This could interact with acidification processes in either a positive or negative way but, whatever the outcome, interpretation of chemical and biological responses to acidic deposition will become more complicated as the dose/response signal becomes noisier.

References

Almer, B., Dickson, W., Ekstrom, C., Hornstrom, E. and Miller, U. (1974). Effects of acidification of Swedish lakes. *Ambio*, **3**, 30–36.

Battarbee, R.W., Maron, B.J., Renberg, I. and Talling, J.F. (1990). *Palaeolimnology and Lake Acidification*. The Royal Society, London.

Battarbee, R.W., Anderson, N.J., Higgitt, S., Oldfield, F., Appleby, P.J., Jones, V.J., Patrick, S.T., Flower, R.J., Kreiser, A., Richardson, N.G., Fritz, S.C., Munro, M.A.R, Rippey, B., Haworth, E.Y., Natkanski, J. and Stevenson, A.C. (1988). *Lake Acidification in the United Kingdom 1800–1986*. ENSIS Publishing. London.

Borslett, B. and Brettum, P. (1989). The genus Isoetes in Scandinavia: An ecological review and perspectives. *Aquatic Botany*, **35**, 223–261.

Catling, P.M., Freedman, B., Stewart, C., Kerekes, J.J. and Lefkovitch. (1986). Aquatic plants of acid lakes in Kejimkujik National Park, Nova Scotia; floristic composition and relation to water chemistry. *Can. J. Bot.*, **64**, 724–729.

Cerny, J. (1995). Recovery of acidified catchments in the extremely polluted Krusne Hory mountains, Czech Republic. *Water, Air and Soil Pollut.*, **85**(2), 589–594.

Charles, D.F. (1985). Relationships between surface sediment diatom assemblages and lakewater characteristics in Adirondack lakes. *Ecology*, **66**, 944–1011.

CLAG (1995). *Critical Loads of Acid Deposition for United Kingdom Fresh Waters*. Critical Loads Advisory Group, Sub-group report on Fresh Waters. ITE, Penicuik.

Cosby, B.J., Hornberger, G.M., Galloway, J.N. and Wright, R.F. (1985). Modelling the effects of acid deposition; assessment of a lumped parameter model of soil water and streamwater chemistry. *Water Res.*, **21**, 51–63.

Davis, R.B., Norton, S.A., Hess, C.T. and Brakke, D.F. (1983). Palaeolimnological reconstruction of the effects of atmospheric deposition of acids and heavy metals on the chemistry and biology of lakes in New England and Norway. *Hydrobiologia*, **103**, 113–124.

Driscoll, C.T., Postek, K.M., Kretser, W. and Raynal, D.J. (1995). Long-term trends in the chemistry of precipitation and lake water in the Adirondack region of New York, USA. *Water, Air and Soil Pollut.*, **85**(2), 583–588.

Eriksson, F.L., Hornstrom, E., Mossberg, P. and Nyberg, P. (1983). Ecological effects of lime treatment of acidified lakes and rivers in Sweden. *Hydrobiology*, **101**, 145–164.

Flower, R.J. (1986). The relationship between surface diatom assemblages and pH in 33 Galloway lakes: some regression models for reconstructing pH and their application to sediment cores. *Hydrobiologia*, **143**, 93–103.

Flower, R.J., Battarbee, R.W. and Appleby, P.G. (1987). The recent palaeolimnology of acid lakes in Galloway, south-west Scotland; diatom analysis, pH trends and the role of afforestation. *J. Ecol.*, **75**, 797–824.

Grahn, O. (1977). Macrophyte succession in Swedish lakes caused by deposition of airborne acid substances. *Water, Air and Soil Pollut.*, **7**, 295–305.

Grahn, O. (1985). Macrophyte biomass and production in Lake Gårdsjön – an acidified clearwater lake in SW Sweden. *Ecological Bulletins*, **37**, 203–212.

Grahn, O., Hultberg, H. and Lander, L. (1974). Oligotrophication – a self-accelerating process in lakes subjected to excessive supplies of acid substances. *Ambio*, **3**, 93–94.

Harriman, R. and Morrison, B.R.S. (1982). The ecology of streams draining forested and non-forested catchments in an area of central Scotland subject to acid deposition. *Hydrobiologia*, **88**, 251–263.

Harriman, R., Morrison, B.R.S., Caines, L.A., Collen, P. and Watt, A.W. (1987). Long-term changes in fish populations of acid streams and lochs in Galloway, south west Scotland. *Water, Air and Soil Pollut.*, **32**, 89–112.

Harriman, R., Likens, G.E., Hultberg, H. and Neal, C. (1994). Influence of management practices in catchments on freshwater acidification: Afforestation in the United Kingdom and

North America. In: *Acidification of Freshwater Ecosystems: Implications for the future*. Eds: C.E.W. Steinberg and R.F. Wright. pp. 83–101. John Wiley & Sons.

Harriman, R. Morrison, B.R.S, Birks, H.J.B Christie, A.E.G, Collen, P. and Watt, A.W. (1995). Long-term chemical and biological trends in Scottish streams and lochs. *Water, Air and Soil Pollut.*, **85**, 701–706.

Harriman, R., Watt, A.W. and Christie, A.E.G. (2000). *Patterns and processes of sea-salt deposition at three upland sites in central Scotland. Verh. Internat. Verein. Limnol.* **27**, 1270–1275.

Hendrey, G.R., Baalsrud, K., Traaen, T.S., Morten, L. and Raddum, G. (1976). Acid precipitation: some hydrobiological changes. *Ambio*, **5**, 224–227.

Henriksen, A., Skjelkvale, B.L., Mannio, J., Wilander, A., Harriman, R., Curtis, C., Jensen, J.P., Fjeld, E. and Moiseenko, T. (1998). North European lake survey, 1995. *Ambio*, **27**, 80–91.

Hesthagen, T., Sevalrud, I.H. and Berger, H.M. (1999). Assessment of damage to fish populations in Norwegian lakes due to acidification. *Ambio*, **28**, 112–117.

Hudson, J.A., Crane, S.B. and Robinson, M. (1997). The impact of the growth of new plantation forestry on evaporation and streamflow in the Llanbrynmair catchments. *Hydrology and Earth System Sciences*, **1**, 463–475.

Jeffries, D.S., Clair, P.J., Dillon, P.J., Papineau, M. and Stainton, M.P. (1995). Trends in surface water acidification at ecological monitoring sites in southeastern Canada. *Water, Air and Soil Pollut.*, **85**, 577–582.

Jeffries, D.S. (1997). *1997 Canadian Acid Rain Assessment Vol. 3, Aquatic Effects* Environment Canada, Ottawa, 1997.

Jenkins, A., Renshaw, M., Helliwell, R.C., Sefton, C., Ferrier, R.C. and Swingewood, P. (1997). Modelling surface water acidification in the UK. *IH Report*, **131**, Institute of Hydrology.

Krug, E.C. and Frink, C.R. (1983). Acid rain on acid soil, a new perspective. *Science*, **221**, 520–525.

Laake, M. (1976). *Effekter av lav pH pa produkjon, nedbrytning og stoffkretslop I littoralsonen*. SNSF Project. Internal Report 29/76. Oslo-As, Norway.

Lazarek, S. (1982). Structure and function of a cyanophytan mat community in an acidified lake. *Canadian Journal of Botany*, **60**, 2235–2240.

Lazarek, S. (1985). Epiphytic algal production in the acidified lake Gårdsjön, SW Sweden. *Ecological Bulletins*, **37**, 213–218.

Lucassen, E.C.H.E.T., Bobbink, R., Oonk, M., Brandrud, T.E. and Roelofs, J.G.M. (1999). The effects of liming and reacidification on the growth of *Juncus bulbosus*: a mesocosm experiment. *Aquat. Bot.*, **64**, 95–103.

Maberly, S.C. and Spence, D.H.N. (1983). Photosynthetic inorganic carbon use by freshwater plants. *J. Ecol.*, **71**, 705–724.

Misra, R.D. (1938). Edaphic factors in the distribution of aquatic plants in the English lakes. *J. Ecol.*, **26**, 41.

Murdoch, P.S., Burns, D.A. and Lawrence, G.B. (1998). Relation of climate change to the acidification of surface waters by nitrogen deposition. *Environ. Sci. Technol*, **32**, 1642–1647.

Nalewajko, C. and Paul, B. (1985). Effects of manipulations of aluminium concentrations and pH on phosphate uptake and photosynthesis of planktonic communities in two precambrian shield lakes. *Canadian Journal of Fisheries and Aquatic Sciences*, **42**, 1946–1953.

NAPAP (1998). *NAPAP Biennial Report to Congress: An Integrated Assessment*. National Acid Precipitation Assessment Programme. Silver Spring, Maryland, USA.

Nielsen, S.L., Gacia, E. and Sand-Jensen, K. (1991). Land plants of amphibious *Littorella uniflora* (L.) Aschers maintain utilisation of CO_2 from the sediment. *Oecologia*, **88**, 258–262.

Palmer, M.A., Bell, S.L. and Butterfield, I. (1992). A botanical classification of standing waters in Britain: applications for conservation and monitoring. *Aquatic Conservation: Marine and Freshwater Ecosystems*, **2**, 125–143.

Pearsall, W.H. (1921). The development of vegetation in the English lakes, considered in relation to the general evolution of glacial lakes and rock-basins. *Proc. R. Soc. Lond. B*, **92**, 259–284.

Raddum, G.G. and Fjellheim, A. (1994). Invertebrate community changes caused by reduced acidification. *Acidification of Freshwater Ecosystems: Implications for the Future*. Eds: C.E.W Steinberg and R.F. Wright, pp. 345–354. John Wiley & Sons Ltd.

Raddum, G.G., Fjellheim, A. and Hesthagen, T. (1988). Monitoring of acidification by use of aquatic organisms. *Verh. Int. Ver. Limnol.*, **23**, 2291–2297.

Raven, P.J. (1989). *Short-term changes in the aquatic macrophyte flora of Loch Fleet, S.W. Scotland following catchment liming with particular reference to sublittoral Sphagnum*. Research Paper 39, Palaeoecology Research Unit. University College, London.

Renberg, I., Hellberg, T. and Nilsson, M. (1985). Effects of acidification on diatom communities as revealed by analyses of lake sediments. *Ecological Bulletins*, **37**, 219–223.

Roberts, D.A., Singer, R. and Boylen, C.W. (1985). The submersed macrophyte communities of Adirondack lakes (New York, USA) of varying degrees of acidity. *Aquat. Bot.*, **21**, 219–235.

Roelofs, J.G.M. (1983). Impact of acidification and eutrophication on macrophyte communities in soft waters in The Netherlands. 1: Field observations. *Aquatic Bot.*, **17**, 139–155.

Roelofs, J.G.M, Schuurkes, J.A.A.R. and Smits, A.J.M. (1984). Impact of acidification and eutrophication on macrophyte communities in soft waters. II Experimental studies. *Aquat. Bot.*, **18**, 389–411.

Rosenqvist, I.T. (1978). Alternative sources for acidification of river water in Norway. *Sci. Tot. Environ.* **10**, 39–49.

Rosseland, B.O. and Staurnes, M. (1994). Physiological mechanisms for toxic effects and resistance to acidic water: An ecophysical and ecotoxicological approach. In *Acidification of Freshwater Ecosystems: Implications for the Future*. Eds: C.E.W Steinberg and R.F. Wright, pp. 227–246. John Wiley & Sons Ltd.

Rosseland, B.O., Eldhuset, T.D. and Staurnes, M. (1990). Environmental effects of aluminium. *Environ. Geochem. Health.* **12**, 17–27.

Roy, R.L. and Campbell, P.G.C. (1997). Decreased toxicity of Al to juvenile Atlantic salmon (*Salmo salar*) in acidic soft water containing natural organic matter: A test of the free-ion model. *Environ. Toxicol. Chem.*, **16**, 1962–1969.

Sand-Jensen, K. and Prahl, C. (1982). Oxygen exchange with the lacunae and across leaves and roots of the submerged vascular macrophyte, *Lobelia dortmanna* L. *New Phytol.*, **91**, 103–120.

Schindler, D.W. (1998). A dim future for boreal waters and landscapes. *Bioscience*, **48**, 157–164.

Schuurkes, J.A.A.R., Kok, C.J. and Den Hartog, C. (1986). Ammonium and nitrate uptake by aquatic plants from poorly buffered and acidified waters. *Aquat. Bot.*, **24**, 131–146.

Skjelkvale, B.L. and Henriksen, A. (1995). Acidification in Norway – status and trends. Surface and groundwater. *Water, Air and Soil Pollut.*, **85**, 629–634.

Skjelkvale, B.L. and Wright, R.F. (1998). Mountain lakes; sensitivity to acid deposition and global climate change. *Ambio*, **27**, 280–286.

Staurnes, M., Kroglund, F. and Rosseland, B.O. (1995) Water quality requirement of Atlantic salmon (Salmo salar) in water undergoing acidification or liming in Norway. *Water, Air and Soil Pollut.*, **85**, 347–352.

Stevenson, A.C., Juggins, S., Birks, H.J.B, Anderson D.S., Anderson, N.J., Battarbee, R.W., Berge, F., Davis, R.B., Flower, R.J., Haworth, E.Y., Jones, V.J., Kingston, J.C., Kreiser, A.M., Line, J.M., Munro, M.A.R and Renberg, I. (1991). *The Surface Water Acidification*

Project Palaeolimnology Programme: modern diatom/lake-water chemistry data-set. ENSIS Ltd, London.

Stoddard, J.L., Jeffries, D.S., Lukewille, A., Clair, T.A., Dillon, P.J., Driscoll, C.T., Forsius, M., Johannessen, M., Kahl, J.S., Kellogg, J.H., Kemp, A., Mannio, J., Monteith, D.T., Murdoch, P.S., Patrick, S., Rebsdorf, A., Skjelkvale, B.L., Stainton, M.P., Traaen T., Van Dam, H., Webster, K.E., Wieting, J. and Wilander, A. (1999). Regional trends in aquatic recovery from acidification in North America and Europe. *Nature*, **401**, 575–578.

Stumm, W. and Morgan, J.J. (1970). *Aquatic Chemistry*. John Wiley & Sons.

Yan, N.D. (1979). Phytoplankton community of an acidified heavy metal contaminated lake near Sudbury, Ontario. *Water, Air and Soil Poll.*, **11**, 43–55.

Yan, N.D. and Stokes, P.M. (1978). Phytoplankton of an acid lake and its response to experimental alteration of pH. *Environmental Conservation*, **5**, 93–100.

17 Effects on bryophytes and lichens

J.W. BATES

Special characteristics

Lichens and bryophytes are radically different types of cryptogamic (non-flowering) organisms. Bryophytes are green land plants of a relatively simple form, whereas lichens are not plants at all, but symbiotic associations between a fungus, the mycobiont, and one or more algae, the photobiont(s). From a practical perspective, however, both are pocket-sized autotrophs that occupy similar habitats. They are often conspicuous organisms on hard substrata like rock and tree bark that are impenetrable to roots, as well as in a range of other habitats where poor environmental conditions (e.g. drought, cold, shade or acidity) offer refuges from vascular plant competition (Bates, 1998).

Bryophytes and lichens mostly have no special organs for water absorption (the *poikilohydric* condition) and at best possess only primitive cuticles. A few bryophytes absorb water from the soil and sustain a modest transpiration stream utilizing internal water-conducting hydroids – the so-called *endohydric* species (Hébant, 1977). However, the majority of bryophytes (*ectohydric* species) and lichens are able to absorb moisture over their entire surface. Numerous taxa in both groups can tolerate frequent and intense desiccation. Waxy coatings exist in many bryophytes, especially over the leaves of endohydric mosses and around the pores of thalloid liverworts (Proctor, 1984). This appears to be as important in preventing flooding of the assimilatory tissue as in resisting evaporative loss. Some lichens also have coatings that impede moisture uptake. Well-known instances include the non-wettable, sorediate surfaces of the lichens *Lecanora conizaeoides*[1] and *Hypogymnia physodes* (O'Hare and Williams, 1975; Ahmadjian, 1993), which may partly explain their high pollution tolerance. The absence of waterproof coatings (and stomata) in a wide range of bryophytes and lichens accounts for their high sensitivities to atmospheric pollutants, whether in the form of gases and particulate matter (dry deposition), dissolved in precipitation (wet deposition), or in cloud and fog aerosols (occult deposition).

In the context of atmospheric pollution the main cryptogamic communities that have been investigated include: epiphytes on the twigs, branches and trunks of trees and shrubs; epiliths on natural rock outcrops and man-made masonry; terricolous

[1] Unwettability in this species has also been attributed to the presence of large quantities of the lichen metabolite fumarprotocetraric acid (Hawksworth and Rose, 1976).

Air Pollution and Plant Life, second edition. Edited by J.N.B. Bell and M. Treshow. ISBN 0 471 49090 3 (HB), 0 471 49091 1 (PB).

(ground-inhabiting) mosses and lichens in boreal forests and tundra; peat-mosses (*Sphagnum* spp.), especially in ombrotrophic (blanket and raised) mires. These aesthetically pleasing and often ecologically important coverings of bryophytes and lichens have been significantly impoverished or even completely destroyed over the past 150 years in regions affected by atmospheric pollution. A column devoted to *Literature on Air Pollution and Lichens*, in the journal *The Lichenologist*, recently reached its 49th article (Henderson, 2000, pers. comm.) and has listed just over 3000 publications. The number of bryological publications is smaller, but not insignificant, and this summary is therefore necessarily selective.

Field evidence for decline with increasing SO_2

Historical evidence

The earliest accounts of the high sensitivities of lichens to atmospheric pollutants appeared around the peak of the Industrial Revolution in western Europe. Grindon (1859) in Manchester and Nylander (1866) in Paris both associated the disappearance of lichens from their respective cities with the grossly polluted town air, smoke and sulphur dioxide then being major components of the pollution. Arnold (1892), commenting on the disappearance of mosses from Munich, appears to have been the first person to attribute the loss of bryophytes specifically to atmospheric pollution (Winner, 1988). The landmark work *Air Pollution and Lichens* (Ferry *et al.*, 1973), conveniently summarized much of the historical evidence of lichen decline with increasing burning of fossil fuels and rising atmospheric concentrations of SO_2. Other useful commentaries have been provided by Gilbert (1970, 1973), Hawksworth and Rose (1976), Nash and Wirth (1988) and Richardson (1992). The decline of bryophytes is less well documented, but valuable summaries of the available evidence have been provided by Rose and Wallace (1974), Barkman (1958, 1969), Greven (1992), Winner (1988) and Adams and Preston (1992).

A well-researched historical example is provided by the successive lichen lists compiled for Epping Forest in south-east England. This lies beyond the outermost suburbs of north-east London but, owing to its position downwind, has long been subjected to urban atmospheric pollutants. Table 17.1 records a spectacular decline in lichen species richness over about a century, from the 1860s when the lichen flora

Table 17.1 Progressive impoverishment of the lichen flora of Epping Forest (UK) in relation to increasing atmospheric pollution emanating from London (after Hawksworth *et al.*, 1973)

Dates	No. species	Habitats studied	Collector(s)
1784–96	55	All	Forster
1865–68	120	Epiphytes only	Crombie
1881–82	86	Epiphytes only	Crombie
1909–19	49	All	Paulson and Thompson
1969–70	28	All	Rose and Pentecost

was becoming well known through the work of Crombie, to the late 1960s when the epiphytic flora had become extremely impoverished.

Forster also recorded several epiphytic bryophytes that have not been re-found in the forest by later recorders, notably *Antitrichia curtipendula*, *Cryphaea heteromalla*, *Orthotrichum striatum* and *Ulota crispa* (Adams and Preston, 1992). By the early 1900s the lichen flora was already very dull and the pollution tolerant species *Lecanora conizaeoides* formed a universal covering on bark. The decline in lichen and bryophyte diversity may not have been caused exclusively by SO_2, as smoke was also an important pollutant, and other environmental changes may also have accompanied increasing pollutant release; however, the main features of the lichen decline resemble those seen in many other instances where SO_2 was undoubtedly the main pollutant (Ferry *et al.*, 1973).

The disappearance of *Sphagnum* (peat-mosses) from the blanket peat of large tracts of the southern Pennine Hills (UK) was also almost certainly a result of former high SO_2 pollution from the industrial cities of Manchester and Sheffield (Ferguson and Lee, 1983a). Stratigraphic analysis of the peats underlying the present day *Eriophorum-Vaccinium* vegetation (Tallis, 1964) showed that *Sphagnum* spp. had disappeared quite suddenly about 200 years previously. Abundant industrial soot deposits at this point in the profile are compelling evidence for this 'pollution' hypothesis, and it is supported by contemporary accounts of the rainwater composition reviewed by Ferguson and Lee (1983a).

In The Netherlands, Greven (1992) investigated historical changes in the bryophyte flora using earlier detailed lists for named localities drawn up by J.J. Barkman and others, and supplemented by his own modern inventories. He accumulated persuasive evidence pointing to a general impoverishment of neutrophile[2] bryophytes and an accompanying increase in acidophilous taxa related to the period of high SO_2. Recent increases in neutrophile bryophytes appear to have accompanied declining SO_2 levels and the ascendancy of nitrogenous pollution.

Luxuriance and fertility

Reductions in luxuriance and fertility of lichens and bryophytes have also accompanied increasing SO_2 concentrations over the past 150 years. Larger fruticose and foliose lichens show this particularly well. Hawksworth and Chapman (1971) noted that *Pseudevernia furfuracea* specimens in the Natural History Museum, London collected before 1890 were often fertile, but those collected afterwards were infertile. Hawksworth *et al.* (1973) recorded reductions in cover and the production of ascocarps by sensitive lichens on approaching contemporary SO_2 sources at sublethal concentrations (Table 17.2). One example is provided by the pendulous epiphyte *Usnea ceratina*. This commonly achieves a length of 25–45 cm in unpolluted parts of southern England, but is less than 10 cm in slightly polluted areas where it takes on a contorted form with fewer lateral lobes.

Historical records also demonstrate a decline in luxuriance and fertility of bryophytes in polluted areas. For instance, the moss *Antitrichia curtipendula* was recorded by

[2] Requiring a circumneutral substratum reaction (pH 5.5–7.0).

Table 17.2 Some epiphytic lichens that failed to produce fungal fruiting bodies (ascocarps) in areas of England with a mean winter SO_2 concentration above 40 $\mu g\ m^{-3}$ (after Hawksworth *et al.*, 1973)

Anaptychia ciliaris	*P. sulcata*
Bryoria fuscescens	*Peltigera canina*
Evernia prunastri	*Physcia adscendens*
Hypogymnia physodes	*Pseudevernia furfuracea*
Lobaria amplissima[1]	*Ramalina farinacea*
L. pulmonaria[1]	*R. fraxinea*
Ochrolechia androgyna	*Sphaerophorus globosus*
Parmelia caperata	*Sticta fuliginosa*[1]
P. crinita[1]	*Usnea ceratina*[1]
P. perlata[1]	*U. subfloridana*
P. reticulata[1]	*Xanthoria candelaria*
P. saxatilis[2]	

[1]Also at concentrations above 30 $\mu g\ m^{-3}$.
[2]May fruit at higher concentrations when growing on stone.

W. Baxter and H. Boswell in Bagley Wood a few kilometres from Oxford (UK). In 1843 it was 'plentifully fruiting', but by 1860 it was 'almost gone', these representing the last records of the species for central southern England (Bates, 1995).

Mapping studies

Matching the distributions of cryptogams with contour maps of pollutant concentrations affords a potentially powerful tool for testing pollution hypotheses. The complexities of mapping have been discussed by Hawksworth (1973). Many such studies have shown strong negative relationships of cryptogams with SO_2 and no relationship with smoke levels.

The first individual to map lichens in relation to air pollution was Haugsjå (1930) who presented maps for 20 species in Oslo showing differences in their inner limits. From studying the co-occurrences of species in different parts of Stockholm, Sernander (1926) recognized distinct *zones* of lichen distribution: the inner 'lichen desert' around large industrial plants or city centres where lichens were absent from tree trunks; outwards from this a 'struggle zone' where lichens were present in small quantity; lastly, the 'normal zone' where lichens were abundant on trees and stone. Similar zone maps were soon made for many other cities. Alterations in the extents of these zones, usually as a result of increasing SO_2 concentrations, have afforded one of the major pieces of evidence for favouring the *toxic gas hypothesis* and rejecting the so-called *drought hypothesis*[3] (LeBlanc and Rao, 1973; Coppins, 1973).

Contemporary and historic information have often been combined in distribution maps for individual species. As Figure 17.1 shows, the lichens *Lobaria pulmonaria* and *Ramalina fraxinea*, and the moss *Antitrichia curtipendula*, have been lost from

[3] This attributed the lack of lichens in towns to a drier microclimate rather than to the poisoning effect of pollutants such as SO_2.

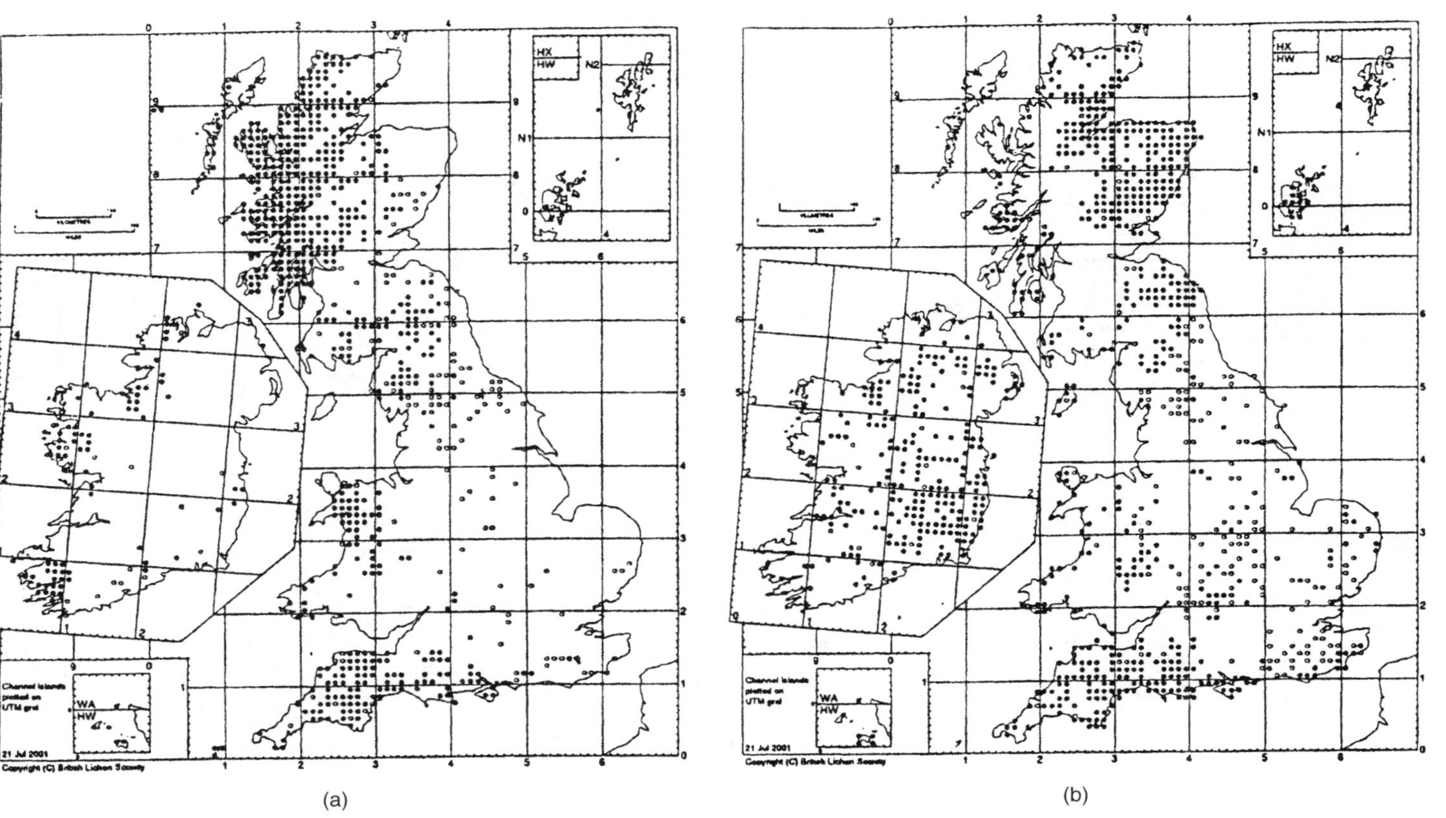

Figure 17.1 Distribution maps, utilizing the 10-km squares of the British and Irish national grids, for three lichens (a, b, d) and a moss (c) showing sensitive responses to industrial and urban SO_2 pollution. (a) *Lobaria pulmonaria*, (b) *Ramalina fraxinea*, (c) *Antitrichia curtipendula*, (d) *Lecanora conizaeoides.* Black dots represent records made in or after 1950, circles denote records made before 1950 or undated. Maps a, b and d are reproduced from the mapping scheme of the British Lichen Society with kind permission of Professor M.R.D. Seaward; map c is reproduced from Hill *et al*. (1994) with the permission of the authors

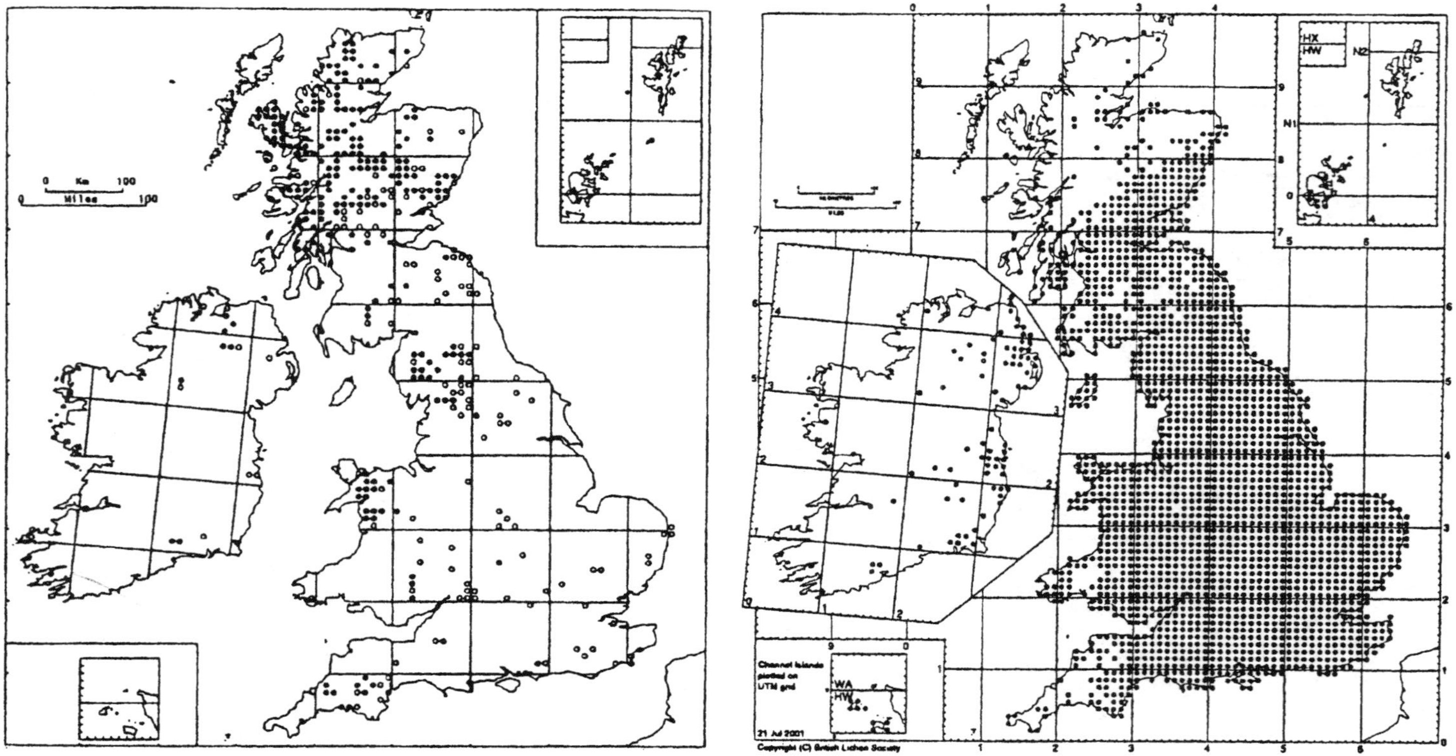

(c)

(d)

Figure 17.1 *(Continued)*

many former sites in central and eastern Britain, evidently owing to SO_2 pollution in these areas. In contrast, the crustose lichen *Lecanora conizaeoides* appears to 'require' SO_2 pollution because it has spread in the most polluted areas since around 1860, and at its peak of abundance in the late 1980s it was absent only from areas of Britain that had been unaffected by SO_2. Such maps invite the definition of 'critical levels' for survival, e.g. *L. pulmonaria* and *R. fraxinea* are absent or decreasing where mean winter levels had exceeded 35 $\mu g\ SO_2\ m^{-3}$ (Seaward and Hitch, 1982). However, many factors can modify the sensitivities of lichens to particular pollutants (Gilbert, 1973; Richardson, 1988). Thus, unless the habitats for mapped cryptogams have been rigorously standardized, attempts to deduce critical levels from distribution maps may be misleading. In establishing relationships between SO_2 levels and cryptogam survival, the least ambiguous circumstances are provided by newly installed point sources in otherwise unpolluted rural regions. The study by Skye (1958) of lichens on trees around the oil-shale works at Kvarntorp, Sweden is a classic example. Numerous (mainly lichen) mapping studies have been undertaken of particular towns and regions, often as bioindication exercises to reveal air pollution zones rather than focusing on the reasons for the responses of individual taxa.

Contemporary community studies

Many investigators have demonstrated the differential sensitivities of lichens and bryophytes around contemporary point sources through studies of the coverage of individual species along transects. One careful investigation of this type involved the sampling of epiphytes along an avenue of regularly spaced *Acer platanoides* trees extending for 1.1 km along a traffic-free passage into the city of Freiburg, Germany (Wirth and Brinckmann, 1977). High winter SO_2 concentrations (80–90 $\mu g\ m^{-3}$) in the city centre were believed to have caused the progressive decline in the total coverage of the 21 lichen and three bryophyte species recorded along the avenue (Figure 17.2a). The common lichens in Figure 17.2b show some contrasting patterns: *Parmelia sulcata* declined progressively towards the city centre; *Hypogymnia physodes* was less sensitive, firstly increasing then decreasing sharply near the centre; *Lecanora conizaeoides* exhibited peak abundance just outside the centre, though it too declined in the centre itself. This and many earlier investigations show the fruticose lichen growth-form to be the most susceptible to SO_2, with foliose lichens intermediate and crustose species the most tolerant. It has been suggested that the increased performance of crustose species such as *L. conizaeoides* under polluted conditions may be partly the result of a release from the competition imposed by normally dominant foliose species (Gilbert, 1973).

Various authors (e.g. Skye, 1968; Gilbert, 1970; Johnsen and Søchting, 1973) have demonstrated the progressive acidification of tree bark and other substrata on approaching SO_2 sources, a result of the overwhelming of inherent buffer properties by excess H^+. However, Gilbert (1970) deduced that this acidification effect was not in itself the primary cause of lichen and bryophyte sensitivity to SO_2.

Gilbert (1970) investigated the importance of microsite factors in determining the occurrence of bryophytes and lichens in several distinct habitats (asbestos roofs, sandstone wall tops, free standing *Fraxinus* trees, short grassland) along an SO_2 gradient around Newcastle upon Tyne, UK (Table 17.3). Shelter reduces the exposure

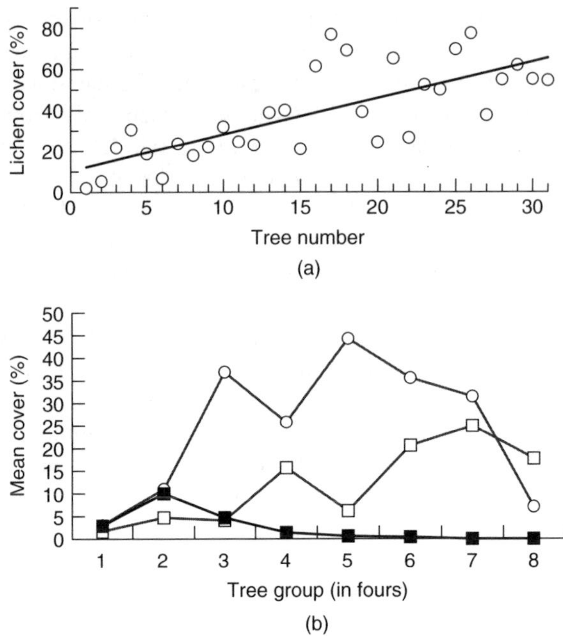

Figure 17.2 Changes in lichen cover of *Acer platanoides* trunks on approaching (from right to left) the centre of Freiburg along Hindenburg Avenue in 1974. (a) Total lichen cover on the south sides of individual sampled trees and fitted regression line. (b) Changes in cover values of three important lichens along the avenue, averaged for each group (A–H) of four successive sampled trees: ○, *Hypogymnia physodes*; ■, *Lecanora conizaeoides*; □, *Parmelia sulcata*. Redrawn from Wirth and Brinckmann (1977)

Table 17.3 The distance (km) westwards from central Newcastle in the late 1960s that 50% of the normal assemblage of lichens and bryophytes was attained in four contrasted habitats (after Gilbert, 1970)

Habitat	Bryophytes	Lichens
Nutrient-rich tree bark	21	16
Sandstone wall tops	12.5	10
Asbestos roofs	9	6.5
Short grassland	8.5	<8.5

to air-borne SO_2 and was seen as a restriction of cryptogams to dense woodland, long grassland, tree bases, bark crevices, walls backed by hedges and the kerbstones of graves in churchyards. Alkaline substrata like asbestos, mortar and limestone markedly modify the ionic forms of SO_2 present in the surface water film, and allow cryptogams to penetrate further into polluted areas. Local nutrient enrichment (e.g. bird droppings on roofs, roadside dust accumulations on sandstone walls and 'nutrient streaks' beneath wounds on tree trunks) all enhanced survival of lichens. These effects might be explicable through increased pH and through specific effects of nutrient elements (e.g. P, K, N) in reducing susceptibility to SO_2.

In North America several facilities generating SO_2 are widely removed from other pollution sources (Winner, 1988). Some, such as the iron-sintering plant at Wawa, Ontario, the copper smelters at Sudbury, Ontario and those at Murdochville, Quebec, have created horrendous scenes of environmental degradation arising from SO_2 and heavy metal pollution. Sulphur dioxide is the only significant pollutant released from the natural gas refineries at Fox Creek, Alberta. Here, Winner and Bewley (1978) found that even terricolous mosses were affected. The normally dominant 'feathermoss' *Pleurozium schreberi* progressively decreased in cover with increasing SO_2 stress nearer the refinery, but the subordinates *Hylocomium splendens* and *Ptilium crista-castrensis* increased in cover closer to the refinery, perhaps in response to lowered competition from *P. schreberi*, before succumbing to the high SO_2 concentrations nearest the plant.

Transplants

The transplantation of specimens from a locality with clean air to sites with different levels of SO_2 pollution was used originally to determine relative sensitivities of cryptogam species, but increasingly the technique has been used to apply realistic fumigation levels in physiological and ultrastructural investigations. Brodo (1961, 1967) popularized a method in which discs of bark with lichen attached were cut out with a punch from donor tree trunks and relocated in holes of identical dimensions in the receiver trees. The performance of the lichens in their new habitats is usually assessed from photographs or tracings of thallus outlines. Variations of this technique have been applied in several studies (LeBlanc and Rao, 1966, Hawksworth, 1973). Recent investigators have often assessed the physiological condition of the transplants after exposure (e.g. Sheridan *et al.*, 1976; Ferry and Coppins, 1979; Holopainen, 1984; Kardish *et al.*, 1987; Garty *et al.*, 1998; Gonzalez *et al.*, 1998; Hufnagel and Türk, 1998; Gonzalez and Pignata, 1999). Transplants involving terricolous and peatland lichens and bryophytes have also been used increasingly in SO_2 research (Kauppi, 1976; Ferguson and Lee, 1983b; Bates, 1993).

Sulphur content

The demonstration of correlations between S contents of cryptogams and measured SO_2 levels provides evidence for the involvement of this pollutant in the destruction of lichens and bryophytes (Hawksworth, 1973). Clear correlations between S content and physiological markers of SO_2 injury have been obtained both in indigenous specimens and in transplants exposed in polluted regions (e.g. Boonpragob and Nash, 1990; von Arb *et al.*, 1990; Palomaki *et al.*, 1992; Gonzalez and Pignata, 1994). In several more recent studies the sulphur isotope discrimination ratio ($\delta^{34}S$) has been used to try to pinpoint sources of sulphur absorbed by cryptogams. This topic is discussed further in Chapter 8.

Physiological and biochemical effects of SO_2

Demonstrating that sensitivity rankings of bryophytes and lichens to SO_2 in the laboratory match those observed under polluted conditions in the field, provides compelling evidence that SO_2 is the major pollutant limiting species distributions in the field. In comparison with the vascular plant literature on SO_2, there have been few studies employing long-term fumigations with realistic concentrations of the pollutant. Accordingly, there are real difficulties in extrapolating results of short-term laboratory experiments to the responses of bryophytes and lichens under field conditions (Nash, 1988), and much reliance is still given to field observations in defining sensitivities. On the positive side, their simple morphology means that non-vascular cryptogams can be conveniently incubated in solutions containing dissolved SO_2 and considerable progress has been made in understanding mechanisms of cellular damage and tolerance of the pollutant.

Responses to long-term fumigations

Usually, cryptogams cannot be fumigated for long periods employing the facilities that are routinely used for potted vascular plants. The shoots or thalli do not readily absorb moisture from soil and quickly dry-out in a moving air stream if they are deprived of the considerable resistances to evaporation imposed in nature by their complex colonial life-forms and specific microsites. Also, many species are shade-adapted and injured by high light intensities.

Field fumigation is a more expensive but probably preferable approach that has, however, rarely been employed. Moser *et al.* (1980) carried out an *in situ* fumigation of tundra lichens (*Cladonia* spp) at Anaktuvuk Pass (Alaska) by releasing SO_2 from a nozzle and measuring its concentration in 0.2-m strips of natural vegetation at 1-m intervals downwind of the point of release. The samples in the nearest two strips (0.3–0.5, 1.0–1.2 m) were killed outright by the unrealistically high SO_2 levels (>1 ppm) received, but those further away showed dose-related photosynthetic reductions. *Cladonia rangiferina* appeared to be more sensitive to SO_2 than *C. stellaris* and *C. cucullata.* Data on epiphytic lichen tolerances to long-term SO_2 (and O_3, see later) fumigations at realistic levels were obtained in the UK during the Liphook Forest Fumigation Project (Bates *et al.*, 1996). Here circular plots of young conifers were subjected to target concentrations of the pollutants by means of computer-controlled gas releases on the upwind perimeters. Three lichens (*Evernia prunastri*, *Hypogymnia physodes*, *Lecanora conizaeoides*) colonized the acid bark of the conifers during the experimental fumigations from 1987–90. *E. prunastri* occurred only at ambient SO_2 concentration (4 ppb) and was therefore the most sensitive of the lichens. *H. physodes* was abundant at 4 pbb, much scarcer at 12 ppb and rare at 20 ppb. In marked contrast, *L. conizaeoides* was scarce at 4 ppb, but had appreciable coverage in the 12 and 20 ppb plots, indicating a positive requirement for SO_2 or some associated factor.

Although the difficulties associated with fumigation can be overcome, it is often simpler to employ sodium bisulphite[4] solution as a dissolved analogue of gaseous SO_2. The main difficulty in employing bisulphite (HSO_3^-) solutions is in equating these to gaseous SO_2 concentrations in the moving air above the vegetation (Nieboer *et al.*, 1977). Ferguson *et al.* (1978) compared the responses of *Sphagnum* spp to soluble

[4] The term 'bisulphite' is here taken to mean any solution containing SO_3^{2-} and HSO_3^- ions irrespective of the terminology used by cited authors.

Table 17.4 A comparison of the effects of sulphur pollutants on the growth of some *Sphagnum* species (after Ferguson *et al.*, 1978)

Species	Percentage reduction in growth caused by pollutant		
	SO_2 (131 μg m^{-3})	HSO_3^- (1 mM)	SO_4^{2-} (5 mM)
S. tenellum	100	100	73.2
S. imbricatum	79.6	89.3	57
S. papillosum	–	85.2	56
S. capillifolium	42.5	84.5	81.7
S. magellanicum	8.9	84.6	56.7
S. recurvum	15.5	68.1	29.6

sulphur pollutants (HSO_3^-, SO_4^{2-}) and gaseous SO_2 treatments of four months' duration. Increasing HSO_3^- concentrations caused an initial rise in growth and chlorophyll content followed by a marked decline at higher levels. Sulphate caused much smaller reductions in growth. Almost the same rankings of the sensitivities of the species were obtained employing gaseous SO_2 and aqueous HSO_3^-, with *S. tenellum* appearing the most and *S. recurvum* the least sensitive taxon (Table 17.4). A closely similar ranking was obtained when Ferguson and Lee (1983b) fumigated five *Sphagnum* species for 12 weeks at the more realistic SO_2 concentration of 65 μg m^{-3}. Most of the information summarized below is derived from comparatively short-term SO_2 fumigations and bisulphite treatments.

Metabolic and ultrastructural effects of SO_2

From comparative investigations Fields (1988) ranked the effects of SO_2 on metabolic processes in lichens in order of decreasing sensitivity: nitrogen fixation > membrane integrity > carbohydrate transfer > photosynthesis > respiration > photosynthetic pigments. In solution, SO_2 gives rise to a mixture of sulphite ions (SO_3^{2-}), bisulphite ions (HSO_3^-) and sulphurous acid (H_2SO_3): sulphite, the least toxic form, predominates at pH 6 and above; the more toxic bisulphite predominates at pH 2–5; highly toxic sulphurous acid becomes increasingly important below pH 4 (Saunders and Wood, 1973). A greater toxicity of dissolved SO_2 with increasing acidity has been clearly demonstrated in lichens (Puckett *et al.*, 1973; Türk and Wirth, 1975), peat-mosses (Ferguson and Lee, 1979) and forest floor mosses (Bharali, 1999).

Early reports of chlorophyll degradation as a result of SO_2 treatment often involved gaseous concentrations far exceeding those experienced in the field. Protons derived from sulphurous acid were believed to bring about phaeophytinization of chlorophyll. Puckett *et al.* (1973) found a striking relationship between chlorophyll loss in lichens and the increasing oxidizing power from HSO_3^- as the pH was progressively lowered. Such drastic effects may represent a late stage in cell injury. For example, Sanz *et al.* (1992) recorded chlorophyll reduction in fruticose lichens only when photosynthesis was damaged beyond recovery and ultrastructural damage was apparent. Nevertheless, pigment losses have been reported in circumstances under field conditions where the cryptogams may have been only moderately stressed by pollutants (e.g. Baxter *et al.*, 1989a; Gonzalez and Pignata, 1994; Silberstein *et al.*, 1996a; Garty *et al.*, 1998).

Baddeley *et al.* (1973) studied the effects of aqueous SO_2 on lichen respiration using an oxygen electrode and found a dose-related depression in all species at 'acceptably low' concentrations. Results were least variable when the measurements were made at a pH realistic for the lichen substratum under consideration. Most lichen studies have found respiration less sensitive than photosynthesis (e.g. Beekley and Hoffman, 1981; Malhotra and Khan, 1983). Thus, Sanz *et al.* (1992) obtained an effect on respiration only at high SO_2 concentrations in *Evernia prunastri* and *Ramalina fraxinea.* In three forest mosses, Winner and Kock (1982) and Winner and Bewley (1983) concluded that dark respiration was unaffected by SO_2, but Baxter *et al.* (1991a) reported an initial suppression of respiration in *Sphagnum cuspidatum* although this was later reversed, probably owing to oxidation of the dissolved pollutant to harmless sulphate (see later).

Measurements of photosynthesis in fully hydrated bryophytes and lichens and, more recently, of the quantum efficiency of photosystem (PS) II have been made by many workers. In a pioneer investigation Hill (1971) compared photosynthesis rates in three epiphytic lichens incubated in sodium bisulphite solutions and obtained a sensitivity order (*Usnea subfloridana* > *Hypogymnia physodes* > *Lecanora conizaeoides*) that agreed well with field sensitivities to supposed SO_2 pollution. A similar comparison of four urban mosses by Inglis and Hill (1974) showed a less clear separation of the species. However, the mosses appeared to be more sensitive to aqueous SO_2 than the lichens, a photosynthetic reduction being noted at 0.02 mM sulphite in the mosses, but only at 0.1 mM in *Usnea subfloridana*, the most sensitive lichen. Ferguson and Lee (1979) reported a similarly high sensitivity of *Sphagnum* spp to bisulphite solutions utilizing both O_2 evolution and $^{14}CO_2$ incorporation to measure photosynthesis. Their results also supported the relatively high SO_2 tolerance of *S. recurvum* revealed in long-term growth experiments (Ferguson *et al.*, 1978). The importance of hydration status was demonstrated in forest mosses by Winner and Bewley (1983). Less SO_2 absorption occurred as the shoots were dried, but some injury to photosynthesis was always apparent upon rehydration unless the mosses had been air dry when exposed to the pollutant. Similar results were obtained with the lichen *Cladonia mitis* by Coxson (1988) and further critical observations on this subject have been made for lichens by Holopainen and Kauppi (1989).

Recently, fluorometer measurements of parameters of chlorophyll fluorescence have become established as a rapid method for assessing the impact of pollutants upon the photoassimilatory system of bryophytes and lichens (Baxter *et al.*, 1989a; Potter *et al.*, 1996; Tuba *et al.*, 1997). A detailed analysis of the relationship between chlorophyll degradation and changes in chlorophyll *a* fluorescence properties was made in the lichen *Parmelia quercina* collected from sites with SO_2 (and O_3) polluted air in northern Spain (Calatayud *et al.*, 1996). Thalli from polluted sites showed slower rates of Q_A (primary quinone acceptor of PSII) reoxidation, a greater proportion of closed PSII reaction centres, lower non-photochemical quenching (NPQ) and photochemical quenching, and a higher quantum efficiency of the remaining open PSII reaction centres. Confirmation of the effects of SO_2 on chlorophyll *a* fluorescence was obtained in fumigation studies with the fruticose lichens *Evernia prunastri* and *Ramalina farinacea* (Deltoro *et al.*, 1999). *E. prunastri* suffered larger reductions in net photosynthesis,

PSII-mediated electron flow and NPQ than *R. farinacea*, and also exhibited decreased activities of antioxidant enzymes.

In lichens, carbohydrate transfer from photobiont to mycobiont is an important function of the symbiosis. Specific transfer compounds, glucose where the photobiont is a cyanobacterium and one of the polyols where it is a green alga, diffuse across the small extracellular gap between the plasma membranes of the two symbionts. Absorption by the fungus presumably involves specific carrier proteins in the hyphal cell membranes. Fields and St Clair (1984) showed that short-term fumigations with 1 and 2 ppm gaseous SO_2 led to decreased carbohydrate transfer in *Collema polycarpon* (glucose) and *Parmelia chlorochroa* (ribitol). As the fumigations also greatly increased electrolyte leakage from the cells it was concluded that the carrier proteins in the cell membranes had probably been denatured. The greater accessibility of the cell membranes than of the chloroplasts and their thylakoids was believed to explain the higher sensitivity of carbohydrate transfer than photosynthesis to SO_2. von Arb and Brunold (1990) also observed greatly reduced release of ^{14}C-photosynthate in samples of *Parmelia sulcata* from the Swiss city of Biel that correlated well with other indicators of pollutant stress.

Leakage of K^+ and other electrolytes has commonly been used to indicate damage to the plasma membrane in cryptogams as a result of SO_2 exposure (e.g. Puckett *et al.*, 1977; Pearson and Henriksson, 1981; Fields and St Clair, 1984; Hart *et al.*, 1988; Silberstein *et al.*, 1996a). The kinetics of this process in lichens has been investigated by Puckett *et al.* (1977).

Nitrogen fixation, known only in lichens with a blue-green algal photobiont, appears to be one of the most sensitive metabolic processes to SO_2 although it has been little studied (Richardson and Nieboer, 1983). In *Stereocaulon paschale* the process was completely inhibited by a 50 mM bisulphite solution at pH 5.8, whereas blocking of photosynthesis required a 500 mM bisulphite solution (Hällgren and Huss, 1975). Vincent (1990) showed proportionately greater inhibition of N_2-fixation in *Peltigera canina* than of respiration and photosynthesis when specimens were placed in the SO_2 (and NO_2) polluted atmosphere of Toulouse (France).

Changes to the internal concentrations of a range of molecules within cryptogams have been noted in SO_2 fumigation and town air exposure studies. These include amino acids (Vincent, 1990; Silberstein *et al.*, 1996b), ATP (Aulio, 1984; Kardish *et al.*, 1987; Garty *et al.*, 1988; Silberstein *et al.*, 1996a), fatty acids and lipids (Beltman *et al.*, 1980; Malhotra and Khan, 1983; Bychek-Guschina *et al.*, 1999), antioxidants (Gonzalez and Pignata, 1994; Silberstein *et al.*, 1996b), proteins (Malhotra and Khan, 1983; Gonzalez and Pignata, 1994) and the release of 'stress' ethylene (Garty *et al.*, 1995).

Ultrastructural effects of SO_2 have been investigated by transmission electron microscopy of fumigated lichens, but few relevant observations, apart from those of early workers, who frequently noted plasmolysis of cells and bleaching of pigments, have been made in bryophytes. Plasmolysis has also been observed in lichen algae (Eversman, 1978), and dying algae cease to autofluoresce as their chlorophyll degrades (Kauppi, 1980). In the *Nostoc* (cyanobacterium) photobiont of *Peltigera canina*, Sharma *et al.* (1983) described a decrease in the number of carboxysomes (locations of RUBISCO) and increased production of cyanophycin granules, the alga's N store.

In specimens of the green algal lichen *Cladonia alpestris* transplanted around a sulphite pulp mill, Ikonen and Kärenlampi (1976) observed swollen and disrupted thylakoids in the algal chloroplasts, accumulations of starch grains and granulation of the stroma. Pyrenoglobuli were absent from such specimens. Subsequent studies (Holopainen and Kärenlampi, 1984; Holopainen and Kauppi, 1989) have revealed a wider range of ultrastructural abnormalities and injury stages in green lichen algae (Table 17.5). In *Peltigera canina* the mycobiont appeared as susceptible to SO_2 injury as the *Nostoc* photobiont, possibly because of the unusually thin-walled nature of the hyphae in this species (Table 17.5). Modenesi (1993) presented striking scanning electron micrographs showing the accumulation of calcium oxalate (weddellite) crystals on the surface of the SO_2 sensitive lichen *Parmotrema reticulatum* as a result of exposure to the pollutant. The same symptom was induced by treatment with paraquat, suggesting that the crystals were a by-product of the detoxification of free radicals.

Mechanisms of SO_2 tolerance

The attractive hypothesis that differential sensitivity is the result of differential uptake of SO_2 has received little support from recent experimental work. Winner and Bewley (1983) could find no differences in SO_2 uptake by three forest mosses that explained their different sensitivities. Indeed, Winner *et al.* (1988) concluded that mosses in all habitats are 100–1000 times stronger sinks for the pollutant than vascular plants. Silberstein *et al.* (1996b) found substantial S uptake at polluted sites by the relatively SO_2-tolerant lichen *Xanthoria parietina*. Gries *et al.* (1997) carefully measured SO_2 uptake by 11 fruticose and foliose lichens in a continuous flow system and concluded that their uptake rates were all closely similar. Moreover, pollutant uptake was identical in thalli that had been killed by heat treatment to that in living thalli. Uptake of SO_2 therefore appears to be a passive phenomenon and tolerant cryptogams presumably ameliorate it after entry as there is no evidence of tolerant enzymes (e.g. Ziegler, 1977).

An observation indicative of amelioration is that inhibition of photosynthesis in lichens by aqueous SO_2 is reversible. Puckett *et al.* (1974) found that the lichen *Umbilicaria muhlenbergii* could overcome several repeated applications of aqueous SO_2 before eventually succumbing (Figure 17.3). Syratt and Wanstall (1969) had earlier suggested that the high tolerance of the epiphytic moss *Dicranoweisia cirrata* to SO_2 was due to a mechanism capable of oxidizing dissolved SO_2 to harmless SO_4^{2-} that employed metabolic energy. Baxter *et al.*, 1989a) found a large difference in the SO_2 tolerances of two populations of *Sphagnum cuspidatum*. Laboratory application of bisulphite inhibited photosynthesis and growth in shoots from the unpolluted population in north Wales, but it had little effect in *S. cuspidatum* from the polluted locality in the southern Pennines. Shoots from the polluted locality were found to promote the rapid oxidation and disappearance of the HSO_3^- from the incubation solution (Figure 17.4), a property that was associated with a higher concentration of transition metal ions (Fe^{3+}, Mn^{2+}, Cu^{2+}) on their cell wall exchange sites (Baxter *et al.*, 1989b, 1991a). These metals, and especially Fe^{3+}, were present as a result of heavy industrial pollution in the past in the southern Pennines, and readily oxidize bisulphite by acting as electron acceptors. This passive amelioration mechanism could be induced in the unpolluted Welsh plants by experimental additions of Fe^{3+} and removed from Pennine plants by

Table 17.5 A summary of ultrastructural changes at different injury stages caused by SO_2 (215 μg m^{-3}) in **(a)** the green-algal photobiont (*Trebouxia*) of *Bryoria capillaris* and *Hypogymnia physodes*, **(b)** the cyanobacterial (*Nostoc*) photobiont of *Peltigera*, and **(c)** the mycobiont of *Peltigera*. The higher injury stages were produced by increasing both the period of fumigation and the moisture content of the lichens during exposure (after Holopainen and Kauppi, 1989)

Type of injury	Stage of injury					
	Normal	Slight	Intermediate		Severe	
	0	1	2	3	4	5
(a)						
Mitochondrial swelling and deformation	–	+	++	++	++	0
Chloroplast envelope folding	–	+	++	++	0	0
Pyrenoglobuli deformation	–	–	+	++	++	0
Appearance of myelin-like figures	–	–	++	++	++	0
Mitochondrial cristae and matric degeneration	–	–	+	++	++	0
Thylakoid folding	–	–	+	++	++	0
Chloroplast stroma granulation	–	–	+	++	++	0
Nuclear degeneration	–	–	+	++	++	0
Cytoplasmic granulation	–	–	–	+	++	0
Vescicle complex rupture	–	–	–	+	++	0
Thylakoid rupture	–	–	–	–	+	++
Cell membrane rupture	–	–	–	–	+	++
Cell contents collapse	–	–	–	–	+	++
(b)						
Cell matrix granulation	–	+	++	++	++	0
Precipitation of cell matrix and storage granules	–	–	+	++	++	0
Thylakoid degeneration	–	–	+	++	++	0
Thylakoid disintegration	–	–	–	+	++	0
Cell contents collapse	–	–	–	–	+	++
(c)						
Mitochondrial swelling and deformation	–	+	++	++	++	0
Mitochondrial matrix degeneration	–	–	+	++	++	0
Nuclear degeneration	–	–	+	++	++	0
Cytoplasmic granulation	–	–	–	+	++	0
Plasmalemmasome and vacuole rupture	–	–	–	+	++	++
Cell membrane rupture	–	–	–	+	++	++
Cell contents granulation and collapse	–	–	–	–	++	++

Key: –, no injury; +, observable injury; ++, severe injury; 0, organelle no longer visible

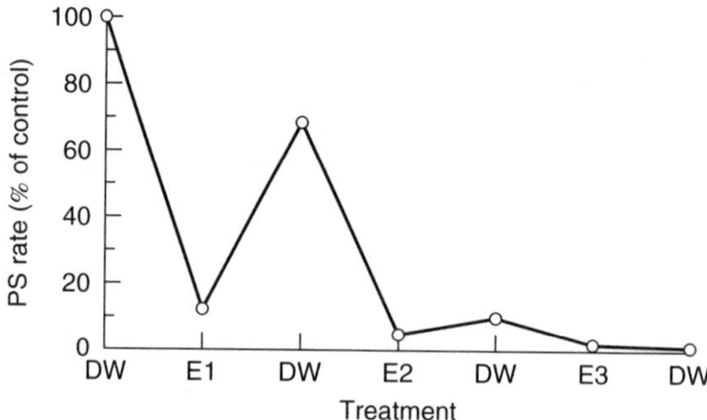

Figure 17.3 The effect on photosynthetic rate (PS) of repeated exposures (E1-E3) to aqueous sulphur dioxide (75 ppm SO_2, pH 3, 15 min) in the lichen *Umbilicaria muhlenbergii*. The initial (control) photosynthetic rate was determined in material hydrated with distilled water (DW). Samples (5 mm discs cut from thalli) were washed thoroughly with distilled water (DW) and maintained fully hydrated for 6 h to permit recovery after each SO_2 exposure. Redrawn from Puckett *et al.* (1974)

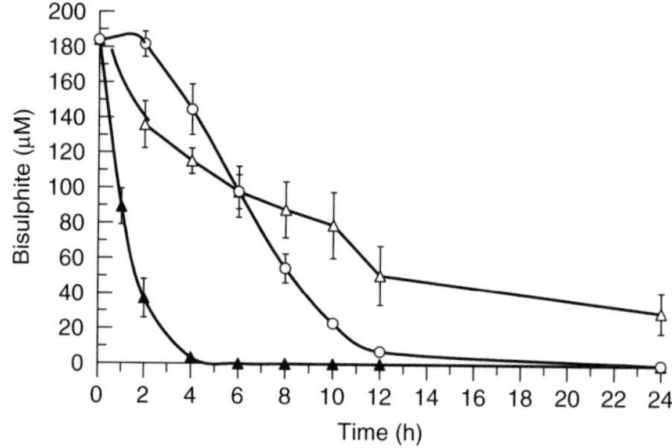

Figure 17.4 Disappearance of bisulphite from an artificial rainwater solution: ▲, in the presence of *Sphagnum cuspidatum* from a polluted site; Δ, in the presence of *S. cuspidatum* from an unpolluted site; ○, in the absence of *Sphagnum*. Graph shows means of three replicates and their standard errors. Redrawn from Baxter *et al.* (1989b)

eluting away Fe^{3+} with the chelating agent EDTA (Baxter *et al.*, 1991b). In a study of three lichens with different SO_2 susceptibilities, Miszalski and Niewiadomska (1993) investigated the hypothesis that bisulphite oxidation was under metabolic control. Oxidation was highest in the moderately SO_2 tolerant *Hypogymnia physodes* and least in the SO_2 sensitive *Usnea filipendula*; however, addition of the metabolic inhibitors KCN and DCMU did not markedly alter the rate of bisulphite oxidation. Thus, again the oxidation appeared to be a passive process that was correlated with the metal contents of the lichens. However, in the terricolous mosses *Rhytidiadelphus triquetrus* and *Pleurozium schreberi* the oxidation of bisulphite has a significant metabolic component (Bharali, 1999) as it proceeded at a much higher rate in the light than in the dark, and it was sensitive to inhibitors of oxidative metabolism. An increased sulphate content of *Xanthoria parietina* in polluted sites was taken as evidence that this lichen has a considerable ability to oxidize SO_2 although whether this occurred actively was not demonstrated (Silberstein *et al.*, 1996b).

In higher plants one of the responses to SO_2 is an increase in activities of the enzymes peroxidase and superoxide dismutase as a defence against increased oxidant formation. However, Lee and Studholme (1992) reported that peroxidase activity in *Sphagnum cuspidatum* from the polluted southern Pennines is significantly lower than at the unpolluted Welsh site. Furthermore, *S. cuspidatum* transplanted from the Welsh to the Pennine site showed a decrease in peroxidase activity whereas it increased in vascular plants. These results suggest that peroxidase is not part of the defence against SO_2 in *Sphagnum*. In contrast, the tolerant lichen *Xanthoria parietina* contained more peroxidase activity than the sensitive *Ramalina duriaei*, and peroxidase activity was higher at polluted compared to unpolluted sites (Silberstein *et al.*, 1996b). *X. parietina* was also shown to increase its peroxidase activity (Figure 17.5) and its content of the antioxidant compound reduced glutathione (GSH) in response to bisulphite additions, whereas *R. duriaei* did not show enhanced peroxidase activity and had no GSH protection. Parietin, the conspicuous yellow lichen metabolite in *X. parietina*, is an anthraquinone derivative that may also behave as an antioxidant.

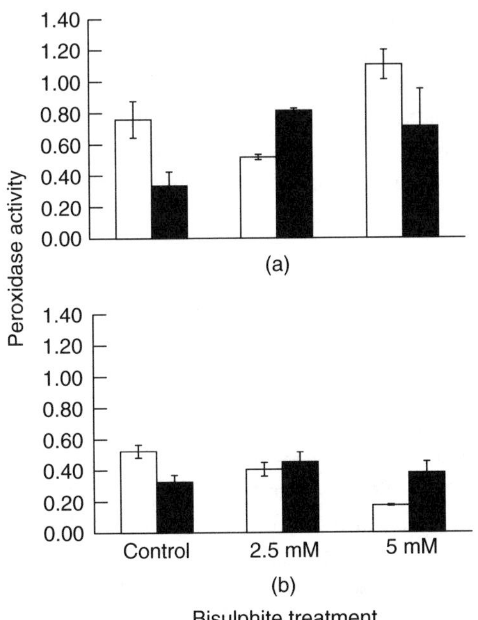

Figure 17.5 Effect of bisulphite treatments on peroxidase activity (ΔOD mg^{-1} protein min^{-1}) in (a) the relatively SO_2-tolerant lichen *Xanthoria parietina*, and (b) the relatively SO_2-sensitive lichen *Ramalina duriaei*. White bars show activity immediately after treatment, black bars show activity 24 h after removal of the bisulphite. Data are means of three replicates and their standard deviations. Redrawn from Silberstein *et al.* (1996b)

Earlier the importance of calcareous substrata in resisting the toxic influences of SO_2 was discussed mainly in terms of their buffer properties towards acid sulphur compounds. Nieboer *et al.* (1979) and Richardson *et al.* (1979) demonstrated that several metal ions (Ca^{2+}, Sr^{2+}, Mg^{2+}, Na^+, K^+) reduced the K^+ leakage and photosynthetic inhibition in SO_2-treated lichens suggesting a more specific ameliorating effect of the metals. In the terricolous moss *Rhytidiadelphus triquetrus*, a species that acts as a calcicole in polluted areas (Bates, 1993), the addition of Ca^{2+} ameliorates photosynthetic inhibition by dissolved SO_2 (Bharali, 1999). The Ca^{2+} ions may protect by strengthening linkages between polar groups in the plasma membrane.

The comparison of *Xanthoria parietina* (tolerant) and *Ramalina duriaei* (sensitive) by Silberstein *et al.* (1996b) is the most complete catalogue of cryptogamic SO_2 defences that has yet been achieved. These include both constitutive and induced mechanisms. The former embrace an inherently greater pH buffer capacity of the thallus in *X. parietina* whilst the latter includes the ability to generate amino acids (proline, arginine), both important in resisting pH decreases induced by the build up of $SO_4{}^{2-}$ from oxidation of SO_2.

Responses to other pollutants

Compared with the overwhelming evidence showing that SO_2 is damaging to cryptogams, field observations indicating that other atmospheric pollutants may be damaging to bryophytes and lichens are modest. The main exceptions to this are provided by airborne fluorides, dealt with below, and wet deposited acidity, which is considered in the context of bryophytes and lichens by Farmer *et al.* (1992) and Bates (2000). This section examines the field and laboratory-based evidence that NO_x, NH_3, O_3 and PAN, and fluorides may be harmful to bryophytes and lichens.

NO_x

Oxides of nitrogen, mainly nitrogen dioxide (NO_2) and nitric oxide (NO), are now among the most important atmospheric pollutants of urban environments (Chapters 2 and 3). In some instances they may accompany high levels of SO_2, whereas in many western European and North American cities the ascendency of NO_x has occurred whilst SO_2 has been declining in importance. As the cryptogamic flora had already been decimated by SO_2 in most regions affected, it is unsurprising that little evidence has been forthcoming that NO_x is itself destructive to bryophytes and lichens. In The Netherlands, van Dobben and ter Braak (1999) identified many species that were sensitive to SO_2, but very few that could be used to predict NO_2 concentrations. Nevertheless, it has long been known that epiphytic lichen floras are poorer on trees by major roads than elsewhere (Ferry *et al.*, 1973), a possible indication of the involvement of NO_x derived from vehicle engines.

A multiple regression study by von Arb *et al.* (1990), utilizing physiological indicators of stress in the lichen *Parmelia sulcata* at urban sites in northern Switzerland, identified strong effects of NO_x and SO_2, but little injury was attributable to O_3. Moreover, a recent investigation of factors limiting the recolonization by epiphytic lichens of trees in central London also identified high NO_x concentrations as the major factor preventing the townwards spread of *Parmelia saxatilis* (Batty, 1997). Angold (1997) investigated heathland sites adjacent to a busy road in the New Forest (south England) and attributed vegetation changes on approaching the road to increasing NO_x concentrations from vehicle exhausts. Podetia of the lichen *Cladonia portentosa* were significantly shorter and had fewer first order branches nearer the road. However, it was not clear whether this was directly attributable to NO_x or a response to the increasing vigour of vascular plants nearer the road.

Early fumigation experiments with bryophytes and lichens have been criticized for utilizing unrealistically high concentrations of NO_2 (Farmer *et al.*, 1992). The effects of more realistic exposures (NO_2 at 122.4 $\mu g\ m^{-3}$) on the endohydric moss *Polytrichum formosum* were varied: growth of existing shoots was firstly stimulated, later these shoots suffered a 46% reduction in number, and there was a 36% reduction of new shoot production (Bell *et al.*, 1992). Morgan *et al.* (1992) examined the effects of fumigations with NO and NO_2 on the nitrate reductase (NR) enzyme in four mosses. NO_2 at 35 ppb stimulated NR activity, probably because the gas dissolves to form mainly nitrate, the enzyme's substrate. In contrast, NO at the same concentration was inhibitory towards NR, presumably through the formation of toxic nitrite ions in the surface water film on the mosses. Earlier evidence of the high toxicity of nitrite ions had been obtained by Marti (1983) working with isolated lichen photobionts, whereas nitrate and sulphate were relatively harmless. Investigations in the author's laboratory have likewise shown a high sensitivity of respiration and photosynthesis in epiphytic lichen thalli (Batty, 1997) and shoots of the terricolous mosses *Pleurozium schreberi* and *Rhytidiadelphus triquetrus* (Bharali, 1999) to nitrite.

NH_3

Ammonia is becoming an increasingly common atmospheric pollutant in rural areas of Europe and in the worst affected countries such as The Netherlands it may represent 50–80% of total nitrogen deposition. The destruction of heathlands, in which bryophytes and lichens may be significant components, in regions badly affected by this pollutant is described in Chapter 12.

In The Netherlands, de Bakker (1989), and van Dobben and de Bakker (1996) noticed an increased frequency from 1970 onwards, a period when SO_2 levels were falling but NH_3 emissions rising, of the so-called 'nitrophilous' lichens[5] (e.g. species of *Candelariella*, *Phaeophyscia*, *Physcia*, *Xanthoria*). These replaced 'acidophiles' (e.g. *Hypogymnia physodes*, *Lecanora conizaeoides*, *Pseudevernia furfuracea*) on the bark of oak (*Quercus*) and beech (*Fagus*). De Bakker (1989) demonstrated that a major effect of NH_3 emissions was to raise the pH of bark that had previously been acidified by SO_2, and he reasoned that 'nitrophilous' epiphytes were actually more dependent on high pH than on the nitrogen supply. Comparable losses of acidophiles and increases in neutrophiles among bryophytes in several habitats have also been attributed to falling SO_2 concentrations and rising NH_3 in The Netherlands (Greven, 1992). Increases in N content, and discoloration and die-back of the heathland lichen *Cladonia portentosa* in

[5] Species that are normally associated with sites enriched by bird droppings.

The Netherlands may also be partly the result of an excessive uptake of NH_3 (Brown, 1992). Recent increases in epiphytic green algae appear to correlate with increased N deposition in northern Europe (Bråkenhielm and Quinhong, 1995; Søchting, 1997; Poikolainen *et al.*, 1998). Algal films can rapidly overgrow slower growing epiphytic lichens and mosses, and deprive them of light (Barkman, 1958), and could therefore have a profound effect on the structure of cryptogamic communities.

There are considerable problems in making realistic fumigations of cryptogams with NH_3, a reactive and highly soluble gas, rarely found at concentrations above 10 ppb (Lee *et al.*, 1998). Ammonia readily dissolves in the surface water on bryophytes and lichens, yielding the ammonium ion (Miller and Brown, 1999). NH_4^+ is adsorbed passively onto the negatively charged exchange sites of the cell walls in lichens and is relatively labile (Brown, Standell and Miller, 1995; Miller and Brown, 1999), apparently passing rapidly from the cell wall to the cell interior. The physiological consequences of increased ammonium ion uptake in bryophytes and lichens are largely unknown. Brown and Tomlinson (1993) noted injury to the lichen *Parmelia sulcata* upon application of ammonium chloride solutions over 17 weeks and demonstrated inhibition of photosynthesis by 10 mM NH_4Cl in the laboratory. Sanchez-Hoyos and Manrique (1995) observed declining levels of antheraxanthin, a xanthophyll pigment that protects the photosynthetic apparatus against photooxidative damage, and increasing chlorophyll *b* in *Ramalina capitata*, as a result of repeated treatments with 10 mM NH_4Cl over 10 days. As this lichen is typically found on ammonia-enriched bird perching stones in Spain it would be expected to be more resistant than other lichens to this treatment. Increasingly, the ammonium ion is detected as a component of total wet deposited nitrogen.

O_3 and PAN

Bryophytes and lichens appear to be less sensitive to ozone than the most sensitive vascular plant indicators, however, rather few investigations into their sensitivities to O_3 have been made. In Europe, O_3 pollution develops mainly in sunny weather in summer (Chapters 3 and 4) when cryptogams are least likely to be metabolically active and susceptible. In California, ozone is a more persistent problem. Sigal and Nash (1983) demonstrated wide-reaching effects of oxidant air pollution on lichens growing upon conifers in the mountains near Los Angeles. Despite these findings, Boonpragob *et al.* (1989) detected appreciable quantities of other dry-deposited pollutants (NO_3^-, NH_4^+, SO_4^{2-}, PO_4^{3-}, H^+, F^-, Cl^-) when transplants of the epiphytic lichen *Ramalina menziesii* were exposed in Los Angeles. These substances accumulate on the pendulous, net-like thallus in dry weather (Boonpragob and Nash, 1990) and are dissolved and absorbed into the cells during rainstorms. Several (NO_3^-, F^-) appeared to be accumulated to toxic levels, and are more probable causes of the extinction of *R. menziesii* in Los Angeles than O_3 (Ross and Nash, 1983; Boonpragob and Nash, 1991). Clear field evidence of bryophyte or lichen damage by O_3 has so far not been forthcoming from studies in Europe (e.g. von Arb *et al.*, 1990; Bates *et al.*, 1996).

Low water solubility may mean that O_3 penetrates lichen thalli and moss shoots rather poorly when they are fully hydrated and potentially most susceptible. Laboratory fumigations of cryptogams with ozone have certainly yielded mixed results. Brown and

Smirnoff (1978) found no suppression of photosynthesis in *Cladonia rangiformis* by 2 ppm O_3 for 30 min and by 6 ppm O_3 for 2.5 h. They speculated that entry of the pollutant was probably slowed by contact with the less susceptible hyphae of the mycobiont, preventing entry to the (presumed) more sensitive photobiont cells. Nash and Sigal (1979) noted a lag before effects of fumigation were noticeable in thalli of *Parmelia sulcata* and the more tolerant *Hypogymnia enteromorpha.* As already noted, *Ramalina menziesii* was more tolerant to O_3 than *Parmelia caperata* (Ross and Nash, 1983). These sensitivity rankings show no relationship with those obtained for SO_2, indicating that quite different mechanisms are involved. Eversman and Sigal (1987) demonstrated photosynthetic decline and alterations to ultrastructure in *Parmelia caperata* and *Umbilicaria mammulata*, but no structural damage was found in *Cladonia arbuscula* or *Trebouxia* (photobiont of *C. stellaris*) by Rosentreter and Ahmadjian (1977). Farmer *et al.* (1992), investigating causes of 'acid rain' damage, fumigated some SO_2-sensitive *Lobaria* species for 6 days with realistic O_3 concentrations but detected no increase in electrolyte leakage or inhibition of photosynthesis. Tarhanen, Holopainen and Oksanen (1997) fumigated four epiphytic lichens at 40, 150 and 300 ppb for 2 and 4 weeks and found a slight decrease rather than increase in K^+ leakage, indicating a lack of membrane damage. However, ultrastructural symptoms of damage to the photobiont (starch accumulation, increased opacity of the pyrenoid) were detected in the more sensitive *Bryoria capillaris* and *Usnea hirta.* Ozone may be responsible for changes in the chlorophyll fluorescence parameters of the lichen *Parmelia quercina* observed by Calatayud *et al.* (1999) upon exposure to town air in Spain. These effects appeared to be primarily a result of impairment of the normal protective mechanisms against photooxidative injury, as feeding with the antioxidant ascorbate stimulated electron flow, photochemical quenching and non-radiative dissipation of energy.

In bryophytes, differential effects of exposure to 3.6, 28.3 and 43.2 ppb O_3 were found for several *Sphagnum* species (Gagnon and Karnosky, 1992). Potter *et al.* (1996) also investigated effects on *Sphagnum* and found the greatest sensitivity in *S. recurvum* (= *S. fallax*), the European peat-moss with the highest tolerance to SO_2. Lee *et al.* (1998) screened 22 common British bryophytes with acute (150 ppb) O_3 exposures and detected reduced photosynthesis or increased membrane leakage in only four species. Germination of spores of the moss *Polytrichum* commune was also unaffected by 150 ppb O_3, although that of two ferns was more seriously affected by these acute exposures (Boseley *et al.*, 1998). It is possible that the effects of the oxidant O_3 may be most important when bryophytes are experiencing conditions like desiccation that increase free radical activity within the cells.

Peroxyacetyl nitrate (PAN) is another secondary photochemical pollutant that is particularly associated with southern California. In preliminary studies, Sigal and Taylor (1979) fumigated three lichens with PAN and detected a reduction in photosynthesis in *Hypogymnia enteromorpha* and *Parmelia sulcata* but not in *Collema nigrescens.* Eversman and Sigal (1984) examined chloroplast ultrastructure of the two PAN-sensitive lichens as damage to thylakoids had been previously discovered in vascular plants exposed to this pollutant. In the photobiont cells of both species, following fumigation with 100 ppb PAN, they found evidence of abnormally high starch accumulation and of reduced pyrenoid area, but no major disruption of thylakoid membranes was noted. The results indicated that the main result of PAN treatment was to cause the algal cells to act as if free-living and stop releasing photosynthate to the mycobiont. The effects of O_3 and PAN on cryptogams merit further investigation particularly in developing tropical countries.

Fluorides

In contrast with other pollutants dealt with in this section, it is well documented that airborne fluorides have a devastating effect on lower plants (Gilbert, 1973; Ahmadjian, 1993). They are associated with specific point sources such as aluminium smelters, brick kilns and certain types of chemical works, but have attracted limited attention (Chapter 9). Early investigations around aluminium factories at Fort William, Scotland (Gilbert, 1971) and Arvida, Canada (LeBlanc *et al.*, 1972) showed similar patterns of bryophyte and lichen impoverishment to those caused by SO_2. Cryptogams were entirely absent from trees, fence posts and rocks nearest these sources, but less affected on the ground. These observations have been extended by more recent work by Roberts *et al.* (1979), Roberts and Thompson (1980) and Davies (1982). Perkins *et al.* (1980) and Perkins and Millar (1987a, b) followed spatial and temporal changes in the lichens of rocks and trees around an aluminium reduction plant that opened in 1970 on Anglesey, North Wales. Their papers record many interesting observations about relative sensitivities of species and show positive responses of the lichens as pollutant release was later reduced.

Little is known about the mechanisms by which fluorides cause injury or are tolerated. Belandria *et al.* (1989) found that on a molal basis, dissolved fluoride (NaF) was much less inhibitory to ascospore germination of four lichens than dissolved SO_2. *Lecanora conizaeoides* was the least affected of the lichens by either pollutant. Using transplants, Palomaki *et al.* (1992) showed that F concentrations in *Bryoria fuscescens*, a lichen epiphyte of Norway spruce, fluctuated seasonally, perhaps due to leaching by precipitation. Distinctive types of ultrastructural damage were observed for SO_2 and F, the latter occurring when thallus F reached 30–40 ppm. Fluoride caused swelling and degradation of the thylakoids in photobiont cells and the appearance of numerous lipid droplets in the stroma (Holopainen and Karenlampi, 1985).

Re-establishment with declining SO_2 levels

The re-establishment of cryptogams in regions where the air was formerly polluted with SO_2 is largely a recent phenomenon applying to a small number of developed countries (Chapter 3) where, through socio-economic changes, clean air legislation and increasing use of low-sulphur fuels, emissions have been greatly reduced. The short history of this topic and the reasons for the varied responses to SO_2 abatement are briefly considered in this section.

Early claims of improvements to epiphytic lichen floras with improving air quality have mostly not been corroborated (Gilbert, 1992). The first convincing demonstration of recovery was by Seaward (e.g. 1976, 1982) and Henderson-Sellers and Seaward

(1988) who painstakingly mapped the spread of the lichen *Lecanora muralis* into central Leeds (UK) on various types of masonry with ameliorating SO_2 pollution. Three separate waves of advance were recognized: (1) a vanguard on asbestos cement roofing of mean pH 9.8, which was colonized as the average SO_2 concentration fell below about 200 $\mu g\ m^{-3}$, (2) a consolidating population on mortar and concrete of mean pH 8.35, some 550 m behind the vanguard; (3) a slow occupation of siliceous masonry (e.g. wall capstones) of mean pH 5.25 occurring hundreds of metres behind the other two waves, but accelerating as pollution levels progressively fell. The importance of substratum type here is in line with what is known about effects of pH on ionic forms of dissolved SO_2 and ameliorating properties of Ca^{2+} ions described above. However, there was also some indication that genotypes of *L. muralis* with elevated SO_2 tolerance may have been involved in certain phases of the colonization.

There have been many more reports, some largely anecdotal, of epiphytes recolonizing former territories (Gilbert, 1992). Showman (1981) described recolonization following air quality improvements around a rurally situated coal-fired power plant in the upper Ohio River valley. Previously injured lichens took about 2 years to re-commence growth, and 4 years elapsed before any recolonization was evident, comparable to the lag observed in the studies of *Lecanora muralis*. From 1973 to 1996 the improvements included an increase from 6 to 20 epiphytic species and *Flavoparmelia* (= *Parmelia*) *caperata*, previously absent, had colonized 27 out of 28 sites (Showman, 1997). Recovery occurring on a large scale over 25 years has also been documented (Gunn *et al.*, 1995) for the area around the Sudbury (Ontario, Canada) metal smelters following a 90% reduction in SO_2 emissions. Not only are sensitive epiphytic lichens now frequent in the area, but acidified lakes are also being recolonized by acid-sensitive invertebrates. Marked changes observed in The Netherlands (de Bakker, 1989; Greven, 1992; van Dobben and de Bakker, 1996) are complicated by the de-acidifying effect that increasing NH_3 deposition has had on tree bark. The return of some lichens to the Jardin du Luxembourg in central Paris (Seaward and Letrouit-Galinou, 1991) is particularly symbolic, as it was here that Nylander (1866) made his initial observations on the sensitivity of lichens to atmospheric pollution.

The situation in and around London has been recorded in some detail and is probably typical of other western European cities (Rose and Hawksworth, 1981; Hawksworth and McManus, 1989). In central London annual mean SO_2 concentrations have declined progressively from above 300 $\mu g\ m^{-3}$ in the mid-1960s to around 20 $\mu g\ m^{-3}$ at the present time, with less spectacular declines in the inner and outer suburbs. Lichen colonization commenced around 1970 and 49 epiphytes had been recorded by 1989. At about the same time discoveries of rare epiphytic mosses were being made in London including *Orthotrichum striatum* on Hampstead Heath in 1989 (Adams and Preston, 1992; also see Bates *et al.*, 1997). Among the most rapid recolonizers were some extremely SO_2 sensitive lichens like *Parmelia caperata* and *P. perlata* that were advancing well ahead of more SO_2 tolerant types. Thus, lichen recolonization did not appear to be an orderly advance inwards of the various lichen zones, but instead consisted of more sporadic appearances of species that were thought to have particularly efficient dispersal mechanisms. These were designated ‘zone skippers’ by Hawksworth and McManus (1989). Gilbert (1992) added the name 'zone dawdlers' for common taxa that reinvaded poorly, supposedly because of dispersal limitations (Table 17.6).

In contrast to these observations, little change occurred over the period 1979–90 in the trunk communities of mature pedunculate oaks (*Quercus robur*). Bates *et al.* (1990) pointed out that many of the records of invaders had been primarily from less-acid bark, such as that of willows (*Salix* spp). Utilizing transplant experiments to overcome inherent dispersal limitations, Batty (1997) confirmed that colonization of mature oaks in central London by *Parmelia caperata* was mainly restricted by the low pH of the bark, itself caused by decades of exposure to high SO_2 levels. Low pH favours production of the more toxic forms of dissolved SO_2. In contrast, colonization by *P. saxatilis* in London appears to be restricted primarily by high NO_x levels. A later summary of data from a transect into London (Bates *et al.*, 2001) confirmed that no significant re-colonization of mature oaks by foliose or fruticose lichens had occurred over the period 1979–2000 despite progressively declining SO_2 levels. However, *Lecanora conizaeoides* had contracted towards the city centre, where the concentration of the pollutant was still appreciable, and it had become extinct in the suburbs and surrounding countryside (Figure 17.6). In the absence of significant macro-lichen invasion, the behaviour of *L. conizaeoides* suggests a direct dependence on elevated SO_2 levels rather than a response to competition as suggested by Gilbert (1973).

Bioindication of SO_2 pollution

A large proportion of the literature on cryptogams concerns their use as 'bioindicators' of polluted zones in towns and cities. The main advantage of using cryptogams is that they are indigenous integrators of local atmospheric conditions and thus more likely to provide information that is relevant to other biota (crops, humans) than chemical monitoring gauges. As highly sensitive organisms they may give warning of effects before

Table 17.6 Recolonization by epiphytic lichens: species that recolonize rapidly ('zone skippers') and those that are extremely slow to recolonize ('zone dawdlers') in the UK with declining SO_2 levels (after Hawksworth and McManus, 1989, and Gilbert, 1992)

Rapid colonizers	Poor colonizers
Candelaria concolor	*Calicium* spp.
Evernia prunastri	*Chaenotheca* spp.
Parmelia caperata	*Chrysothrix candelaris*
Parmelia perlata	*Diploicia canescens*
Parmelia revoluta	*Hypocenomyce scalaris*
P. *subrudecta*	*Lecanactis abietina*
Physcia aipolia	*Opegrapha vulgata*
Ramalina farinacea	*Parmelia saxatilis*
Usnea subfloridana	*Parmeliopsis ambigua*
Xanthoria polycarpa	*Pertusaria amara*

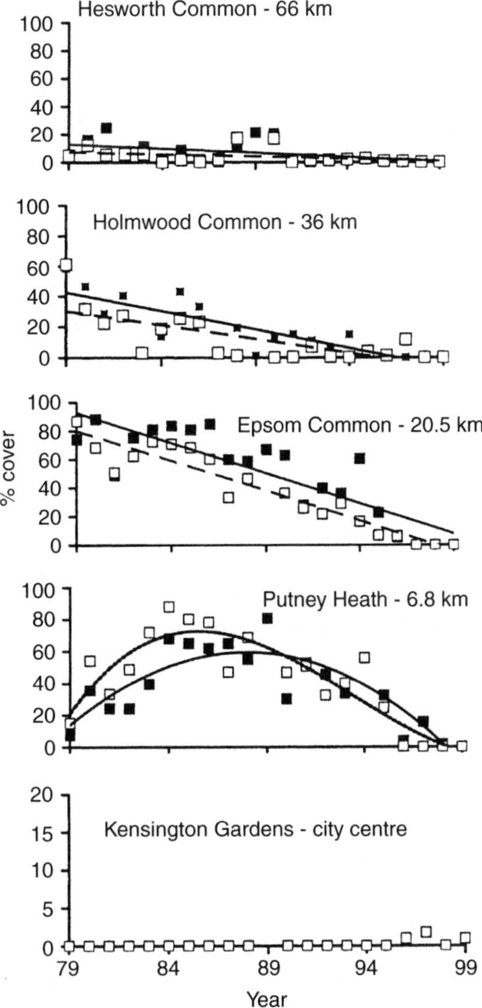

Figure 17.6 Annual changes in the mean cover of *Lecanora conizaeoides* at two heights on trunks of oak at five stations on a transect extending S.S.W. from central London. The lichen has become extinct at the four outermost sites as SO_2 levels have declined, but it has recently colonized the bases of oaks in the city centre. The distance of each station from the city centre is given after the name. □, height 45 cm; ■, height 150 cm. Redrawn from Bates *et al.* (2001)

they reach concentrations that harm larger organisms. The potential usefulness of lichen bioindication in human health, was demonstrated most forcibly by a study (Cislaghi and Nimis, 1997) of the diversity of epiphytic lichens in the Veneto region of northeastern Italy. Lichen biodiversity correlated strongly with levels of lung cancer mortality in young (under 55 years) resident men. The correlation results from underlying causal relationships between levels of atmospheric pollutants (mainly SO_2) and lichen biodiversity on the one hand, and between levels of atmospheric pollutants (probably mainly carcinogens emitted with SO_2) and lung cancer on the other (Cislaghi and Nimis, 1997). Several specialist texts deal with lichen bioindication and biomonitoring methods and the reader is referred to these for further details: Hawksworth and Rose (1976); Nash and Wirth (1988); Richardson (1992); Gries (1996); Nimis *et al.* (2002, in press).

Among the many physiological processes and molecular indicators of SO_2 injury, several have been exploited as potential biomonitoring tools. In lichens, Silberstein *et al.* (1996a) emphasized the need to use an indicator that represented the entire thallus, not just the photobiont, and favoured ATP content. Lurie and Garty (1991) and Garty *et al.* (1995) have promoted ethylene production by lichens as a possible biomonitoring tool for air pollutants. However, the processes that trigger ethylene production in lichens may be complex involving heavy metals as well as SO_2, therefore, this approach may only be valid in particular situations.

References

Adams, K.J. and Preston, C.D. (1992). Evidence for the effects of atmospheric pollution on bryophytes from national and local recording. In: Harding, P.T., ed. *Biological recording of changes in British wildlife*. ITE Symposium no. 26. London: HMSO, 31–43.

Ahmadjian, V. (1993). *The lichen symbiosis*. New York: John Wiley & Sons.

Angold, P.G. (1997). The impact of a road upon adjacent heathland vegetation: effects on plant species composition. *Journal of Applied Ecology* **34**, 409–417.

Arnold. (1892). Zur Lichenenflora von München. *Ber. Bayer. Bot. Ges.* **2**, 2–76.

Aulio, K. (1984). The effects of sulphuric compounds on the ATP content of the peat moss *Sphagnum fuscum*. *Chemosphere* **13**, 915–918.

Baddeley, M.S., Ferry, B.W. and Finegan, E.J. (1973). Sulphur dioxide and respiration in lichens. In: Ferry, B.W., Baddeley, M.S. and Hawksworth, D.L., eds. *Air pollution and lichens*. London: Athlone Press, 299–313.

Barkman, J.J. (1958). *Phytosociology and ecology of cryptogamic epiphytes*. Assen: Van Gorcum.

Barkman. (1969). The influence of air pollution on bryophytes and lichens. In: *Air pollution, proceedings of the First European Congress on the influence of air pollution on plants and animals*. Wageningen: Centre for Agriculture Publishing and Documentation, 197–209.

Bates, J.W. (1993). Regional calcicoly in the moss *Rhytidiadelphus triquetrus*: survival and chemistry of transplants at a formerly SO_2-polluted site with acid soil. *Annals of Botany* **72**, 449–455.

Bates, J.W. (1995). A bryophyte flora of Berkshire. *Journal of Bryology* **18**, 503–620.

Bates, J.W. (1998). Is 'life-form' a useful concept in bryophyte ecology? *Oikos* **82**, 223–237.

Bates, J.W. (2000). Mineral nutrition, substratum ecology and pollution. In: Shaw, A.J. and Goffinet, B.G., eds. *Bryophyte biology*. New York: Cambridge University Press, 248–311.

Bates, J.W., Bell, J.N.B. and Farmer, A.M. (1990). Epiphyte recolonisation of oaks along a gradient of air pollution in south-east England, 1979–1990. *Environmental Pollution* **68**, 81–99.

Bates, J.W., Bell, J.N.B. and Massara, A.C. (2001). Loss of *Lecanora conizaeoides* and other fluctuations of epiphytes on oak in S.E. England over 21 years with declining SO_2 concentrations. *Atmospheric Environment* **35**, 2557–2568.

Bates, J.W., McNee, P.J. and McLeod, A.R. (1996). Effects of sulphur dioxide and ozone on lichen colonization of conifers in the Liphook Forest Fumigation Project. *New Phytologist* **132**, 653–660.

Bates, J.W., Preston, C.D., Proctor, M.C.F., Hodgetts, N.G. and Perry, A.R. (1997). Occurrence of epiphytic bryophytes in a 'tetrad' transect across southern Britain. 1. Geographical trends in abundance and evidence of change. *Journal of Bryology* **19**, 685–714.

Batty, K. (1997). *Population dynamics of epiphytic lichens in relation to changing air quality*. PhD thesis. Imperial College, London.

Baxter, R., Emes, M.J. and Lee, J.A. (1989a). Effects of the bisulphite ion on growth and photosynthesis in *Sphagnum cuspidatum* Hoffm. *New Phytologist* **111**, 457–462.

Baxter, R., Emes, M.J. and Lee, J.A. (1989b). The relationship between extracellular metal accumulation and bisulphite tolerance in *Sphagnum cuspidatum* Hoffm. *New Phytologist* **111**, 463–472.

Baxter, R., Emes, M.J. and Lee, J.A. (1991a). Short term effects of bisulphite on pollution-tolerant and pollution sensitive populations of *Sphagnum cuspidatum* Ehrh. (ex Hoffm.). *New Phytologist* **118**, 425–431.

Baxter, R., Emes, M.J. and Lee, J.A. (1991b). Transition metals and the ability of *Sphagnum* to withstand the phytotoxic effects of the bisulphite ion. *New Phytologist* **118**, 433–439.

Beekley, P.K. and Hoffman, G.R. (1981). Effects of sulfur dioxide fumigation on photosynthesis, respiration, and chlorophyll content of selected lichens. *Bryologist* **84**, 379–390.

Belandria, G., Asta, J. and Nurit, F. (1989). Effects of sulphur dioxide and fluoride on ascospore germination of several lichens. *Lichenologist* **21**, 79–86.

Bell, S., Ashenden, T.W. and Rafarel, C.R. (1992). Effects of rural roadside levels of nitrogen dioxide on *Polytrichum formosum* Hedw. *Environmental Pollution, Series A*, **76**, 11–14.

Beltman, I.H., de Kok, L.J., Kuiper, P.J.C. and van Hasselt, P.R. (1980). Fatty acid composition and chlorophyll content of epiphytic lichens and a possible relation to their sensitivity to air pollution. *Oikos* **35**, 321–326.

Bharali, B. (1999). *Ecophysiological effects of atmospheric pollutants on terricolous mosses*. PhD thesis. Imperial College, London.

Boonpragob, K. and Nash, III T.H. (1990). Seasonal variation of elemental status in the lichen *Ramalina menziesii* Tayl. from two sites in southern California: evidence for dry deposition accumulation. *Environmental and Experimental Botany* **30**, 415–428.

Boonpragob, K. and Nash, III T.H. (1991). Physiological responses of the lichen *Ramalina menziesii* Tayl. to the Los Angeles urban environment. *Environmental and Experimental Botany* **31**, 229–238.

Boonpragob, K., Nash, III T.H. and Fox C.A. (1989). Seasonal deposition patterns of acidic ions and ammonium to the lichen *Ramalina menziesii* Tayl. in southern California. *Environmental and Experimental Botany* **29**, 187–197.

Boseley, A., Petersen, R. and Rebbeck, J. (1998). The resistance of the moss *Polytrichum commune* to acute exposure of simulated acid rain or ozone compared to two fern species: spore germination. *Bryologist* **101**, 512–518.

Bråkenhielm, S. and Quinghong, L. (1995). Spatial and temporal variability of algal and lichen epiphytes on trees in relation to pollutant deposition in Sweden. *Water, Air and Soil Pollution* **79**, 61–74.

Brodo, I.M. (1961). Transplant experiments with corticolous lichens using a new technique. *Ecology* **42**, 838–841.

Brodo, I.M. (1967). Lichen growth and cities: a study on Long Island. *Bryologist* **69**, 427–449.

Brown, D.H. (1992). Impact of agriculture on bryophytes and lichens. In: Bates, J.W. Farmer, A.M., eds. *Bryophytes and lichens in a changing environment*. Oxford: Clarendon Press, 259–283.

Brown, D.H. and Smirnoff, N. (1978). Observations on the effect of ozone on *Cladonia rangiformis. Lichenologist* **10**, 91–94.

Brown, D.H., Standell, C.J. and Miller, J.E. (1995). Effects of agricultural chemicals on lichens. *Cryptogamic Botany* **5**, 220–223.

Brown, D.H. and Tomlinson, H. (1993). Effects of nitrogen salts on lichen physiology. *Bibliotheca Lichenologica* **53**, 27–34.

Bychek-Guschina, I.A., Kotlova, E.R. and Heipieper, H. (1999). Effects of sulfur dioxide on lichen lipids and fatty acids. *Biochemistry, Moscow* **64**, 61–65.

Calatayud, A., Deltoro, V.I., Abadia, A., Abadia, J. and Barreno, E. (1999). Effects of ascorbate feeding on chlorophyll fluoresence and xanthophyll cycle components in the lichen *Parmelia quercina* (Willd.) Vainio exposed to atmospheric pollutants. *Physiologia Plantarum* **105**, 679–684.

Calatayud, A., Sanz, M.J., Calvo, E., Barreno, E. and del Valle-Tascon, S. (1996). Chlorophyll *a* fluorescence and chlorophyll content in *Parmelia quercina* thalli from a polluted region of northern Castellón (Spain). *Lichenologist* **28**, 49–65.

Cislaghi, C. and Nimis, P.L. (1997). Lichens, air pollution and lung cancer. *Nature* **387**, 463–464.

Coppins, B.J. (1973). The 'Drought Hypothesis'. In: Ferry, B.W., Baddeley, M.S and Hawksworth, D.L., eds. *Air pollution and lichens*. London: Athlone Press, 124–142.

Coxson, D.S. (1988). Recovery of net photosynthesis and dark respiration on rehydration of the lichen *Cladonia mitis*, and the influence of prior exposure to sulphur dioxide while desiccated. *New Phytologist* **108**, 483–487.

Davies, F.B.M. (1982). Accumulation of fluoride by *Xanthoria parietina* growing in the vicinity of the Bedfordshire brickfields. *Environmental Pollution Series A* **29**, 189–196.

de Bakker, A.J. (1989). Effects of ammonia emission on epiphytic lichen vegetation. *Acta Botanica Neerlandica* **38**, 337–342.

Deltoro, V.I., Gimeno, C., Calatayud, A. and Barreno, E. (1999). Effects of SO_2 fumigations on CO_2 gas exchange, chlorophyll *a* fluorescence emission and antioxidant enzymes in the lichens *Evernia prunastri* and *Ramalina farinacea. Physiologia Plantarum* **105**, 648–654.

Eversman, S. (1978). Effects of low-level SO_2 on *Usnea hirta* and *Parmelia chlorochroa. Bryologist* **81**, 368–377.

Eversman, S. and Sigal, L.L. (1984). Ultrastructural effects of peroxyacetyl nitrate (PAN) on two lichen species. *Bryologist* **87**, 112–118.

Eversman, S. and Sigal, L.L. (1987). Effects of SO_3, O_3, and SO_2 and O_3 in combination on photosynthesis and ultrastructure of two lichen species. *Canadian Journal of Botany* **65**, 1806–1818.

Farmer, A.M., Bates, J.W. and Bell, J.N.B. (1992). Ecophysiological effects of acid rain on bryophytes and lichens. In: Bates, J.W. and Farmer, A.M., eds. *Bryophytes and lichens in a changing environment*. Oxford: Clarendon Press, 284–313.

Ferguson, P. and Lee, J.A. (1979). The effects of bisulphite and sulphate upon photosynthesis in *Sphagnum. New Phytologist* **82**, 703–712.

Ferguson, P. and Lee, J.A. (1983a). Past and present sulphur pollution in the southern Pennines. *Atmospheric Environment* **6**, 1131–1137.

Ferguson, P. and Lee, J.A. (1983b). The growth of *Sphagnum* species in the southern Pennines. *Journal of Bryology* **12**, 579–586.

Ferguson, P., Lee, J.A. and Bell, J.N.B. (1978). Effects of sulphur pollutants on the growth of *Sphagnum* species. *Environmental Pollution, Series A* **16**, 151–162.

Ferry, B.W., Baddeley, M.S. and Hawksworth, D.L., eds. (1973). *Air pollution and lichens*. London: Athlone Press.

Ferry, B.W. and Coppins, B.J. (1979). Lichen transplant experiments and air pollution studies. *Lichenologist* **11**, 63–73.

Fields, R.D. (1988). Physiological responses of lichens to air pollutant fumigations. In: Nash, T.H., Wirth, V., eds. *Lichens, bryophytes and air quality*. Berlin: J. Cramer, 175–200.

Fields, R.D. and St. Clair, L.L. (1984). The effects of SO_2 on photosynthesis and carbohydrate transfer in the two lichens: *Collema polycarpon* and *Parmelia chlorochroa*. *Bryologist* **87**, 297–301.

Gagnon, Z.E. and Karnosky, D.F. (1992). Physiological response of three species of *Sphagnum* to ozone exposure. *Journal of Bryology* **17**, 81–91.

Garty, J., Cohen, Y. and Kroog, N. (1998). Airborne elements, cell membranes, and chlorophyll in transplanted lichens. *Journal of Environmental Quality* **27**, 973–979.

Garty, J., Kardish, N., Hagemeyer, J. and Ronen, R. (1988). Correlation between the concentration of adenosine tri-phosphate, chlorophyll degradation and the amounts of airborne heavy metals and sulphur in transplanted lichen. *Archives of Environmental Contamination and Toxicology* **17**, 601–611.

Garty, J., Kauppi, M. and Kauppi, A. (1995). Differential responses of certain lichen species to sulfur-containing solutions under acidic conditions as expressed by the production of stress-ethylene. *Environmental Research* **69**, 132–143.

Gilbert, O.L. (1970). Further studies on the effect of sulphur dioxide on lichens and bryophytes. *New Phytologist* **69**, 605–627.

Gilbert, O.L. (1971). The effects of airborne fluorides on lichens. *Lichenologist* **5**, 26–32.

Gilbert, O.L. (1973). The effect of airborne fluorides. In: Ferry, B.W., Baddeley, M.S. and Hawksworth, D.L., eds. *Air pollution and lichens*. London: Athlone Press, 176–191.

Gilbert, O.L. (1992). Lichen reinvasion with declining air pollution. In: Bates, J.W. Farmer, A.M., eds. *Bryophytes and lichens in a changing environment*. Oxford: Clarendon Press, 159–177.

Gonzalez, C.M. and Pignata, M.L. (1994). The influence of air pollution on soluble proteins, chlorophyll degradation, MDA, sulfur and heavy metals in a transplanted lichen. *Chemistry and Ecology* **9**, 105–113.

Gonzalez, C.M. and Pignata, M.L. (1999). Effect of pollutants emitted by different urban-industrial sources on the chemical response of the transplanted *Ramalina ecklonii* (Spreng.) Mey. & Flot. *Toxicological and Environmental Chemistry* **69**, 61–73.

Gonzalez, C.M., Orellana, L.C., Casanovas, S.S. and Pignata, M.L. (1998). Environmental conditions and chemical response of a transplanted lichen to an urban area. *Journal of Environmental Management* **53**, 73–81.

Greven, H.C. (1992). *Changes in the Dutch bryophyte flora and air pollution*. Dissertationes Botanicae, Band 194. Berlin & Stuttgart: J. Cramer.

Gries, C. (1996). Lichens as indicators of air pollution. In: Nash, III T.H., ed. *Lichen biology*. Cambridge: Cambridge University Press, 240–254.

Gries, C., Sanz, M.-J., Romagni, J.G., Goldsmith, S., Kuhn, U., Kesselemeier, J. and Nash, III T.H. (1997). The uptake of gaseous sulphur dioxide by non-gelatinous lichens. *New Phytologist* **135**, 595–602.

Grindon, L.H. (1859). *The Manchester flora*. London: W. White.

Gunn, J., Keller, W., Negusanti, J., Potvin, R., Beckett, P. and Winterhalder, K. (1995). Ecosystem recovery after emission reductions: Sudbury, Canada. *Water, Air and Soil Pollution* **85**, 1783–1788.

Hällgren, J.-E. and Huss, K. (1975). Effects of SO_2 on photosynthesis and nitrogen fixation. *Physiologia Plantarum* **34**, 171–176.

Hart, R., Webb, P.G., Biggs, R.H. and Portier, K.M. (1988). The use of lichen fumigation studies to evaluate the effects of new emission sources on Class 1 areas. *JAPC* **38**, 144–147.

Haugsjå. (1930). Über den Einfluß der Stadt Oslo auf die Flechten-vegetation der Bäume. *Nyt. Mag. Naturvidensk* **68**, 1–116.

Hawksworth, D.L. (1973). Mapping studies. In: Ferry, B.W., Baddeley, M.S. and Hawksworth, D.L., eds. *Air pollution and lichens*. London: Athlone Press, 38–76.

Hawksworth, D.L. and Chapman, D.S. (1971). *Pseudevernia furfuracea* (L.) Zopf and its chemical races in the British Isles. *Lichenologist* **5**, 51–58.

Hawksworth, D.L. and McManus, P.M. (1989). Lichen recolonization in London under conditions of rapidly falling sulphur dioxide levels, and the concept of zone skipping. *Botanical Journal of the Linnean Society* **100**, 99–109.

Hawksworth, D.L. and Rose, F. (1976). *Lichens as pollution monitors*. London: Arnold.

Hawksworth, D.L., Rose, F. and Coppins, B.J. (1973). Changes in the lichen flora of England and Wales attributable to pollution of the air by sulphur dioxide. In: Ferry, B.W., Baddeley, M.S. and Hawksworth, D.L., eds. *Air pollution and lichens*. London: Athlone Press, 330–367.

Hébant, C. (1977). *The conducting tissues of bryophytes*. Bryophytorum Bibliotheca, 10. Vaduz: J. Cramer.

Henderson, A. (2000). Literature on air pollution and lichens XLIX. *Lichenologist* **32**, 89–102.

Henderson-Sellers, A. and Seaward, M.R.D. (1979). Monitoring lichen reinvasion of ameliorating environments. *Environmental Pollution* **19**, 207–213.

Hill, D.J. (1971). Experimental study of the effect of sulphite on lichens with reference to atmospheric pollution. *New Phytologist* **70**, 831–836.

Hill, M.O., Preston, C.D. and Smith, A.J.E. (1994). Atlas of the bryophytes of Britain and Ireland. Volume 3. Mosses (*Diplolepideae*). Colchester: Harley Books.

Holopainen, T.H. (1984). Cellular injuries in epiphytic lichens transplanted to air polluted areas. *Nordic Journal of Botany* **4**, 393–408.

Holopainen, T.H. and Kärenlampi, L. (1984). Injuries to lichen ultrastructure caused by sulphur dioxide fumigations. *New Phytologist* **98**, 285–294.

Holopainen, T.H. and Kärenlampi, L. (1985). Characteristic ultrastructural symptoms caused in lichens by experimental exposure to nitrogen compounds and fluorides. *Annales Botanici Fennici* **22**, 333–342.

Holopainen, T.H. and Kauppi, M. (1989). A comparison of light, fluorescence and electron microscopic observations in assessing the SO_2 injury of lichens under different moisture conditions. *Lichenologist* **21**, 119–134.

Hufnagel, G. and Türk, R. (1998). Monitoring of air pollutants by exposed lichens in Salzburg (Austria). *Sauteria* **9**, 281–288.

Ikonen, S. and Kärenlampi, L. (1976). Physiological and structural changes in reindeer lichens transplanted around a sulphite pulp mill. In: Kärenlampi, L., ed. *Plant damage caused by air pollution*. Kuopio: University of Kuopio, 37–45.

Inglis, F. and Hill, D.J. (1974). The effect of sulphite and fluoride on carbon dioxide uptake by mosses in the light. *New Phytologist* **73**, 1207–1213.

Johnsen, I. and Søchting, U. (1973). Influence of air pollution on the epiphytic lichen vegetation and bark properties of deciduous trees in the Copenhagen area. *Oikos* **24**, 344–351.

Kardish, N., Ronen, R., Bubrick, P. and Garty, J. (1987). The influence of air pollution on the concentration of ATP and on chlorophyll degradation in the lichen *Ramalina duriaei* (De Not.) Bagl. *New Phytologist* **106**, 697–706.

Kauppi, M. (1976). Fruticose lichen transplant technique for air pollution experiments. *Flora* (Jena) **165**, 407–414.

Kauppi, M. (1980). Fluorescence microscopy and microfluorimetry for the examination of pollution damage in lichens. *Annales Botanici Fennici* **17**, 163–173.

LeBlanc, F. and Rao, D.N. (1966). Réaction de quelques lichens et mousses épiphytiques à l'anhydride sulfureux dans la région de Sudbury, Ontario. *Bryologist* **69**, 338–346.

LeBlanc, F. and Rao, D.N. (1973). Evaluation of the pollution and drought hypotheses in relation to lichens and bryophytes in urban environments. *Bryologist* **76**, 1–19.

LeBlanc, F., Rao, D.N. and Comeau, G. (1972). Indices of atmospheric purity and fluoride pollution pattern in Arvida, Quebec. *Canadian Journal of Botany* **50**, 991–998.

Lee, J.A. and Studholme, C.J. (1992). Responses of *Sphagnum* species to polluted environments. In: Bates, J.W. and Farmer, A.M., eds. *Bryophytes and lichens in a changing environment*. Oxford: Clarendon Press, 314–332.

Lee, J.A., Caporn, S.J.M., Carroll, J., Foot, J.P., Johnson, D., Potter, L. and Taylor, A.F.S. (1998). Effects of ozone and atmospheric nitrogen deposition on bryophytes. In: Bates, J.W., Ashton, N.W. Duckett, J.G., eds. *Bryology for the twenty-first century*. Leeds: Maney and The British Bryological Society, 331–341.

Lurie, S. and Garty, J. (1991). Ethylene production by the lichen *Ramalina duriaei*. *Annals of Botany* **68**, 317–319.

Malhotra, S.S. and Khan, A.A. (1983). Sensitivity to SO_2 of various metabolic processes in an epiphytic lichen, *Evernia mesomorpha*. *Biochemie und Physiologie der Pflanzen* **178**, 121–130.

Marti, J. (1983). Sensitivity of lichen phycobionts to dissolved air pollutants. *Canadian Journal of Botany* **61**, 1647–1653.

Miller, J.E. and Brown, D.H. (1999). Studies of ammonia uptake and loss by lichens. *Lichenologist* **31**, 85–93.

Miszalski, Z. and Niewiadomska, E. (1993). Comparison of sulphite oxidation mechanisms in three lichen species. *New Phytologist* **123**, 345–349.

Modenesi, P. (1993). An SEM study of injury symptoms in *Parmotrema reticulatum* treated with paraquat or growing in sulphur dioxide-polluted air. *Lichenologist* **25**, 423–433.

Morgan, S.M., Lee, J.A. and Ashenden, T.W. (1992). Effects of nitrogen oxides on nitrate assimilation in bryophytes. *New Phytologist* **120**, 89–97.

Moser, T.J., Nash, III T.H. and Clark, W.D. (1980). Effects of a long-term field sulfur dioxide fumigation on Arctic caribou forage lichens. *Canadian Journal of Botany* **58**, 2235–2240.

Nash, III T.H. (1988). Correlating fumigation studies with field effects, In: Nash, III T.H. and Wirth, S., eds. *Lichens, bryophytes and air quality*. Bibliotheca Lichenologia, 30. Berlin: J. Cramer, 201–216.

Nash, III T.H. and Sigal, L.L. (1979). Gross photosynthetic response of lichens to short-term ozone fumigations. *Bryologist* **82**, 280–285.

Nash, III T.H. and Wirth, V., eds. (1988). *Lichens, bryophytes and air quality*. Berlin: J. Cramer.

Nieboer, E., Richardson, D.H.S., Lavoie, P. and Padovan, D. (1979). The role of metal-ion binding in modifying the toxic effects of sulphur dioxide on the lichen *Umbilicaria muhlenbergii*. I. Potassium efflux studies. *New Phytologist* **82**, 621–632.

Nieboer, E., Tomassini, F.D., Puckett, K.J. and Richardson, D.H.S. (1977). A model for the relationship between gaseous and aqueous concentrations of sulphur dioxide in lichen exposure studies. *New Phytologist* **79**, 157–162.

Nimis, P.L., Wolseley, P.A. and Scheidegger, C., eds. (2002). *Lichen monitoring*. Kluwer, in press.

Nylander, W. (1866). Les lichens du Jardin du Luxembourg. *Bulletin de la Société Botanique de France* **13**, 364–372.

O'Hare, G.P. and Williams, P. (1975). Some effects of sulphur dioxide flow on lichens. *Lichenologist* **7**, 116–120.

Palomaki, V., Tynnyrinen, S. and Holopainen, T. (1992). Lichen transplantation in monitoring fluoride and sulfur deposition in the surroundings of a fertilizer plant and a strip mine at Siilinjarvi. *Annales Botanici Fennici* **29**, 25–34.

Pearson, L.C. and Henriksson, E. (1981). Air pollution damage to cell membranes in lichens. II. Laboratory experiments. *Bryologist* **84**, 515–520.

Perkins, D.F., Millar, R.O. and Neep, P.E. (1980). Accumulation of airborne fluoride by lichens in the vicinity of an aluminium reduction plant. *Environmental Pollution Series A* **21**, 155–168.

Perkins, D.F. and Millar, R.O. (1987a). Effects of airborne fluoride emissions near an aluminium works in Wales. Part 1. Corticolous lichens growing on broadleaved trees. *Environmental Pollution* **47**, 63–78.

Perkins, D.F. and Millar, R.O. (1987b). Effects of airborne fluoride emissions near an aluminium works in Wales. Part 2. Saxicolous lichens growing on rocks and walls. *Environmental Pollution* **48**, 185–196.

Poikolainen, J., Lippo, H., Hongisto, M., Kubin, E., Mikkola, K. and Lindgren, M. (1998). On the abundance of epiphytic green algae in relation to the nitrogen concentrations of biomonitors and nitrogen deposition in Finland. *Environmental Pollution* **102**, 85–92.

Potter, L., Foot, J.P., Caporn, S.J.M. and Lee, J.A. (1996). Responses of four *Sphagnum* species to acute ozone fumigation. *Journal of Bryology* **19**, 19–32.

Proctor, M.C.F. (1984). Structure and ecological adaptation. In: Dyer, A.F. and Duckett, J.G., eds. *The experimental biology of bryophytes*. London: Academic Press, 9–37.

Puckett, K.J., Nieboer, E., Flora, W.P. and Richardson, D.H.S. (1973). Sulphur dioxide: its effect on photosynthetic ^{14}C fixation in lichens and suggested mechanisms of phytotoxicity. *New Phytologist* **72**, 141–154.

Puckett, K.J., Richardson, D.H.S., Flora, W.P. and Nieboer, E. (1974). Photosynthetic ^{14}C fixation by the lichen *Umbilicaria muhlenbergii* (Ach.) Tuck. following short exposures to aqueous sulphur dioxide. *New Phytologist* **73**, 1183–1192.

Puckett, K.J., Tomassini, F.D., Nieboer, E. and Richardson, D.H.S. (1977). Potassium efflux by lichen thalli following exposure to aqueous sulfur-dioxide. *New Phytologist* **79**, 135–145.

Richardson, D.H.S. (1992). *Pollution monitoring with lichens*. Naturalists' Handbooks 19. Slough: Richmond Publishing.

Richardson, D.H.S. (1988). Understanding the pollution sensitivity of lichens. *Botanical Journal of the Linnean Society* **96**, 31–43.

Richardson, D.H.S. and Nieboer, E. (1983). Ecophysiological responses of lichens to sulphur dioxide. *Journal of the Hattori Botanical Laboratory* **54**, 331–351.

Richardson, D.H.S., Nieboer, E., Lavoie, P. and Padovan, D. (1979). The role of metal-ion binding in modifying the toxic effects of sulphur dioxide on the lichen *Umbilicaria muhlenbergii*. II. C^{14}-fixation studies. *New Phytologist* **82**, 633–643.

Roberts, B.A. and Thompson, L.K. (1980). Lichens as indicators of fluoride emission from a phosphorus plant, Long Harbour, Newfoundland, Canada. *Canadian Journal of Botany* **58**, 2218–2228.

Roberts, B.A., Thompson, L.K. and Sidhu, S.S. (1979). Terrestrial bryophytes as indicators of fluoride emissions from a phosphorus plant, Long Harbour, Newfoundland, Canada. *Canadian Journal of Botany* **57**, 1583–1590.

Rose, C.I. and Hawksworth, D.L. (1981). Lichen recolonization in London's cleaner air. *Nature* **289**, 289–292.

Rose, F. and Wallace, E.C. (1974). Changes in the bryophyte flora of Britain. In: Hawksworth, D.L., ed. *The changing flora and fauna of Britain*. London: Academic Press, 27–46.

Rosentreter, R. and Ahmadjian, V. (1977). Effect of ozone on the lichen *Cladonia arbuscula* and the *Trebouxia* phycobiont of *Cladina stellaris*. *Bryologist* **80**, 600–605.

Ross, L.J. and Nash, III T.H. (1983). Effect of ozone on gross photosynthesis of lichens. *Environmental and Experimental Botany* **23**, 71–77.

Sanchez-Hoyos, M.A. and Manrique, E. (1995). Effect of nitrate and ammonium ions on the pigment content (xanthophylls, carotenes and chlorophylls) of *Ramalina capitata*. *Lichenologist* **27**, 155–160.

Sanz, M.J., Gries, C. and Nash, III T.H. (1992). Dose-response relationships for SO_2 fumigations in the lichen *Evernia prunastri* (L.) Ach. *New Phytologist* **122**, 313–319.

Saunders, P.J.W. and Wood, C.M. (1973). Sulphur dioxide in the environment: its production, dispersal and fate. In: Ferry, B.W., Baddeley, M.S. and Hawksworth, D.L., eds. *Lichens and air pollution*. London: Athlone Press, 6–37.

Seaward, M.R.D. (1976). Performance of *Lecanora muralis* in an urban environment. In: Brown, D.H., Hawksworth, D.L. and Bailey, R.H., eds. *Lichenology: progress and problems*. London: Academic Press, 323–357.

Seaward, M.R.D. (1982). Lichen ecology of changing urban environments. In: Bornkamm, R., Lee, J.A. and Seaward, M.R.D., eds. *Urban ecology: 2nd European ecological symposium*. Oxford: Blackwell Scientific, 181–189.

Seaward, M.R.D. and Hitch, C.J.B. (1982). *Atlas of the lichens of the British Isles*. Volume 1. Cambridge: Institute of Terrestrial Ecology.

Seaward, M.R.D. and Letrouit-Galinou, M.A. (1991). Lichens return to the Jardin du Luxembourg after an absence of almost a century. *Lichenologist* **23**, 118–126.

Sernander, R. (1926). *Stockhoms natur*. Uppsala: Almquist & Wiksells.

Sharma, P., Bergman, B., Hällbom, L. and von Hofsten, A. (1983). Ultrastructural changes of *Nostoc of Peltigera canina* in presence of SO_2. *New Phytologist* **92**, 573–579.

Sheridan, R.P., Sanderson, C. and Kerr, R. (1976). Effects of pulpmill emissions on lichens in the Missoula valley, Montana. *Bryologist* **79**, 248–252.

Showman, R.E. (1981). Lichen recolonization following air quality improvement. *Bryologist* **84**, 492–497.

Showman, R.E. (1997). Continuing lichen recolonization in the upper Ohio River valley. *Bryologist* **100**, 478–481.

Sigal, L.L. and Nash, III T.H. (1983). Lichen communities on conifers in southern California mountains: an ecological survey relative to oxidant air pollution. *Ecology* **64**, 1343–1354.

Sigal, L.L. and Taylor, O.C. (1979). Preliminary studies of the gross photosynthetic response of lichens to peroxyacetylnitrate fumigations. *Bryologist* **82**, 564–575.

Silberstein, L., Siegel, B.Z., Siegel, S.M., Mukhtar, A. and Galun, M. (1996a). Comparative studies on *Xanthoria parietina*, a pollution-resistant lichen, and *Ramalina duriaei*, a sensitive species. I. Effects of air pollution on physiological processes. *Lichenologist* **28**, 355–365.

Silberstein, L., Siegel, B.Z., Siegel, S.M., Mukhtar, A. and Galun, M. (1996b). Comparative studies on *Xanthoria parietina*, a pollution-resistant lichen, and *Ramalina duriaei*, a sensitive species. II. Evaluation of possible air pollution-protection mechanisms. *Lichenologist* **28**, 367–383.

Skye, E. (1958). Luftföroreningars inverkan på busk- och bladlavfloran kring skifferoljeverket i Närkes Kvarntorp. *Svensk Botaniser Tiddskrift* **52**, 133–190.

Skye, E. (1968). Lichens and air pollution. *Acta Phytogeographica Suecica* **52**, 1–123.

Søchting, U. (1997). Epiphyllic cover on spruce needles in Denmark. *Annales Botanici Fennici* **34**, 157–164.

Syratt, W.J. and Wanstall, P.J. (1969). The effects of sulphur dioxide on epiphytic bryophytes. In: *Proceedings of the First European Congress on the Influence of Air Pollution on Plants and Animals*. Wageningen: Centre for Agricultural Publishing and Documentation, 79–85.

Tallis, J.H. (1964). Studies on the Southern Pennine peats. II. The behaviour of Sphagnum. *Journal of Ecology* **52**, 345–353.

Tarhanen, S., Holopainen, T. and Oksanen, J. (1997). Ultrastructural changes and electrolyte leakage from ozone fumigated epiphytic lichens. *Annals of Botany* **80**, 611–621.

Tuba, Z., Csintalan, Z., Badacsonyi, A. and Proctor, M.C.F. (1997). Chlorophyll fluorescence as an exploratory tool for ecophysiological studies on mosses and other small poikilohydric plants. *Journal of Bryology* **19**, 401–407.

Türk, R. and Wirth, V. (1975). The pH dependence of SO_2 damage to lichens. *Oecologia* **19**, 285–291.

van Dobben, H.F. and de Bakker, A.J. (1996). Re-mapping epiphytic lichen biodiversity in The Netherlands: effects of decreasing SO_2 and increasing NH_3. *Acta Botanica Neerlandica* **45**, 55–71.

van Dobben, H.F. and ter Braak, C.J.F. (1999). Ranking of epiphytic lichen sensitivity to air pollution using survey data: a comparison of indicator scales. *Lichenologist* **31**, 27–39.

Vincent, J.P. (1990). Influence d'une atmosphere urbaine sur les differentes fonctions d'une espèce lichenique. Etude 'in situ' et en chambrettes experimentales. *Science of the Total Environment* **95**, 167–180.

von Arb, C. and Brunold, C. (1990). Lichen physiology and air pollution. I. Physiological responses of in situ *Parmelia sulcata* among air pollution zones within Biel, Switzerland. *Canadian Journal of Botany* **68**, 35–42.

von Arb, C., Mueller, C., Amman, K. and Brunold, C. (1990). Lichen physiology and air pollution. II. Statistical analysis of the correlation between SO_2, NO_2, NO and O_3, and chlorophyll content, net photosynthesis, sulphate uptake and protein synthesis of *Parmelia sulcata* Tayl. *New Phytologist* **115**, 431–438.

Winner, W.E. (1988). Responses of bryophytes to air pollution. In: Nash, III T.H. and Wirth, V., eds. *Lichens, bryophytes and air quality*. Berlin: J. Cramer, 141–173.

Winner, W.E., Atkinson, C.J. and Nash, III T.H. (1988). Comparisons of SO_2 absorption capacities of mosses, lichens, and vascular plants in diverse habitats. In: Nash III, T.H. and Wirth, V., eds. *Lichens, bryophytes and air quality*. Berlin: J. Cramer, 217–230.

Winner, W.E. and Bewley, J.D. (1978). Contrasts between bryophyte and vascular plant synecological responses in an SO_2-stressed white spruce association in Central Alberta. *Oecologia* (Berlin) **33**, 311–325.

Winner, W.E. and Bewley, J.D. (1983). Photosynthesis and respiration of feather mosses fumigated at different hydration levels with SO_2. *Canadian Journal of Botany* **61**, 1456–1461.

Winner, W.E. and Kock, G.W. (1982). Water relations and SO_2 resistance of mosses. *Journal of the Hattori Botanical Laboratory* **52**, 431–440.

Wirth, V. and Brinckmann, B. (1977). Statistical analysis of the lichen vegetation of an avenue in Freiburg (South-West Germany), with regard to injurious anthropogenous influences. *Oecologia* (Berlin) **28**, 87–101.

Ziegler, I. (1977). Sulfite action on ribulosediphosphate carboxylase in the lichen *Pseudevernia furfuracea*. *Oecologia* **29**, 63–66.

18 Modification of plant response by environmental conditions

G. MILLS

Introduction

The growth, development and reproductive potential of a plant are closely coupled to the environmental conditions in which it lives. Where one or more of the essential conditions for growth (light, water, nutrients etc.) are in short supply or in excess, the plant may exhibit a 'stress' response such as anthocyanin pigmentation of the leaves, reduced growth, or leaf and/or fruit abscission. Whether the plant is adversely affected by stress or healthy, the conditions in which it lives will influence both the uptake (flux) of a gaseous pollutant via the stomata and the ability of the plant to detoxify the pollutant and repair any damage it has induced. For example, some environmental conditions cause stomatal closure and effectively protect the plant against pollutant damage, whilst in other conditions, the closing mechanism of the guard cells is damaged, leading to increased pollutant uptake and consequent effect. By studying the modifying influence of the environment on plant responses to pollutants it becomes possible to predict how plants will respond in the cocktail of environmental conditions associated with the open field. Such information is of particular relevance for evaluation of the impacts of global climate change and for the modelling of pollutant effects across continents.

To provide a comprehensive review of all of the literature on the effects of environmental conditions on the separate responses of trees, crops, natural vegetation and lower plants, to all of the major pollutants would require far more space than is available here. In the first part of this chapter the influences of the most important climatic and soil factors on the responses to the gaseous pollutants O_3, NO_x, NH_3 and SO_2 are discussed. For each factor, the pollutant and vegetation groups that are of greatest relevance are considered in the most detail. Later in the chapter, the ways in which multiple co-occurring factors influence responses to pollutants are examined using the increased sensitivity of plants to SO_2 during the winter as an example.

Modifying effect of individual environmental factors

Many environmental factors influence the magnitude of responses of plants by altering the flux of pollutants into the plant. As pollutant flux is considered in detail in

Air Pollution and Plant Life, second edition. Edited by J.N.B. Bell and M. Treshow. ISBN 0 471 49090 3 (HB), 0 471 49091 1 (PB).

Chapter 4, this chapter will focus on the consequences for the plant of altered flux, rather than the mechanisms by which flux is altered. The conditions associated with maximum pollutant flux via the stomata of most temperate plants can be summarised as: high humidity (low vapour pressure deficit), non-limiting water supply, saturating light levels, temperatures in the range 20–32 °C and moderate wind speed (Grüters *et al.*, 1994; Grünhage and Jager, 1994; Grünhage *et al.*, 1999). Environmental conditions can also modify responses to pollutants by influencing growth and energy availability for repair mechanisms. The net effect of altered flux and/or growth is manifested as a changed response to the pollutant relative to that expected in optimum conditions for plant growth. The key conditions that modify the responses to pollutants have been separated into climatic and soil factors for ease of discussion.

Climatic factors

Temperature

Temperature affects the rate of metabolic processes in plants such as photosynthesis, respiration and growth as well as the flux of pollutants to the plant. However, its modifying influence on the photosynthetic response to O_3, SO_2 and NO_x appears to be both species and pollutant specific. For example, the sensitivity of photosynthesis to SO_2 increased with increasing temperature in barley, rape, bean and oats but decreased with increasing temperature in tobacco and rice (Black, 1982).

Comparisons of the responses to pollutants in 'summer' versus 'winter' temperature regimes have demonstrated the modifying influence of this factor. For instance, Foot *et al.* (1996) reported negative effects of ozone on the root growth of heather (*Calluna vulgaris*) during winter exposure to ozone (mean 6.8 °C) but not during summer exposure (mean 12.3 °C). Using the same exposure facilities, the growth of the bryophyte *Polytrichum commune* was also found to be significantly reduced by ozone exposure in winter (mean 6.4 °C) but not summer conditions (mean 15 °C) (Potter *et al.*, 1996). However, the reverse occurred for *Sphagnum recurvum*, showing that generalisations cannot be made. Nevertheless, controlled environment studies do offer some support to the suggestion of increased sensitivity to ozone at lower temperatures. For example, by changing the day/night temperature regime from 20/16 °C to 15/11 °C, the sensitivity of *Phleum pratense* to 80 ppb O_3 was doubled (Mortensen and Nilsen, 1992). Kleier *et al.* (1998), found a similar effect when rapid cycling brassica (*Brassica rapa* L.) was exposed to ozone at root temperatures of 13 °C and 18 °C. They postulated that at the higher root temperature, plants were able to compensate for O_3-induced reductions in photosynthesis and could thus avoid a decrease in biomass and reproductive output.

Several researchers have taken advantage of naturally occurring periods of temperature stress to study possible modifying effects on the responses of plants to pollutants. For example, a period of high temperatures (35–40 °C) coincided with exposure of wheat to 110 ppb ozone during anthesis (Meyer *et al.*, 1997), a growth stage that is known to be particularly sensitive to O_3 (Soja *et al.*, in press). Reductions in the quantum yield of photosynthesis were greater during this period of heat stress than in another set of wheat plants exposed to ozone at more moderate temperatures (20–25 °C) during anthesis. In a different study, periods of very low temperatures (down to −6 °C and <0 °C for 2 weeks) occurred during long-term exposure of heather (*Calluna vulgaris*) to ozone at 70 ppb in open-top chambers (Foot *et al.*, 1997). Severe damage (greater than 50% of the shoots dark brown or blackened) was recorded on 35% of the plants grown in ozone, whereas injury was either absent or more limited on the filtered-air plants. No such injury had been detected prior to the period of sub-zero temperatures. Enhancement of injury during the winter has been more widely reported for SO_2 than for NO_x and O_3. The causes are considered later as an example of how several interacting factors can influence responses to pollutants.

Humidity

The humidity of the air surrounding the leaves of a plant has a profound effect on stomatal conductance and hence the flux of pollutants into the plant (see Chapter 4). Thus, the effects of humidity are most clearly demonstrated when measuring short-term responses such as changes in net photosynthesis and the development of visible injury on leaves. Most recent research has focussed on the influence of humidity on the response to ozone. This is because the highest ozone concentrations often coincide with high vapour pressure deficit, VPD (= low relative humidity), and therefore occur when stomatal conductance is relatively low (Wieser and Havranek, 1993; Thoene *et al.*, 1996). Data from the ICP Vegetation, an international research programme monitoring the effects of ambient ozone on plants in Europe and the USA, can be used to illustrate the significance of this co-occurrence. When the data from all 35 sites were combined, it emerged that at lower VPD less ozone was required to cause chlorotic injury on white clover (*Trifolium repens*) than at higher VPD (Figure 18.1, Benton *et al.*, 2000). For example, the extent of injury at the second 28 d harvest of the 1999 growth season was the same on clover grown at a site in Italy as at one in Sweden (Mills, unpublished). However, injury first occurred in harvest interval 1–2 following a 5 d AOT40[1] of 1330 ppb.h and VPD of 1.57 kPa at the Italian site and after a much lower 5 d AOT40 of 405 ppb.h at the more humid Swedish site, where the corresponding VPD was 0.33 kPa. Thus, the same amount of injury was caused by one-third as much ozone at the more humid site. The strong influence of VPD on the extent of ozone injury development has also been shown in laboratory experiments and biomonitoring with other clover species (Balls *et al.*, 1996, Tonneijck and van Dijk, 1997) and with tobacco in Italy (Biondi *et al.*, 1992) and Spain (Ribas *et al.*, 1998). Concordant results were obtained in open-top chamber studies: VPD was the most important non-plant factor influencing the extent of leaf injury on ozone-sensitive grassland species (Bungener *et al.*, 1999a) and bean (Thompson *et al.*, 1992).

Although fewer studies have been conducted for other gaseous pollutants, their potential for impact is also partially controlled by the effect of VPD on stomatal conductance. For example, Kropff *et al.* (1990) showed that the photosynthetic rate of field bean (*Vicia faba*) was reduced more by SO_2 at low VPD than at high VPD due to a higher rate of uptake.

[1] AOT40 (accumulated dose over a threshold of 40 ppb) is the sum of the differences between the hourly mean ozone concentration (in ppb) and 40 ppb for each hour when the concentration exceeds 40 ppb, accumulated during daylight hours. Units ppb.h.

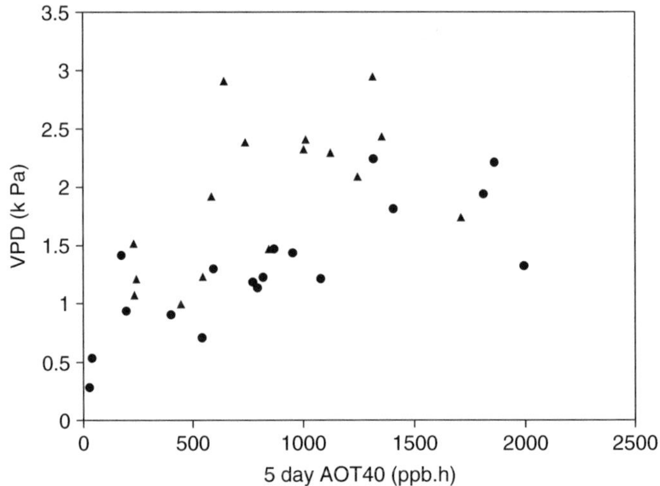

Figure 18.1 The AOT40 (accumulated during daylight hours) and the mean vapour pressure deficit (0930–1630h) during the 5 days preceding the appearance of ozone injury on the leaves of *Trifolium* species grown outdoors at experimental sites in 15 countries of Europe. The data are from the 1995 (▲) and 1996 (●) summer experiments of the ICP Vegetation. Redrawn from Benton *et al.*, 2000. Reprinted from *Agric. Ecosys. Environ.*, **78**, Benton *et al.*, 'An international cooperative programme indicates the widespread occurrence of ozone injury on crops', 19–30, 2000, with permission of Elsevier Science

Light

Light levels can modify the response to pollutants via instantaneous effects on the rates of stomatal conductance and photosynthesis, or by longer term effects on the growth and development of plants. Farage and Long (1999) reported a typical response where the light-saturated rate of photosynthesis (A_{sat}) of wheat (*Triticum aestivum*) was 35% lower in the 80 ppb O_3 treatment than in the 5 ppb control. Although it cannot easily be determined whether the light level changed the response to ozone or vice versa, such combinations of effects are important in tree canopies, where light levels can vary from full sunlight at the top of the canopy down to deep shade for the lower leaves. Tjoelker *et al.* (1995) exposed individual branches of an 18 m tall stand of sugar maple (*Acer saccharum*) to ozone at 95 ppb and monitored the response in branches along a light gradient of 14.5 to 0.7 mol m^{-2} d^{-1}. The shaded leaves exhibited greater ozone-induced decreases in A_{sat} than the sun-leaves (Figure 18.2, Tjoelker *et al.*, 1995). This effect was not caused by a difference in stomatal conductance and thus was not flux dependent. Compared with the sun-leaves, the shaded leaves exhibited greater proportional decreases in A_{sat}, and had lower chlorophyll contents and quantum efficiencies when exposed to ozone than the controls. In an earlier study, sugar maple seedlings were more sensitive to 99–115 ppb ozone in growth chambers when grown in shade conditions rather than full sunlight, with effects such as reduced root growth, decreased leaf area:plant dry weight and increased leaf respiration being noted (Tjoelker *et al.*, 1993). More recently, Mortensen (1999) showed that there was

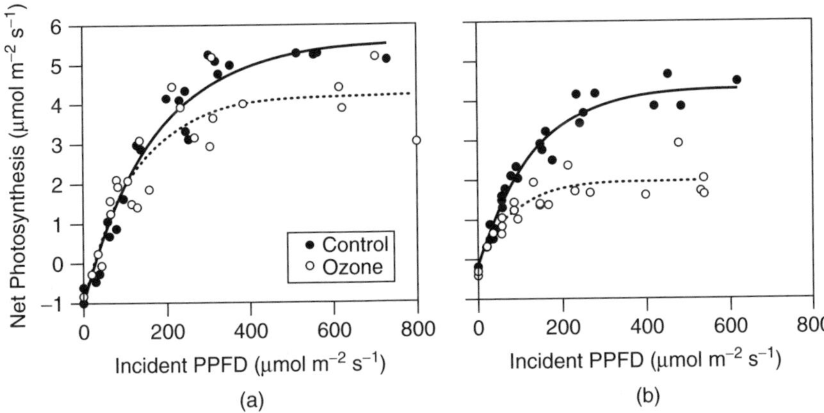

Figure 18.2 The effect of ozone on the light response of photosynthesis in leaves of (a) a sun branch (PPFD 6.9 mol m^{-2} d^{-1}) and (b) a deep shade branch (PPFD 0.8 mol m^{-2} d^{-1}) exposed to ozone (○) compared to controls (•) after 56 d of ozone exposure ($n = 4$ leaves for each branch and each treatment). The non-linear regressions differed between ozone-treated and control leaves within sun ($P = 0.003$) and shade ($P < 0.001$) leaf groups. Redrawn from Tjoelker *et al.*, 1995. Reproduced by permission of Blackwell Science Ltd

greater injury on mountain birch (*Betula pubescens*) when exposed to 78 ppb of ozone in 50% shade than there was on the seedlings grown in full sunlight.

The modifying influence of light on the responses of plants to SO_2 during the winter is considered later.

Wind speed

The primary influence of wind speed on pollutant action is to alter the flux of pollutants to the plants by altering the diffusion of the gases between the atmosphere and the leaf surface (boundary layer resistance). In an early study, Ashenden and Mansfield (1977) demonstrated the importance of this factor by growing plants in wind tunnels in which pollutants were added into the air stream before it flowed over the test plants. *Lolium perenne* was exposed to 117 ppb SO_2 for 28 d at wind speeds of 10 and 25 m min^{-1}. At the lower wind speed there was a high boundary layer resistance and no effects of SO_2 were detected. However, at the higher wind speed (a low boundary layer resistance), a 33% reduction in the dry weight of the shoots of the *L. perenne* and visible injury were evident. In more recent studies, the rate of air changes per minute has been set in closed and open-top exposure chambers to levels that ensure low boundary layer resistance, and any 'inhibitory effect' of low wind speed on pollutant flux is thus avoided.

Soil factors

Soil water content

Mechanisms to reduce water loss such as partial stomatal closure and reduced rates of expansion of new leaves are induced in leaves in the very early stages of soil drying following 'root-to-shoot' signalling by plant hormones, principally abscisic acid (see review by Mansfield, 1998). It clearly follows that the moisture content of the soil is a key determinant of stomatal conductance and hence the flux of pollutants into a plant. Such effects have been studied in the field and laboratory by growing plants in conditions of water restriction and comparing the responses with those of normally watered plants.

Open-top chamber experiments in the USA and Europe have shown that several crops are less sensitive to ozone when grown with limited availability of soil moisture (e.g. wheat, Fuhrer, 1996; cotton, Temple *et al.*, 1988a; bean, Moser *et al.*, 1988). Typical reductions in the effect of ozone on yield or biomass reduction are shown in Figure 18.3 (Fuhrer, 1996). Tingey and Hogsett (1985) showed that such effects were primarily due to reduced uptake through partially closed stomata. By artificially inducing stomatal closure in well-watered plants using abscisic acid, the plants could be protected against ozone damage; if the stomata were subsequently opened artificially by treating the plants with fusicoccin, ozone sensitivity was restored. Other reported 'beneficial' effects of drought stress coinciding with ozone treatment include reduced effects of O_3 on photosynthesis (Temple *et al.*, 1988b; Vozzo *et al.*, 1995); reduced water consumption (Kobayashi *et al.*, 1993) and decreases in visible injury (Pell *et al.*, 1993; Vandermeiren *et al.*, 1995). In soybean, however, Heggestad *et al.* (1985) showed that a high soil moisture deficit (SMD) is detrimental when combined with ozone treatment because prolonged ozone exposure reduced root growth resulting in an inability to respond to a lower water table. A yield reduction of 25% was reported when O_3 and SMD treatments were combined compared with 5% and 4% reductions when the two treatments were applied singly.

Ozone-drought interaction studies for tree species have shown that drought-induced stomatal closure was partially inhibited by ozone, resulting in increased water loss (e.g. Norway spruce, Karlsson *et al.*, 1995; beech, Pearson and Mansfield, 1993). Sap

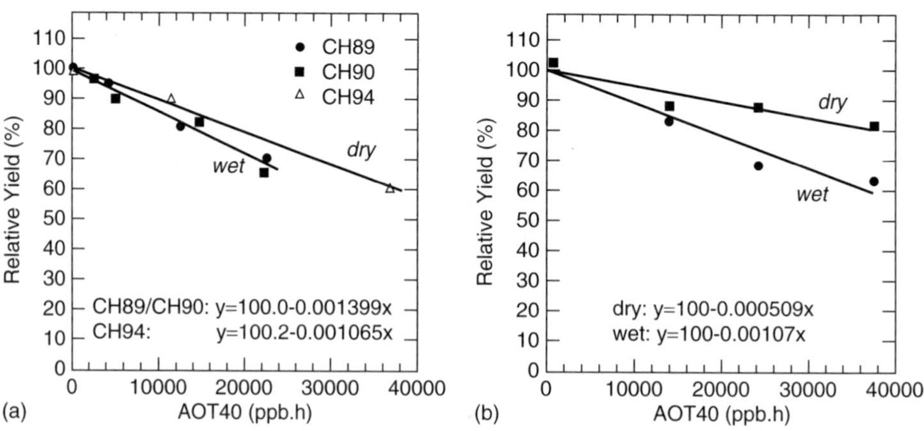

Figure 18.3 Exposure–response relationship for (a) grain yield of wheat and (b) biomass of white clover, exposed in open-top chambers under irrigated (wet) or non-irrigated (dry) conditions. Redrawn from Fuhrer, 1996. Reproduced by permission of the author

flow and water use in ash (*Fraxinus excelsior*) are greater during ozone episodes (Wiltshire, Unsworth and Wright, 1994). Skärby *et al.* (1998) concluded that, over a whole growing season, trees exposed to ozone during a drought might lose more water and use more energy than well-watered trees, and thus are likely to show an enhanced decline in growth and vitality.

In contrast to trees and crops, no clear patterns have yet emerged for water stress:ozone interactions in natural vegetation. Field observations showed that ozone injury on natural vegetation in Ohio and Indiana was widespread in a wet year with moderate ozone levels, but virtually absent in a dry year with high ozone (Showman, 1991). Furthermore, there have been few records of visible injury on wild plants growing during the very dry summer months in Mediterranean areas, suggesting that drought conditions are 'protecting' plants from high ambient ozone episodes (Davison and Barnes, 1998). These hypotheses were tested in an open-top chamber study of 24 species of semi-natural grassland (Bungener *et al.*, 1999b). The authors expected to find that those species sensitive to reduced irrigation would be protected against O_3 effects on productivity when watering was reduced. However, only two (*Trifolium repens* and *Dactylis glomerata*) of the nine species that fell into this category responded in this way. Three species (*Onobrychis sativa, Lotus corniculatus* and *Carum carvi*) were more sensitive to ozone in the dry treatment, and no difference in response to ozone occurred in the other four species. Bungener and colleagues hypothesised that in some species, the response to drought increases ozone sensitivity, leading to enhanced effects of ozone on shoot growth.

Fewer studies have been performed on the effects of soil moisture on the responses to other pollutants since such interactions are less likely to occur in the field than ozone–drought interactions. Lucas (1990) exposed Timothy grass (*Phleum pratense*) to combinations of SO_2 and NO_2 (30, 60 or 90 ppb of each) for 40 d in solardomes. Decreased partitioning to the roots and increased water loss were evident in the two highest pollutant mixture treatments when water was withheld for 23 d. More water consumption occurred in the polluted plants even though their rate of growth was slower. Imposition of severe drought stress on individual leaves by excision from the plant failed to cause complete stomatal closure and led to enhanced water loss from the polluted leaves (Lucas, 1990). Several such effects in excised leaves have been reported (see review by Mansfield, 1998) and are considered to be due to damaged stomatal mechanisms including damage to the adjacent epidermal cells that leads to a loss of mechanical resistance against the guard cells.

Soil nutrient status

Uptake of macronutrients (N, P, K, Mg, Ca, S) and micronutrients (e.g. Mn, Zn, Cu etc.) from the soil is essential for healthy plant growth. Some nutrients, including S and N, can also be acquired from the atmosphere in gaseous form by stomatal conductance. When nutrient availability is limited, resource allocation, leaf development, canopy structure, reproductive development and rates of senescence can all be altered. Hence, an interaction with air pollutants is likely.

It might be expected that those gaseous pollutants that contain plant nutrients (e.g. NO_2, SO_2, NH_3) will have a beneficial effect on the plant when occurring in conditions of poor nutrient availability. Some studies have shown this to be the case. For example, in a controlled environment study, Rowland *et al.* (1987) studied this possibility by exposing plants to a relatively high concentration of NO_2 (300 ppb) whilst controlling nitrate supply to the roots of hydroponically grown barley plants. With 0 or 0.01 mM of nitrate in the nutrient solution, NO_2 significantly increased the shoot:root ratio above that of the filtered-air controls; at 0.1, 0.5, 1.0 and 10 mM nitrate, no effect of NO_2 on the shoots or roots were noted. On a field scale, maximum growth reductions due to SO_2 (150 ppb, 4 h d^{-1}, 5 d per week for 45 d) were observed in wheat plants grown directly in an agricultural soil in open-top chambers without added fertiliser whereas no effects of SO_2 were noted in plants grown with optimal levels of NPK (Agrawal and Verma, 1997).

An Australian study attempted to separate out the effects of NO_x and SO_2 as nutrient sources (Murray *et al.*, 1994). Nitrogen- and sulphur-deficient white clover (*Trifolium repens*) were exposed to NO, NO_2 and SO_2 combinations using open-top chambers. Exposure to 29 ppb NO and NO_2 for 4 h d^{-1} for 149 d halved the shoot weight of white clover and clearly was not beneficial. Similarly, exposure to increasing SO_2 concentrations (80–388 ppb, 4 h d^{-1}, 149 d) caused a linear decrease in growth even though there was a substantial increase in the S content of leaves. When exposed to all three pollutants, the detrimental effects on plant growth were additive even though the nutrient contents of the leaves increased. It was concluded that NO_2 may have partially ameliorated the problems of nitrogen deficiency, leading to increased stomatal conductance and thus increased flux of SO_2 into the plant with a consequent increase in apparent SO_2 toxicity at each SO_2 concentration.

Each of the studies described so far have used relatively high pollutant concentrations similar to the level experienced close to a local pollutant source. Working at lower concentrations of SO_2, Rantanen *et al.* (1994), showed that 30 ppb each of SO_2 and NO_2 had a greater effect on two-year-old seedlings of Norway spruce when combined with limited availability of K, Ca and Mg. However, this effect was transitory and only occurred during the rapid growth phrase in early spring. Root responses and overall nutrient demand were unaffected by the end of the 158 d exposure.

Overall, there is some evidence that high concentrations of SO_2 and/or NO_2 have a beneficial effect on plants grown in conditions of nutrient stress. However, effects of lower levels of the pollutants can be more pronounced when nutrient availability is limited.

Most experiments on the effects of ozone on crops have been carried out in non-limiting nutrient conditions since the use of fertilisers to optimise nutrient availability is widespread in agriculture. Where nutrient interactions have been investigated in crops, they tend to have been studied at the physiological level using high ozone concentrations and a wide range of nutrient availability (e.g. Pell *et al.*, 1990), although some OTC experiments have been conducted (e.g. Fangmeier *et al.*, 1997). Ozone impacts are more likely to be influenced by nutrient status in partially managed or unmanaged forests and areas of natural/semi-natural vegetation where plants are often growing in nutrient-limiting conditions. Such interactions are of greater ecological significance and are considered in more detail than nutrient:ozone interactions in crops.

In several tree species, ozone effects are less pronounced when nutrients are limiting to growth. Ozone stress may be tolerated at low nutrients owing to compensatory mechanisms and defence metabolism (Rantanen *et al.*, 1994; Greitner *et al.*, 1994) and increased root growth relative to that of the shoots (Tjoelker and Luxmoore, 1991, abs; Pell *et al.*, 1995). However, in one species, birch (*Betula pendula* Roth), some studies have shown that ozone has a greater impact when nutrients are limiting. Landolt *et al.* (1997) showed that well-fertilised birch had enhanced Rubisco activity and was better able to repair ozone injury than the low-fertiliser treatment. Similarly, Paakkonen and Holopainen (1995) found that sufficient nitrogen supply conferred ozone resistance on birch by enhanced leaf biomass production, decreased leaf loss and increased root weight. Other studies with this species have also indicated that ozone effects are more pronounced when nutrient levels are low (e.g. Schmutz *et al.*, 1995; Matyssek *et al.*, 1997).

The possibility of an influence of nutrient status on the response to ozone in wild plants has been investigated comprehensively for *Plantago major* (Whitfield *et al.*, 1996, 1998). As with birch, the effects of ozone were more severe in root-restricted, nutrient-deficient plants. For example, the mean relative growth rate of an ozone-sensitive line of *P.major* was greater in nutrient-deficient soil than in nutrient-rich soil following exposure to ozone for 4 weeks at 70 ppb for 7 h d^{-1} (Davison and Barnes, 1998). By the final harvest, ozone treatment reduced seed production by two-thirds in nutrient-limited soil, but had no significant effect in nutrient-rich soil. This enhanced effect of ozone occurred even though nutrient deficiency reduced the stomatal conductance of the sensitive line and hence the flux of ozone into the leaves. The ecological implications of such an enhanced effect when nutrient availability is low are described by Davison and Barnes (1998).

Salinity

In areas of the world where irrigation is routinely used, soil salinity can have a major adverse effect on crop productivity. For example, it has been estimated that salinity affects an average of 11% of irrigated land in India (Singh, 1992) and up to 26% of the arable land in parts of Pakistan (Ahmad, 1990). The possibility of interactions between salinity and ozone has been investigated for five cultivars of rice (Welfare *et al.*, 1996). The majority of interactions were additive, with antagonistic effects on the growth, assimilation and nutrient status of some of the cultivars being noted. Olszyk *et al.* (1988) also found that the interactions between ozone and salinity were additive for alfalfa. In contrast, a series of experiments with wheat and soybean have shown that low levels of salinity protect plants against SO_2 effects on growth and injury by decreasing stomatal conductance (Qifu and Murray, 1991, 1993; Huang *et al.*, 1994; Ma and Murray, 1994). However, when pre-treated with high salinity levels, soybean was severely injured by subsequent SO_2 exposure even though the stomata were closed (Ma and Murray, 1994).

Increased sensitivity to SO_2 in winter conditions: an example of multi-factor interactions

Although the factors have been considered individually so far, in ambient conditions a plant responds to pollutant exposure in a mixture of environmental conditions. Not only will a factor influence pollutant effect, it will also affect the response of the plants to other environmental factors. Enhanced sensitivity to SO_2 during the winter has been reported, and will be used as an example to show how multiple factors combine to influence the response to an individual pollutant.

Much of the work on SO_2 effects that was conducted during the 1970s and 1980s tested the hypothesis that 'SO_2 has a greater impact on vegetation during winter conditions than in summer conditions'. As the highest SO_2 concentrations are generally found in the winter, the existence of such an interaction has important ecological and economic implications. Experiments were either conducted in closed chambers where usually one or two factors (e.g. temperature and light intensity) were varied, or in the open field, where plants were subjected to the naturally occurring mixture of climatic conditions associated with a northern European winter (short days, low light, high humidity and low temperature).

One of the early investigations that suggested that there might be increased sensitivity during the winter months was performed by Davies (1980). Timothy grass (*Phleum pratense*) was exposed to 120 ppb of SO_2 for 5 weeks with either low irradiance/short day length or high irradiance/long daylength. Shoot and root weights were reduced by 50% in 'winter' light conditions, but were unaffected in 'summer' light conditions. Jones and Mansfield (1982) confirmed that the magnitude of response to SO_2 was dependant on the growth rate of the plants, with slower growing plants being of greatest sensitivity. Thus, in environmental conditions leading to slow growth (such as low light and temperature), the effects of SO_2 on the shoots were more pronounced.

The photosynthetic process was also found to be more sensitive to SO_2 during winter than summer conditions. For example, Mansfield and Jones, (1985) found that 4–6 weeks of exposure of *Phleum pratense* to 110 ppb SO_2 reduced net photosynthesis when the plants were grown at low intensities (135 μmol m^{-2} s^{-1}), but not at medium intensities (360 μmol m^{-2} s^{-1}), even though uptake was 50% greater at the higher light level. Similarly, Kropff *et al.* (1990) demonstrated an enhanced reduction in the photosynthetic rate of *Vicia faba* by SO_2 at 8 °C compared to 18 °C. The enhanced sensitivity to SO_2 during 'winter' conditions appears to be explained by reduced detoxification of sulphite (formed as SO_2 dissolves in the moisture of the apoplast) into much less toxic sulphate due to a lack of production of respiratory substrate at low temperatures (Jones and Mansfield, 1982; Darrall, 1989; Kropff *et al.*, 1990). In addition, low VPD (high humidity) also associated with winter conditions, results in increased flux of SO_2 into the leaf with a concomitant enhanced effect on photosynthesis (Kropff *et al.*, 1990).

The winter sensitivity hypothesis was further tested by exposing plants to ambient or near ambient conditions over a long time period. In one such experiment, the grass *Poa pratensis* was exposed in hemispherical greenhouses to 62 ppb of SO_2 from November through to the following September (Whitmore and Mansfield, 1983). Relative to the filtered air controls, the growth of *P. pratensis* was significantly reduced in late winter, but recovered during late spring, leading to no difference in the dry weight per plant at the end of the experiment. The winter growth depression was greater than additive when 62 ppb of NO_2 was combined with the 62 ppb SO_2 treatment. Using the same exposure facilities, Pande and Mansfield (1985) showed similar growth reduction, followed by recovery in winter barley treated with SO_2 (and NO_2). The winter sensitivity hypothesis was also tested in natural climatic conditions by exposing field-grown plants to SO_2 using a field-release system in which SO_2 was released upwind of a commercial crop of winter barley (Baker *et al.*, 1986). Growth reductions at the highest mean SO_2 concentrations (44–73 ppb, seasonal average) were greatest during the winter months, with recovery occurring during the spring and summer. In an earlier experiment, pronounced leaf injury occurred in SO_2-exposed winter wheat following a period of sub-zero temperatures and heavy snow fall (Baker *et al.*, 1982).

Thus, there is strong experimental evidence that the impact of SO_2 is greater in winter conditions when growth is restricted than in summer conditions when plants are actively growing. Experiments performed in the open field have confirmed the predictions made from closed-chamber experiments.

Conclusions

The focus of this chapter mirrors the research emphases of the previous two decades, namely SO_2 research in the 1980s and O_3 research in the 1990s. As knowledge increased on the responses to these pollutants in optimum growing conditions, interest in plant responses in less than optimum conditions followed. Less research has been conducted on the modifying influence of environmental conditions on plant responses to N-pollutants, and thus there has been less to report.

For each pollutant, researchers have considered how the conditions that are most likely to apply at the time of exposure might influence response. Studies have shown that SO_2 activity is more pronounced in the winter, when plant growth and metabolism are slow. This observation is of considerable relevance for European and North American exposures to the pollutant. However, in recent times there has been increasing concern about high levels of SO_2 in the developing countries of the world (Mansfield, 1999). Clearly, more information is now needed on how plants indigenous to, or grown in, these countries respond to SO_2 in the high temperatures that are typical of these areas. In ozone research, the emphasis has been on quantifying the modifying influence of climatic conditions and soil moisture deficit on ozone uptake via the stomata. Monitoring of effects such as changes in photosynthetic rate and the extent of ozone injury on the foliage has clearly demonstrated that soil and climatic conditions are modifying short-term responses to the pollutant. The challenge now is to link changes in flux to longer term effects such as growth and yield reductions. This would allow potential effects on vegetation to be modelled over large geographical areas such as Europe and North America. Recent interest in NH_3 pollution has led to an increase in research into plant effects. However, more information is needed on how its effects are modified by environmental conditions.

The examples presented in this chapter have shown that environmental conditions influence pollutant flux by an effect on stomatal aperture. Instantaneous, but transient, effects on flux are exerted by factors such as vapour pressure deficit, temperature, wind speed and light, whereas effects of soil moisture deficit and nutrient status are longer lasting. For a given flux, the response to the pollutant is then modified by the influence of environmental conditions such as temperature and nutrient availability on growth and energy production for repair processes. Thus, it can be concluded that the net effect of the pollutant on the plant results from a complex web of interactions between the plant and its environment before, during and after exposure to the pollutant.

References

Agrawal, M. and Verma, M. (1997). Amelioration of sulphur dioxide phytotoxicity in wheat cultivars by modifying NPK nutrients. *Journal of Environmental Management* **49**, 231–244.

Ahmad, S. (1990). Soil salinity and water management. In: *Proceedings of the Indo-Pak Workshop on Soil Salinity and Water Management*. PARC, Islamabad, 3–18.

Ashenden, T.W. and Mansfield, T.A. (1977). Influence of wind speed on the sensitivity of ryegrass to SO_2. *Journal of Experimental Botany* **28**, 729–735.

Baker, C.K., Colls, J.J., Fullwood, A.E. and Seaton, G.G.R. (1986). Depression of growth and yield in winter barley exposed to sulphur dioxide in the field. *New Phytologist* **104**, 233–241.

Baker, C.K., Unsworth, M.H. and Greenwood, P. (1982). Leaf injury on wheat plants exposed in the field in winter to SO_2. *Nature* **299**, 239–240.

Balls, G.R., Palmer-Brown, D. and Sanders, G.E. (1996). Investigating microclimate influences on ozone injury in clover (*Trifolium subterraneum*) using artificial neural networks. *New Phytologist* **132**, 271–280.

Benton, J., Fuhrer, J., Gimeno, B.S., Skärby, L., Palmer-Brown, D., Ball, G., Roadknight, C. and Mills, G. (2000). An international cooperative programme indicates the widespread occurrence of ozone injury on crops. *Agriculture, Ecosystems and Environment* **78**, 19–30.

Biondi, F., Mignanego, L. and Schenone, G. (1992). Correlation between environmental parameters and leaf injury in *Nicotiana tabacum* L. cv Bel–W3. *Environmental Monitoring and Assessment* **22**, 73–87.

Black, V.J. (1982). Effects of sulphur dioxide on physiological processes in plants. In: Unsworth, M.H. and Ormrod, D.P. (Eds). *Effects of gaseous air pollution in agriculture and horticulture*, Butterworth Scientific, London, pp. 67–91.

Bungener, P., Balls, G.R., Nussbaum, S., Geissmann, M., Grub, A. and Fuhrer, J. (1999a). Leaf injury characteristics of grassland species exposed to ozone in relation to soil moisture condition and vapour pressure deficit. *New Phytologist* **142**, 271–282.

Bungener, P., Nussbaum, S., Grub, A. and Fuhrer, J. (1999b). Growth response of grassland species to ozone in relation to soil moisture conditions and plant strategy. *New Phytologist* **142**, 283–293.

Darrall, N.M. (1989). The effect of air pollutants on physiological processes in plants. *Plant, Cell and Environment* **12**, 1–30.

Davies, T. (1980). Grasses are more sensitive to SO_2 pollution in conditions of low irradiance and short days. *Nature* **284**, 483–485.

Davison, A.W. and Barnes, J.D. (1998). Effects of ozone on wild plants. *New Phytologist* **139**, 135–151.

Fangmeier, A., Grüters, U., Jäger, H.-J. and Vermehren, B. (1997). Effects of elevated CO_2, nitrogen supply and tropospheric ozone on spring wheat–II. Nutrients (N, P, K, S, Ca, Mg, Fe, Mn, Zn). *Environmental Pollution* **96**, 43–59.

Farage, P.K. and Long, S.P. (1999). The effects of O_3 fumigation during leaf development on photosynthesis of wheat and pea: An *in vivo* analysis. *Photosynthesis Research* **59**, 1–7.

Foot, J.P., Caporn, S.J.M., Lee, J.A. and Ashenden, T.W. (1996). The effect of long-term ozone fumigation on the growth, physiology and frost sensitivity of *Calluna vulgaris*. *New Phytologist* **133**, 503–511.

Foot, J.P., Caporn, S.J.M., Lee, J.A. and Ashenden, T.W. (1997). Evidence that ozone exposure increases the susceptibility of plants to natural frosting episodes. *New Phytologist* **135**, 369–374.

Fuhrer, J. (1996). The critical level for effects of ozone on crops, and the transfer to mapping. In: *Critical levels for ozone in Europe: Testing and finalizing the concepts*. UN-ECE Workshop report. University of Kuopio, Department of Ecology and Environmental Science, 27–43.

Greitner, C.S., Pell, E.J. and Winner, W.E. (1994). Analysis of aspen foliage exposed to multiple stresses – ozone, nitrogen deficiency and drought. *New Phytologist* **127**, 579–589.

Grünhage, L. and Jäger, H.J. (1994). Influence of the atmospheric conductivity on the ozone exposure of plants under ambient conditions: considerations for establishing ozone standards to project vegetation. *Environmental Pollution* **85**, 125–129.

Grünhage, L., Jäger, H.J., Haenel, H.-D., Lopmeier, F.J. and Hanewald, K. (1999). The European critical levels for ozone; improving their usage. *Environmental Pollution* **105**, 163–173.

Grüters, U., Fangmeier, A. and Jäger, H.J. (1994). Modelling stomatal responses of spring wheat (*Triticum aestivum*) to ozone and different levels of water supply. *Environmental Pollution* 141–149.

Heggestad, H.E., Gish, T.J., Lee, E.H., Bennett, J.H. and Douglass, L.W. (1985). Interaction of soil moisture stress and ambient ozone on growth and yields of soybeans. *Phytopathology* **75**, 472–477.

Huang, L.B., Murray, F. and Yang, X.H. (1994). Interaction between mild salinity and sublethal SO_2 pollution on wheat (*Triticum aestivum* cultivar Wilgoyne (Ciano Gallo)). 2. Accumulation of sulfur and ions. *Agriculture, Ecosystems and Environment* **47**, 335–351.

Jones, T. and Mansfield, T.A. (1982). The effect of SO_2 on growth and development of seedlings of *Phleum pratense* under different light and temperature environments. *Environmental Pollution* **27**, 57–71.

Karlsson, P.E., Medin, E.L., Selldén, G., Selldén, G.F., Wallin, G., Ottoson, S. and Skärby, L. (1995). Ozone and drought stress: Interactive effects on the growth and physiology of Norway spruce (*Picea abies* L. Karst). *Water, Air and Soil Pollution* **85**, 1325–1330.

Kleier, C., Farnsworth, B. and Winner, W. (1998). Biomass, reproductive output, and physiological responses of rapid-cycling Brassica (*Brassica rapa*) to ozone and modified root temperature. *New Phytologist* **139**, 657–664.

Kobayashi, K., Miller, J.E., Flagler, R.B. and Heck, W.W. (1993). Model analysis of interactive effects of ozone and water stress on the yield of soybean. *Environmental Pollution* **82**, 39–45.

Kropff, M.J., Smeets, W.L.M., Meijer, E.M.J., Van der Zalm, A.J.A. and Bakx, E.J. (1990). Effects of sulphur dioxide on leaf photosynthesis: the role of temperature and humidity. *Physiologia Plantarum* **80**, 655–661.

Landolt, W., Günthardt-Georg, M.S., Pfenninger, I., Einig, W., Hampp, R., Maurer, S. and Matyssek, R. (1997). Effect of fertilization on ozone-induced changes in the metabolism of birch (*Betula pendula*) leaves. *New Phytologist* **137**, 389–397.

Lucas, P.W. (1990). The effects of prior exposure to sulphur dioxide and nitrogen dioxide on the water relations of Timothy grass (*Phleum pratense*) under drought conditions. *Environmental Pollution* **66**, 117–138.

Ma, Q.F. and Murray, F. (1994). Responses to sequential exposure to SO_2 and salinity in soybean (*Glycine max* L.). *Water, Air and Soil Pollution* **73**, 143–155.

Mansfield, T.A. (1998). Stomata and plant water relations: Does air pollution create problems? *Environmental Pollution* **101**, 1–11.

Mansfield, T.A. (1999). SO_2 pollution: a bygone problem or a continuing hazard? In: *Physiological Plant Ecology*. (Eds) M.C. Press, J.D. Scholes and M.G. Barker. Blackwell Science, pp. 219–240.

Mansfield, T.A. and Jones, T. (1985). Growth/environment interactions in SO_2 responses of grasses. In: *Sulphur Dioxide and Vegetation*. Stanford University Press, California, pp. 332–345.

Matyssek, R., Maurer, S., Gunthardt-Georg, M.S., Landolt, W., Saurer, M. and Polle, A. (1997). Nutrition determines the 'strategy' of *Betula pendula* for coping with ozone stress. *Phyton-Annales Rei Botanicae* **37**, 157–167.

Meyer, U., Köllner, B. and Willenbrink, J. (1997). Physiological changes in agricultural crops induced by different ambient ozone exposure regimes I. Effects on photosynthesis and assimilate allocation in spring wheat. *New Phytologist* **136**, 645–652.

Mortensen, L.M. (1999). Foliar injuries caused by ozone in *Betula pubescens* Ehrh. and *Phleum pratense* L. as influenced by climatic conditions before and during O_3 exposure. *Acta Agriculturae Scandinavica Section B-Soil and Plant Science* **49**, 44–49.

Mortensen, L.M. and Nilsen, J. (1992). Effects of ozone and temperature on growth of several wild plant species. *Norwegian Journal of Agricultural Sciences* **6**, 195–204.

Moser, T.J., Tingey, D.T., Rodecap, K.D., Rossi, D.J. and Clark, C.S. (1988). Drought stress applied during the reproductive phase reduced ozone induced effects in bush bean. In: (Eds) Heck, W.E., Taylor, O.C. and Tingey, D.T. *Assessment of Crop Loss from Air Pollutants*. Elsevier Applied Science, London, pp. 345–364.

Murray, F., Monk, R., Clarke, K. and Qifu, M. (1994). Growth responses of N and S deficient white clover and burr medic to SO_2, NO and NO_2. *Agriculture, Ecosystems and Environment* **50**, 113–121.

Olszyk, D.M., Maas, E.V., Kats, G. and Francois, L.E. (1988). Soil salinity and ambient ozone – lack of stress interaction for field-grown alfalfa. *Journal of Environmental Quality* **17**, 299–304.

Paakkonen, E. and Holopainen, T. (1995). Influence of nitrogen supply on the response of clones of birch (*Betula pendula* Roth) to ozone. *New Phytologist* **129**, 595–603.

Pande, P.C. and Mansfield, T.A. (1985). Responses of winter barley to SO_2 and NO_2 alone and in combination. *Environmental Pollution* **39**, 281–291.

Pearson, M. and Mansfield, T.A. (1993). Interacting effects of ozone and water stress on the stomatal resistance of beech (*Fagus sylvatica*). *New Phytologist* **123**, 351–358.

Pell, E.J., Sinn, J.P., Eckhardt, N., Vinten-Johansen, C., Winner, W.E. and Mooney, H.A. (1993). Response of radish to multiple stresses II: Influence of season and genotype on plant response to ozone and soil moisture deficit. *New Phytologist* **123**, 153–163.

Pell, E.J., Sinn, J.P. and Johansen, C.V. (1995). Nitrogen supply as a limiting factor determining the sensitivity of *Populus tremuloides* Michx to ozone stress. *New Phytologist* **130**, 437–446.

Pell, E.J., Winner, W.E., Vinten-Johansen, C. and Mooney, H.A. (1990). Response of radish to multiple stresses: I. Physiological and growth responses to changes in ozone and nitrogen. *New Phytologist* **115**, 439–446.

Potter, L., Foot, J.P., Caporn, S.J.M. and Lee, J.A. (1996). The effects of long-term elevated ozone concentrations on the growth and photosynthesis of *Sphagnum recurvum* and *Polytrichum commune*. *New Phytologist* **134**, 649–656.

Qifu, M. and Murray, F. (1991). Soil-salinity modifies SO_2 sensitivity in soybean. *New Phytologist* **119**, 269–274.

Qifu, M. and Murray, F. (1993). Effects of SO_2 and salinity on nitrogenase activity, nitrogen concentration and growth of young soybean plants. *Environmental and Experimental Botany* **33**, 529–567.

Rantanen, L., Palomaki, V., Harrison, A.F., Lucas, P.W. and Mansfield, T.A. (1994). Interactions between combined exposure to SO_2 and NO_2 and nutrient status of trees: effects on nutrient content and uptake, growth, needle ultrastructure and pigments. *New Phytologist* **128**, 689–701.

Ribas, A., Filella, I., Gimeno, B.S. and Peñuelas, J. (1998). Evaluation of tobacco cultivars as bioindicators and biomonitors of ozone phytotoxical levels in Catalonia. *Water, Air and Soil Pollution* **107**, 347–365.

Rowland, A.J., Drew, M.C. and Wellburn, A.R. (1987). Foliar entry and incorporation of atmospheric nitrogen dioxide into barley plants of different nitrogen status. *New Phytologist* **107**, 357–371.

Schmutz, P., Bucher, J.B., Gunthardt-Georg, M. S., Tarjan, D. and Landolt, W. (1995). Response of poplar to ozone alone and in combination with NO_2 at different nitrogen fertilization levels. *Phyton-Annales Rei Botanicae* **35**, 269–289.

Showman, R.E. (1991). A comparison of ozone injury to vegetation during moist and drought years. *Journal of the Air and Waste Management Association* **41**, 63–64.

Singh, N.T. (1992). Dry land salinity in the Indo-Pakistan sub-continent. In: *Degradation and restoration of arid lands*. Texas Technical University, Lubbock, 179–248.

Skärby, L., Ro-Poulsen, H., Wellburn, F.A.M. and Sheppard, L.J. (1998). Impacts of ozone on forests: a European perspective. *New Phytologist* **139**, 109–122.

Soja, G., Barnes, J., Vandermeiren, K., Pleijel, H. and Mills, G. (in press). Phenological weighting of ozone exposures in the calculation of critical levels for non-woody species. *Environmental Pollution*.

Temple, P.J., Kupper, R.S., Lennox, R.L. and Rohr, K. (1988a). Physiological and growth responses of differentially irrigated cotton to ozone. *Environmental Pollution* **53**, 255–263.

Temple, P.J., Benoit, L.F., Lennox, R.W., Reagan, C.A. and Taylor, O.C. (1988b). Combined effects of ozone and water stress on alfalfa growth and yield. *Journal of Environmental Quality* **17**, 108–113.

Thoene, B., Rennenberg, H. and Weber, P. (1996). Absorption of atmospheric NO_2 by spruce (*Picea abies*) trees 2. Parameterization of NO_2 fluxes by controlled dynamic chamber experiments. *New Phytologist* **134**, 257–266.

Thompson, C.R., Kats, G., Olszyk, D.M. and Adams, C.J. (1992). Humidity as a modifier of vegetation responses to ozone: Design and testing of a humidification system for open-top field chambers. *Journal of the Air Waste Management Association* **42**, 1063–1066.

Tingey, D.T. and Hogsett, W.E. (1985). Water stress reduces ozone injury via a stomatal mechanism. *Plant Physiology* **77**, 944–947.

Tjoelker, M.G. and Luxmoore, R.J. (1991). Soil-nitrogen and chronic ozone stress influence physiology, growth and nutrient status of *Pinus taeda* L. and *Liriodendron tulipfera* L. seedlings. *New Phytologist* **119**, 69–81.

Tjoelker, M.G., Volin, J.C., Oleksyn, J. and Reich, P.B. (1993). Light environment alters response to ozone stress in seedlings of *Acer saccharum* Marsh. and hybrid *Populus* L. *New Phytologist* **124**, 627–636.

Tjoelker, M.G., Volin, J.C., Oleskyn, J. and Reich, P.B. (1995). Interaction of ozone pollution and light effects on photosynthesis in a forest canopy experiment. *Plant, Cell and Environment* **18**, 895–905.

Tonneijck, A.E.G. and Van Dijk, C.J. (1997). Assessing effects of ambient ozone on injury and growth of *Trifolium subterraneum* at four rural sites in the Netherlands with ethylenediurea (EDU). *Agriculture, Ecosystems and Environment* **65**, 79–88.

Vandermeiren, K., De Temmerman, L. and Hookham, N. (1995). Ozone sensitivity of *Phaseolus vulgaris* in relation to cultivar differences, growth stage and growing conditions. *Journal of Water, Air and Soil Pollution* **85**, 1455–1460.

Vozzo, S.F., Miller, J.E., Pursley, W.A. and Heagle, A.S. (1995). Effects of ozone and soil water deficit on field-grown soybean: I. Leaf gas exchange. *Journal of Environmental Quality* **24**, 663–670.

Welfare, K., Flowers, T.J., Taylor, G. and Yeo, A.R. (1996). Additive and antagonistic effects of ozone and salinity on the growth, ion contents and gas exchange of five varieties of rice (*Oryza sativa* L.). *Environmental Pollution* **92**, 257–266.

Whitfield, C.P., Davison, A.W. and Ashenden, T.W. (1996). Interactive effects of ozone and soil volume on *Plantago major*. *New Phytologist* **134**, 287–294.

Whitfield, C.P., Davison, A.W. and Ashenden, T.W. (1998). The effects of nutrient limitation on the response of *Plantago major* to ozone. *New Phytologist* **140**, 219–230.

Whitmore, M.E. and Mansfield, T.A. (1983). Effects of long-term exposures to SO_2 and NO_2 on *Poa pratensis* and other grasses. *Environmental Pollution* **31**, 217–235.

Wieser, G. and Havranek, W.M. (1993). Ozone uptake in the sun and shade crown of spruce – quantifying the physiological-effects of ozone exposure. *Trees – Structure and Function* **7**, 227–232.

Wiltshire, J.J.J., Unsworth, M.H. and Wright, C.J. (1994). Seasonal changes in water use of ash trees exposed to ozone episodes. *New Phytologist* **127**, 349–354.

19 Air pollutant–abiotic stress interactions

A.W. DAVISON AND J.D. BARNES

For more than a century, scientists have investigated the effects of specific pollutants individually, often under controlled conditions, and mostly in the absence of other stresses. Yet, there has long been an awareness that the physical environment can alter plant responses to pollutants and vice versa. It is therefore surprising that it is only in the past 20 years the potential importance of interactions between pollutants and the common abiotic stresses (cold, heat, water deficit, wind and extremes of light) has been recognized. There are two major reasons for this: (i) technical considerations such as a lack of facilities for imposing controlled stress and (ii) the fact that the majority of research until relatively recently focussed on summer-grown, annual crops. This gave little incentive for research on interactions because the incidence of stress is minimized in crops through good husbandry i.e. by growing drought/cold resistant varieties, irrigation, fertilization, and the use of shade screens and windbreaks to protect vulnerable crops. It is only since the advent of research on pasture species, forest trees and wild plants, that the potential importance of interactions between air pollutants and abiotic stresses has become recognized. In this chapter, we focus on winter injury and soil moisture deficit to illustrate the ways in which pollutants can modify plant responses to abiotic stress and vice versa. We use case studies involving SO_2, acid deposition and nitrogen to highlight the potential importance of interactions and discuss some of the problems involved in the investigation and interpretation of pollution effects in the field.

Importance of abiotic stress

A glance at a global map of biomes or crop distribution illustrates the overriding importance of the physical environment in determining the geographical range of vegetation. Each species can tolerate a range of temperature, light, water and wind, but the amplitude of the range and the nature of the response to extreme temperatures or developing water deficit depends on the developmental stage of the plant, nutritional status and the incidence of pests, diseases or other stresses. In short, responses to change in the physical environment are not fixed, they are very variable, even within a species. Indeed, farmers exploit this variation to minimize crop losses. Nevertheless, losses remain significant. Winter frosts, for example, are responsible for losses in agricultural crops estimated at over a billion dollars per year in the USA alone at 1975 prices (White

Air Pollution and Plant Life, second edition. Edited by J.N.B. Bell and M. Treshow. ISBN 0 471 49090 3 (HB), 0 471 49091 1 (PB).

and Hass, 1975). This draws clear attention to the potential economic importance of pollution–stress interactions.

Plants survive extreme conditions through a variety of morphological and physiological adaptations. Indeed, those that grow in seasonally stressed environments commonly sense changes in the environment that signal the approach of winter or the onset of drought and acclimate (or harden) as the degree of stress intensifies. After the stress they 'de-harden'. This acclimation process is known to involve profound alterations in membrane chemistry and properties, the up-regulation of anti-oxidant defences and the accumulation of osmoregulatory and cryoprotectant compounds, all of which demand changes in resource and energy allocation. Since pollutants are known to alter resource allocation there is a *prima facia* case for considering that they might also modify the way in which plants react to abiotic stresses. Furthermore, all the major air pollutants (SO_2, NO_x and ozone), along with the reactive oxygen species generated during their metabolism, are known to react with cell membrane constituents (e.g. amino acids, proteins, unsaturated fatty acids and sulphydryl groups). As membranes house many of the enzymes and transport systems associated with key metabolic activities, it is conceivable that pollutants can also influence acclimation processes directly.

SO_2 and winter stress

Sulphur emissions have fallen dramatically in most industrialized western cities in the last two decades of the twentieth century but before that, ground-level concentrations of SO_2 were extremely high and there is good evidence of direct effects of the pollutant and interactions with abiotic stresses.

The atmospheric concentration of SO_2 varies both in time and space, reflecting variations in energy demand with season, dispersion and the rate of deposition and chemical transformation. In the UK, before sulphur emissions were reduced, demand for energy was greatest during winter so atmospheric concentrations of SO_2 were commonly twice as high as in summer (Figure 19.1). Superimposed upon these seasonal variations were diurnal changes in concentration of the order of 10–20% of the daily mean due to increased energy demand at night and meteorological conditions. As a consequence, SO_2 concentrations were commonly correlated with temperatures below 10 °C.

There has been anecdotal evidence for over a century that the effects of SO_2 pollution are unexpectedly severe during the winter, implying some form of interaction. Cohen and Ruston (1925), who were pioneers in the experimental study of air pollution, were amongst the first to demonstrate the phenomenon. Working with vegetation in the heavily polluted city of Leeds, in northern England, they made several important observations. For example, they noted that several winter hardy species that were planted-out in autumn, including wallflower (*Erysimum cheiri*) and cabbage (*Brassica oleracea*), did not survive in the most polluted parts of the city whereas they did in the suburbs. They also showed that the common shrub, privet (*Ligustrum ovalifolium*), stayed evergreen in rural areas but lost its leaves progressively earlier in the winter in the more polluted parts of the city. This was a common phenomenon in many northern

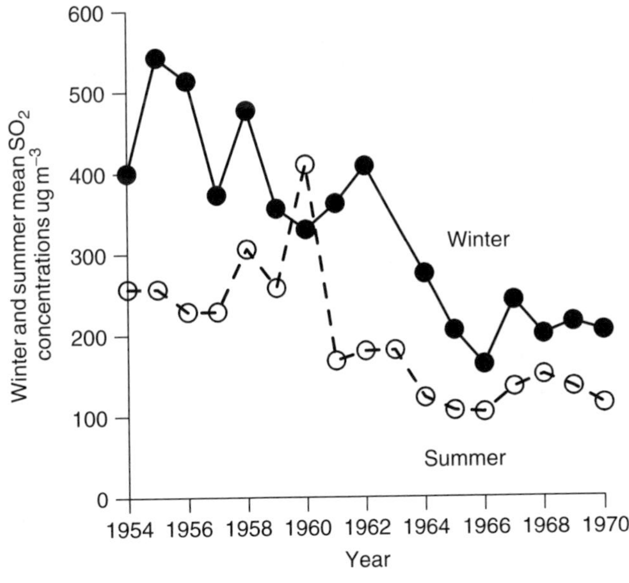

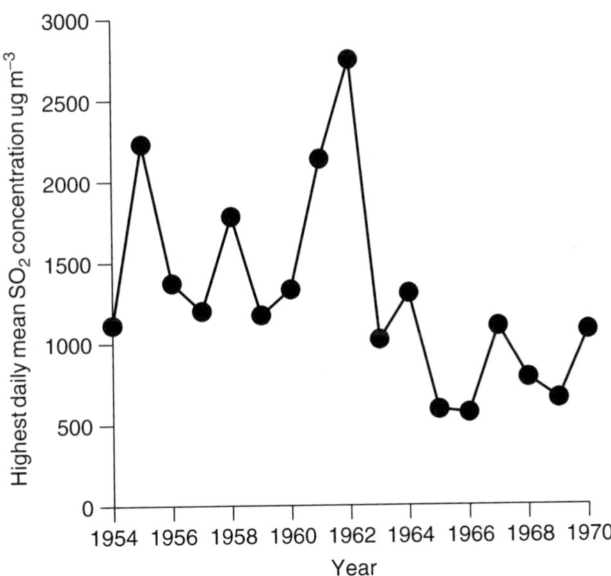

Figure 19.1 Illustration of winter and summer mean SO_2 concentrations in Sheffield city centre from 1954 to 1970. The upper graph shows that in the 1950s winter mean concentrations were very high and about twice those in summer. From the early 1960s concentrations fell with the introduction of smoke control orders. The lower graph shows the highest daily mean concentration in each year. In the 1962/63 winter episode, which had a peak daily mean of 2700 $\mu g\ m^{-3}$, all the *Coleus* plants *inside* a greenhouse at the University were totally defoliated overnight. (Data from Warren Spring, 1976.)

British cities until SO_2 levels were reduced between 1960 and the 1980s. Although the observations are anecdotal, the frequent association between damage, SO_2 and cold winter weather provides a strong indication that the combination of pollution and the winter environment may be considerably greater than the sum of the individual effects.

Most early attempts to investigate the effects of SO_2 by fumigating plants under semi-controlled conditions were performed in North America (Thomas, 1951). The principal aim was to estimate the effects of emissions from smelters on summer-grown annual crops. As smelter stacks produced high concentrations of SO_2 at ground level for short periods of time, most experiments mimicked this, using concentrations that would be considered very high today. Interactions with relative humidity, light and temperatures above 5 °C were investigated in some studies (cited by Thomas, 1951) and the most important of these was humidity. Low humidity greatly reduced the effects of SO_2 because of stomatal closure. Probably because of the focus on summer crops, interactions with temperatures <5 °C were not investigated.

In contrast, in the UK, much of the SO_2 originated from coal used in domestic heating and was released at low levels, so it produced a blanket of relatively high concentrations over very long periods of time, especially during the winter. When experimental work started in the UK, fumigation patterns reflected this and interest centred on perennial grasses. In addition, most studies continued through the winter months when temperatures were low and SO_2 concentrations were at their highest (Bleasdale, 1952, 1973; Bell and Clough, 1973; Horsman *et al.*, 1979; Jones and Mansfield, 1982; Whitmore and Mansfield, 1983). This work indicated that there were significant effects of SO_2 at much lower concentrations than reported from the USA and gradually it was recognized that abiotic environmental factors were probably of prime importance in governing plant responses to SO_2 pollution (Bell and Clough, 1973; Ashenden and Mansfield, 1977; Bell *et al.*, 1979; Black, 1982). This view was confirmed by Davies (1980). She examined the effect of light on the response of timothy grass (*Phleum pratense*) to SO_2. When plants were exposed to 343 μg $SO_2 m^{-3}$ (=120 ppb) in two day lengths and two irradiances there were no effects on plants grown in long days and higher irradiance. However, SO_2-treated plants grown in short days and at low irradiance, exhibited marked reductions (30–60%) in shoot and root yields. Davies' explanation was that plants grown and exposed at low levels of irradiance exhibited a reduced capacity to detoxify the solution products of SO_2. She also pointed out that plants rarely grow under unstressed conditions in the field, so the effects of pollution might be grossly underestimated by the majority of experiments conducted under controlled (i.e. favourable) conditions.

At about the same time, researchers operating in the Czechoslovakian mountains began to provide evidence that SO_2 could affect the extent of winter injury (i.e. frost damage, desiccation and/or pigment loss) suffered by forest trees. Transplantation experiments performed with saplings of Norway spruce (*Picea abies*) revealed severe damage during the winter at an SO_2-polluted site, whilst trees transferred to clean sites remained healthy (Materna, 1974, 1984). Subsequent field observations led Materna (1984) to the conclusion that exposure to long-term mean concentrations greater than 20 μg SO_2 m^{-3} could increase susceptibility to frost damage and result in die-back. The effect increased with altitude and it was concluded that the damage observed represented the integrated action of air pollution and other stresses. Similar conclusions were drawn by Finnish researchers examining the agents responsible for the damage to conifers growing in and around industrialized towns subject to a complex mixture of pollutants. Damage was noted to be much greater in winter than summer, and transplantation experiments with Scots pine (*Pinus sylvestris*) and Norway spruce indicated that the chief cause of mortality was pollutant damage during the winter (Havas, 1971; Huttunen, 1978). Measurements of osmotic potential, pH, buffering capacity, electrical conductivity and non-structural carbohydrate content of spruce needles sampled at two-week intervals showed that near a fertilizer factory pollution affected physiological characteristics associated with frost resistance. The implication was that some combination of SO_2, NO_x and HF reduced frost resistance (Huttunen *et al.*, 1981a,b). Strong supporting evidence that exposure to elevated levels of SO_2 can alter frost resistance was produced through controlled environment studies. Keller (1978a,b), employing Norway spruce trees grown in cabinets out of doors, showed that exposure to elevated levels of SO_2 resulted in increased sensitivity to freezing in spring, a finding subsequently confirmed by Feiler *et al.* (Michael *et al.*, 1982; Feiler, 1985). However, the data collected by Davison and Bailey (1982) probably afford the clearest demonstration that SO_2 predisposes plants to freezing injury. Working with perennial ryegrass (*Lolium perenne*) these authors showed that post-freezing survival was reduced by exposure to SO_2, with the strength of the interaction dependent upon mineral nutrition (Figure 19.2). The interaction was greatest in plants supplied with abundant nitrogen and sulphur.

Subsequent experiments performed by the same authors (reported in Davison and Barnes, 1986) revealed that freezing resistance could be reduced significantly by exposure to elevated levels of SO_2 for only two weeks during the hardening phase of the experiment. Field verification of Davison and Bailey's findings came during an experiment on winter wheat (*Triticum aestivum* cv. Bounty) in which a free air exposure system was used to deliver elevated levels of the pollutant to a standing crop in the field (Baker *et al.*, 1982). Severe leaf injury, restricted to fumigated plants, developed after a period when temperatures fell to $-9.2\,^{\circ}C$ following a mild spell.

Although experiments conducted over the last 20 years provide strong evidence that SO_2 reduces freezing resistance, the economic and/or ecological significance of such an interaction is more difficult to assess. In long-lived species such as trees, damage to needles or branches can be quantified to derive estimates of the economic costs involved, but the ecological importance is more problematic. Although the extent of frost damage induced by the pollutant may be severe, plants may compensate during later growth and thus effects on final growth/yield maybe minimal (see Baker *et al.*, 1982). Effects on pasture species are more complicated because of the potential for natural selection. The first indication that this might happen came from the observation that the common varieties of *Lolium perenne* were SO_2-sensitive but populations established in areas of high SO_2 were resistant (Bell and Clough, 1973). This was supported by similar observations around the city of Liverpool (Horsman *et al.*, 1978) and the demonstration of differences in SO_2 tolerance in *Geranium carolinianum* near a power plant in the USA (Taylor and Murdy, 1975). Rapid evolutionary change

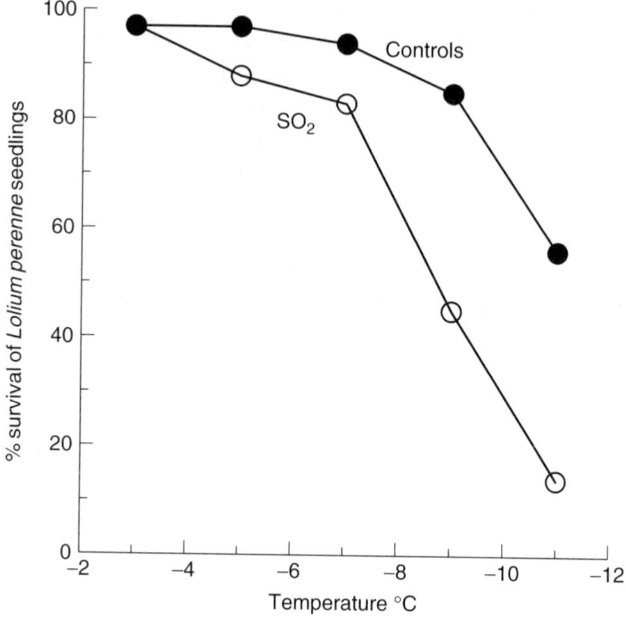

Figure 19.2 Effect of three weeks' exposure to 250 μg SO_2 m^{-3} with two weeks' hardening, on the frost tolerance of perennial ryegrass (*Lolium perenne*). (Redrawn from Davison and Bailey, 1982.)

was convincingly confirmed by an extensive study of sown grasses in Manchester by Bell and co-workers (Wilson and Bell, 1985). They demonstrated that after sowing, populations rapidly evolved resistance in a polluted part of the city, but when SO_2 concentrations dropped, resistance was lost. Davison and Bailey (1982) suggested that SO_2–frost interaction might accelerate this selection process. If this is the case, economic losses might be experienced for only the first season or two after the re-sowing of pastures, until resistant genotypes predominate. However, the agronomic quality of selected populations has not been assessed.

Acid deposition, forest decline and winter injury

It is instructive to compare research on this topic in Europe with that in the USA. European work will be considered first.

The effects of acid deposition were first investigated almost 100 years ago (Cohen and Ruston, 1925), but there was a resurgence of interest in North America and Europe in the 1970s and 1980s following the realization that air pollution was causing acidification of soil and fresh water in many regions, including pristine areas remote from immediate sources of acidifying gases. The principal cause was initially identified as sulphur deposition but the importance of nitrogen began to be recognized in the 1980s, especially as sulphur emissions decreased. Reports began to appear, particularly in

Germany, of new-type forest damage (*neuartige Waldschäden*) based upon the visible symptoms observed in some areas. The news spread after the magazine *Der Spiegel* reported that forests were being damaged on an unprecedented scale (Blank *et al.*, 1988) and soon the idea that many Central European forests were 'dying' became entrenched in the literature and in the popular press. Dire predictions of decreasing forest health and *Waldsterben* ('forest death') stimulated much research and many theories as to the causes. The hypotheses that evolved, based almost entirely on field observations or experiments conducted on young trees grown under (semi-) controlled conditions, incorporated a role for interactions involving biotic and abiotic factors such as drought, frost and pollutants (reviewed by Schulze and Freer-Smith, 1990; Sandermann *et al.*, 1997; Chappelka and Freer-Smith, 1995; Skärby *et al.*, 1998. See also Chapters 15 and 18).

Much attention was paid to suggestions that O_3–frost and/or O_3–drought interactions might be responsible for triggering the simultaneous decline of trees in disparate regions because the symptoms linked with tree decline were recorded after a series of particularly harsh winters and dry summers (Bosch and Rehfuess, 1988). Early reports revealed that conifers exposed to elevated levels of O_3 exhibited more damage after freezing than controls (Brown *et al.*, 1987; Barnes and Davison, 1988; Chappelka *et al.*, 1990). Subsequent investigations suggested that the observed effects may have been largely due to enhanced sensitivity to winter desiccation and/or photoinhibition rather than shifts in freezing tolerance *per se*, since evidence supporting effects of O_3 on the timing and extent of frost hardiness in conifers is rather weak (see reviews Barnes *et al.*, 1996; Skärby *et al.*, 1998). In contrast, there is strong experimental evidence that elevated levels of O_3 impair stomatal function, resulting in stomatal sluggishness and less control over transpiration under field conditions (Barnes *et al.*, 1990; Wiltshire *et al.*, 1994). This can increase the risk of summer drought and predispose foliage to desiccation during the winter (see Wellburn *et al.*, 1997). These are subtle effects that could damage a proportion of tree foliage. The significance of such interactions might be expected to depend upon the proportion and vitality of the remaining foliage, and the capacity of the tree to compensate.

In Europe, concerns about forest health were, and still are, largely based on surveys in which crown condition is assessed visually (see Chapter 15). Such surveys of the visible condition of trees can be useful if observers are trained, work to rigidly standardized criteria and sample sizes are large, but even in those circumstances, interpretation is difficult. Unfortunately many of the earlier surveys fell short of these standards. In the 1990s reports began to appear criticizing them, their interpretation and the idea that there was a widespread forest decline (Innes, 1992; Skelly and Innes, 1994; Kandler and Innes, 1995). Since then it has become clear that the predicted widespread forest death, *Waldsterben*, has not materialized. The strongest evidence that pollution is not the cause of a general decline comes from objective measurements of the volume of living trees (i.e. the growing stock) and growth (i.e. increase in stem wood volume). In most European forests both the growing stock and growth have increased in recent decades (Kauppi, Mielikäinen and Kuusela, 1992; Spiecker *et al.*, 1996; Spiecker, 1999). The causes of this increase are not clear (Spiecker, 1999) but it adds weight to the conclusion that pollution is not causing widespread decline.

The effects of pollution occur only at specific locations where there are known to be high pollutant concentrations (Kauppi *et al.*, 1992). This illustrates the considerable difficulties in interpreting field data and argues for caution in extrapolating from what *can* happen in controlled environments or open-top chambers to what actually happens in the field. Whilst it is conceivable that pollutant–abiotic stress interactions are predisposing or contributing factors to potential decline in the growth and vigour of trees in some parts of Europe, their role and importance on a wider scale remain unclear.

Red spruce decline and acid deposition in North Eastern USA

The eastern North American species, red spruce (*Picea rubens*) showed a dramatic decline at high altitude sites in the Appalachians between 1965 and 1982 (Johnson, 1992). Needles became red-brown in winter followed by defoliation and crown dieback. The younger foliage showed most damage and the symptoms strongly resembled winter injury (Curry and Church, 1952; Evans, 1986). Experiments performed by Sheppard, Smith and Cannell (1989) and DeHayes *et al.* (1990) have shown that the species rarely hardens sufficiently for all individuals to withstand mid-winter temperatures without damage. Sheppard and Pfanz (2000) have commented that this absence of a safety margin between hardening capacity and minimum winter temperatures may single out this species as being particularly vulnerable to pollutant–winter stress interactions.

Johnson *et al.* (1988, 1996) observed that the beginning of the decline in red spruce vitality corresponded with a decade of unusually cold winters. This also coincided with a peak in sulphur emissions and wet sulphur deposition in the 1960s (Johnson *et al.*, 1996). The same research team also noted that there was a spatial correlation between spruce decline, frequency of dead stems, cloud exposure and SO_4^{-2} ion concentrations. This led Johnson and McLaughlin (1986) to postulate that the combination of wet deposited N and S compounds, and photochemical oxidants, might predispose trees growing at high altitude to frost, drought and/or other climatic perturbations. At about the same time, Nihlgård (1985) suggested that additional N received from cloud and rain might increase growth at the expense of cryoprotection. Subsequently, Sheppard (1994) collated the available information and produced a linear relationship between the loss of mid-winter hardiness and sulphur deposition. Thus, the available experimental data suggest a strong link between the decline of red spruce, winter injury and some components of acidic deposition.

Many experiments using open-top chambers and other semi-controlled facilities showed that acid deposition can alter frost resistance but that does not, of course, prove a cause–effect relationship between decline and acidic deposition. Furthermore, Sheppard and Pfanz (2000) pointed out that few experiments have fully simulated the conditions that occur in the field, particularly the length of time needles are exposed to cloud, the number of wetting and drying cycles, timing of exposures and frost events. Consequently, the most compelling evidence comes from field manipulations such as those of DeHayes *et al.* (1991) and Vann *et al.* (1992). DeHayes *et al.* (1991) transplanted juvenile trees to a high altitude, polluted site. Some were exposed to ambient cloud water and others were protected, then frost hardiness was assessed. Hardiness of protected saplings was found to be 3–5 °C greater in late autumn and winter than those exposed to ambient pollution. This is a potentially significant loss in frost hardiness in a species that has little safety margin between its maximum hardening capacity and minimum winter temperatures. Vann *et al.* (1992) used a different approach. They enclosed branches of healthy trees in aerated Teflon bags for 12 weeks during the growing season. Branches were exposed to: charcoal-filtered air (i.e. containing neither cloudwater nor pollutant gases); ambient gases but no cloud water; filtered air with deionized water mist; or ambient cloud water plus pollutant gases. Frost resistance was tested in January, over 3 months after the bags were removed. Branches exposed to clean air and mist were most hardy and there was a 10 °C difference in frost hardiness between the treatments, showing that typical ambient mist could predispose trees to frost damage. After reviewing the work on red spruce, Sheppard and Pfanz (2000) concluded that frost/pollutant interactions probably make a significant contribution to the decline of this slow-growing species, at high altitude sites in the Appalachian mountains.

The mechanism underlying the action of acid mist on frost hardiness remains unclear (see Sheppard, 1994). One problem is that most experiments used a simulated mist that contained all of the components of ambient acid mist (hydrogen ions, SO_4^{2-}, NH_4^+ and NO_3^-). Consequently, it is impossible to separate the effects of individual components. In one of the few exceptions, Cape *et al.* (1991) used pair-wise combinations of SO_4^{2-}, NH_4^+ and NO_3^-, and the mixture, and found clear evidence that SO_4^{2-} and ammonium are the most important components affecting frost hardiness of red spruce. They suggested that the NH_4^+ exacerbated the effects of the SO_4^{2-} possibly by effects on SO_4^{2-} uptake.

The lesson from a comparison between research into forest decline in Europe and North America is that pollution *can* affect response to abiotic stress and play an important part in decline, but its importance varies in individual cases. It must not be assumed that effects that occur in controlled environments actually do occur or have any significance in the field.

Water deficit-pollutant interactions

This section will focus on interactions between SO_2, NO_2, and water deficit. SO_2 and NO_2 are considered together because the pollutants often co-occur because they both arise from burning of fossil fuels. Interactions involving ozone are covered in Chapter 18. On a global basis, plant productivity is limited more by water availability than by any other factor (Kramer, 1983). There has long been experimental evidence that the partial stomatal closure caused by soil moisture deficit can reduce the sensitivity of plants to gaseous air pollutants (Macdowall, 1965; Tingey and Hogsett, 1985). Conversely, ozone, which often causes stomatal closure, can alter the effects of water stress (see Chapter 18). However, more recently, research has shown that exposure to low concentrations of SO_2 (as low as 20 ppb) can impair the ability of leaves to conserve effectively and regulate their water loss – a finding of potential importance in both an ecological and agronomic context. Several laboratory-based experiments pay

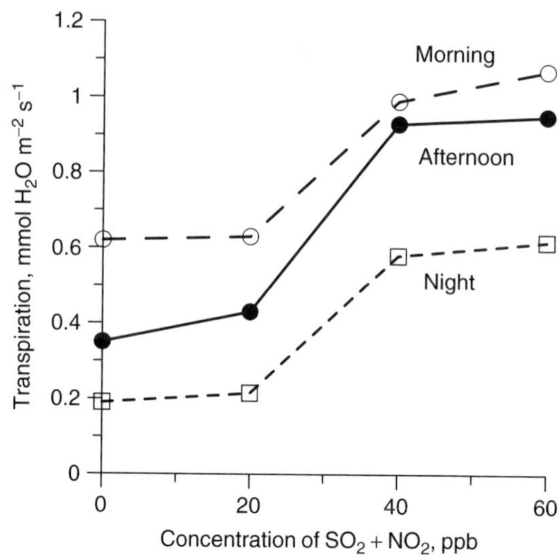

Figure 19.3 Transpiration rates (mmol H_20 m^{-2} s^{-1}) of birch (*Betula pubescens*) after exposure to clean air or 20, 40, or 60 pbb each of $SO_2 + NO_2$ for one month. (Redrawn from Neighbour *et al.*, 1988.)

testimony to the fact that exposure to SO_2 (and/or NO_2) results in stomatal dysfunction (see reviews by Mansfield and Pearson, 1996; Mansfield, 1998). This is illustrated in Figure 19.3 which shows the transpiration rate of birch (*Betula pubescens*) after exposure to clean air or mixtures containing 20, 40 or 60 ppb each of SO_2 and NO_2 (Neighbour *et al.*, 1988). There was much greater water loss in plants that had either 40 or 60 ppb of the pollutants. Electron microscopy showed that in plants exposed to the pollutants there were patches of stomata with abnormally wide open stomata and damaged nearby epidermal cells. Using the simple technique of drying detached leaves showed that the rate of stomatal closure was also greatly affected, even by 40 ppb $SO_2 + NO_2$. Investigation by Black and Black (1979) indicates that the effect is due, in part, to damage selectively sustained by subsidiary cells (or the epidermal cells that perform the same function if there are not anatomically distinct subsidiary cells). The loss in turgor of the affected cells prevents hydrostatic stomatal closure. However, other factors are known to play a role (see Robinson *et al.*, 1998). For example, the work of Atkinson *et al.* (1991) showed that in leaves of barley (*Hordeum vulgare*) exposed to 35 ppb SO_2 + 35 ppb NO_2 there was a small reduction of stomatal opening. However, when abscisic acid (ABA) was applied in the transpiration stream, stomata that were exposed to the pollutants responded less rapidly, resulting in greater water loss. Mansfield and Pearson (1996) commented that well-watered plants with little endogenous ABA may show different reactions to $SO_2 + NO_2$ from mildly stressed plants that have elevated ABA.

There is a great deal of evidence that gaseous air pollutants reduce resource allocation to the roots so the root-to-shoot ratio decreases (Bell, 1982; Cooley and Manning, 1987; Darrall, 1989; Chappelka and Freer-Smith, 1995; Taylor and Ferris, 1996). In 24 SO_2

experiments on grasses summarized by Bell (1982), 21 showed a significant reduction in root growth and in 10 there was a change in root-to-shoot ratio. This has obvious implications for the effects of water stress. One of the first reports relating to SO_2 was that of Tingey *et al.* (1971) who found that roots were more sensitive to 50 ppb SO_2 than shoots, and that root length and diameter were also affected. Jones and Mansfield (1982), using a maximum of 120 ppb SO_2 and timothy grass (*Phleum pratense*) showed that assimilate transport to the roots was affected and that effects on root growth could be detected before those on the shoots. In general, SO_2 reduces root-to-shoot ratio, NO_2 has little effect on its own but there may be interactive effects between SO_2 and NO_2 (Bell, 1982; Darrall, 1989; Taylor and Ferris, 1996).

Considering that there is clear evidence of effects of SO_2 or $SO_2 + NO_2$ on stomatal function and on root-to-shoot ratio, it is surprising that there has been very little experimental investigation of the potential effects on response to water stress. One of the few papers to report on this is Mansfield *et al.* (1988). As Figure 19.4 shows, well-watered plants of *Phleum pratense* had a relative growth rate averaging about 0.180 week^{-1} but when water was withheld, it fell with the concentration of $SO_2 + NO_2$. Even relatively low concentrations of the pollutants altered the rate of depletion of soil water, as well as the pattern with depth (Figure 19.5). The extent to which these effects were due to stomatal dysfunction and altered root growth/morphology is not known. Clearly, there

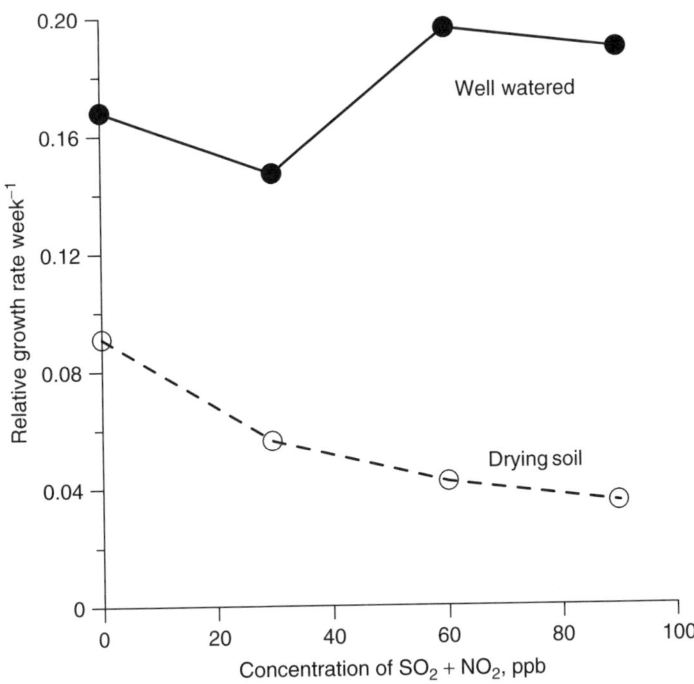

Figure 19.4 Effects on the mean relative growth rate (week^{-1}) of *Phleum pratense* exposed for 40 days to $SO_2 + NO_2$, followed by 23 days in which plants were watered or water was withheld. (Data from Mansfield *et al.*, 1988.)

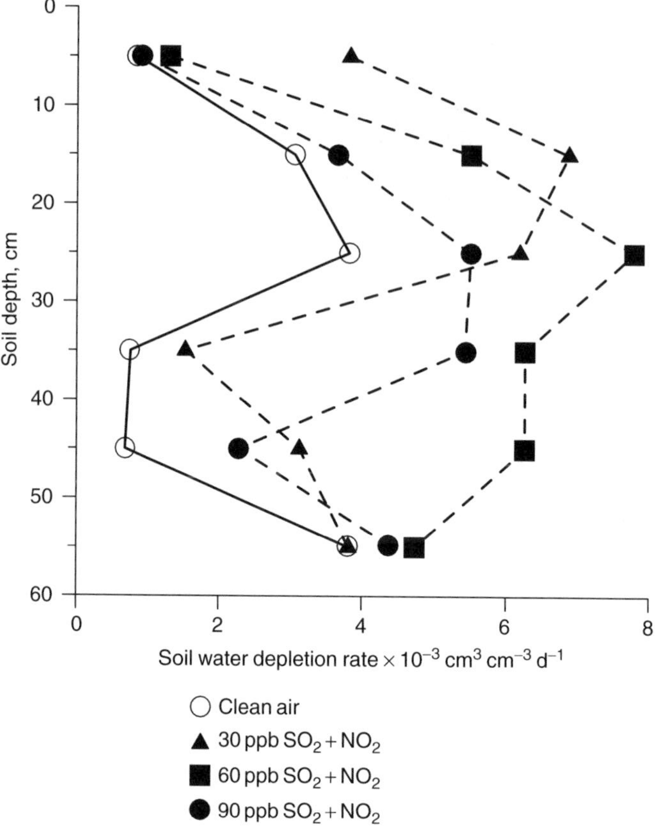

Figure 19.5 Rates of soil water depletion with depth under unwatered plants of *Phleum pratense* exposed to $SO_2 + NO_2$ (Redrawn from Mansfield *et al.*, 1988.)

are potentially important interactions between the pollutants and response to water stress but the question that remains to be answered is the extent to which these effects, at present-day SO_2 and NO_2 concentrations, affect plant performance in the field.

Nitrogen enrichment and winter damage

Over the past few decades, there has been a well-documented increase in the emission of nitrogen-containing compounds, notably nitric oxide (NO) and nitrogen dioxide (NO_2) arising from combustion, and ammonia (NH_3) which mostly originates from animal production. As a consequence nitrogen deposition has also increased to rates ranging from a background of < 5 to >60 kg ha^{-1} y^{-1} in the worst places. The potential effects of nitrogen enrichment are far reaching (Vitousek *et al.*, 1997). Ecological aspects are considered in detail in Chapter 12.

There have been reports of increased nitrogen fertilization altering plant response to both drought and cold for almost a century (see summary in Barnes *et al.*, 1996). The reasons for this are increased leaf production and transpiration surface, usually at the expense of root growth, and prolongation or stimulation of bud burst at times when plants should be cold-hardened. In order to assess the potential for such interactions it is necessary to distinguish between the effects of gaseous pollutants (principally NO_2 and NH_3) and deposited nitrogen (NO_3^- and NH_4^+).

There has been little published on the effects of NO_2 on freezing tolerance but Renner (1990) found no effect of high concentrations on the freezing resistance of ryegrass. Similarly, Freer-Smith and Mansfield (1987) found only marginal effects of 30 ppb $NO_2+/-SO_2$ on the frost hardiness of Norway spruce. However, bud break was earlier, which could make the trees sensitive to late frosts. More recently, Caporn *et al.* (2000) reported that 40 ppb of $SO_2 + NO_2$ reduced the frost resistance of shoots of heather, *Calluna vulgaris*. Exposing the plants for 8 months over the growing season increased plant growth and root/shoot ratio. Frost tolerance, as measured by electrolyte leakage, was reduced, whether exposure was during the hardening or de-hardening period. Although 40 ppb is much lower than most earlier work, the concentrations are still high compared with those found in rural habitats in Europe and the USA so it would be very desirable to repeat the experiment using lower concentrations.

The literature on NH_3 is sparse but more conclusive. Field observations suggest that NH_3 can reduce frost hardiness. For example, Pietilä *et al.* (1991) reported dieback of pine near fur farms in winter while De Temmerman *et al.* (1987) noted that frost injury was most pronounced near NH_3 sources. Experimental support was provided by open-top chamber studies which showed that ammonia at 105 $\mu g\ m^{-3}$ (typical of areas close to animal units) increased electrolyte leakage of Scots pine needles exposed to freezing temperatures (Dueck *et al.*, 1990). Exposure to NH_3 and then to temperatures lower than $-10\,°C$ led to more severe damage in early autumn and spring than in winter. The effects were thought to be due to prolongation of growth delaying winter hardening. Dueck (1990) used several NH_3 concentrations, some in combination with SO_2. There was no effect at -4 or $-7\,°C$ but it started at $-10\,°C$ and the effects of the two pollutants were synergistic. In the case of *Calluna*, fumigated with 100 $\mu g\ m^{-3}$ NH_3 for 4–7 months, frost sensitivity was unaffected in autumn but it was in February, just before the start of growth (Van der Eerden *et al.*, 1991). Although there are reports of NH_3 not having any effect on frost hardiness (e.g. Clement *et al.*, 1995) the body of evidence supports the view that there are important effects at NH_3 concentrations that occur near sources. Conversely, there is no evidence of effects at the typical rural concentrations of $<1\ \mu g\ m^{-3}$.

The potential deleterious effects of deposited nitrogen were first realized in the 1980s. One of the first suggestions was that it might extend the growing season and change the synchrony between hardening and the occurrence of frost, which would predispose vegetation to winter injury (Nihlgård, 1985, Friedland *et al.*, 1985). The extensive literature on fertilizers has repeated suggestions that nitrogen may predispose plants to greater risk of winter injury (see Barnes *et al.*, 1996) but it is conflicting and difficult to interpret in relation to deposited nitrogen. After reviewing it, Barnes *et al.* (1996)

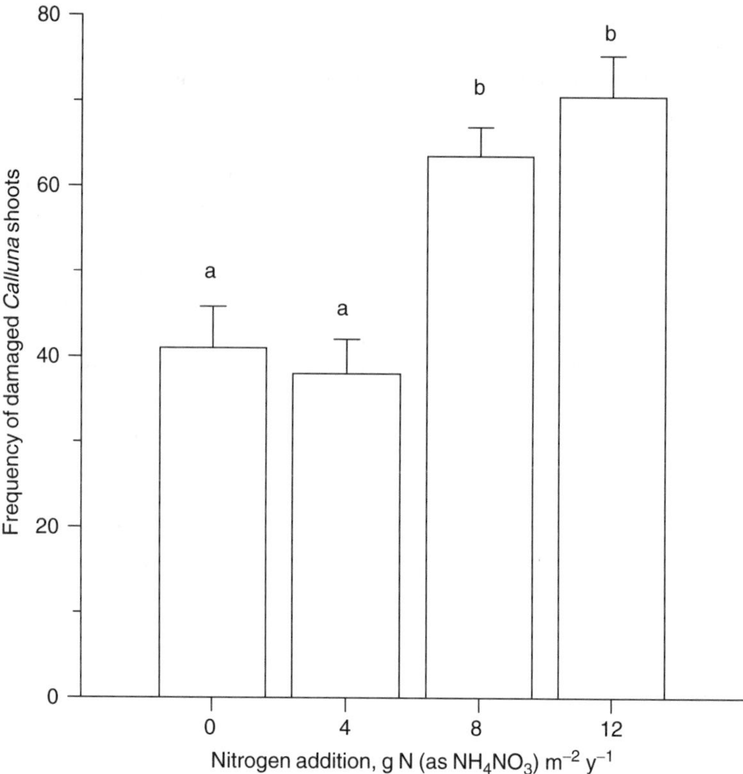

Figure 19.6 Effects of long-term addition of NH_4NO_3 on the % frequency of winter damaged *Calluna* shoots. Columns sharing a letter are not significantly different. (Redrawn from Carroll *et al.,* 1999.) Reproduced by permission of Carroll *et al.*, 'The effect of increased deposition of atmospheric nitrogen on *Callum vulgaris* in upland Britain' in *New Phytologist*, Vol. 141, pp. 423–431, Blackwell Science

concluded that cause–effect relationships for the effects of nitrogen on frost hardiness were unsubstantiated (at least for woody species). The exception appeared to be in relation to *Calluna*. Heathland dominated by heather is an important part of the oceanic European landscape and it has been reducing in area for several years (see Chapter 12). The causes of the loss and the conversion of heath to grass are complex but there is evidence that an interaction between deposited nitrogen and winter injury is involved. Bobbink and Heil (1993) concluded that 'Although the significance of frost sensitivity of *Calluna* plants is still uncertain at ambient nitrogen and sulphur loads, high levels of ammonia or ammonium sulphate seem to influence this phenomenon'. Since then there have been some important observations made in long-term experiments. Using low rates of nitrogen addition (7.7 and 15.4 kg ha^{-1} y^{-1}) Power *et al.* (1998) found no effect on frost sensitivity of *Calluna* in early winter but a slightly decreased hardiness in April, coupled with accelerated spring bud burst. Lee and Caporn (1998) and Carroll *et al.* (1999) used higher rates of application but their data are compatible with

Power *et al.* (1998). They showed that between 1990 and 1994 addition of NH_4NO_3 to *Calluna* moorland produced a marked dose-related increase in shoot extension, canopy height and litter accumulation. However, since 1994 there has been no dose response in shoot extension but an increase in susceptibility to late winter injury (Figure 19.6). The damage was characterized by browning of shoot tips in early to late spring. This work, although using high rates of nitrogen application, adds to the evidence that nitrogen–winter stress interactions are involved in the decline of *Calluna*.

References

Ashenden, T.W. and Mansfield, T.A. (1977). Influence of wind speed on the sensitivity of ryegrass to SO_2. *Journal of Experimental Botany* **28**: 729–735.

Atkinson, C.J., Wookey, P.A. and Mansfield, T.A. (1991). Atmospheric pollution and the sensitivity of stomata on barley leaves to abscisic acid and carbon dioxide. *New Phytologist* **117**: 535–541.

Baker, C.K., Unsworth, M.H. and Greenwood, P. (1982). Leaf injury on wheat plants exposed in the field in winter to SO_2. *Nature* **299**: 149–151.

Barnes, J.D. and Davison, A.W. (1988). The influence of ozone on the winter hardiness of Norway spruce (*Picea abies* [L.] Karst.). *New Phytologist* **108**: 159–166.

Barnes, J.D., Eamus, D., Davison, A.W., Ro-Poulsen and Mortensen, L. (1990). Persistent effects of ozone on needle water loss and wettability in Norway spruce. *Environmental Pollution* **63**: 345–363.

Barnes, J.D., Hull, M.R. and Davison, A.W. (1996). Impacts of air pollutants and rising CO_2 in winter time In: *Plant Growth and Air Pollution* (ed. by M Yunus, M Iqbal). London, Wiley, pp. 135–166.

Bell, J.N.B. (1982). Sulphur dioxide and the growth of grasses. In: *Effects of Gaseous Air Pollution in Agriculture and Horticulture* (ed. by M.H. Unsworth and D.P. Ormrod). Butterworth Scientific, London, pp. 225–246.

Bell, J.N.B., Rutter, A.J. and Relton, J. (1979). Studies on the effects of low levels sulphur dioxide on the growth of *Lolium perenne* L. *New Phytologist* **83**: 627–643.

Bell, J.N.B. and Clough, W.S. (1973). Depression in yield of rye grass exposed to SO_2. *Nature* **241**: 47–49.

Black, V.J. (1982). Effects of sulphur dioxide on physiological processes in plants. In: *Effects of Gaseous Air Pollution in Agriculture and Horticulture*. (ed. by M.H. Unsworth and D.P. Ormrod). Butterworth Scientific, London, pp. 67–91.

Black, C.R. and Black, V.J. (1979). The effects of low concentrations of sulphur dioxide on stomatal conductance and epidermal cell survival in field bean (*Vicia faba* L.). *Journal of Experimental Botany* **30**: 291–298.

Blank, L.W., Roberts, T.M. and Skeffington, R.A. (1988). New perspectives on forest decline. *Nature* **336**: 27–30.

Bleasdale, J.K.A. (1952). Atmospheric pollution and plant growth. *Nature* **169**: 376–377.

Bleasdale, J.K.A. (1973). Effects of coal smoke pollution gases on the growth of ryegrass (*Lolium perenne* L.). *Environmental Pollution* **5**: 275–280.

Bobbink, R. and Heil, G.W. (1993). Atmospheric deposition of sulphur gas and nitrogen in heathland ecosystems. In: *Heathlands: Patterns and Processes in a Changing Environment* (ed. by R. Aerts and G.W. Heil). Kluwer Academic Publishers, pp. 25–50.

Bosch, C. and Rehfuess, K.E. (1988). On the significance of frost phenomena with the novel forest damage. *Forstwissenschaftliches Centralblatt* **107**: 123–130.

Brown, K.A., Roberts, T.M. and Blank, L.W. (1987). Interaction between ozone and cold sensitivity in Norway spruce: a factor contributing to the forest decline in Central Europe? *New Phytologist* **105**: 149–155.

Cape, J.N., Leith, I.D., Fowler, D., Murray, M.B., Sheppard, L.J., Eamus, D. and Wilson, R.H.F. (1991). Sulphate and ammonium in mist impair the frost hardening of red spruce seedlings. *New Phytologist* **188**: 119–126.

Caporn, S.J.M., Ashenden, T.W. and Lee, J.A. (2000). The effect of exposure to NO_2 and SO_2 on frost hardiness of *Calluna vulgaris*. *Environmental and Experimental Botany* **43**: 111–119.

Carroll, J.A., Caporn, S.J.M., Cawley, L., Read, D.J. and Lee, J.A. (1999). The effect of increased deposition of atmospheric nitrogen on *Calluna vulgaris* in upland Britain. *New Phytologist* **141**: 423–431.

Chappelka, A.H. and Freer-Smith P.H. (1995). Predisposition of trees by air pollutants to low temperatures and moisture stress. *Environmental Pollution* **87**: 105–117.

Chappelka, A.H., Kush, J.S., Meldahl, R.S. and Lockaby, B.G. (1990). An ozone-low temperature interaction in Loblolly pine (*Pinus taeda* L.). *New Phytologist* **114**: 721–726.

Clement, J.M.A.M., Venema, J.H. and Van Hasselt, P.R. (1995). Short-term exposure to atmospheric ammonia does not affect low temperature hardiness of winter wheat. *New Phytologist* **131**: 345–351.

Cohen, J.B. and Ruston, A.G. (1925). *Smoke, a Study of Town Air*. 2nd edition. Edward Arnold, London.

Cooley, D.R. and Manning, W.J. (1987). The impact of ozone on assimilate partitioning in plants. *Environmental Pollution* **47**: 95–113.

Curry, J.R. and Church, T.W. (1952). Observations on winter drying of conifers in the Adirondacks. *Journal of Forestry* **50**: 114–116.

Darrall, N.M. (1989). The effects of air pollution on physiological processes in plants. *Plant, Cell and Environment* **12**: 1–30.

Davies, T. (1980). Grasses more sensitive to SO_2 pollution in conditions of low irradiance and short days. *Nature* **284**: 483–485.

Davison, A.W. and Bailey, I.F. (1982). SO_2 pollution reduces the freezing resistance of ryegrass. *Nature* **297**: 400–402.

Davison, A.W. and Barnes, J.D. (1986). Effects of winter stress on pollutant responses. In: *How are the Effects of Air Pollutants on Agricultural Crops Influenced by the Interaction with Other Limiting Factors?* Air Pollution Research Report CEC DGXII, Brussels, pp. 16–32.

DeHayes, D.H., Waite, C.E., Ingle, M.A. and Williams, M.W. (1990). Winter injury susceptibility and cold tolerance of current and year-old needles of red spruce trees from several provenances. *Forest Science* **36**: 982–994.

DeHayes, D.H., Thornton, F.C., Waite, C.E. and Ingle, M.A. (1991). Ambient cloud deposition reduces cold tolerance of red spruce seedlings. *Canadian Journal of Forest Research* **21**: 1292–1295.

De Temmerman, L.D., Ronse, A., Van den Cruys A. and Meeus-Verdinne, K. (1987). Ammonia and pine die-back in Belgium. In: *Air Pollution and Ecosystems. Proceedings International Symposium, Grenoble, France* (ed. by P. Mathy). Reidel, Dordrecht, pp. 774–779.

Dueck, T.A. (1990). Effect of ammonia and sulphur dioxide on the survival and growth of *Calluna vulgaris* (L.) Hull seedlings. *Functional Ecology* **4**: 109–116.

Evans, L.S. (1986). Proposed mechanisms of initial injury – causing apical die back in red spruce at high elevation in eastern North America. *Canadian Journal of Forestry Research* **16**: 1113–1116.

Feiler, S. (1985). Influence of SO_2 on membrane permeability and consequences on frost sensitivity of spruce (*Picea abies* [L.] Karst.). *Flora* **177**: 217–226.

Freer-Smith, P.H. and Mansfield, T.A. (1987). The combined effects of low temperature and $SO_2 + NO_2$ pollution on the new season's growth and water relations of *Picea sitchensis*. *New Phytologist* **106**: 225–237.

Friedland, A.J., Hawley, G.J. and Gregory, R.A. (1985). Investigations of nitrogen as a possible contributor to red spruce (*Picea rubens* Sarg.) decline. In: *Proceedings of Symposium on the Effects of Air Pollutants on Forest Ecosystems*. University of Minnesota Press, Minneapolis, pp. 83–107.

Havas, P.J. (1971). Injury in pines growing in the vicinity of a chemical processing plant in Northern Finland. *Acta Forest. Fenn.* **121**: 1–21.

Horsman, D.C., Roberts, T.M. and Bradshaw, A.D. (1978). Evolution of sulphur dioxide tolerance in perennial ryegrass. *Nature* **276**: 493–494.

Horsman, D.C., Roberts, T.M., Lambert, M. and Bradshaw, A.D. (1979). Studies on the effect of sulphur dioxide on perennial ryegrass (*Lolium perenne* L.) I. Characteristics of the fumigation system and preliminary results. *Journal of Experimental Botany* **30**: 485–493.

Huttunen, S. (1978). Effects of air pollution on provenances of Scots pine and Norway spruce in Northern Finland. *Silva Fennica* **12**: 1–16.

Huttunen, S., Havas, P. and Laine, K. (1981a) Effects of air pollutants on the wintertime water economy of the Scots pine (*Pinus sylvestris*). *Holarctic Ecology* **4**: 94–101.

Huttunen, S., Kärenlampi, L. and Kolari, K. (1981b) Changes in osmotic potential and some related physiological variables in the needles of polluted Norway spruce (*Picea abies*). *Annales Botanici Fennicae* **18**: 63–71.

Innes, J.L. (1992). Forest decline. *Progress in Physical Geography* **16**: 1–64.

Johnson, A.H. (1992). The role of biotic stresses in the decline of red spruce in high elevation forests of the Eastern United States. *Annual Review of Phytopathology* **30**: 349–367.

Johnson, A.H., Cook, E.R. and Siccama, T.G. (1988). Climate and red spruce growth and decline in the northern Appalachians. *Proceedings of the National Academy of Sciences of the United States of America* **85**: 5369–5373.

Johnson, A.H., DeHayes, D.H. and Siccama, T.G. (1996). Role of acid deposition in the decline of red spruce (*Picea rubens* Sarg.) in the montane forests of the north eastern USA. In: *Forest Trees and Palms: Disease and Control* (ed. by S.P. Raychauduri, K. Maramarosch), Science Publishers, New Hampshire, pp. 49–71.

Johnson, A.H. and McLaughlin, S.B. (1986). *The Nature and Timing of the Deterioration of Red Spruce in the Northern Appalachian Mountains from Acid Deposition, Long Term Trends*. National Academy Press, Washington, DC.

Jones, T. and Mansfield, T.A. (1982). The effect of SO_2 on growth and development seedlings of *Phleum pratense* under different light and temperature environments. *Environmental Pollution* **27**: 57–71.

Kandler, O. and Innes, J.L. (1995). Air pollution and forest decline in Central Europe. *Environmental Pollution* **90**: 171–180.

Kauppi, P.E., Mielikainen, K. and Kuusela, K. (1992). Biomass and carbon budget of European forests, 1971–1990. *Science* **256**: 70–74.

Keller, T. (1978a) Frostschaden als forge einer latenten. Immisionschadigung. *Staub.* **38**: 24–26.

Keller, T. (1978b) Wintertime atmospheric pollutants – do they affect the performance of deciduous trees in the ensuing growing season? *Environmental Pollution* **16**: 243–247.

Kramer, P.J. (1983). *Water Relations of Plants*. Academic Press New York.

Lee, J.A. and Caporn, S.J.M. (1998). Ecological effects of atmospheric reactive nitrogen deposition on semi-natural terrestrial ecosystems. *New Phytologist* **139**: 127–134.

Macdowall, F.D.H. (1965). Predisposition of tobacco to ozone damage. *Canadian Journal of Plant Science* **45**: 1–12.

Mansfield, T.A. (1998). Stomata and plant water relations: does air pollution create problems? *Environmental Pollution* **101**: 1–11.

Mansfield, T.A. and Pearson, M. (1996). Disturbances in stomatal behaviour in plants exposed to air pollution. In: *Plant Growth and Air Pollution* (ed. by M Yunus, M Iqbal). Wiley, London, pp. 179–193.

Mansfield, T.A., Wright, E.A., Lucas, P.W. and Cottam, D.A. (1988). Interactions between air pollutants and water stress. In: *Air Pollution and Plant Metabolism* (ed. by S. Schulte-Hostede, N.M. Darrall, L.W. Blank, A.R. Wellburn), Elsevier, London, pp. 288–306.

Materna, J. (1974). Einfluss der SO_2-immissionen auf Fichtenpflanzen in Wintermonaten. *IXth Int. Tagung uber Luftverunreinigung und Forstwirtschaft*, Marianske Lazne, Czechoslovakia, pp. 14–26.

Materna, J. (1984). Impact of atmospheric pollution on natural ecosystems. In: *Air Pollution and Plant Life* (ed. by M. Treshow), Wiley, London, pp. 376–416.

Michael, G., Feiler, S., Ranft, H. and Tesche, M. (1982). Der Einfluss von Schwefeldioxid und Frost auf Fichten (*Picea abies* L. Karst.) *Flora* **172**: 317–326.

Neighbour, E.A., Cottam, D.A. and Mansfield, T.A. (1988). Effects of sulphur dioxide and nitrogen dioxide on the control of water loss by birch (*Betula spp.*). *New Phytologist* **108**: 149–157.

Nihlgård, B. (1985). The ammonium hypothesis. An additional explanation for forest dieback in Europe. *Ambio* **14**: 2–8.

Pietilä, M., Lahdesmaki, P., Pietilainen, P., Ferm, A., Hytonen, J. and Patila, A. (1991). High nitrogen deposition causes changes in amino acid concentrations and protein spectra in needles of Scots pine (*Pinus sylvestris*). *Environmental Pollution* **72**: 103–115.

Power, S.A., Ashmore, M.R., Cousins, D.A. and Sheppard, L.J. (1998). Effects of nitrogen deposition on the stress sensitivity of *Calluna vulgaris*. *New Phytologist* **138**: 663–673.

Renner, C.J. (1990). Interactive effects of sulphur dioxide, nitrogen dioxide and the winter environment on *Lolium perenne*. PhD Thesis, Newcastle University.

Robinson, M.F., Heath, J. and Mansfield, T.A. (1998). Disturbances in stomatal behaviour caused by air pollutants. *Journal of Experimental Botany* **49**: 461–469.

Sandermann, H., Wellburn, A.R. and Heath, R.L. (1997). Forest decline and ozone: synopsis. In. *Forest Decline and Ozone* (ed. by H. Sandermann, A.R. Wellburn, R.L. Heath). Springer-Verlag, Berlin Heidelberg. Ecological Studies **127**: pp. 369–377.

Schulze, E.D. and Freer-Smith, P.H. (1990). An evaluation of forest decline based on field observations focussed on Norway spruce, *Picea abies*. *Proceedings of the Royal Society of Edinburgh Section B Biological Sciences* **97**: 155–168.

Sheppard, L.J. (1994). Causal mechanisms by which sulphate, nitrate and acidity influence frost hardiness in red spruce. Review and hypothesis. *New Phytologist* **127**: 69–82.

Sheppard, L. and Pfanz, H. (2000). Impacts of air pollutant on cold hardiness. In: *Conifer Cold Hardiness* (ed. by F. Bigras, S. Columbo), Kluwer Academic, The Netherlands, pp. 305–333.

Sheppard, L.J., Smith, R.I. and Cannell, M.G.R. (1989). Frost hardiness of *Picea rubens* growing in spruce decline regions of the Appalachians. *Tree Physiology* **5**: 23–27.

Skelly, J.M. and Innes, J.M. (1994). Waldsterben in the forests of Central Europe and Eastern North America: fantasy or reality? *Plant Disease* **78**: 1021–1032.

Skärby, L., Ro-Poulsen, H., Wellburn, F.A.M. and Sheppard, L.J. (1998). Impacts of ozone on forests: a European perspective. *New Phytologist* **139**: 109–122.

Spiecker, H. (1999). Overview of recent growth trends in European forests. *Water, Soil and Air Pollution* **116**: 33–46.

Spiecker, H., Mielikäinen, K., Köhl, M. and Skovsgaard, J.P. (eds.) (1996). *European Forest Institute Research Report No. 5*. Springer-Verlag, Berlin.

Taylor, G.E. and Murdy, W.H. (1975). Population differentiation of an annual plant species, *Geranium carolinianum*, in response to sulphur dioxide. *Botanical Gazette* **136**: 212–215.

Taylor, G. and Ferris, R. (1996). Influence of air pollution on root physiology and growth. In: *Plant Growth and Air Pollution* (ed. by M. Yunus, M. Iqbal), Wiley, London, pp. 375–393.

Thomas, M.D. (1951). Gas damage to plants. *Annual Revue of Plant Physiology* **2**: 293–322.

Tingey, D.T., Heck, W.W. and Reinert, R.A. (1971). Effects of low concentrations of ozone and sulfur dioxide on foliage, growth and yield of radish. *Journal of the American Society of Horticultural Science* **96**: 369–371.

Tingey, D.T. and Hogsett, W.E. (1985). Water stress reduces ozone injury via a stomatal mechanism. *Plant Physiology* **77**: 944–947.

Van der Eerden, L.J., Dueck, T.A., Berdowski, J.J.M., Greven, H. and Van Dobben, H.F. (1991). Influence of NH_3 and $(NH_4)_2SO_4$ on heathland vegetation. *Acta Botanica Neerlandica* **40**: 281–297.

Vann, D.R., Strimbeck, G.R. and Johnson, A.H. (1992). Effects of ambient levels of airborne chemicals on freezing resistance of red spruce foliage. *Forest Ecology Management* **51**: 67–79.

Vitousek, P.M., Aber, J., Howarth, R.W., Likens, G.E., Matson, P.A., Schindler, D.W., Schlesinger, W.H. and Tilman, G.D. (1997). Human alteration of the global nitrogen cycle: sources and consequences. *Ecological Applications* **7**: 737–750.

Warren Spring Laboratory (1976). *National Survey of Air Pollution 1961–1971. Volume 4*. HMSO London.

Wellburn, A.R., Barnes, J.D., Lucas, P.W., McLeod, A.R. and Mansfield, T.A. (1997). Controlled ozone exposures and field observations of ozone effects in the UK. In: *Forest Decline and Ozone: A Comparison of Controlled Chamber and Field Experiments*. Ecological Studies Series Vol. 127 (ed. by H. Sandermann, A.R. Wellburn, R.L. Heath), Springer-Verlag, Heidelberg, pp. 201–236.

White, G.F. and Hass, J.E. (1975). *Assessment of research on natural hazards*. MIT Press.

Whitmore, M.E. and Mansfield, T.A. (1983). Effects of long-term exposures to SO_2 and NO_2 on *Poa pratensis* and other grasses. *Environmental Pollution* **31**: 217–235.

Wilson, G.B. and Bell, J.N.B. (1985). Studies on the tolerance to SO_2 of grass populations in polluted areas. III. Investigations on the rate of development of tolerance. *New Phytologist* **100**: 63–77.

Wiltshire, J.J., Unsworth, M.H. and Wright, C.J. (1994). Seasonal changes in water use of ash trees exposed to ozone episodes. *New Phytologist* **127**: 349–354.

20 Effects of air pollutants on biotic stress

W. FLÜCKIGER, S. BRAUN AND E. HILTBRUNNER

Introduction

In their effect on plants, air pollutants are interacting with other environmental abiotic and biotic stress factors in a complex way. In particular, host plant–parasite relationships can be influenced considerably. In the context of forest damages and yield losses due to air pollution parasites are often involved. Air pollution as an environmental stress factor affecting host–parasite relationships has previously been discussed by several authors (Heagle, 1973; Treshow, 1975; Laurence, 1981; Hughes and Laurence, 1984; Riemer and Whittaker, 1989; Heliövaara and Väisänen, 1993).

Pollutants can act either directly or indirectly. They can be toxic to the parasite or alter its behaviour or its metabolism. Indirect effects comprise shifts in the abundance of the food plant and changes in plant surfaces, as well as variations in the host plant quality. Pollutants can also lead to alterations at higher trophic levels by affecting predators, parasitoids and pathogens (Riemer and Whittaker, 1989). Although there have been many field surveys and experimental studies carried out over the years, the complex mechanisms of changed host plant–parasite interactions under pollution stress are still not well understood.

The current knowledge on changes in plant–parasite interactions in the presence of air pollutants has mainly been gained from investigations on the effects of fluorides, SO_2, NO_x and to a lesser extent of O_3. Up to now, the majority of field studies have been carried out as gradient investigations around local sources of industrial pollutants, particularly SO_2, NO_x and fluorides. However, in Northern/Central Europe and North America the significance of SO_2 as the most important pollutant has continuously decreased as a consequence of successful abatement strategies (Shannon, 1999; Tarrassón, 1998).

On the other hand, since the 1950s nitrogen deposition in Europe has generally increased (Goulding and Blake, 1993) (Figure 20.1). Actual nitrogen loads in Western/-Central Europe exceed the critical loads for nitrogen for various ecosystems such as forests, heathlands, mesotrophic fens, ombrotrophic bogs, species-rich grassland (UN/ECE, 1996; Posch *et al.*, 1997). An exceedance of critical loads for N according to UN/ECE (1996) may result in shifts in plant communities towards eutrophication (Bobbink *et al.*, 1992; Lee and Caporn, 1998) and nutrient imbalances such as increasing N and decreasing P concentrations (Flückiger and Braun, 1998). (See also Chapter 12.) Insect herbivores are considered to be limited frequently by the availability

Air Pollution and Plant Life, second edition. Edited by J.N.B. Bell and M. Treshow. ISBN 0 471 49090 3 (HB), 0 471 49091 1 (PB).

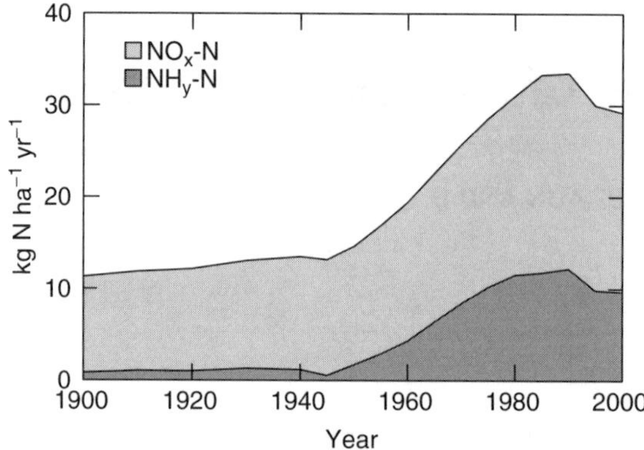

Figure 20.1 Historical development of average N-deposition in Swiss forests, calculated from emission data and transport models (Flückiger and Braun, 1998). Reprinted from *Environmental Pollution*, vol. 102, Flückiger *et al.*, 'Nitrogen deposition in Swiss forests' fig. 2 from pp. 69–76 (1998), with permission from Elsevier Science

of nitrogen. Increased nitrogen in the foliage and nutrient imbalances, respectively, may lead to enhanced susceptibility of trees to parasites. The significance of reduced and oxidized nitrogenous compounds for soil acidification is increasing due to decreasing importance of sulphur compounds. There has been conspicuous acidification of forest soils over the past decades (Falkengren-Grerup, 1987; Berggren *et al.*, 1992; Fitze *et al.*, 1991). Furthermore, nitrogen oxide emissions contribute to the formation of tropospheric ozone. Current ozone concentrations in Europe represent a widespread risk for forest trees and may influence insect performance. Nevertheless, field studies on ozone effects on host plants and herbivores are still rare.

Until recently, there has been little interest in the impacts of elevated CO_2 as the most important 'greenhouse gas' on plant–parasite relationships. It is thought that the productivity of plants may grow under elevated CO_2, but the quality of the host plant may decline because of an increase of carbon-based allelochemicals and a decrease in foliar nitrogen concentration (Webber *et al.*, 1994; Hättenschwiler and Körner, 1996; Stiling *et al.*, 1999).

Interactions between insect herbivores, pathogens and host plants

Insect responses vary considerably with duration and magnitude of the pollution stress. The response also depends on the host plant, insect species and the developmental state of the organism. It is often assumed that the typical interaction between insects and pollutants is due to changes of the host plant quality; the resulting benefits to the insects lead to further harm to the plant. This assumption is possibly too simple. For instance, the damage caused by the leaf hopper *Ossannissonola callosa* to seedlings of sycamore (*Acer pseudoplatanus*) increased the overnight uptake rate of SO_2 by the plants (Warrington *et al.*, 1989). This means that there is a potential for a synergistic interaction between herbivores and pollutants in their adverse effects on plants.

Insects

Field observations

Changes in insect population densities near to pollution sources, mainly sources of fluorides, SO_2, and NO_x, have been observed for about 170 years. The first outbreak of an insect species in forests in relation to air pollution was described by Beling in 1831 (Beling, 1865, cit. in Cramer, 1951). He reported an outbreak of the tortricid moth *Epinotia tedella* in a smoke-damaged young Norway spruce stand in the Harz mountains. Gerlach (1898) and Gerlach (1907) named the beetles *Pissodes harcyniae* and *Pissodes scabricollis* 'smoke weevils' (German: *Rauchrüssler*), as pest outbreaks were widespread in smoke-damaged spruce stands in Saxony.

It is generally believed that sucking insects show higher infestation rates with increasing pollution levels. Data collected by the Rothamsted Insect Survey in the United Kingdom revealed that the abundance of 39 aphid species was positively correlated with SO_2 concentrations. For 34 aphid species no correlations were found and only one species was negatively related to by SO_2 (Houlden *et al.*, in Brown 1995). *Aphis pomi* on *Crataegus sp.* produced the highest population density in the most polluted zone in the central reserve of a motorway. Along the verge and out to a distance of up to 200 m, respectively, the population was significantly smaller (Flückiger *et al.*, 1978). Similar observations were made by Przybylski (1979) who found that the number of aphids was greater close to a motorway. Other insect groups are influenced by SO_2 and NO_x as well. The population level of pine bark bug *Aradus cinnamomeus* was lowest closest to NO_x and SO_2 emitting factories, highest at a distance of 1 to 2 km and then decreased again to background levels with increasing distance from the factories (Heliövaara and Väisänen, 1986). The pine bud moth *Exoteleia dodecella* on Scots pine and the needle miner *Epinotia tedella* on Norway spruce produced the highest population density at a distance of 1.35 and 1.53 km, respectively, from a SO_2 emitting pulp mill in Finland (Oksanen *et al.*, 1996).

Experimental evidence of interactions

Direct effects on insects

There have been only a few studies showing direct effects of air pollutants on insects. Often unrealistic high pollution concentrations were used. Honeybees and male sweat bees (*Lasioglossum zephyrum*) which had been exposed to SO_2 had a reduced brood rearing and flight activity (Hillman and Benton, 1972; Ginevan *et al.*, 1980). Reduced flight activity of male bees might result in a less successful mating behaviour. Feir and Hale (1983) found that the milkweed bug *Oncopeltus fasciatus* exhibited a reduced growth, a smaller final size at maturity and a lower egg production when exposed to a mixture of SO_2, NO, NO_2, CO, O_3 and hydrocarbons at levels comparable to the air pollution concentrations in US cities.

Indirect effects on insects: fumigation experiments

Fumigation experiments with SO_2 and NO_x. In accordance with field observations, a number of fumigation experiments showed that many insect species develop faster and better on their food plants which had been exposed to air pollutants, mainly SO_2 and NO_x. Dohmen *et al.* (1984) reported an increased mean relative growth rate (MRGR) of *Aphis fabae* fed on *Vicia faba* previously fumigated with 400 $\mu g\ m^{-3}$ SO_2 or NO_2 for 7 days. Direct fumigation of aphids on artificial diets had no effect on MRGR. Houlden *et al.* (1990) assessed the MRGR of eight aphid species on six agricultural crops. Prefumigation with 100 nl l^{-1} SO_2 or NO_2 led to an increase in MRGR by between 7% and 75% in all but two cases. Only *Acyrthosiphon pisum* feeding on *Vicia faba* showed a decrease in MRGR. However, feeding the same aphid species on pea fumigated with 90 and 110 ppb SO_2 led to an increase in MRGR by up to 11%. At higher concentrations, MRGR declined again and at 200 ppb SO_2 the growth rate dropped below the rate found on the control plants (Warrington, 1987).

The conifer aphids, *Elatobium abietum, Schizolachnus pineti* and *Cinara pilicornis* were also promoted in growth (increased MRGR) when their host plants had been exposed to SO_2 and NO_x (McNeill *et al.*, 1987; Warrington and Whittaker, 1990, Brown *et al.*, 1993). In chamber experiments with charcoal/Purafil-filtered and non-filtered air along a very busy motorway near Basle (NO_x concentrations up to 1500 $\mu g\ m^{-3}$), the population development of *Aphis pomi* on *Crataegus* sp. and of *Aphis fabae* on *Viburnum opulus* and *Phaseolus vulgaris*, respectively, was significantly enhanced under ambient air conditions (Braun and Flückiger, 1985; Bolsinger and Flückiger, 1984).

Fumigation experiments with ozone. The effects of ozone on aphid performance are less consistent. Studies revealed positive effects of ozone on aphid growth in eight cases, negative effects in four cases and no effect in seven cases (Heliövaara and Väisänen, 1993; Holopainen *et al.*, 1994; Summers *et al.*, 1994). Feeding on O_3-treated food plants significantly enlarged the MRGR of the aphid *Rhopalosiphum padi* (Warrington, 1989). Exposure of Sitka spruce to ozone induced a positive effect on the MRGR of *Cinara pilicornis* when the rearing temperature was 20 °C. Higher temperatures (>23 °C) led to a decrease in the growth rate again (McNeill and Whittaker, 1990). Likewise, the cottonwood aphid, *Chaitophorus populicola*, showed no growth response on cottonwood exposed to 200 ppb O_3 for 5 hours (Coleman and Jones, 1988a). Ozone-rich ambient air (peaks up to 120 ppb O_3) increased the population growth rate of *Phyllaphis fagi* on beech compared to charcoal/Purafil-filtered air, but decreased the growth rate of *Aphis fabae* on bean (Braun and Flückiger, 1989) (Table 20.1).

In contrast to several shoot aphids which can be promoted in growth on O_3- or SO_2- treated host plants, the aphids *Pachypappa vesicalis, Pachypappa tremulae* and *Prociphilus xylostei* feeding on spruce roots seemed to be inhibited when spruce trees were subjected to SO_2- or O_3-exposure (Salt and Whittaker, 1995).

Jeffords and Endress (1984) demonstrated that a moderate prefumigation of oak seedlings with 90 ppb O_3 made the oak leaves less palatable for the gipsy moth *Lymantria dispar*, compared to ambient air (35 ppb O_3) or to high ozone pretreatment (135 ppb O_3). After ozone fumigation of soybean plants and leaves, larvae of the Mexican bean beetle, *Epilachna varivestis*, weighed more and developed faster (Chappelka *et al.*, 1988; Lee *et al.*, 1988). Willow leaf beetles, *Plagiodera versicolor*,

Table 20.1 Peak population size of *Phyllaphis fagi* on *Fagus sylvatica* (left) and of *Aphis fabae* on *Phaseolus vulgaris* (right) in fumigation experiments performed in different years and amino acid concentrations in a phloem exudate taken in 1987. From Braun and Flückiger (1989). Ozone dose was calculated as daylight dose, above a threshold of 40 ppb. (AOT40)

	Peak population size (aphids/plant)						O_3 AOT40 ppm. h
	Phyllaphis fagi on *Fagus sylvatica*			*Aphis fabae* on *Phaseolus vulgaris*			
Year	Ambient air	Filtered air	*p*	Ambient air	Filtered air	*p*	
1985				520 ± 218	924 ± 392	0.01	13.0
1986	129 ± 101	73 ± 60	0.01	1628 ± 675	1923 ± 522	n.s.	12.8
1987	175 ± 129	100 ± 125	0.01	77 ± 66	152 ± 52	0.01	10.3
	Sum of amino acid concentrations ($\mu g\ g^{-1}$ carbohydrate, root transformed)						
	Fagus sylvatica			*Phaseolus vulgaris*			
Year	Ambient air	Filtered air	*p*	Ambient air	Filtered air	*p*	
1987	4.53 ± 1.81	3.15 ± 0.45	0.01	10.69 ± 5.47	10.55 ± 3.8	n.s.	10.3

preferred to feed and to consume more on cottonwood leaf foliage that had been previously exposed to ozone (Coleman and Jones, 1988b). In a dual choice feeding test the beech weevil, *Rhynchaenus fagi*, preferred leaf discs from beech saplings exposed to O_3 rich (up to 115 ppb O_3) ambient air in 65.5% of the cases (Hiltbrunner and Flückiger, 1992).

Fumigation experiments with CO_2. It is assumed that there may be indirect effects of elevated CO_2 on plant–insect interactions because increased CO_2 frequently reduces the nitrogen concentration in the leaves and hence insects may be forced to consume more to maintain their nitrogen intake. Up to now, 61 plant–herbivore relationships under the impact of elevated CO_2 have been investigated. Although different feeding guilds such as leaf-chewing, leaf-miners and sucking insects were included in these studies, the results suggest that the effects of elevated CO_2 are small and rather negative with regard to insect performances (Watt *et al.*, 1998; Bezemer and Jones, 1998). Generally, leaf-chewers seem to compensate for the decreased nitrogen levels in the plant by increasing their food consumption. Phloem feeders and whole cell feeders are the only insects that show a positive CO_2 response (Whittaker, 1999). Stiling *et al.* (1999) provided evidence that the negative effects of elevated CO_2 on leaf miners could mainly be explained by a higher parasitoid attack.

Nitrogen and acid deposition. Increased N in the foliage and nutrient imbalances may lead to increased susceptibility of plants to parasites. The increase in the population development of cereal aphids *Sitobion avenae* and *Metapolophium dirhodum* on wheat plants induced by N fertilization of 120–200 kg N ha yr^{-1} is a well-known example for host-mediated effects of nitrogen on crop plants (Hanisch, 1980). During an outbreak of the heather beetle *Lochmaea suturalis*, N-fertilized *Calluna* plants were more severely damaged (Heil and Diemont, 1983). Growth of beetle larvae was improved by higher N concentration in *Calluna* leaves (Brunsting and Heil, 1985). The scale insect, *Fiorna externa*, artificially infested on young eastern hemlock trees fertilized with 17.5 kg N (NH_4NO_3) per tree, revealed a substantial rise in the survival rate of the nymphs, in number of females with eggs and in number of eggs per female (McClure, 1983). In a N-fertilization experiment (0–160 kg N) in afforestation plots the population size of natural infestations of *Phyllaphis fagi* on beech, *Cinara pilicornis* and *Sacchiphantes abietis* on Norway spruce was positively correlated with the N-treatment levels (Flückiger and Braun, 1999a).

Acid deposition may produce an acidification of the leaf surface (Hoffman *et al.*, 1980). Klingauf (1982) observed that *Acyrthosiphon pisum* uses the pH gradient between the plant surface (slightly acidic) and phloem sap (slightly alkaline) to localize the inserting place of the stylet. Hence, acid deposition on the plant surface might result in an increased aphid settling by changes of the plant surface. Growth rates of *Schizolachnus pineti*, *Eulachnus agilis* and *Cinara pini* were promoted by 12, 20 and 100%, respectively, by an acid mist treatment on pine trees (*Pinus sylvestris*) (Kidd, 1990). On the other hand, the population development of *Phyllaphis fagi* on beech and of *Aphis fabae* on bean was inhibited when host plants had been subjected to acidic mist of pH 2.6 and 3.6 compared to pH 5.6 (Braun and Flückiger, 1989).

Heliövaara *et al.* (1992) studied the effects of simulated acid deposition on the egg survival of the European sawfly *Neodiprion sertifer*. Overwintering egg clusters on Scots pine needles were sprayed with solutions of pH 2, 3, 4, 5, 6 and distilled water. The proportion of hatched larvae increased with increasing acidity. Similar results were obtained from eggs without needles reared in petri dishes, indicating that the acid treatment induced changes in the eggs.

Predators and parasitoids

Predation and parasitism are important regulating factors of insect communities and hence influence the equilibrium in plant–insect relationships. However, many reports have postulated that populations of predators and parasitoids might be reduced or at least altered by air pollutants (Alstad *et al.*, 1982).

A given population of western pine beetle killed more pine trees and showed a higher growth rate in pine stands in the San Bernardino mountains (California) that had a higher proportion of oxidant damaged trees compared to stands with a lower proportion of damaged trees. Dahlsten *et al.* (1997) explained this difference between the pine stands by a higher number of predators and parasitoids in healthier trees.

The braconid parasitoid, *Asobara talida*, and the larvae of *Drosophila sabobscura* as hosts were used in a closed chamber experiment with short-term fumigation of O_3, SO_2 and NO_2 (100 nl l^{-1}). The results demonstrated that the proportion of parasitized hosts and the searching efficiency were both significantly reduced under O_3 fumigation, but not by SO_2 or NO_2 fumigation (Gate *et al.*, 1995; Figure 20.2). Ruohomaki *et al.* (1996) observed no change in parasitism rates of the lepidopteran, *Epirrita autumnata*, on birch along three air pollution gradients of mainly heavy metals and SO_2 emissions from smelters, indicating that parasitoids were sensitive to the same extent to

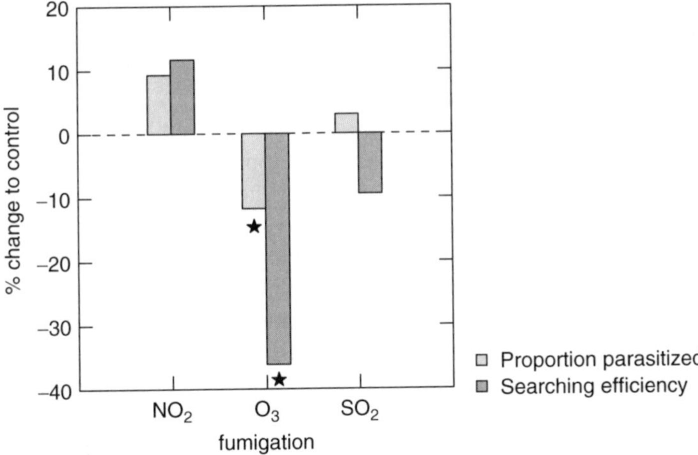

Figure 20.2 Proportion of *Drosophila* larvae parasitized by *Asobara tabida* and searching efficiency of the parasitoid during fumigation with 100 nl l^{-1} O_3, NO_2 or SO_2 (6 chambers, 200 host larvae, 20 parasitoids in each chamber). Both parameters are significantly affected by O_3 fumigation ($p < 0.05$; indicated by asterisks). After Gate *et al.* (1995)

air pollutants as were their host insects. A decreased incidence of predators and parasitoids was apparent in a gradient study with *Aphis pomi* on *Crataegus* sp. close to a motorway. The number of parasitoids and predators such as *Coccinellidae*, *Syrphidae* and *Cecidomyidae* was 8.7 per 10 000 aphids close to the motorway compared to 21.3 at the control site at 200 m distance from the motorway (Braun and Flückiger, 1984).

After infection with the nuclear polyhedrosis virus, the survival rate of the European sawfly *Neodiprion sertifer* was significantly higher when sawflies were fed with needles taken from pine trees treated with simulated acid rain (pH 3) compared to control trees. This means that the acid treatment reduced the susceptibility of the sawflies to the virus disease (Neuvonen *et al.*, 1990). However, Heliövaara *et al.* (1992) could not confirm these results in a pollution gradient study.

Biochemical evidence of interactions

Gaseous air pollutants induce many biochemical changes in plants. In particular, the concentrations of reducing sugars and amino acids increase in plants exposed to SO_2 and O_3 (Koziol and Jordan, 1978; Hughes and Laurence, 1984; Dohmen *et al.*, 1990). Qualitative and quantitative shifts in primary plant metabolites can strongly affect insect phagostimulation, growth and reproduction (Dixon, 1976; McNeill and Southwood, 1978; Schafellner *et al.*, 1999). Above all, an enhanced availability of soluble nitrogen in plants frequently improves growth and reproduction and finally, triggers outbreak of pests (White, 1974, 1993).

One of the first physiological processes to react to air pollution is phloem transport (Spence *et al.*, 1990). It is, therefore, not surprising that many of the changed plant–pest interactions described up to now refer to phloem feeders, and it may also explain the above-mentioned difference between shoot and root aphids. In a comparison of phloem composition found in a field fumigation study with *Viburnum opulus* and *Phaseolus vulgaris* under high NO_x pollution with artificial diets made up of corresponding amino acid concentrations, Bolsinger and Flückiger (1989) were able to show that the raised concentrations of soluble amides and amino acids in the motorway plants were relevant for the development of *Aphis fabae*. The larvae weights were higher, the MRGR increased and the larval developmental time was shorter. Kidd (1990) identified negative and positive effects of amino acids and phenolic compounds in the phloem sap on growth parameters of *Cinara pilicornis*. He assumed that the number and the relative amount of individual amino acids were more important than the overall concentration of the amino acids. The faster development of the tomato pin-worm *Keiferia lycopersicella* on ozone-treated plants could be partly explained by higher concentrations of soluble proteins in stressed plants (Trumble *et al.*, 1987).

Carbohydrates represent the most important energy source for herbivores, especially for later larval insect stages (Otto and Geyer, 1970). There is strong evidence for air pollution impacts via this route, with Mexican bean beetle, *Epilachna varivestis*, which was found to have a feeding preference for SO_2 fumigated soybean leaves, which was positively correlated with pollutant induced changes in sugar concentrations (Hughes and Voland, 1988). Female beetles grew faster by 16% and consumed more by 18% of the ozonated bean foliage which had higher concentrations of non-structural carbohydrates (Lee *et al.*, 1988).

In co-evolution of plant–insect relationships, plants have developed various mechanisms to attract or to repel insects or/and to compensate for the insect damage (Harborne, 1977; Rhoades, 1983; Wold and Marquis, 1997). Secondary plant metabolites may be changed by physical or/and chemical stress factors such as air pollutants (Jones and Pell, 1981; Hurwitz *et al.*, 1979; Jordan *et al.*, 1991). It is often forgotten that plants under stress may concurrently produce increased amounts of repellents and attractants. *Pinus resinosa* and *Betula papyrifera* stressed by SO_2, O_3 or drought formed more ethane (Kimmerer and Kozlowski, 1982) which acts as an efficient repellent to the wood borer beetle *Monochanus alternatus* (Sumimoto *et al.*, 1975). On the other hand, more ethanol was also produced in these stressed plants leading to an increased attraction of beetles such as *Agrilus bilineatus* (Wargo and Montgomery, 1983), *Gnathotrichus sulcatus* (Cade *et al.*, 1970) and *Trypodendron lineatum* (Moeck, 1970). Furthermore, Dunn *et al.* (1990) showed that oaks with relatively low winter starch reserves were more likely to be attacked by *Agrilus bilineatus* in the following summer, indicating that less energy was available for formation of allelochemicals.

In California, Stark and Cobb (1969) provided evidence that *Pinus ponderosa* exposed to oxidants in the San Bernardino mountains was preferably attacked and killed by the western pine beetles *Dendroctonus brevicomis* and *Dendroctonus ponderosa*. Cobb *et al.* (1968a) suggested that ozone-induced shifts in the concentration of reducing sugars could have consequences for terpene synthesis and therefore contribute to the lower resin exudation pressure and hence to the resistance to the bark beetle attack.

Nematodes

Fischer and Führer (1990) investigated the effect of soil acidification on the entomopathogenic nematode *Steinernema kraussei* and its potential host, the nymphs of *Cephalcia abietis*. The nematode forms infective stages that can persist in the soil. Also the nymphs of *C. abietis* stay at least for one year in the soil and are therefore exposed to parasitism by the nematode. The nymphs are killed by a symbiotic, insect-pathogenic bacterium *Xenorhabdus* sp. which is released directly by the nematode into the insect. The host insect dies and the nematode feeds on the bacteria proliferating on the dead insect. Extremely acidic soil conditions reduce the number of nematodes continuously and the regulation of the host population is hampered.

Nematodes are not only affected as predators in plant–insect relationships but also as primary consumers. Occurrence of *Pratylenchus penetrans* was increased when soybean plants were fumigated with non-injuring SO_2 concentrations (250 ppb during 3 weeks, 4 h per day) (Weber *et al.*, 1979). Young tomato plants infested with the same nematode species were more sensitive to a pollutant mixture of SO_2 and O_3, and the combination of the pollutants enhanced the reproduction rate of the nematodes (Shew *et al.*, 1982). On the other hand, ozone exposure of host plants led to substantial decreases in reproduction rates of *Heterodera glycinus, Paratrichodera minor* (both on soybean), *Aphelenchoides fragaria* (on begonia leaves), but only to a slight decrease of *Pratylenchus penetrans* (Brewer, 1979; Weber *et al.*, 1979).

Interactions between pathogens, plants and air pollutants

There is still little fundamental knowledge on host–pathogen interactions under air pollution stress. Pollutants may have direct effects on spore germination of pathogens, hyphal growth and fungal development. Virulence of the pathogen may be promoted or impaired directly, sometimes only during a certain period in the fungal life cycle. As shown already for insects, air pollutants may influence pathogens directly or indirectly by changing the susceptibility of the host plant (Treshow, 1975; Weinstein, 1977). In the case of foliar pathogenic fungi, the plant surface is of predominant importance for a successful infection as it represents the place where the first contact and subsequent reactions between the pathogen and the plant take place. Therefore, the use of fumigation chambers and the resulting induced microclimatic changes represent a strong limitation in pathogen research.

During adhesion of spores and early steps of infection, numerous biochemical signal reactions between the fungus and the host plant occur (formation of phytoalexins, elicitors, etc.) (Laurence, 1981; Huttunen, 1984; Boller, 1995). Pathogens induce different kinds of plant defence reactions that may finally result in a local (e.g. hypersensitive reactions by formation of reactive oxygen species) or systematic plant resistance to the pathogen (Dixon *et al.*, 1994).

Air pollutants may interact in the infection process by modification of the host plant surfaces such as changes in wettability and surface chemistry, as well as changes in foliar exudates and leachates (Blakeman, 1971). On the other hand, many fungal organisms colonize plant tissues as endophytes without apparently affecting the host plant (Petrini, 1986; Petrini, 1991). Some studies indicate that endophytes may serve as antagonists to pathogenic fungi (Clay, 1991). Even pathogens may occur in a latent form in the host tissues for a considerable period. Air pollutants may affect species composition of endophyte communities and cause changes in the life form of the endophytic fungus. The increased isolation frequency of the endophyte *Rhizosphaera kalkhoffi* from Sitka spruce needles growing in SO_2 polluted sites was attributed to its SO_2 tolerance (Magan and Smith, 1996). Up to now, these aspects of plant–endophyte–pathogen interactions have hardly been investigated.

Fungi

Field observations

Field observations of altered pathogen incidence in relation to industrial air pollution go back to the 1920s. Stoklasa (1923) reported increased damage of spruce due to *Chrysomyxa ledi* in the neighbourhood of a smoke emitting plant in Germany. Köck (1935) observed that the fungus *Microsphaera alphitoides* causing powdery mildew on oak was absent in the vicinity of a paper mill in Austria. Flückiger and Oertli (1978) confirmed these observations close to a motorway. A survey of forest ecosystems in Montana and Washington around metal smelters with high SO_2 emission revealed a suppression of the number of pathogenic fungi, particularly of foliage infecting fungi. This was especially true for *Coleosporium solidaginis* on logdepole pine, *Cronartium harknessii, Cronartium comandrae* and *Lophodermium pinastri* on ponderosa and lodgepole pine, *Hyphodermella laricis* on larch, *Melampsora albertensis* on aspen and *Melampsora occidentalis* on black cottonwood and *Pucciniastrum pastulatum* on fir (Scheffer and Hedgecock, 1955). In the zone with highest SO_2 concentrations, these fungi did not occur. In the zone with moderate pollution and tree injuries, fungi were most abundant. Linzon (1978) noted that *Cronartium ribicola* causing blister rust on white pine was almost absent in forests up to 40 km in the main wind direction from the Sudbury smelters (Ontario). On the other hand, Jancarik (1961) observed a higher incidence of *Lophodermium piceae* on spruce needles damaged by SO_2. *Lophodermium pinastri* was quite widespread in areas with average values of 30 to 60 $\mu g\ SO_2\ m^{-3}$ (Huttunen, 1984). A survey of wood decaying fungi in SO_2 polluted industrial areas in former Czechoslovakia and Poland revealed that the occurrence of certain fungal species such as *Schizophyllum commune, Polyporus versicolor, Armillaria mellea, Heterobasidion annosum, Nectria cinnabarina, Melampsora pinitorqua, Stereum pini* and *Trametes pini* was increased over some distance from the pollution sources (Jancarik, 1961; Grzywacz and Wazny, 1973). Similar observations were done by Velagic-Habul *et al.* (1991) in Bosnia-Herzegovina.

Ozone has been reported to weaken trees in natural forest stands and therefore to increase susceptibility of trees to invasive plant pathogens. Along the Blue Ridge Parkway in Virginia, *Pinus strobus* injured by ozone was more subjected to root disease caused by *Verticichodella procera* than visibly undamaged trees (Lackner, 1983). In the San Bernardino mountains *Heterobasidion annosum* invaded more readily freshly cut stumps and roots, respectively, of Jeffrey and ponderosa pines when trees had been previously injured by ozone. Correspondingly, higher infection rate was obtained by an experimental ozone fumigation of young inoculated ponderosa and Jeffrey pine seedlings with the same fungal strains (James *et al.*, 1980a, 1980b).

Experimental investigations on plant–fungus relationships

Fumigation experiments with SO_2

In accordance with field observations, fumigation experiments reveal that the occurrence and fungal development as well as the symptoms on the plants caused by pathogenic fungi can be promoted or inhibited by SO_2 fumigation. The majority of experiments were carried out with crop or horticultural plants. A reduced spore germination, penetration, decreased incidence and, above all, less severe disease development of *Microsphaera alni* on lilac leaves resulted from SO_2 fumigation (Hibben and Taylor, 1975). Similar results were gained by SO_2 fumigation for *Uromyces phaseoli* on pinto beans (130 ppb SO_2) (Weinstein *et al.*, 1975), *Puccinia graminis* on wheat and *Helminthosporium maydis* on corn (100 ppb SO_2 for 110 h) (Laurence *et al.*, 1979). A fungal species that may benefit from SO_2 fumigation is *Sphaerotheca fuliginea* which causes powdery mildew on cucumber. This species showed an enhanced host plant colonization and conidial spore germination when plants had been exposed to 143 μg SO_2 m^{-3}. Higher SO_2 suppressed the fungal growth (Khan *et al.*, 1998). After 5 days of inoculation, more lesions of *Scirrhia acicola* occurred on Scots pine needles when pine seedlings had been fumigated with 200 ppb SO_2 (Weidensaul and Darling, 1979).

Fumigation experiments with O_3

Likewise, ozone fumigation may promote or inhibit pathogenic fungi. Pretreatment of young wheat plants with 87 ppb O_3 (for 3 days) or with 107 ppb (7 days) followed by inoculation with uredospores of *Puccinia recondita* resulted in a great reduction in the number of rust pustules on the ozone fumigated host plants (Dohmen, 1987). Von Tiedemann *et al.* (1990), on the other hand, noted an increased leaf attack caused by *Septoria nodorum*, *Septoria tritici* on wheat and by *Gerlacha nivalis* on wheat and on barley. Disease development was promoted both on leaves with and without visible ozone injury symptoms. Uredia production of *Melampsora medusae* on O_3-treated cottonwood was significantly reduced (Coleman *et al.*, 1987). In contrast, Beare *et al.* (1999a) showed that the incidence and severity of *Melampsora* leaf rust infection on poplar was increased significantly by O_3. They also reported O_3-enhanced conidiospore germination (150–200 nl l^{-1} over 5–6.5 h) and increased number of germ tubes produced (150 nl l^{-1} over 8 h) by *Marssonina tremulae* (Beare *et al.*, 1999b). Prefumigation of poplar host plants increased significantly the size of lesions on old leaves but decreased it on young leaves.

The location of the pathogen development caused may be crucial for the type of interaction. Twig pathogens seem to profit from changes by air pollution. Carey and Kelley (1994) reported that canker production was increased after inoculation with *Fusarium subglutinans* when pine seedlings had been subjected to elevated ozone levels (2.5 fold ambient). Conidial germination of *Apiognomonia veneta* growing *in vitro* was directly inhibited by ozone fumigation (50 and 100 ppb O_3) and accordingly less leaf blight symptoms on London planes were observed when plants were exposed to ozone-rich ambient air compared to filtered air conditions. In contrast, twig lesions were significantly promoted under ozone-rich ambient air after mycelium inoculation with *A. veneta* (von Sury and Flückiger, 1991). The incidence of blossom blight in cherry trees was significantly suppressed when trees had been exposed to ozone-rich ambient air. Cherry trees were infected by spraying *Monilia laxa* spores in full bloom (Tamm, 1994). However, twig lesions were significantly enlarged after mycelium inoculation (Flückiger and Braun, 1999b).

It has been suggested that mycorrhizal fungi protect roots from pathogens (Sylvia and Sinclair, 1983; Azcón-Aguilar and Barea, 1992; Unestam and Damm, 1994). Stress factors including air pollutants may affect formation and function aspects of the symbiotic relationship. Ozone caused a significant decrease in ectomycorrhizas on roots of white birch and white pine seedlings (Keane and Manning, 1987). Bonello *et al.* (1993) reported an improved resistance against root diseases in the presence of the mycorrhizal partner *Hegeloma crustuliniforme*. Ozone exposure (200 ppb 8 h) increased the disease incidence of *Heterobasidion annosum*, but mycorrhizal infection prevented completely this negative effect.

Fumigation with CO_2

Very few studies on effects of elevated CO_2 on pathogens have been conducted. Klironomos *et al.* (1997) found that twice-ambient CO_2 fumigation led to a fourfold increase in airborne fungal propagules. A reduced mildew infection by *Erysiphe graminis* on winter wheat was observed when plants were grown in elevated CO_2 concentrations. The disease reduction could partly be explained by decreased shoot nitrogen concentration under elevated CO_2 (Thompson *et al.*, 1993). Von Tiedemann and Firsching (1998) investigated the combined effects of elevated ozone and carbon dioxide concentrations on the wheat leaf rust *Puccinia recondita*. Ozone reduced substantially the disease severity. These disease-inhibiting effects of ozone were partly compensated under elevated CO_2 as enrichment with carbon dioxide leading to increased total carbohydrate content in the leaves (Figure 20.3).

Nitrogen and acid deposition

Correspondingly to the changes in host-plant–insect relationships, there are some indications that N deposition affects host–pathogen relationships in natural and semi-natural ecosystems. Increased nitrogen deposition rates into forests are reflected by raised foliar N concentration and changed nutrient ratios.

Roelofs *et al.* (1985) reported on a promoted fungal infestation of *Pinus nigra* in Dutch forests with high nitrogen deposition. Pine trees with high N status in the needles were significantly more infected by *Brunchorstia pineae* and *Sphaeropsis sapinea*. *Sphaeropsis sapinea* caused larger necroses on the bark of *Pinus nigra* when trees had been treated with high amounts of nitrogen, but when potassium was added, lesions

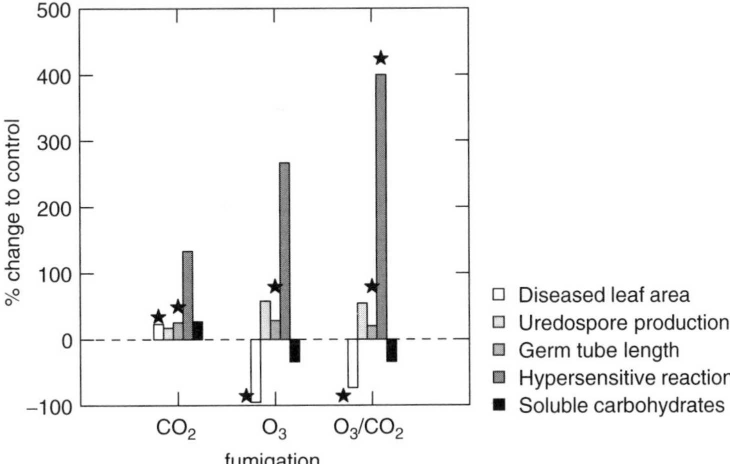

Figure 20.3 Development of diseased wheat leaf area of *Puccinia recondita* f. sp. *tritici*, hypersensitive reaction and soluble carbohydrates in the leaves as well as spore production and germ tube length of the fungus. Data from von Tiedemann and Firsching (1998) (8 chambers, 20 plants per chamber, third leaf stage, 60 days post seeding). Significant differences to the control ($p < 0.05$) are indicated by asterisks. Fumigation conditions were 80 ppb O_3 and/or 650 ppb CO_2. The hypersensitive reaction seems to be responsible for the decreased rust development after O_3 fumigation. Reproduced by permission of *Journal of Plant Diseases and Protection*

remained substantially smaller (De Kam *et al.*, 1991). Beech trees with high N/K ratios in the foliage were more heavily attacked by *Nectria ditissima* in twigs (Flückiger *et al.*, 1986). Experimental N-additions revealed a significant positive correlation between the foliar N/K ratio and the shoot dieback after mycelium inoculation with *Nectria ditissima* (Flückiger and Braun, 1998). A stimulation of the mycelium growth of *Heterobasidion annosum* was found in the outer sapwood tissue of Norway spruce fertilized with nitrogen (Alcubilla *et al.*, 1987). Fertilization with ammonium nitrate in various afforestation plots with young beech and spruce trees resulted in increased pathogen attacks by *Apiognomonia errabunda* and *Phomopsis sp.* on beech and *Botrytis cinerea* on spruce, respectively. The infection rates were significantly increased already by the addition of 10 kg N ha^{-1} yr^{-1} after 5–6 years of continuous fertilization. Disease severity was strongly related to the N/P and N/K ratios in the foliage (Flückiger and Braun, 1999a) (Figure 20.4).

As previously mentioned, mycorrhizal fungi are able to protect roots against root pathogens, for instance *Phytophthora*-caused root diseases (Marx, 1969; Branzanti *et al.*, 1999). Enhanced N loads induce alterations in the mycorrhizal symbiosis, mainly due to an inhibition of the fungal partner (Termorshuizen and Ket, 1991; Wallenda and Kottke, 1998) and therefore affect the susceptibility to pathogens. Following N fertilization the following root diseases were promoted significantly: the root rot of citrus trees caused by *Phytophthora parasitica* (Klotz *et al.*, 1958), the little leaf disease of

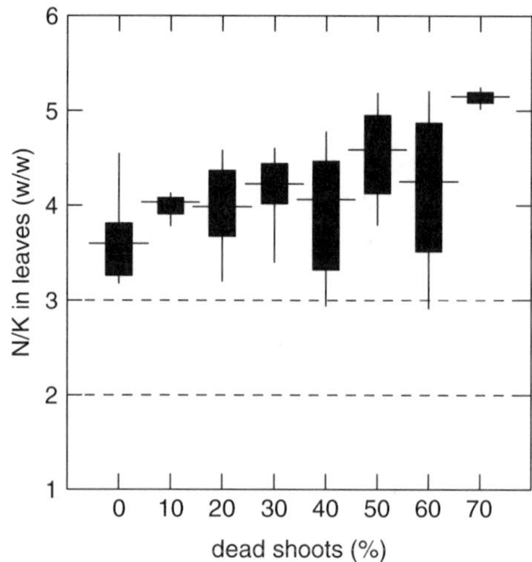

Figure 20.4 Nutrient ratios in the leaves of beech saplings (*Fagus sylvatica*) in an afforestation experiment with different extents of shoot dieback caused by *Apiognomonia errabunda*. Trees were treated five seasons of fertilization with 0, 10, 20, 40, 80 and 160 kg N ha^{-1} a^{-1} in the form of NH_4NO_3, respectively (12 trees per N level). The linear trend is significant at $p < 0.01$ (Flückiger and Braun, 1999a). Even trees without dead shoots have overoptimal N/K ratios. The dashed line indicates classification values after various references cited in Van den Burg (1990). Reproduced with permission from Kluwer Academic Publishers, *Water, Air and Soil Pollution*. vol. 116, pp. 99–110, Figure 6

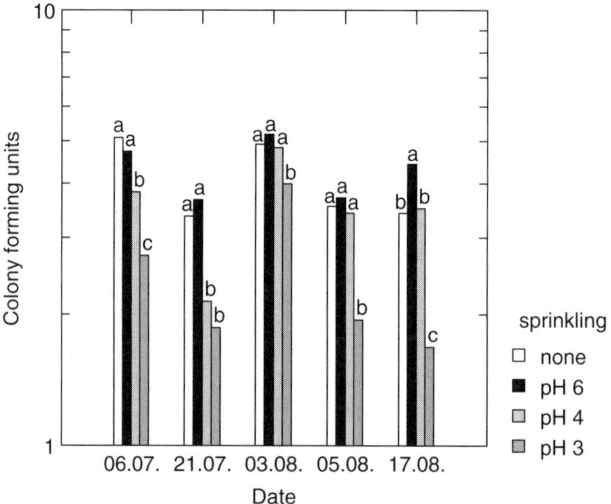

Figure 20.5 Colony-forming units of epiphytic fungi on pine needles (mainly *Aureobasidium pullulans* and *Cladosporium*). after four different sprinkling treatments in the fourth treatment season (5 trees per treatment). Means followed by a different character are significantly different (Ranta, 1990). Reprinted from *Environmental Pollution*, vol. 67, Ranta *et al.*, 'Effects of simulated acid...', fig. 2 from pp. 349–359 (1990) with permission from Elsevier Science

pine by *Phytophthora cinnamoni* (Newhook and Podger, 1972) as well as the root and crown rot of apple trees caused by *Phytophthora cactorum* (Utkhede and Smith, 1995).

Experiments involving effects of acid deposition showed that some pathogenic fungi are negatively affected by addition of acidity. Application of simulated rain of pH 3.2 led to a 86% restriction of teliospore production by *Cronartium fusiforme* on willow oak and to a reduction by 29% in the affected leaf area of kidney bean plants by *Uromyces phaseoli* (Shriner, 1977, 1978). Kvist and Barklund (1984) reported enhanced spore germination of the Scleroderris canker disease *Gremeniella abietina* when conifers had been treated with acid precipitation. Barklund and Unestam (1988) interpreted the improved development of *G. abietina* as a consequence of negative effects of acid mist on the phyllosphere microflora. Von Sury and Flückiger (1993) identified a significant increase in conidial spore germination of *Apiognomonia veneta* after treatment with acid mist of pH 3.0. In acid mist experiments, the foliar colonization by *Aureobasidium pullulans*, which is a known antagonist to many pathogens, was significantly decreased (Ranta, 1990) (Figure 20.5).

Bacteria and virus diseases

There have been only a few investigations that have studied bacterial and viral diseases of plants under the influence of air pollutants. Most of these studies were conducted with crop plants in the presence of SO_2 or O_3.

Besides fungal saprophytes on leaf surfaces, phyllospheric bacteria may be affected by air pollutants. Bacterial communities on leaf surfaces are highly dynamic. Plant exudates are readily used by bacteria and antifungal substances on the leaf surface are often of bacterial origin. Khanna (1986) found that the bacterial population was markedly reduced in the phyllosphere of tropical trees growing at SO_2 polluted sites, although fungal populations on the same trees were promoted under SO_2 pollution. *Xanthomonas phaseoli* was suppressed on SO_2 exposed soybean (100 ppb SO_2) 5 days prior or post inoculation (Laurence and Aluisio, 1981). Likewise, ozone has been suggested to reduce the development of some foliar bacterial diseases. The number of lesions caused by *Pseudomonas glycine* on soybean dropped when leaves had been treated with 80 or 250 ppb O_3 (Laurence and Wood, 1978). Shafer (1992) investigated the effects of simulated acid rain on microorganisms in the rhizosphere. In this study microbial biomass densities, including bacteria, were increased 3 to 8-fold when acid rain was applied directly to the soil. No such effects were obtained when foliage was treated with acidic solutions.

Virus infections may change the sensitivity of plants to ozone. Tobacco plants were less sensitive to ozone when they had been infected by tobacco mosaic virus (Bisessar and Temple, 1977) or by tobacco etch virus (Moyer and Smith, 1975). Similar observations were made with the viral agents, tobacco mosaic virus, alfalfa mosaic virus and tobacco ring spot virus on pinto bean, common bean and soybean plants, respectively (Brennan, 1975; Davis and Smith, 1974, 1976; Wargo *et al.*, 1978). In contrast, the tobacco streak virus enhanced the sensitivity of tobacco plants to ozone injury (Reinert and Gooding, 1978).

Factors contributing to altered host–pathogen relationships

Phyllosphere

Leaf surfaces are of ecological importance in the infection processes of many pathogens. They may be eroded and weathered by air pollutants and thus enhance the capability of the pathogens to penetrate the leaf surface layers directly. Weathered surfaces exhibit an increased wettability. Subsequently, a higher number of water-bound fungal propagules are retained on the leaf surface and the penetration process is alleviated (Shriner and Cowling, 1980). Various studies demonstrate that the phyllosphere microbial communities are affected in the presence of air pollutants (Magan and McLeod, 1991). Phyllosphere saprophytes can have an antagonistic effect on foliar pathogens and function as natural biological control organisms. Any disruption in the balance between phyllosphere saprophytes and pathogens brought about by air pollutants might result in a change in pathogen activity and disease development (Magan *et al.*, 1995; Heagle, 1973; Manning, 1976). Yeasts are known to be antagonistic to foliar pathogens and they are particularly sensitive to SO_2 (Fokkema, 1978; Dowding, 1986). Rapeseed plants that had been inoculated with the pathogen *Alternaria brassicae* as well as with a mixture of two yeasts, *Sporobolomyces roseus* and *Cryptococcus laurenttii*, were exposed to up to 125 ppb SO_2. Depressed infection rate by the pathogen occurred in presence of the yeasts at low SO_2 concentrations (<50 ppb SO_2). At SO_2 concentrations above 100 ppb pathogenic infection was markedly increased, suggesting that low SO_2 concentrations enable yeasts to resist *Alternaria* infections. High SO_2 levels suppressed saprophytic yeasts and plants were sufficiently stressed to allow a rapid, successful infection by the pathogen (Bos, 1986). The hyperparasite *Hyalodendron album*, which can inhibit spore germination and appressorium formation of *Microsphaera alni*, was found to be more sensitive to O_3 and SO_2 than the pathogen (Hibben and Taylor, 1975).

Water relations

Air pollutants such as SO_2 and O_3 are known to modify stomatal functioning and thus the water relations in plants (Black and Unsworth, 1980; Pearson and Mansfield, 1993). This may be of relevance for leaf-infecting fungi which enter the plant via the stomatal apertures. Changes in root growth patterns by ozone (decreased carbon allocation to the roots (Spence *et al.*, 1990) and by soil acidification (Puhe, 1994) may increase the drought susceptibility of plants. Modification in turgor pressure and osmopotential is likely to affect fungi and their development in the host plant as shown for wood decay fungi (Boddy, 1992).

Biochemical evidence of interactions

As emphasized earlier, various changes in plant–insect relationships may be related to biochemical shifts in the plants induced under air pollution stress. Similar biochemical modifications which are induced by air pollution are responsible for alterations in plant–pathogen relationships. They comprise quantitative and qualitative changes in primary (reducing sugars, amino acids) and secondary plant metabolites (phenols, terpenoids, alkaloids). Plants exude sugars, amino acids, proteins and phenolics through the cuticle on the leaf surface. Among them are fungi-stimulating or -suppressing compounds (Blakeman, 1971; Juniper, 1991). Increased permeability of the leaf surface due to air pollutants may also result in an increased exudation of antifungal substances such as terpenoids and phenolics. Gallic acid, present in the leaf surface of many plant species, may restrict colonization by pathogenic fungi (Dix, 1979).

Several studies have outlined the growth inhibition of *Heterobasidion annosum* by atmospheres saturated with resin volatiles. Resistance to *H. annosum* was positively correlated to the resin volatilization (Cobb *et al.*, 1968b; Gibbs, 1972). Trees growing on dry sites yielded less resin and were severely attacked by *H. annosum* (Gibbs, 1968). Specific resin acids were highly fungistatic to *Dothistroma pini* (Franich *et al.*, 1983). *Pinus radiata* fertilized with nitrogen was significantly more strongly attacked by *D. pini*. The fungal infection level was strongly correlated with increased arginine concentration in the needles (Lambert, 1986).

Pathogens may profit from higher concentrations of soluble nitrogen compounds in the plant tissues by higher growth rates, but also by using them and their energy equivalents, respectively, in the degradation of phenolic compounds which are synthesized as inhibitory compounds by the host plant (Tainter and Baker, 1996).

Various stress factors such as hypoxia, drought and air pollutants (SO_2, O_3) produce an increase in ethanol concentrations in roots of woody plants (Kimmerer and Kozlowski, 1982). Wargo and Montgomery (1983) showed that artificial injection of ethanol into the roots stimulated growth of *Armillaria mellea*. It is assumed that sugars, amino acids and ethanol released by roots may attract fungal spores and act as stimulating agents also to pathogenic fungi (Stolzy and Sojka, 1984).

Potential economic impact

There is only limited information on the economic impact of interactions between air pollution and parasite and pest attacks. The pollutant concentrations often do not cause visible effects on plants, but yield reductions at ambient O_3 levels have been identified for the major crop species in different countries (Heck *et al.*, 1983; Fuhrer *et al.*, 1992).

The effect of pollutants may be sufficient to alter the host–parasite relationship in such a way that the performance of insects and pathogens is enhanced and hence causes increased economic impact. Kidd (1990) demonstrated that the damage on spruce foliage treated with acid mist and infested with pine aphids was significantly increased

Table 20.2 Estimates of impact on barley yield of SO_2 interactions with fungal pathogens in an open-air fumigation system. Data are changes in yield compared with control (Mansfield, 1989)

	14–21 ppb SO_2	28–32 ppb SO_2	38–48 ppb SO_2
Powdery mildew	−3%	−2%	−4%
Leaf blotch	+1%	+1.5%	+1.2%
Net economic impact (£ ha^{-1})	−16	−4	−23

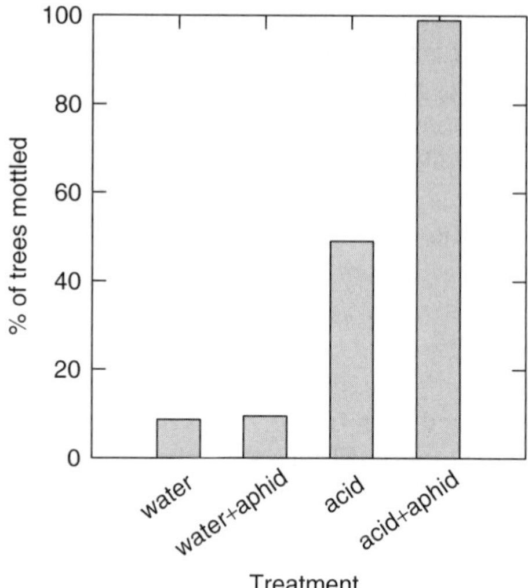

Figure 20.6 Proportion of Scots pine trees showing symptoms of chlorotic mottling in an acid mist treatment (700 μeq l^{-1} H_2SO_4 and 315 μeq l^{-1} HNO_3, pH 3) with and without artificial inoculation with the aphids *Eulachnis agilis*, *Cinara pini* or *Schizolachnus pineti* (10 trees per treatment, inoculation with 5 or 4 apterous, adult aphids per tree). From Kidd (1990). Reproduced by permission of Kidd, 'The effects of simulated acidmist on the growth rates of confer aphids and the implications for tree health' in the *Journal of Applied Entomology* (1990) issue 110, pp. 524–529, Blackwell Wissenschafts-Verlag

compared to each treatment alone (Figure 20.6). Warrington *et al.* (1987) showed that an increase in aphid populations of *Acyrthosiphon pisum* on *Pisum sativum* fumigated with 50 or 80 ppb SO_2 caused a 10% reduction in the yield of peas compared to aphid infested control plants. Mansfield (1989) looked at the effect of powdery mildew (*Erysiphe graminis*) and leaf blight (*Rhynchosporium album*) on barley fumigated with different SO_2 concentrations in an open-field fumigation system and estimated the impact on barley yield loss (Table 20.2). Although the SO_2 effect was opposite on the two fungi, there was a significant economic loss overall.

Concluding remarks and future prospects

There is strong evidence of changes in insect and pathogen performance induced by air pollutants, but the understanding of the ecological significance of such alterations is still limited. Most of the experiments performed have only looked at the response of the pests and pathogens and usually only at a limited number of indices of performance which is not enough to permit conclusions to be drawn at the system level. Much more research is needed to draw conclusions on the ecological and economic significance of the host plant–parasite interactions. Research on plant–insect–pathogen relations in polluted environments has hardly begun. Far too little attention has been paid to predators and parasitoids. Information on changes in phyllosphere microorganisms due to pollution and the effect of pollutants on endophytes is almost non-existent. The long-term effect of soil acidification and eutrophication as a result of air pollution on decomposition and mineralization processes should be examined as well. More attention should be addressed to tropical and subtropical ecosystems with high species richness. Finally, more studies are needed combining fumigation experiments with *in vitro* experiments in order to elucidate the biochemical basis of changes in parasite performance under pollution stress.

References

Alcubilla, M., von Aufsess, H., Cerny, G., Heibl, R. and Rehfuess, K.E. (1987). Stickstoffdüngungsversuche in einer Fichtenwuchsstockung (*Picea abies* Karst.) auf devastierter Kalkmergel-Rendzina. II Verlauf künstlicher Inokulationen mit *Heterobasidion annosum* (Fr.) Bref. und Pilzhemmwirkung von Bast und Holz im Biotest. *Forstwiss.Cbl.* **106**, 27–44.

Alstad, D.N., Edmunds, G.F. and Weinstein, L.H. (1982). Effects of air pollutants on insects populations. *Ann. Rev. Entomol.* **27**, 369–384.

Azcón-Aguilar, C. and Barea, J.M. (1992). Interactions between mycorrhizal fungi and other rhizosphere microorganisms. In: *Mycorrhizal functioning: an integrative plant-fungal process*, ed. Allen, M.F., pp. 163–198. Routledge, Chapman & Hall, London.

Barklund, P. and Unestam, T. (1988). Infection experiments with *Gremeniella abietina* on seedlings of Norway spruce and Scots pine. *Eur. J. For. Path.* **18**, 409–420.

Beare, J.A., Archer, S.A. and Bell, J.N.B. (1999b). *Marssonina* leafspot disease of poplar under elevated ozone: pre-fumigated host and in vitro studies. *Environ. Pollut.* **105**, 409–417.

Beare, J.A., Archer, S.A. and Bell, J.N.B. (1999a). Effects of *Melampsora* leaf rust disease and chronic ozone exposure on poplar. *Environ. Pollut.* **105**, 419–426.

Berggren, D., Bergkvist, B., Falkengren-Grerup, U. and Tyler, G. (1992). Influence of acid deposition on metal solubility and fluxes in natural soils. In: *Chemical Climatology and Geomedical Problems*, eds. Låg, J., pp. 141–154. Norwegian Academy of Sci. Letters, Otta, Norway.

Bezemer, T.M. and Jones, T.H. (1998). Plant-insect herbivore interactions in elevated atmospheric CO_2: Quantitative analyses and guild effects. *Oikos* **82**, 212–222.

Bisessar, S. and Temple, P.J. (1977). Reduced ozone injury on virus-infected tobacco in the field. *Plant Disease Reptr.* **61**, 961–963.

Black, V.S. and Unsworth, M.H. (1980). Stomatal responses to sulphur dioxide and vapour pressure deficit. *J. Exp. Bot.* **31**, 667–677.

Blakeman, J.P. (1971). The chemical environment of the leaf surface in relation to growth of pathogenic fungi. In: *Ecology of Leaf Surface Microorganisms*, eds. Preece, T.F. and Dickinson, Ch., pp. 255–268. Academic Press, New York.

Bobbink, R., Boxman, D., Fremstad, E., Heil, G., Houdijk, A. and Roelofs, J. (1992). Critical Loads for nitrogen eutrophication of terrestrial and wetland ecosystems based upon changes in vegetation and fauna. In: *Critical Loads for Nitrogen*, eds. Grennfelt, P. and Thörnelöf, E., pp. 111–159.

Boddy, L. (1992). Microenvironmental aspects of the xylem defenses to wood decay fungi. In: *Defense Mechanisms of Woody Plants against Fungus*, ed. Blanchette, R.A., pp. 96–132. Springer, Berlin.

Boller, T. (1995). Chemoperception of microbial signals in plant cells. *Ann. Rev. Plant Physiol. Plant Microbiol.* **46**, 189–214.

Bolsinger, M. and Flückiger, W. (1984). Effect of air pollution at a motorway on the infestation of *Viburnum opulus* by *Aphis fabae*. *Eur. J. For. Path.* **14**, 256–260.

Bolsinger, M. and Flückiger, W. (1989). Ambient air-pollution induced changes in amino-acid pattern of the phloem sap in host plants – relevance to aphid infestation. *Environ. Pollut.* **56**, 209–216.

Bonello, P., Heller, W. and Sandermann Jr, H. (1993). Ozone effects on root-disease susceptibility and defence responses in mycorrhizal and non-mycorrhizal seedlings of Scots pine (*Pinus sylvestris* L.). *New Phytol.* **124**, 653–663.

Bos, M. (1986). Effect of SO_2 on the antagonism between saprophytic phyllosphere organisms and the pathogen *Alternaria brassicae* on agar plates and summer rape. Institute for Plant Pathology Research, Wageningen, Holland.

Branzanti, M.B., Rocca, E. and Pisi, A. (1999). Effect of ectomycorrhizal fungi on chestnut ink disease. *Mycorrhiza* **9**, 103–109.

Braun, S. and Flückiger, W. (1984). Increased population of the aphid *Aphis pomi* at a motorway. Part 1: Field evaluation. *Environ. Pollut. (Ser. A.)* **33**, 107–120.

Braun, S. and Flückiger, W. (1985). Increased population of the aphid *Aphis pomi* at a motorway. 3. The effect of exhaust-gases. *Environ. Pollut (A)* **39**, 183–192.

Braun, S. and Flückiger, W. (1989). Effect of ambient ozone and acid mist on aphid development. *Environ. Pollut.* **56**, 177–187.

Brennan, E. (1975). On exclusion as the mechanism of ozone resistance of virus-infected plants. *Phytopathology* **65**, 1054–1055.

Brewer, J.B. (1979). Interactions of ozone, sulfur dioxide and *Pratylenchus* species on tomato and soybean. MSc thesis. North Carolina State University Raleigh, North Carolina.

Brown, V.C. (1995). Insect herbivores and gaseous air pollutants–current knowledge and predictions. In: *Insects in a Changing Environment*, eds. Harrington, R. and Stork, N.E., pp. 219–249. Academic Press, London.

Brown, V.C., Ashmore, M.R. and McNeill, S. (1993). Experimental investigations of the effects of air-pollutants on aphids on coniferous trees. *Forstw. Cbl.* **112**, 128–132.

Brunsting, A.M.H. and Heil, G.W. (1985). The role of nutrients in the interaction between a herbivorous beetle and some competing plant species in heathlands. *Oikos* **44**, 23–26.

Cade, S.C., Hruthfiord, B.F. and Gara, R.J. (1970). Identification of a primary attractant for *Gnathotrichus sulcatus isolated* from western hemlock logs. *J. Econ. Entomol.* **63**, 1014–1015.

Carey, W.A. and Kelley, W.D. (1994). Interaction of ozone exposure and *Fusarium subglutinans* inoculation on growth and disease development of loblolly pine seedlings. *Environ. Pollut.* **84**, 35–43.

Chappelka, A.H., Kraemer, M.E. and Mebrahtu, T. (1988). Effects of ozone on soybean resistance to Mexican bean beetle feeding preference and larval antibiosis. *Environ. Pollut.* **53**, 418–419.

Clay, K. (1991). Endophytes as antagonists of plant pests. In: *Microbial Ecology of Leaves*, eds. Andrews, J.H. and Hirano, S.S., pp. 331–357. Brock/Springer Verlag, New York.

Cobb, F.W., Kristic, M. and Zavarin, E. (1968b). Inhibitory effects of volatile oleoresin components on *Fomes annosus* and four *Ceratocystis* species. *Phytopathology* **58**, 1327–1335.

Cobb, F.W., Wood, D.L., Stark, R.W. and Parmeter, J.R. (1968a). Theory on the relationship between oxidant injury and bark beetle infestation. *Hilgardia* **39**, 141–152.

Coleman, J.S., Jones, C.G. and Smith, W.H. (1987). The effect of ozone on cottonwood leaf rust interactions: independence of abiotic stress, genotype and leaf ontogeny. *Can. J. Bot.* **65**, 949–953.

Coleman, J.S. and Jones, G.C. (1988a). Acute ozone stress on the eastern cottonwood (*Populus deltoides* Bartr.) and the pest potential of the aphid *Chaitophorus populicola* Thomas (Homoptera: Aphididae). *Environ. Entom.* **17**, 207–212.

Coleman, J.S. and Jones, G.C. (1988b). Plant stress and insect performance: Cottonwood, ozone and a leaf beetle. *Oecologia* **76**, 57–61.

Cramer, H.H. (1951). Die geographischen Grundlagen des Massenwechsels von *Epinotia tedella*. *Forst. Cbl.* **70**, 42–53.

Dahlsten, D.L., Rowney, D.L. and Kickert, R.N. (1997). Effects of oxidant air pollutants on western pine beetle (Coleoptera: Scolytidae) populations in southern California. *Environ. Pollut.* **96**, 415–423.

Davis, D.D. and Smith, S.H. (1974). Reduction of ozone-sensitivity of pinto bean by bean common mosaic virus. *Phytopathology* **64**, 383–385.

Davis, D.D. and Smith, S.H. (1976). Reduction of ozone sensitivity of pinto bean by virus reduced local lesions. *Plant Disease Reptr.* **60**, 31–34.

De Kam, M., Versteegen, C.M., Van den Burg, J. and Van der Werf, D.C. (1991). Effects of fertilization with ammonium sulphate and potassium sulphate on the development of *Sphaeropsis sapinea* in Corsican pine. *Neth. J. Pl. Path.* **97**, 265–274.

Dix, N.J. (1979). Inhibition of fungi by gallic acid in relation to growth on leaves and litter. *Trans. Brit. Mycol. Soc.* **73**, 329–336.

Dixon, A.F.G. (1976). *Biologie der Blattläuse*. Fischer Verlag, Stuttgart.

Dixon, R.A., Harrison, M.J. and Lamb, L.J. (1994). Early events in the activation of plant defense responses. *Ann. Rev. Phytopathol.* **32**, 479–501.

Dohmen, G.P. (1987). Secondary effects of air pollution: ozone decreases brown rust disease potential in wheat. *Environ. Pollut.* **43**, 189–194.

Dohmen, G.P., Koppers, A. and Langebartels, C. (1990). Biochemical response of Norway spruce (*Picea abies* (L.) Karst.) towards 14-month exposure to ozone and acid mist: Effects on amino acid, glutathione and polyamine titers. *Environ. Pollut.* **64**, 375–384.

Dohmen, G.P., McNeill, S. and Bell, J.N.B. (1984). Air pollution increases *Aphis fabae* pest potential. *Nature* **307**, 53.

Dowding, P. (1986). Leaf yeasts as indicators of air pollution. In: *Microbiology of the Phyllosphere*, eds. Fokkema, N.J. and van den Heuvel, J., pp. 121–136. Cambridge University Press, Cambridge.

Dunn, J.P., Potter, D.A. and Kimmerer, T.W. (1990). Carbohydrate reserves, radial growth and mechanisms of resistance of oak trees to phloem-boring insects. *Oecologia* **83**, 458–468.

Falkengren-Grerup, U. (1987). Long-term changes in pH of forest soils in southern Sweden. *Environ. Pollut.* **43**, 79–90.

Feir, D. and Hale, R. (1983). Growth and reproduction of an insect model in controlled mixtures of air pollutants. *Int. J. Environ. Studies* **20**, 223–228.

Fischer, P. and Führer, E. (1990). Effect of soil acidity on the entomophilic nematode *Steinernema kraussei* Steiner. *Biol. Fert. Soils* **9**, 174.

Fitze, P., Burri, A., Achermann, M., Kuhn, N. and Polomski, J. (1991). Bodenversauerung. *Bulletin BGS* **15**, 41–52.

Flückiger, W. and Braun, S. (1998). Nitrogen deposition in Swiss forests and its possible relevance for leaf nutrient status, parasite attacks and soil acidification. *Environ. Pollut.* **102**, 69–76.

Flückiger, W. and Braun, S. (1999a). Nitrogen and its effects on growth, nutrient status and parasite attacks in beech and Norway Spruce. *Water, Air and Soil Poll.* **116**, 99–110.

Flückiger, W. and Braun, S. (1999b). Stress factors of urban trees and their relevance for vigour and predisposition for parasite attacks. *Acta Horticulturae* **496**, 325–334.

Flückiger, W., Braun, S., Flückiger-Keller, H., Leonardi, S., Asche, N., Bühler, U. and Lier, M. (1986). Untersuchungen über Waldschäden in festen Buchenbeobachtungsflächen der Kantone

Basel-Landschaft, Basel-Stadt, Aargau, Solothurn, Bern, Zürich und Zug. *Schweiz. Z. Forstwesen* **137**, 917–1010.

Flückiger, W. and Oertli, J.J. (1978). Der Einfluss verkehrsbedingter Luftverunreinigungen auf den Befall der Eiche durch *Microsphaera alphitoides*. *Phytopathol. Ztsch.* **93**, 363–366.

Flückiger, W., Oertli, J.J. and Baltensweiler, W. (1978). Observations of an aphid infestation on hawthorn in the vicinity of a motorway. *Naturwiss.* **65**, 654–655.

Fokkema, N.J. (1978). Fungal antagonism in the phyllosphere. *Ann. Appl. Biol.* **89**, 115–119.

Franich, R.A., Gadget, P.M. and Shain, L. (1983). Fungistatic effects of *Pinus radiata* needle epicuticular fatty and resin acids on *Dothistroma pini*. *Physiol. Plant. Pathol.* **23**, 183–195.

Fuhrer, J., Grandjean Grimm, A., Tschannen, W. and Shariat-Madari, H. (1992). The response of spring wheat (*Triticum aestivum* L.) to ozone at higher elevations. *New Phytol.* **121**, 211–219.

Gate, I.M., McNeill, S. and Ashmore, M.R. (1995). Effects of air pollution on the searching behaviour of an insect parasitoid. *Water, Air and Soil Poll.* **85**, 1425–1430.

Gerlach, C. (1898). Beitrag zur Lebensweise unserer beiden Harzrüsselkäfer *Pissodes harcyniae* und *P. scabricollis*. *Forstl. Naturw. Zeitsch.* **7**, 137–147.

Gerlach, C. (1907). Einige charakteristische Merkmale und Beweismittel für das Vorhandensein von Rauchschäden in Fichtenbeständen auf Grund eigener Beobachtungen und Erfahrungen. *Allg. Forst- und Jagdz.* **83**, 375–382.

Gibbs, J.N. (1968). Resin and the resistance of conifers to *Fomes annosus*. *Ann. Bot.* **32**, 649–665.

Gibbs, J.N. (1972). Tolerance of *Fomes annosum* to pine oleoresins and pinosylvins. *Eur. J. For. Path.* **2**, 147–157.

Ginevan, M.E., Lane, D.D. and Greenberg, L. (1980). Ambient air concentrations of sulphur dioxide affect flight activity in bees. *Proc. Natl. Acad. Sci. USA* **77**, 5631–5633.

Goulding, K.W.T. and Blake, L. (1993). Testing the PROFILE model on long-term data. In: *Critical Loads: Concepts and Applications. Proceedings of a Conference 12–14 Feb. 1992, Grange over Sands*, eds. Hornung, M. and Skeffington, R.A., pp. 68–73. Institute of Terrestrial Ecology, London HMSO.

Grzywacz, A. and Wazny, J. (1973). The impact of industrial air pollutants on the occurrence of several important pathogenic fungi of forest trees in Poland. *Eur. J. For. Path.* **3**, 129–141.

Hanisch, H.C.H. (1980). Untersuchungen zum Einfluss unterschiedlich hoher Stickstoffdüngung zu Weizen auf die Populationsentwicklung von Getreideblattläusen. *Ztsch. Pflanzenkr. Pfl. schutz* **87**, 546–556.

Harborne, J.B. (1977). *Introduction to Ecological Biochemistry*. Academic Press, London.

Hättenschwiler, S. and Körner, Ch. (1996). System-level adjustments to elevated CO_2 in model spruce ecosystems. *Global Change Biology* **2**, 377–387.

Heagle, A.S. (1973). Interactions between air pollutants and plant parasites. *Ann. Rev. Phytopathol.* **11**, 365–388.

Heck, W.W., Adams, R.M., Cure, W.W., Heagle, A.S., Heggestad, H.E., Kohut, R.J., Kress, L.W., Rawlings, J.O. and Taylor, O.C. (1983). A reassessment of crop loss from ozone. *Environ. Sci. & Technol.* **12**, 572–581.

Heil, G.W. and Diemont, W.H. (1983). Raised nutrient levels change heathland into grassland. *Vegetatio* **53**, 113–120.

Heliövaara, K. and Väisänen, R. (1986). Industrial air pollution and the pine bark bug *Aradus cinnamomeus* Panz. (Het., Arachidae). *J. Appl. Entomol.* **101**, 469–478.

Heliövaara, K. and Väisänen, R. (1993). *Insects and Pollution*. CRC Press, Boca Raton.

Heliövaara, K., Väisänen, R. and Varama, M. (1992). Acidic precipitation increases egg survival in *Neodiprion sertifer*. *Entomol. Exp. Appl.* **62**, 55.

Hibben, C.R. and Taylor, M.P. (1975). Ozone and sulphur dioxide effects on the lilac powdery mildew fungus. *Environ. Pollut.* **9**, 107–114.

Hillman, R.C. and Benton, A.W. (1972). Biological effects of air pollution on insects, emphasizing the reactions of the honeybee (*Apis mellifera* L.) to sulphur dioxide. *J. Elisha Mitchell Sci. Soc.* **88**, 195.

Hiltbrunner, E. and Flückiger, W. (1992). Altered feeding preference of beech weevil *Rhynchaenus fagi* L. for beech foliage under ambient air pollution. *Environ. Pollut.* **75**, 333–336.

Hoffman, W.A., Lindberg, S.E. and Tuner, R.R. (1980). Precipitation acidity – the role of the forest canopy in acid exchange. *J. Environ. Quality.* **9**, 95–100.

Holopainen, J.K., Braun, S. and Flückiger, W. (1994). The response of spruce shoot aphid *Cinara pilicornis* Hartig to ambient and filtered air at two elevations and pollution climates. *Environ. Pollut.* **86**, 233–238.

Houlden, G., McNeill, S., Aminu-Kano, M. and Bell, J.N.B. (1990). Air pollution and agricultural aphid pests. 1. Fumigation experiments with SO_2 and NO_2. *Environ. Pollut.* **67**, 305–314.

Hughes, P.R. and Laurence, J.A. (1984). Relationships of biochemical effects of air pollutants on plants to environmental problems: insect and microbial interactions. In: *Gaseous Air Pollutants and Plant Metabolism*, eds. Koziol, W.J. and Whatley, F.R., pp. 361–377. Butterworths, London.

Hughes, P.R. and Voland, M.L. (1988). Increases in feeding stimulants as the primary mechanism by which SO_2 enhances performance of Mexican bean beetle on soybean. *Entomol. Exp. Appl.* **48**, 257–262.

Hurwitz, B., Pell, E.J. and Sherwood, R.T. (1979). Status of coumestrol and 4,7-hydroxyflavone in alfalfa foliage exposed to ozone. *Phytopathol.* **69**, 810–813.

Huttunen, S. (1984). Interactions of disease and other stress factors with air pollution. In: *Air Pollution and Plant Life*, ed. Treshow, M., pp. 321–356. Wiley, New York.

James, R.L., Cobb, F.W., Wilcox, W.W., Miller, P.R. and Parmeter, J.R. (1980b). Effects of oxidant air pollution on susceptibility of pine roots to Fomes *annosus. Phytopathology* **70**, 560–563.

James, R.L., Cobb, F.W., Wilcox, W.W. and Rowney, D.L. (1980a). Effects of photochemical oxidant injury of ponderosa and Jeffrey pines on susceptibility of sapwood and freshly cut stumps to *Fomes annosus. Phytopathology* **70**, 704–708.

Jancarik, V. (1961). Vyskyt drevo kaznych hub u kourem poskozovane oblasti Krusnych hor. *Lesnictvi* **7**, 667–692.

Jeffords, M.R. and Endress, A.C. (1984). Possible role of ozone in the defoliation by the gypsy moth (Lepidoptera: *Lymantriidae*). *Environ. Entomol.* **13**, 1249–1252.

Jones, J.V. and Pell, E.J. (1981). The influence of ozone on the presence of isoflavones in alfalfa foliage. *J. Air Pollut. Cont. Assoc.* **31**, 885–886.

Jordan, D.N., Green, T.H., Chappelka, A.H., Lockaby, B.G., Muldahl, R.S. and Gjerstad, D.H. (1991). Response of total tannins and phenolics in loblolly pine foliage exposed to ozone and acid rain. *J. Chem. Ecol.* **17**, 505–513.

Juniper, B.E. (1991). The leaf from the inside and the outside: a microbe's perspective. In: *Microbial Ecology of Leaves*, eds. Andrews, J.H. and Hirano, S.S., pp. 21–42. Brock/Springer Verlag, New York.

Keane, K.D. and Manning, W.J. (1987). Effects of ozone and simulated acid rain and ozone and sulfur dioxide on mycorrhizal formation in paper-birch and white pine. In: *Acid Rain: scientific and Technical Advances*, eds. Perry, R., Harrison, R.M., Bell, J.N.B. and Lester, J.N., pp. 608–613. Selper, London.

Khan, M.R., Khan, M.W. and Pasha, M.J. (1998). Effects of sulfur dioxide on the development of powdery mildew of cucumber. *Environ. Exp. Bot.* **40**, 265–273.

Khanna, K.K. (1986). Phyllosphere microflora in certain plants in relation to air pollution. *Environ. Pollut.* **42**, 191–200.

Kidd, N.A.C. (1990). The effects of simulated acid mist on the growth rates of conifer aphids and the implications for tree health. *J. Appl. Entomol.* **110**, 524–529.

Kimmerer, T.W. and Kozlowski, T.T. (1982). Ethylene, ethane, acetaldehyde and ethanol production by plants under stress. *Plant Physiol.* **69**, 840–847.

Klingauf, F.A.J. (1982). Breeding for resistance to aphids Faba bean improvement. In: *Proceedings of the Faba Bean Conference Cairo, Egypt, 7–11 March 1981*, eds. Hawtin, G. and Webb, C., pp. 285–295. Matinus-Nijhoff, The Hague.

Klironomos, J.N., Rillig, M.C., Allen, M.F., Zak, D.R., Pregitzer, K.S. and Kubiske, M.E. (1997). Increased levels of airborne fungal spores in response to *Populus tremuloides* grown under elevated atmospheric CO_2. *Can. J. Bot.* **75**, 1670–1673.

Klotz, L.J., De Wolfe, T.A. and Wong, P.P. (1958). Decay of fibrous roots of citrus. *Phytopathology* **48**, 616–622.

Koziol, M.J. and Jordan, C.F. (1978). Changes in carbohydrate levels in red kidney bean (*Phaseolus vulgaris* L.) exposed to sulphur dioxide. *J. Exp. Bot.* **29**, 1037–1043.

Köck, G. (1935). Eichenmehltau und Rauchgasschäden. *Ztsch. Pflanzenkr.* **45**, 44–45.

Kvist, K. and Barklund, P. (1984). Air pollution problems in Swedish forests. *Forstw. Cbl.* **103**, 74–82.

Lackner, A.L. (1983). Root disease and insect infestations on air pollution sensitive *Pinus strobus* and studies of pathogenicity of *Verticicladiella procera. Plant Disease* **67**, 679–681.

Lambert, M.J. (1986). Sulphur and nitrogen nutrition and their interactive effects on *Dothistroma* infection in *Pinus radiata. Can. J. For. Res.* **16**, 1055–1062.

Laurence, J.A. (1981). Effects of air pollutants on plant-pathogen interactions. *Ztsch. Pflanzenkr. Pfl. schutz* **87**, 156–172.

Laurence, J.A. and Aluisio, A.L. (1981). Effects of sulphur dioxide on expansion of lesions caused by *Corynebacterium nebraskense* in maize and by *Xanthomonas phaseoli* var. *sojensis* in soybean. *Phytopathology* **71**, 445–448.

Laurence, J.A., Weinstein, L.H., McCune, M.S. and Aluisio, A.L. (1979). Effects of sulfur dioxide on southern corn leaf blight of maize and stem rust of wheat. *Plant Disease Reptr.* **63**, 975–978.

Laurence, J.A. and Wood, F.A. (1978). Effects of ozone on infection of soybean by *Pseudomonas glycine. Phytopathology* **68**, 441–445.

Lee, E.H., Wu, Y., Barrows, E.M. and Mulchi, C.L. (1988). Air pollution, plants and insects: growth and feeding preferences of the Mexican bean beetle on bean foliage stress by sulphur dioxide or ozone. *Environ. Pollut.* 1–4, 442.

Lee, J.A. and Caporn, S.J.M. (1998). Ecological effects of atmospheric reactive nitrogen deposition on semi-natural terrestrial ecosystems. *New Phytol.* **139**, 127–134.

Linzon, S.N. (1978). Effects of airborne sulphur pollutants on plants. In: *Sulphur in the Environment: Part II. Ecological Impacts*, ed. Nriagu, S.O., pp. 109–162. Wiley, New York.

Magan, N., Kirkwood, I.A., McLeod, A.R. and Smith, M.K. (1995). Effect of open-air fumigation with sulphur dioxide and ozone on phyllosphere and endophytic fungi of conifer needles. *Plant, Cell Environ.* **18**, 291–302.

Magan, N. and McLeod, A.R. (1991). Effects of atmospheric pollutants on phyllosphere microbial communities. In: *Microbial Ecology in Leaves*, eds. Andrews, J.H. and Hirano, S.S., pp. 379–400. Brock/Springer Verlag, New York.

Magan, N. and Smith, M.K. (1996). Isolation of the endophytes *Lophodermium piceae* and *Rhizosphaera kalkhoffii* from Sitka spruce needles in poor and good growth sites and in vitro effects of environmental factors. *Phyton* **36**, 103–110.

Manning, W.J. (1976). The influence of ozone on plant surface microfloras. In: *Microbiology of aerial plant surfaces*, eds. Dickinson, C.H. and Preece, T.F., pp. 159–172. Academic Press, London.

Mansfield, P.J. (1989). *Interactions of Atmospheric Sulphur Dioxide with Fungal Diseases of Winter Barley*. Ph D Thesis, University of London.

Marx, D.H. (1969). The influence of ectotrophic mycorrhizal fungi on the resistance of pine roots to pathogenic infections. I Antagonism of mycorrhizal fungi to root pathogenic fungi and soil bacteria. *Phytopathology* **59**, 153–163.

McClure, M.S. (1983). Competition between herbivores and increased resource heterogeneity. In: *Variable Plants and Herbivores in Natural and Managed Systems.*, eds. Denno, R.F. and McClure, M.S., pp. 125–153. Academic Press, London.

McNeill, S., Aminu-Kano, M., Houlden, G., Bullock, J.M., Citrone, S. and Bell, J.N.B. (1987). The interaction between air pollution and sucking insects. In: *Acid Rain – Scientific and technical advances*, eds. Perry, R., Harrison, R.M., Bell, J.N.B. and Lester, J.N., pp. 602–607. Selper, London.

McNeill, S. and Southwood, T.R.E. (1978). The role of nitrogen in the development of insect plant relationships. In: *Biochemical Aspects of Plant and Animal Coevolution*, ed. Harborne, J.B., pp. 77–98. Academic Press, London.

McNeill, S. and Whittaker, J.B. (1990). Air pollution and tree-dwelling aphids. In: *Population Dynamics of Forest Insects*, eds. Watt, A.D., Leather, S.R., Hunter, M.B. and Kidd, N.A.C., pp. 195–208. Intercept, Andover.

Moeck, H.A. (1970). Ethanol as a primary attractant for the ambrosia beetle *Trypodendron lineatum* (Coleoptera: Scolytidae). *Can. Entom.* **102**, 985–995.

Moyer, J.W. and Smith, S.H. (1975). Oxidant injury reduction on tobacco induced by tobacco etch virus infection. *Environ. Pollut.* **9**, 103–106.

Neuvonen, S., Saikkonen, K. and Haukioja, E. (1990). Simulated acid rain reduces the susceptibility of the European pine sawfly (*Neodiprion sertifer*) to its nuclear polyhedrosis virus. *Oecologia* **83**, 209–212.

Newhook, F.S. and Podger, F.D. (1972). The role of *Phytophthora cinnamomi* in Australian and New Zealand forests. *Ann. Rev. Phytopathol.* **10**, 299–326.

Oksanen, J., Holopainen, J.K., Nerg, A. and Holopainen, T. (1996). Levels of damage of Scots pine and Norway spruce caused by needle miners along SO_2 gradient. *Ecography* **19**, 229–236.

Otto, D. and Geyer, W. (1970). Zur Bedeutung des Zuckergehaltes der Nahrung für die Entwicklung nadelfressender Kieferninsekten. *Arch. Forstwes.* **19**, 135–150.

Pearson, M. and Mansfield, T.A. (1993). Interacting effects of ozone and water stress on the stomatal resistance of beech (*Fagus sylvatica* L.). *New Phytol.* **123**, 351–358.

Petrini, O. (1986). Taxonomy of endophytic fungi of aerial plant tissues. In: *Microbiology of the Phyllosphere*, eds. Fokkema, N.J. and van den Heuvel, J., pp. 175–187. Cambridge University Press, Cambridge.

Petrini, O. (1991). Fungal endophytes of tree leaves. In: *Microbial Ecology of Leaves*, eds. Andrews, J.H. and Hirano, S.S., pp. 179–197. Springer Verlag, New York.

Posch, M., Hettelingh, J.-P., de Smet, P.A.M. and Downing, R.J. (1997). Calculation and mapping of critical thresholds in Europe. Status report 1997. *RIVM Report* 259 101 007. Bilthoven, Netherlands, National Institute of Public Health and the Environment, Convention of Long-Range Transboundary Air Pollution on the United Nations Economic Commission for Europe.

Przybylski, Z. (1979). The effect of automobile exhaust gases on the arthropods of cultivated plants, meadows and orchards. *Environ. Pollut. (A)* **19**, 157–161.

Puhe, J. (1994). Die Wurzelentwicklung der Fichte (*Picea abies* (L.) Karst.) bei unterschiedlichen chemischen Bodenbedingungen. *Berichte des Forschungszentrums Waldökosysteme Reihe A* **108**, Göttingen.

Ranta, H.M. (1990). Effect of simulated acid rain on quantity of epiphytic microfungi on Scots pine (*Pinus sylvestris* L.) needles. *Environ. Pollut.* **67**, 349–359.

Reinert, R.A. and Gooding, G.V. (1978). Effect of ozone and tobacco streak virus alone and in combination on *Nicotiana tabacum*. *Phytopathology* **68**, 15–17.

Rhoades, D.F. (1983). Herbivore population dynamics and plant chemistry. In: *Variable Plant and Herbivores in Natural and Managed Systems*, eds. Denno, R.F. and McClure, M.S., pp. 155–220. Academic Press, New York.

Riemer, J. and Whittaker, J.B. (1989). Air pollution and insect herbivores: observed interactions and possible mechanisms. In: *Insect-Plant Interactions*, eds. Bernays, E.A., pp. 73–105. CRC Press, Boca Raton.

Roelofs, J.G., Kempers, A.J., Houdijk, A.L. and Jansen, J. (1985). The effects of air-borne ammonium sulphate on *Pinus nigra* var. *maritima* in the Netherlands. *Plant and Soil* **84**, 45-56.

Ruohomaki, K., Kaitaniemi, P., Kozlov, M., Tammaru, T. and Haukioja, E. (1996). Density and performance of *Epirrita autumnata* (Lepidoptera: Geometridae) along three air pollution gradients in Northern Europe. *J. Appl. Ecol.* **33**, 773–784.

Salt, D.T. and Whittaker, J.B. (1995). Populations of root-feeding aphids in the Liphook forest fumigation experiment. *Plant, Cell Environ.* **18**, 321–325.

Schafellner, C., Berger, R., Dermutz, A., Fueher, E. and Mattanovich, J. (1999). Relationship between foliar chemistry and susceptibility of Norway spruce (Pinaceae) to *Pristiphora abietina* (Hymenoptera: Tenthredinidae). *Can. Entomol.* **131**, 373–385.

Scheffer, T.C. and Hedgecock, G.G. (1955). Injury to northwestern forest trees by sulphur dioxide from smelters. US Department of Agriculture, Forest Service, Tech. Bull. 1117 Washington DC.

Shafer, S.R. (1992). Responses of microbial populations in the rhizosphere to deposition of simulated acidic rain onto foliage and/or soil. *Environ. Pollut.* **76**, 267–278.

Shannon, J.D. (1999). Regional trends in wet deposition of sulfate in the United States and SO_2 emissions from 1980 through 1995. *Atmos. Environ.* **33**, 807–816.

Shew, B.B., Reinert, R.A. and Barker, K.R. (1982). Response of tomatoes to ozone, sulfur dioxide and infection by *Pratylenchus penetrans*. *Phytopathology* **72**, 822–826.

Shriner, D.S. (1978). Effects of simulated acid rain on host-parasite interactions in plant diseases. *Phytopathology* **68**, 213–218.

Shriner, D.S. (1977). Effects of simulated rain acidified with sulfuric acid on host-parasite interactions. *Water, Air and Soil Poll.* **8**, 9–14.

Shriner, D.S. and Cowling, E.B. (1980). Effects of rainfall acidification on plant pathogens. In: *Effects of Acid Precipitation on Terrestrial Ecosystems*, eds. Hutchinson, T.C. and Haves, M., pp. 435–442. Plenum Press, New York.

Spence, R.D., Rykiel, E.J. and Sharpe, P.J. (1990). Ozone alters carbon allocation in Loblolly pine: Assessment with carbon-11 labelling. *Environ. Pollut.* **64**, 93–106.

Stark, R.W. and Cobb, F.W. (1969). Smog injury, root diseases and bark beetle damage in ponderosa pine. *Calif. Agric.* **23**, 13–15.

Stiling, P., Rossi, A.M., Hungate, B., Dijkstra, P., Hinkle, C.R., Knott, W.M. and Drake, B. (1999). Decreased leaf-miner abundance in elevated CO_2: Reduced leaf quality and increased parasitoid attack. *Ecological Appl.* **9**, 240–244.

Stoklasa, L. (1923). *Die Beschädigung der Vegetation dutch Rauchgase und Fabrikexhaltationen*. Urban und Schwarzenberg, Berlin, Wien.

Stolzy, L.H. and Sojka, R.E. (1984). Effects of flooding on plant disease. In: *Flooding and Plant Growth*, ed. Kozlowski, T.T., pp. 221–264. Academic Press, New York.

Sumimoto, M., Shiraga, M. and Kando, T. (1975). Ethane in pine needles preventing the feeding of the beetle *Monochamus alternatus*. *J. Ins. Physiol.* **21**, 713–722.

Summers, C.G., Retzlaff, W.A. and Stephenson, S. (1994). The effect of ozone on the mean relative growth rate of *Diuraphis noxia* (Mordvilko) (Homoptera: Aphididae). *J. Agricult. Entomol.* **11**, 181–187.

Sylvia, D.M. and Sinclair, W.A. (1983). Phenolic compounds and resistance to fungal pathogens induced in primary roots of Douglas fir seedlings by the ectomycorrhizal fungus *Laccaria laccata*. *Phytopathology* **73**, 390–397.

Tainter, F.H. and Baker, F.A. (1996). *Principles of Forest Pathology*. John Wiley & Sons, Canada.

Tamm, L. (1994). *Epidemiological aspects of sweet cherry blossoms blight caused by* Monilinia laxa. PhD Thesis, University Basel.

Tarrassón, L. (1998). Transboundary acidifying air pollution in Europe (International cooperative programme for the monitoring and evaluation of long range transmission of air pollutants in Europe). Meteorological synthesizing Centre-West.

Termorshuizen, A.J. and Ket, P.C. (1991). Effects of ammonium and nitrate on mycorrhizal seedlings of *Pinus sylvestris*. *Eur. J. For. Path.* **21**, 404–413.

Thompson, G.B., Brown, J.K.M. and Woodward, F.I. (1993). The effects of host carbon dioxide, nitrogen and water supply on the infection of wheat by powdery mildew and aphids. *Plant, Cell Environ.* **16**, 687–694.

Treshow, M. (1975). Interactions of air pollutants and plant diseases. In: *Responses of Plants to Air Pollution*, eds. Mudd, J.B. and Kozlowski, T.T., pp. 307–334. Academic Press, London.

Trumble, J.T., Hare, J.D., Musselmann, R.C. and McCool, P.M. (1987). Ozone induced changes in host plant suitability: Interactions of *Keiferia lycopersicella* and *Lycopersicon esculentum*. *J. Chem. Ecol.* **13**, 203–218.

UN/ECE (1996). Manual on methodologies for mapping critical loads/levels and geographical areas where they are exceeded, revised version June 1996. *Convention on Long-range Transboundary Air Pollution*. Berlin, Umweltbundesamt.

Unestam, T. and Damm, E. (1994). Biological control of seedling diseases by ectomycorrhizae. *INRA editions/les colloques* **68**.

Utkhede, R.S. and Smith, E.M. (1995). Effect of nitrogen form and application method on incidence and severity of *Phytophthora* crown and root rot of apple trees. *Eur. J. Plant. Path.* **101**, 283–289.

Van den Burg, J. (1990). *Foliar analysis for determination of tree nutrient status – a compilation of literature data*. Rijksinstituut voor Onderzoek in de Bos- en Landschapsbouw 'De Dorschkamp', Wageningen.

Velagic-Habul, E., Lazarev, V. and Custovic, H. (1991). Evaluation of emission of SO_2 and occurrence of pathogenic fungi of forest tree species. *Plant Protection* **42**, 153–164.

von Sury, R. and Flückiger, W. (1991). Effects of air pollution and water stress on the leaf blight and twig cankers of London planes (*Platanus x acerifolia* (Ait.) Willd.) caused by *Apiognomonia veneta* (Sacc. + Speg.) Höhn. *New Phytol.* **118**, 397–405.

von Sury, R. and Flückiger, W. (1993). Effects of acid, or ammonium-enriched, artificial mist on leaf blight of London plane [*Platanus x acerifolia* (Ait) Willd.] and on the behaviour of the causal fungus, *Apiognomonia veneta* (Sacc. + Speg.) Höhn. *New Phytol.* **124**, 447–454.

von Tiedemann, A. and Firsching, K.H. (1998). Combined whole-season effects of elevated ozone and carbon dioxide concentrations on a simulated wheat leaf rust (*Puccinia recondita* f. sp. *tritici*) epidemic. *Ztsch. Pflanzenkr. Pfl. schutz* **105**, 555–566.

von Tiedemann, A., Ostländer, P., Firsching, K.H. and Fehrmann, H. (1990). Ozone episodes in southern lower Saxony (FRG) and their impact on the susceptibility of cereals to fungal pathogens. *Environ. Pollut.* **67**, 43–59.

Wallenda, T. and Kottke, I. (1998). Nitrogen deposition and ectomycorrhizas. *New Phytol.* **139**, 169–187.

Wargo, P.M. and Montgomery, M.E. (1983). Colonization by *Armillaria mellea* and *Agrilus bilineatus* of oaks injected with ethanol. *For. Sci.* **29**, 848–857.

Wargo, R.H., Pell, E.J. and Smith, S.H. (1978). Induced resistance to ozone injury of soybean by tobacco ring spot virus. *Phytopathology* **68**, 715–719.

Warrington, S. (1987). Relationship between SO_2 dose and growth of the aphid *Acyrthosiphon pisum* on peas. *Environ. Pollut.* **43**, 155–162.

Warrington, S. (1989). Ozone enhances the growth rate of cereal aphids. *Agric. Ecosys. Environ.* **26**, 65–68.

Warrington, S., Cottam, D.A. and Whittaker, J.B. (1989). Effects of insect damage on photosynthesis, transpiration and SO_2 uptake by sycamore. *Oecologia* **80**, 139.

Warrington, S., Mansfield, T.A. and Whittaker, J.B. (1987). Effect of SO_2 on the reproduction of pea aphids *Acyrthosiphon pisum* and the impact of SO_2 and aphids on the growth and yield of peas. *Environ. Pollut.* **48**, 285–294.

Warrington, S. and Whittaker, J.B. (1990). Interactions between Sitka spruce, the green spruce aphid, sulphur dioxide pollution and drought. *Environ. Pollut.* **65**, 363–370.

Watt, A.D., Flückiger, W., Leith, I.D. and Lindsay, E. (1998). Atmospheric pollution, elevated CO_2 and spruce aphids. *Forestry Commission T.P.* **24**. Edinburgh.

Webber, A.N., Nie, G.Y. and Long, S.P. (1994). Acclimation of photosynthetic proteins to rising atmospheric CO_2. *Photosynthesis Research* **39**, 413–425.

Weber, D.F., Reinert, R.A. and Barker, K.R. (1979). Ozone and sulfur dioxide effects on reproduction and host-parasite relationships of selected plant-parasitic nematodes. *Phytopathology* **69**, 624–628.

Weidensaul, T.C. and Darling, S.L. (1979). Effects of ozone and sulphur dioxide on the host-pathogen relationship of Scots pine and *Scirrhia acicola*. *Phytopathology* **69**, 939–941.

Weinstein, L.H. (1977). Fluoride and plant life. *J. Occup. Medicine* **19**, 49–78.

Weinstein, L.H., McCune, M.S., Aluisio, A.L. and Van Leuken, P. (1975). The effects of sulphur dioxide on the incidence and severity of bean rust and early blight of tomato. *Environ. Pollut.* **9**, 145–155.

White, T.C.R. (1974). A hypothesis to explain outbreaks of looper caterpillars with special reference to populations of *Selidosema suavis* in a plantation of *Pinus radiata* in New Zealand. *Oecologia* **16**, 279–301.

White, T.C.R. (1993). *The Inadequate Environment. Nitrogen and the Abundance of Animals*. Springer Verlag, New York.

Whittaker, J.B. (1999). Impacts and responses at population level of herbivorous insects to elevated CO_2. *Eur. J. Entomol.* **96**, 149–156.

Wold, E.N. and Marquis, R.J. (1997). Induced defense in white oak: Effects on herbivores and consequences for the plant. *Ecology* **78**, 1356–1369.

21 Effects of air pollutants in developing countries

F.M. MARSHALL

Air pollution trends in developing countries – concentrations and distributions

Increasing energy demands associated with economic growth and industrialisation in Asia, Africa and Latin America have resulted in dramatic increases in air pollution emissions. Problems are compounded by rapid and poorly planned industrial growth in developing countries, the close proximity of industrial complexes and thermal power plants to residential areas (Singh, 1995) and the fact that air pollution control in developing countries is often inadequate for technical and economic reasons. Air pollution kills more than 2.7 million people annually, with more than 90 percent of these deaths in developing countries and two-thirds of them in Asia (UNDP, 1998). Thus it is not surprising that most attention to date has focused on the direct impact of these industrial and urban emissions on human health. However, very little is known about pollutant concentrations in many suburban and rural areas, where there may be significant indirect impacts of air pollution on human health, through reduced crop yields, food quality and income.

Sulphur dioxide, one of the major phytotoxic primary pollutants, is emitted mainly from the combustion of coal and fuel oil, with increased emissions associated with the rapidly increasing energy demands in many developing countries. For example, Asian energy demand is doubling every 12 years, and 80 percent of the demand is met by burning fossil fuels, mainly coal (van Aardenne *et al.*, 1999). As a result, SO_2 emissions in Asia are predicted to increase from 34×10^6 tonnes in 1990 to 110×10^6 tonnes by 2020 (van Aardenne *et al.*, 1999). In China, coal burning alone accounted for 72 percent of total energy consumption in 1998, causing more than half of the country's SO_2 emissions (Yanjia and Kebin, 1999). China is now the leading emitter of SO_2 in the world. Coal-based power generation has also greatly increased in India over the last decade and now accounts for 64 percent of electricity generation (Agrawal and Singh, 2000). Smelters are another important, but more localised source of sulphur dioxide.

Road traffic is a relatively minor contributor in terms of national emissions of SO_2 in the developed world due to the low sulphur content of fuel. However, this is not the case in many developing countries which have much higher sulphur levels in diesel

Air Pollution and Plant Life, second edition. Edited by J.N.B. Bell and M. Treshow. ISBN 0 471 49090 3 (HB), 0 471 49091 1 (PB).

fuel. For example, a recent study (Kitada and Azad, 1998) in Dhaka has shown that motor traffic contributes over 50 percent of overall sulphur dioxide emissions and that mean concentrations around the city can reach up to c. 200 $\mu g\ m^{-3}$. In this and many other developing country cities the number of motor vehicles is also increasing rapidly.

Traffic also plays the major role in emissions of NO_x, with nitric oxide as the principal primary pollutant but being rapidly oxidised to NO_2. All high temperature combustion processes result in the emission of NO_x, with thermal power stations representing the other major source. A global increase in NO_x emissions from 40×10^6 tonnes in the mid-1980s to $55–66 \times 10^6$ tonnes per year by 2025 has been predicted (Lee *et al.*, 1997), with substantially higher percentage increases in some developing countries, such as China. These increases in NO_x are predicted to cause widespread increases in O_3 levels across the developing world.

Urban air pollution problems are exacerbated in developing countries by the higher age of vehicles, poor vehicle maintenance, overloading of buses and trucks, poor fuel quality and poorly planned and maintained narrow streets and roads which worsen congestion (Faiz and Sturm, 2000). For example, in India the Central Pollution Control Board has estimated that 67 percent of Delhi's pollution load is due to vehicles registered before 1991; furthermore, a projected 80 percent of all vehicles in the year 2000 are two or three wheelers with inefficient two stroke engines (Singh, 1995) and two wheeler production continues to grow (by 20 percent per annum). The number of vehicles in India has increased from 11 million to 21 million between 1986 and 1991 (Singh, 1995), causing considerable traffic congestion and the increasing urban population remains almost totally dependent on road transport.[1]

Ozone levels can be elevated above the natural background of about 40 ppb through a complex series of photochemical reactions also involving both NO_x and volatile organic compounds (VOCs). Motor vehicles, particularly inefficient and poorly tuned engines characteristic of developing countries, are the major source of VOCs. Furthermore the high temperatures and high light intensity characteristic of many developing country cities favour the production of ozone. Although the precursors for ozone are produced in cities, the levels of this secondary pollutant are often higher on the outskirts of the city, due to local destruction by NO at ground level within the city (UK PORG, 1993). Ozone is a particular cause for concern because elevated ozone levels can be widespread over rural agricultural areas, particularly downwind of cities (UK PORG, 1993). There has been very little coordinated monitoring of ozone levels in rural areas of developing countries, but the limited data available are consistent with this, indicating phytotoxic levels in a number of important agricultural areas (Ashmore and Marshall, 1999).

Agriculture in the developing world

Agriculture plays a crucial role in ensuring food security and in the economic growth of developing countries. The need to feed rapidly growing populations has been an

[1] There has been some recent success in addressing this issue in Delhi through the removal of commercial vehicles over 15 years old, and conversion of taxis, buses and autorickshaws to run on compressed natural gas.

important driving force behind efforts to increase agricultural production in developing countries (Conway and Barbier, 1990). Increased food production is of particular significance for the poor, the vast majority of whom depend directly on agriculture or other forms of natural resource exploitation for at least part of their livelihoods (Cox *et al.*, 1998). More than 850 million people in the developing world are currently undernourished (Carney, 1999). In addition, the poor may have little or no choice of area in which to farm and are more likely to be living and farming close to industrial pollution sources. Thus, alleviation of constraints on agricultural productivity, including air pollution, remains a high priority for poverty reduction efforts.

While the impacts of air pollution, and particularly ozone, on agriculture in North America and western Europe have received considerable attention, there has been little recognition of this issue in the developing countries of Asia, Africa, and Latin America. Here the social and economic significance of air pollution impacts on agriculture may be much greater, bearing in mind the considerable importance of national agricultural production to maintain food security and earn foreign exchange, However, air pollution control in general is argued for in the face of limited resources and a general desire to promote industrial development. It is therefore extremely important to establish local field-based evidence to demonstrate the true costs and benefits of air pollution abatement. This would include both direct and indirect (via agriculture) impacts of air pollution on human health. Air pollutants may also have adverse impacts on forestry in developing countries, but this has been studied even less than for agricultural crops. A useful review of available information has recently been published by Innes and Haron (2000).

Direct evidence of adverse impacts of air pollution on crops in developing countries

Ozone

Major field studies on the direct effects of ambient O_3 on crop yield have been carried out in India, Pakistan, Mexico and Egypt (reviewed in Ashmore and Marshall, 1999). Many of these studies have used anti-ozonant chemicals applied to the foliage or as a soil drench. This technique is relatively easy to employ and less costly than chamber filtration systems (see review of field techniques by Bell and Marshall, 2000).

The anti-ozonant, ethylenediurea (EDU) provides up to 100 percent protection against O_3-induced visible injury and yield reduction. The earliest work with EDU in developing countries was carried out by Bambawale (1986a, 1989) on local potato cultivars in the Indian Punjab. He demonstrated that leaf spot could be eliminated by application of EDU, this being the first demonstration of the phytotoxicity of O_3 in the field in the Indian subcontinent. Bambawale (1986b) also demonstrated a similar phenomenon at another Indian north western plains location in Uttar Pradesh.

Some EDU work outside Mexico City was also reported in the same year. Here there has been considerable concern about the direct health impacts of ozone arising from motor vehicles, because the climate and topography are particularly conducive to high levels of ozone production. Ozone concentrations in the city have reached 400 ppb and frequently exceed 150 ppb. This is also one of the few countries where impacts on both crops and trees have also been assessed. Laguette Rey *et al.* (1986) applied EDU to two cultivars of *Phaseolus* bean in Montecillos and found a major yield reduction of 29 percent in the untreated as compared to the EDU treated plants in one cultivar, but the yield reduction in the other cultivar was only 4 percent. A number of other studies from both industrialised and developing countries have shown significant differences in sensitivity of cultivars to ozone (Ashmore and Marshall, 1999), which suggests the potential to select resistant cultivars for use in polluted areas.

More recently, an important experiment was conducted by Wahid *et al.* (2001) in Pakistan. Here the protective effect of EDU on soybean (*Glycine max*) was assessed at a suburban site, a remote rural site and a rural roadside site around the city of Lahore in the pre-monsoon growing season. The decrease in seed weight of the nontreated plants as compared to the untreated plants was 53, 65 and 74 percent for the suburban, remote rural and rural roadside sites, respectively. Oxidant concentrations were also higher at the rural sites. The results suggest that ozone may have widespread impacts on crop yields in the Punjab, which is the most important agricultural area in Pakistan.

In Egypt agriculture is effectively confined to the Nile Delta where there are major air pollution sources and climatic conditions highly conducive to ozone production. Some rural O_3 data are available for Egypt, which indicate mean concentrations of >75 ppb in the Delta during the summer (Farag *et al.*, 1993). Hassan *et al.* (1995) grew local cultivars of radish and turnip with and without EDU at two sites: one a suburban area of Alexandria and the second in the village of Abbis, 30 km to the south in the Nile Delta agricultural region. The 6 h mean oxidant concentrations recorded in February and March were 55 ppb in Alexandria and 67 ppb in Abbis. The effects of EDU on yield reflected this in that the harvested dry weight of radish was reduced by 30 percent in the untreated plants at Abbis and by 24 percent in Alexandria. Turnip is much less sensitive to ozone, and showed a 17 percent reduction in harvested dry weight of the untreated plants in Abbis, but no effect in Alexandria.

Recently further studies have been conducted into the effects of O_3 in the field in India. Varshney and Rout (1998) exposed tomato plants at three sites in Delhi, with and without EDU treatment. In the untreated plants the shoot and root dry weight were reduced by 26 percent and 18 percent, respectively.

A very limited number of open-top chamber filtration studies have been carried out in developing countries. The most important series of experiments was again in the outskirts of Lahore in Pakistan using two local cultivars each of winter wheat and rice repeated for two successive years (Maggs *et al.*, 1995; Wahid *et al.*, 1995a,b). The crops were grown in open-top chambers ventilated with ambient or charcoal-filtered air and were subject to local cultivation practices. The four experiments all showed significant reductions in yield in the unfiltered air treatment as compared to filtered air, ranging from 34 to 46 percent. The concentrations of sulphur dioxide at this site were negligible, but a series of closed chamber fumigation studies were carried out to assess the relative contributions of ozone and nitrogen dioxide to the observed yield reductions (Maggs, 1996). These studies, using the same cultivars showed no effect of NO_2 either alone or in combination with O_3, indicating that the yield reductions recorded at Lahore were caused by ambient O_3 alone. These findings were supported by the later study using EDU with soybean (Wahid *et al.*, 2001)

Sulphur dioxide

A large number of chamber and field studies have been conducted to investigate the impact of sulphur dioxide on crops in developing countries, particularly in India. Wheat, which appears to be particularly sensitive as compared to other major crop plants, has been studied most extensively. However, the majority of these studies are closed chamber fumigations involving abnormally high concentrations of SO_2, with only a limited number of the fumigations utilising ambient levels of the pollutant. Field experiments are generally transect studies away from point sources, again due to the complex and costly nature of open-top fumigations systems which have been widely adopted in industrialised countries. In these transect studies plant material is usually grown in standard soils and containers and exposed for all or part of the growing season along gradients of the pollutant. Results can be complicated by the presence of other pollutants, notably NO_x. One important transect study was reported by Singh *et al.* (1990) in which a local wheat cultivar was grown at different distances from a coal-fired power station in Uttar Pradesh, India. This generated a dose–response relationship, which was subsequently used by Ashmore and Marshall (1997) in an illustrative local risk assessment to estimate that farmers operating within 10 km of Indian power stations might experience yield losses of 10–60 percent, depending on the scale of emissions. Thus it is apparent that location of industries in rural areas may present a serious threat to crop production on a local scale.

Despite evidence of increasing SO_2 emissions in developing countries, the experimental evidence indicates that phytotoxic impacts on agriculture are largely associated with point sources, and would affect a localised zone in the vicinity of industries located in rural areas. However, this ignores the issue of urban and peri-urban agriculture, which is exposed in many places to high levels of both SO_2 and other pollutants. Urban and peri-urban agriculture is often neglected by policy makers and planners, but there is strong evidence that it plays an increasing role in food supply of rapidly urbanising populations in the developing world and a very important role in the nutrition of the urban poor (UNDP, 1996). A recent research programme carried out in the Indian cities of Delhi and Varanasi has confirmed the importance of such agriculture both for the local economy and as a source of food security, direct income generation and employment for the poor (Marshall *et al.*, 2000). This study included a series of transect experiments to assess the impact of air pollution on four important local crops, palak/spinach beet *(Beta vulgaris)* and moong bean *(Vigna radiata)* grown during the summer season and wheat *(Triticum aestivum)* and mustard *(Brassica campestris)* during the winter. The plants were grown in containers in standard soils and sites with air pollution levels varying from relatively clean to highly polluted. There were negative correlations ($p < 0.05$) with mean SO_2 concentrations and the yield of both moong bean and palak in Varanasi. The leaf dry weight for palak ranged from 0.66 g plant^{-1} at the site with the lowest SO_2 (40 μg m^{-3}) to 0.18 g plant^{-1} at the site with highest SO_2 (110 μg m^{-3}) (M. Agrawal, pers. comm. Reported in Marshall *et al.*, 2000). In the winter experiment there were again negative correlations with SO_2 concentrations in both cities, but this was only significant ($p < 0.05$) for wheat in Varanasi where there was a 23 percent difference in grain yield between the sites with the highest (141 $\mu g\ m^{-3}$) and lowest (35 $\mu g\ m^{-3}$) SO_2 concentrations (C.K. Varshney, M. Agrawal, pers. comm.)

Nitrogen dioxide

Very little research has been carried out into the impacts of nitrogen dioxide on crop yields in developing countries. Although the study by Maggs *et al.* (1995) indicated that ambient levels of NO_x were not having an impact on yields of wheat and rice at suburban sites outside Lahore, this pollutant may indeed be important in and around urban areas. The impact of NO_x was also considered in the study described above in which the effect of SO_2 on four crops in Delhi and Varanasi were investigated. In Varanasi the NO_2 concentrations during the summer ranged from 22–112 $\mu g\ m^{-3}$ across 10 sites, and there was a negative correlation with moong bean yield ($p < 0.05$) (Marshall *et al.*, 2000). In winter, wheat yield was significantly negatively correlated ($p < 0.05$) with NO_2 concentrations ranging from 31 to 105 $\mu g\ m^{-3}$ (Marshall *et al.*, 2000). In Delhi the yields of both mustard and wheat were negatively correlated with NO which ranged from 79 and 197 $\mu g\ m^{-3}$. (Marshall *et al.*, 2000). The transect study in Varanasi also raised the possibility that urban air pollution was having an impact on the nutritional quality, in addition to the yield of crops. The results showed significant negative relationships with carbohydrate and energy contents and both SO_2 and NO_2 for moong bean and wheat (M. Agrawal, pers. comm.).

Transect studies and filtration studies are important experimental tools, but care must be taken in interpreting the results, due to the possibility that pollutants can have a combined effect. It has been long recognised that NO_2 and SO_2 can have synergistic or additive interactions (Ashenden and Mansfield, 1978), and that O_3 can also influence the impact of these pollutants (Bell 1984). In some cases antagonistic interactions have also been observed (Bell, 1984).

Priorities for further research

The limited field experimental data described above clearly indicate that significant crop losses may be occurring in a number of important agricultural areas in the developing world, with ozone the major cause for concern. However, this issue is little recognised, and resources available to investigate it are limited. It is therefore important to be able to identify and illustrate geographical areas where there is a high risk of major crop losses in order to target further research efforts.

The evidence so far indicates that there is a need for monitoring of pollutant levels in agricultural areas, particularly around major cities and in the vicinity of industries. In the absence of any regular pollution monitoring to date, global emissions inventories and the application of a critical levels approach can be used to indicate those areas of the developing world where pollutant levels are likely to be sufficient to cause significant crop yield losses. This will assist the process of identifying priority areas for monitoring programmes and crop field studies.

Increases in NO_x are predicted to cause widespread increases in O_3 levels across the developing world. Chameides *et al.* (1994) estimated that an increase would take place in the area of cereal crops subjected to O_3 concentrations above the threshold for yield reductions from 9–35 percent of world area to 30–75 percent by 2025. Ashmore and Marshall (1997) used nitrogen dioxide emissions projections for 2025 (from Lee *et al.*, 1997) to estimate areas at high, medium and low risk of crop losses due to ozone. The estimates of ozone formation were based on the study of Chameides *et al.* (1994). The results of the risk assessment indicated that the majority of the cultivated area of India, China and South Africa would fall into the high risk category (a possible yield reduction of 15 percent or more), whilst many other countries such as Egypt, Malaysia, Bangladesh and the Philippines are predicted to have large areas of land in the high risk category. More recently Chameides *et al.* (1999) modelled O_3 production in China and concluded that concentrations were at that time sufficiently high as to reduce yield of winter wheat production over a large part of the country, notably in the Yangtze Delta, the south east Yellow River Delta, the eastern Pearl River Delta and Sichuan. The critical level used was based on European studies with wheat, which is not particularly sensitive to ozone as compared with other extensively studied crops such as soybean and field bean (*Phaseolus*). This supports the hypothesis that although the impacts of ozone have been little studied in the developing world, this pollutant may be causing widespread yield losses on a range of important crop species. Significant impacts of ozone in such countries, where the need to increase food production to meet the requirements of growing populations, and to earn foreign exchange, is often vital, could be of much greater economic or social importance than in those parts of the world currently experiencing agricultural surpluses.

When the risk assessment is repeated for sulphur dioxide, using emissions projections from the Stockholm Environment Institute for 2025 (Ashmore and Marshall 1997, reported in Marshall *et al.*, 1997), there is a very similar pattern, although there are more easily distinguishable areas of high risk relating to specific industrial units such as in the South African Highveld or the Nile Delta in Egypt. Sulphur dioxide may pose a significant, but more localised threat to crop production, than is the case with ozone.

Whilst it has been well established that large differences do exist between species and cultivars, the dose–response data and critical levels used to assess the risk to agriculture in developing countries are by necessity based on the results of experimental studies from North America and Europe. It is important to establish field-based studies to assess the impact of air pollution under local agricultural conditions in the developing world. One factor affecting the sensitivity to pollution will be climate. For example, the stomatal uptake of pollutants will be influenced by humidity. Agricultural practices can also have a major impact. Stomatal uptake will be enhanced when crops are irrigated as for example in the major wheat and rice growing areas of the Punjab, whilst rain fed crops will tend to have a more limited uptake. Other factors such as intercropping, fertilisers, soil salinity and pesticides can all influence the uptake and/or physiological impact of pollutants. If the influence of these factors was understood more fully they would offer the potential to ameliorate the effects of air pollution through changes to agronomic practices.

In the case of ozone, the impacts on crop yield at a given pollution level are rather higher in a number of experimental studies than would have been predicted from established dose–response data. Indeed a comparison of experimental studies from around the world suggests that cultivars from developing countries may be more sensitive to ozone (Ashmore and Marshall, 1997). However, in the absence of comparable dose–response relationships, it is difficult to separate a systematic difference in cultivar sensitivity to ozone from the effects of agronomic practices. If systematic differences can be found between cultivars this raises the possibility of ameliorating the impacts of air pollution by selecting resistant cultivars.

Recommendations to ameliorate the impacts of air pollution will in general require an initial investment from the resource-poor farmers and/or governments of developing countries. It is therefore important to assess the impacts of air pollution on crop yield and quality and the costs of addressing them, relative to other more recognised threats to crops in the developing world such as pests, diseases and soil salinity. This raises the issue of another major research priority, that is to assess the impact of air pollution on increasing the susceptibility of crops to pest and disease damage. Whilst there is considerable evidence from industrialised countries of ambient levels of air pollution altering the performance of both insect pests and fungal pathogens (Bell *et al*., 1993) (see Chapter 20) there is very little published research from the developing world, where pests and diseases often represent a much greater threat. There are, however, a limited number of studies that strongly suggest that the performance of some important chewing and sucking insect pests may be enhanced by ambient levels of air pollution. One recent field study based in Varanasi, India (Davies, 2000) demonstrated an increased performance of the mustard aphid *(Lipaphis erysimi)* on mustard and of the diamond-backed moth *(Plutella xylostella)* on cabbage in terms of a range of growth and reproductive parameters, including increase in fecundity and reproductive period at more polluted sites in and around the city.

Acknowledgements

I am very grateful for the contributions and support of colleagues at Imperial College, particularly Nigel Bell and Craig Davies, and to Dr Madhoolika Agrawal and other friends and colleagues in India.

References

Agrawal, M. and Singh, J. (2000). Impact of coal power plant emissions on foliar elemental concentrations in plants in a low rainfall tropical region. *Environmental Monitoring and Assessment* **60**, 261–282.

Ashenden, T.W. and Mansfield, T.A. (1978). Extreme pollution sensitivity of grasses when SO_2 and NO_2 are present in the atmosphere together. *Nature, London*., **273**, 142–143.

Ashmore, M.R. and Marshall, F.M. (1999). Ozone impacts on agriculture: an issue of global concern. *Advances in Botanical Research*, **29**, 32–52.

Ashmore, M.R. and Marshall, F.M. (1997). *The Impacts and Costs of Air Pollution on Agriculture in Developing Countries*. Final technical report submitted to the Department for

International Development, Environment Research Programme. Project No. ERP6289. Imperial College of Science, Technology and Medicine, London.

Bambawale, O.M. (1986a). Evidence of ozone injury to a crop plant in India. *Atmospheric Environment*, **20**, 1501–1503.

Bambawale, O.M. (1986b). Detection and severity of ozone injury on potato crop in north western plains of India. In: *Proceedings of National Seminar on Environmental Pollution Control and Monitoring*, October 1986, CS10, Chandigor, pp. 145–154.

Bambawale, O.M. (1989). Control of ozone injury on potato. *Indian Phytopathology*, **42**, 509–513.

Bell, J.N.B. (1984). Direct effects of air pollution on plants. In: Troyanowsky, C. (ed.). *Air Pollution and Plants*. VCH, Weinheim, pp. 116–127.

Bell, J.N.B., McNeill, S., Houlden, G., Brown, V.C. and Mansfield, P.J. (1993). Atmospheric change: effect on plant pests and diseases. *Parasitology*, **106**, S11–S24.

Bell, J.N.B. and Marshall, F.M. (2000). Field studies on impacts of air pollution on agricultural crops. In: Agrawal, S.B. and Agrawal, M. (eds.) *Environmental Pollution Plant Response*. Lewis Publishers, Boca Raton, pp. 99–110.

Carney, D. (1999). *Approaches to sustainable livelihoods for the rural poor*. ODI Poverty Briefing No 3, February 1999. Overseas Development Institute, London.

Chameides, W.L., Kasibhatla, P.S., Yienger, J. and Levy, H. (1994). Growth of continental scale metro-agro-plexes, regional ozone pollution and world food production. *Science*, **264**, 74–77.

Chameides, W.L., Li, X., Tang, X., Zhou, X., Luo, C., Kiang, C.S., St. John, J., Saylor, R.D., Liu, S.C., Lam, K.S., Wang, T. and Giorgi, F. (1999). Is ozone pollution affecting crop yields in China? *Geophysical Research Letters*, **26**, 867–870.

Conway, G.R. and Barbier, E.B. (1990). *After the Green Revolution: Sustainable Agriculture for Development*. Earthscan Publications, London.

Cox, A., Farrington, J. and Gilling, J. (1998) *Reaching the Poor? Developing a poverty screen for agricultural research proposals*. ODI Working Paper 112, November 1998. Overseas Development Institute, London.

Davies, C. (2000). Air Pollution and Agricultural Insect Pests in Urban and Peri-urban Areas of India: A Case Study of Varanasi. PhD Thesis, University of London.

Faiz, A. and Sturm, P.J. (2000). New directions: air pollution and road traffic in developing countries. *Atmospheric Environment* **34**, 4745–4746.

Farag, S.A., Rizk, H.F.S., El-Bahnasaway, R.M. and Meleigy, M.I. (1993). The effect of pesticides on surface ozone concentrations. *International Journal of Environmental Education and Information*, **12**, 217–224.

Hassan, I.A., Ashmore, M.R. and Bell, J.N.B. (1995). Effect of ozone on radish and turnip under Egyptian field conditions. *Environmental Pollution*, **89**, 107–114.

Innes, J.L. and Haron, A.H. (2000). *Air Pollution and the Forests of Developing and Rapidly Industrialising Regions*. Report No. 4 of the IUFRO Task Force on Environmental Change. CABI Publishers, Wallingford.

Kitada, T. and Azad, A.K. (1998). Study on the air pollution control system in Dhaka, Bangladesh. *Environmental Technology* **19**, 443–459.

Laguette Rey, H.D., de Bauer, L.I., Shibata, J.K. and Mendoza, N.M. (1986). Impacto de los oxidante ambtales en el cultivo de frigol, en Montecillos, estado de Mexico *Centro do Fitopatologia*, **66**, 83–95.

Lee, D.S., Kohler, I., Grobler, E., Rohrer, F., Sausen, K., Gallardo-Klenner, L., Olivier, J.G.J., Dentener, F.J. and Bowman, A.F. (1997). Estimations of global NO_x emissions and their uncertainties. *Atmospheric Environment* **31**, 1735–1749.

Maggs, R. (1996). The Effects of Ozone and Nitrogen Dioxide on Pakistan Wheat *(Triticum aestivum)* and Rice *(Oryza sativa)* Cultivars. PhD Thesis, University of London.

Maggs, R., Wahid, A., Shamsi, S.R.A. and Ashmore, M.R. (1995). Effects of ambient air pollution on wheat and rice yield in Pakistan. *Water, Air and Soil Pollution*, **85**, 1311–1316.

Marshall, F.M., te Lintelo, D.J.H., Wildig, Z., Stonehouse, J., Bell, J.N.B., Ashmore, M.R., and Batty, K. (2000). *The Impacts and Policy Implications of Air Pollution on Crops in Developing Countries*. Final Technical Report. Department for International Development, Environment Research Programme. R6992. Imperial College of Science, Technology and Medicine, London.

Marshall, F.M., Ashmore, M.R. and Hinchcliffe, F. (1997). *A Hidden Threat to Food Production: Air Pollution and Agriculture in the Developing World*. Gatekeeper Series No. SA73. International Institute for Environment and Development, London.

Singh, O.N. (1995). Local air pollution in India. *Pure and Applied Chemistry* **67**: 1462–1465.

Singh, J.S., Singh, K.P. and Agrawal, M. (1990). *Environmental Degradation of Obra-Renukoot-Singrauli Areas and its Impact on Natural and Derived Ecosystems*. Project report submitted to Ministry of Environment and Forests, Government of India, 4/167/84/MAB/EN – 2RE.

UK PORG (1993). *Ozone in the United Kingdom 1993*. United Kingdom Photoxidant Review Group, 3rd Report. Department of the Environment, London.

UNDP (1996). *Urban Agriculture: Food, Jobs and Sustainable Cities*. United Nations Development Program, New York.

UNDP (1998). *Human Development Report 1998*. United Nations Development Program. Oxford University Press, New York.

Varshney, C.K. and Rout, C. (1998). Ethylene diurea (EDU) protection against ozone injury in tomato plants at Delhi. *Bulletin of Environmental Contamination and Toxicology*, **61**, 188–193.

Van Aardenne, J.A., Carmichael, G.R., Levy, H., Streets, D. and Hordijk, L. (1999). Anthropogenic NO_x emissions in Asia in the period 1990–2020. *Atmospheric Environment*, **33**: 633–646.

Wahid, A., Maggs, R., Shamsi, S.R.A., Bell, J.N.B. and Ashmore, M.R. (1995a). Air pollution and its impact on wheat yield in the Pakistan Punjab. *Environmental Pollution*, **88**, 147–154.

Wahid, A., Maggs, R., Shamsi, S.R.A., Bell, J.N.B. and Ashmore, M.R. (1995b). Effects of air pollution on rice yield in the Pakistan Punjab. *Environmental Pollution*, **90**, 323–329.

Wahid, A., Shamsi, S.R.A., Milne, E., Marshall, F.M. and Ashmore, M.R. (2001). Effects of oxidants on soybean growth and yield in the Pakistan Punjab. *Environmental Pollution*, **113**, 271–280.

Yanjia, W. and Kebin, H. (1999). The air pollution picture in China. *IEEE Spectrum 1999* **36**(12): 55–58.

22 Air quality guidelines and their role in pollution control policy

M.R. ASHMORE

Introduction

Complete control of atmospheric emissions is impossible, and the costs of decreasing pollution emissions typically rise greatly as the degree of removal increases. Thus a key question for policy-makers is the extent to which pollutant emissions should be reduced in order to protect biodiversity, forest health and crop production. Ideally, it would be possible to define air quality standards for biodiversity, and for crop and forest production, as is done for human health, and to devise cost-effective emission control programmes to ensure that these standards are met. However, the process of defining such standards is challenging when considering the impacts on vegetation, because of the range of species, and of climatic and edaphic conditions, which need to be considered. This chapter provides an overview of these issues and describes some of the methods which have been used to define guidelines for effects on vegetation. Rather than provide a comprehensive account of all the air quality guidelines which have been proposed, the chapter will concentrate on three major pollutants which have been considered in detail in earlier chapters: ozone (Chapter 6); sulphur dioxide (Chapter 8) and nitrogen deposition (Chapter 12).

It is important to appreciate that, because the values which are proposed for air quality guidelines may have significant political and economic implications, the process of evaluating scientific data to identify guideline values needs to be considered in the context of policy development. The political structures and goals within which guidelines are derived may vary in ways which have a major influence on the methods through which values are derived. Air pollution control policy has, for obvious reasons, been primarily focussed on the protection of human health. Protection of the natural environment in general, and of vegetation in particular, has been a secondary policy consideration. The major exception to this has been the development of international agreements on the control of emissions of transboundary pollutants within Europe, which have been negotiated on the basis of reducing damage due to the acidification of soils and freshwaters. In 1999, this approach was extended to the effects of nitrogen deposition and ozone under the Gothenburg Protocol. Because Europe is the region of the world where the development of air quality guidelines for the natural environment

Air Pollution and Plant Life, second edition. Edited by J.N.B. Bell and M. Treshow. ISBN 0 471 49090 3 (HB), 0 471 49091 1 (PB).

has been most directly linked to pollution control policy, it will be the prime focus of this chapter.

Definitions

Before proceeding further, it is useful to consider the terminology used in this field, and in particular to distinguish the terms air quality standard and air quality guideline, and to relate these terms to other concepts used in pollution control policy.

Air quality guidelines were described by the World Health Organisation (WHO, 1987), in the context of effects on human health, as values which indicate the levels of a chemical at which no adverse effect is expected, based on current best scientific judgement. Their aim was identified as being to provide a basis for protecting public health, and to provide background information and guidance for use in policy discussions. It is important to distinguish such guidelines from air quality standards, which set the value within a legally enforceable framework. Standards may be more relaxed than guideline values for social, economic or political reasons. A similar distinction, but with different jargon, is found in the structure of EU air quality policy. This defines 'limit values' which are mandatory for member states and must be met within a given number of years; and 'guide values' which aim to provide higher levels of protection but are not mandatory (Edwards, 1998).

In Europe, a further concept, that of the critical load, has been introduced and applied, defined as 'a quantitative estimate of an exposure to one or more pollutants below which significant harmful effects on specified sensitive elements of the environment do not occur, according to present knowledge' (Bull, 1991). The critical load refers to the pollutant input in all forms of deposition, but, analogously, critical levels, defined in terms of atmospheric concentrations over a given averaging time, have also been defined for pollutant gases. In broad terms, critical loads and levels have the characteristics of air quality guidelines. It is important to appreciate that the critical

Table 22.1 Comparison between conventional ecotoxicological guidelines and critical loads/levels (modified from Bull, 1995). Reproduced by permission of Kluwer Academic Publishers, *Water, Air and Soil Pollution*, vol. 85, from Bull, 'Critical loads ...', table 1, p. 203

Conventional approach	Critical loads and levels
Objectives based on laboratory tests	Field-based studies used to establish values
Lethal effects are the usual response used	Range of chemical or biological criteria may be used
Objectives set well below known effects to provide margin of safety	Objectives set as close to known effect levels as possible
No beneficial effects assumed to occur at any level	Beneficial effects (e.g. increased growth rates) may be considered
Environmental damage from exceedance usually occurs over a short time	Environmental damage from exceedance may occur over years or decades, and may be cumulative

load concept is based on a concept of long-term sustainability, and does not relate to current levels of impact – the concern is rather over the cumulative effects of deposition, often over decades. Thus, where current deposition rates exceed critical loads, action to reduce emission rates may be justified in order to prevent future damage, even though field studies demonstrate no adverse effects of current deposition rates. Critical loads were originally developed in the context of soil and freshwater acidification, which lie outside the scope of this chapter; Cresser (2000) provides a recent critical review of the development and application of critical loads in this context.

Table 22.1, modified from Bull (1995), summarises the key conceptual differences between the critical load approach and conventional ecotoxicological objectives. The key differences are that critical loads aim to deal with long-term effects at an ecosystem level, while conventional ecotoxicological approaches rely on short-term laboratory tests on individual test organisms, with the application of safety factors to account for the uncertainty of extrapolation to long-term ecosystem impacts.

Basic issues in defining guidelines

There are four key steps in defining an air quality guideline:

1. Identifying the appropriate sensitive species or ecosystem on which to base the guideline.
2. Identifying the response parameter of concern and deciding what size of change in this parameter can be judged as adverse.
3. Identifying the method of characterising pollutant exposure.
4. Using appropriate experimental data, field data or models to determine the threshold pollutant exposure for the identified species and response parameter.

It is important to appreciate the difficulties in defining such thresholds, in terms of pollutant deposition or concentration, for adverse effects on vegetation. These difficulties arise from three major sources:

1. There are conceptual problems in defining what is meant by an 'adverse effect', both scientifically and in terms of societal judgement. For example, are small shifts in species composition of urban lichen communities, or changes in species composition in the microfauna of a woodland floor which do not affect nutrient cycling, or small changes in the genetic composition of populations, really 'significant harmful effects'? This issue often needs to be addressed in areas where other human impacts, e.g. through changes in land use, have already changed substantially the structure and composition of plant communities.
2. While the adverse effects of air pollution may be clear in situations where they are present in high concentrations and have dramatic effects, as a 'threshold' concentration is approached, the effects become gradually more subtle, and are difficult to detect in field observations. Furthermore, in many cases, the threshold concentration may be close to the natural background concentration of the pollutant

or there may be no obvious threshold. Existing experimental methods may be adequate for assessing direct effects on single organisms, but assessing the long-term effects of air pollution on complex plant communities is beyond the scope of current experimental approaches.

3. Effects on vegetation of chronic exposures to air pollution may be the result of the long-term accumulation of pollutants or long-term chemical changes in soils and vegetation. There is no obvious experimental method of directly testing the consequences of these cumulative effects. The alternative approach, which has been adopted to define critical loads of acidity to prevent long-term damage to ecosystems, is to define a critical chemical concentration for biological effects in the relevant medium (typically soil or freshwater), and then to develop mathematical models to estimate the level of atmospheric deposition which would eventually lead to this chemical criterion being exceeded.

Relationship to exposure–response relationships

There is a close relationship between an air quality guideline or critical level and exposure–response relationships. Sanders *et al.* (1995) identify four generic exposure–response relationships from the literature (Figure 22.1), which need consideration in the context of defining air quality guidelines. Curve (a) is an idealised relationship which provides a clear threshold exposure which could be used in setting a guideline value. However, it should be recognised that, even if this curve did represent the generic response of vegetation to atmospheric pollution, then there would still be problems raised by the fact there will be a family of such relationships, depending on factors

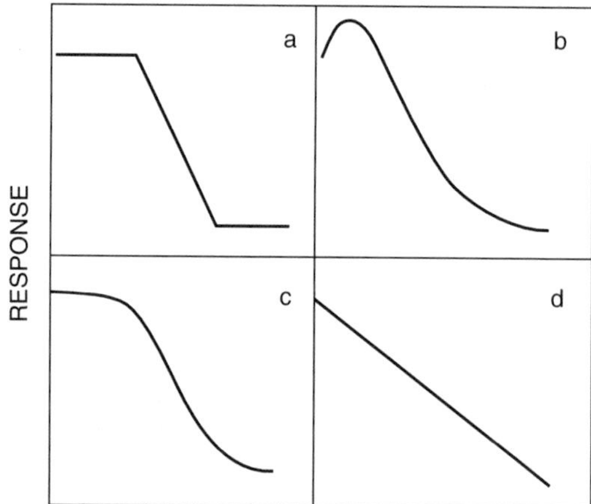

Figure 22.1 Types of response curves for biological response to air pollutants: (a) clear step change in response above a threshold; (b) stimulation at low concentrations; (c) asymptotic change of response; (d) linear response. Source: Sanders *et al.* (1995). Reproduced with permission from Kluwer Academic Publishers, *Water, Air and Soil Pollution*, **85**, 4, table 1 from Sanders *et al.* p. 193

such as species, genotype and growth conditions. Curves (b), (c) and (d) are more typical of relationships for air pollution impacts on vegetation, and specific examples are included in the following sections. Each raises specific issues about methods of setting air quality criteria. In the case of (b), the issue is how to deal with the stimulatory effect of low concentrations. From a production perspective, it might be argued that a guideline should be set only to prevent adverse effects. However, this stimulation in one response criterion may be associated with other potentially adverse effects, for example on sensitivity to drought or cold stress. Furthermore, the time element may need to be considered; while atmospheric nitrogen deposition may initially stimulate plant growth, for example, over a longer time period it may lead to nutrient imbalances or changes in competitive balance. In the case of curves (c) and (d), the issues relate rather to the problem of defining the appropriate threshold; is complete protection required, or is it possible to define some acceptable degree of impact, based on its biological or statistical significance? Analogously, the setting of air quality guidelines for many carcinogens recognises the lack of a threshold and is based on an assessment of the concentration or exposure corresponding to a defined level of risk.

There is a further aspect to the relationship between exposure–response relationships and air quality guidelines which is of considerable importance. Policy evaluation is increasingly concerned to quantify the benefits of investment in measures to reduce pollutant emissions, either through formal cost-benefit analysis or through a more pragmatic assessment of whether the scale of the cost is commensurate with the scale of the benefit. Response curves such as those in Figure 22.1 might be used to define the benefits of different levels of emission control for agricultural production; however, the problems of both quantifying and valuing long-term benefits in terms of biodiversity mean that a precautionary approach may be more appropriate than one based on formal cost-benefit analysis.

Development of air quality guidelines

Europe as a continent is characterised by the large number of individual countries, many of relatively small area, with high densities of pollutant emissions. This fact, together with the recognition that the long-range transport of air pollution did not respect international boundaries has meant that international fora have played a central role in assessing the need for, and the values of, air quality guidelines and standards to protect vegetation. There are three organisations which have played a major role in developing and applying air quality guidelines in Europe. The World Health Organisation produced a set of 'Air Quality Guidelines for Europe' in the mid-1980s (WHO, 1987); these have been revised over the past five years, and a new version of this publication has just been published (WHO, 2000). Although these documents concentrate on air quality guidelines to protect human health, guidelines have also been developed for vegetation. The process of scientific review leading to the revised WHO guidelines has been influential in the development of air quality objectives within the European Union, which has the power to set air quality limit values which are binding on all member states. The EU has recently issued a Directive which sets limit values for certain pollutants to protect vegetation or ecosystems; these limit values apply only in rural background areas, and not in urban or industrial areas where the focus of policy measures will be effects on human health. Individual European countries may also develop their own national guidelines or standards for vegetation; if these countries are EU members, these guidelines cannot be more relaxed than those set by the EU. Finally, the United Nations Economic Commission for Europe (UN/ECE), which covers all countries of Europe, has played an important role in the international negotiation of emission reductions based on targets for long-term environmental protection, through the Convention on Long-Range Transboundary Air Pollution. This has provided the framework for the development of critical loads and levels.

Although the United States and Canada are members of UN/ECE, air quality guidelines for vegetation have evolved in North America primarily in a national, rather than an international, framework. In the United States, National Ambient Air Quality Standards are set by the national Environmental Protection Agency; these distinguish primary standards – set to protect human health – and secondary standards set to protect wider societal welfare, including vegetation. An extensive process of detailed scientific review, policy assessment and Congressional review is involved in developing the recommended standards. The position in Canada is somewhat different, in that the National Ambient Air Quality Objectives (NAAQOs), unlike the NAAQS in the US, are not legally enforced, Three types of objectives are set: maximum desirable, acceptable and tolerable concentrations.

Tables 22.2 and 22.3 summarise the critical loads and levels which have been adopted within the framework of UN/ECE for the three pollutants – sulphur dioxide, ozone and nitrogen deposition – and which are discussed in more detail in the remainder of this chapter. Other air quality guidelines for these pollutants have been proposed and are in use in other policy contexts and in other countries. It is important to emphasise that the two tables, and the following text, are not intended to summarise recommended values for application elsewhere in the world; rather the processes involved in establishing these values are used to illustrate the broader issues of establishing air quality guidelines for vegetation which were discussed in the earlier sections of this chapter.

Table 22.2 Summary of critical levels for sulphur dioxide and ozone adopted by UN/ECE

	Sulphur dioxide	Ozone
Nature of exposure index	Annual or winter mean ($\mu g\ m^{-3}$)	Cumulative seasonal exposure above 40 ppb during daylight hours (ppm.h)
Agriculture	30	3 (over 3 months)
Forests	20 15 (harsh climates)	10 (over 6 months)
Semi-natural vegetation	20 15 (harsh climates)	3 (over 3 months)
Lichens	10	–

Table 22.3 Summary of proposed critical loads (kg N ha^{-1} yr^{-1}) for effects of nitrogen deposition on vegetation (adapted from Bobbink and Roelofs, 1995). Reproduced by permission of Kluwer Academic Publishers, *Water, Air and Soil Pollution*, vol. 85, from Bobbink *et al.*, 'Nitrogen critical loads. . ., table on p. 2417

Ecosystem	Critical load	Effect of exceedance
Forests		
Acidic coniferous forests	7–20***	Changes in ground flora and mycorrhizas
Acidic deciduous forests	10–20**	Changes in ground flora and mycorrhizas
Calcareous forests	15–20*	Changes in ground flora
Forests in humid climates	5–10*	Decline in lichens
Heathlands		
Lowland dry heaths	15–20***	Transition from heather to grass
Lowland wet heaths	17–22**	Transition from heather to grass
Upland *Calluna* heaths	10–20*	Decline in heather and moss dominance
Arctic and alpine heaths	5–15*	Decline in lichens, mosses and evergreen dwarf shrubs
Grasslands and wetlands		
Calcareous grasslands	15–35**	Increase in tall grasses, altered diversity
Montane grasslands	10–15*	Increase in tall grasses, altered diversity
Shallow soft-water bodies	5–10***	Decline in isoetid species
Ombrotrophic bogs	5–10**	Decline in typical mosses, increase in tall grasses

Quality of evidence supporting the proposed critical load:
***reliable
**quite reliable
*uncertain

Ozone

Ozone is the pollutant for which the derivation of air quality guidelines has been most directly linked to exposure–response relationships. European approaches to devising air quality guidelines for ozone have differed significantly from those in the US, which have been based on seasonal mean concentrations. The aim in Europe was to develop an exposure index which reflected the cumulative nature of ozone impacts and which incorporated the capacity of vegetation to detoxify ozone entering the plant. Empirical analysis of experimental data from open-top chamber studies carried out across Europe, as part of a wider EU programme (Jäger *et al.*, 1993), established that a linear relationship with yield reduction in wheat could be demonstrated using an exposure index termed AOT40 (Accumulated exposure Over a Threshold concentration of 40 ppb): (Figure 22.2). The fact that this relationship is based on a combination of data from experiments in several countries indicates its wide relevance across northern and central Europe (Fuhrer *et al.*, 1997). The AOT40 index for crops is calculated by summing the differences between the actual concentration and 40 ppb for each daylight hour during a growing season when the concentration exceeds 40 ppb. Using the relationship illustrated in Figure 22.2, which is equivalent to curve (d) in Figure 22.1, a critical level could be set for any chosen value of yield loss; in practice, a yield loss of 5 percent

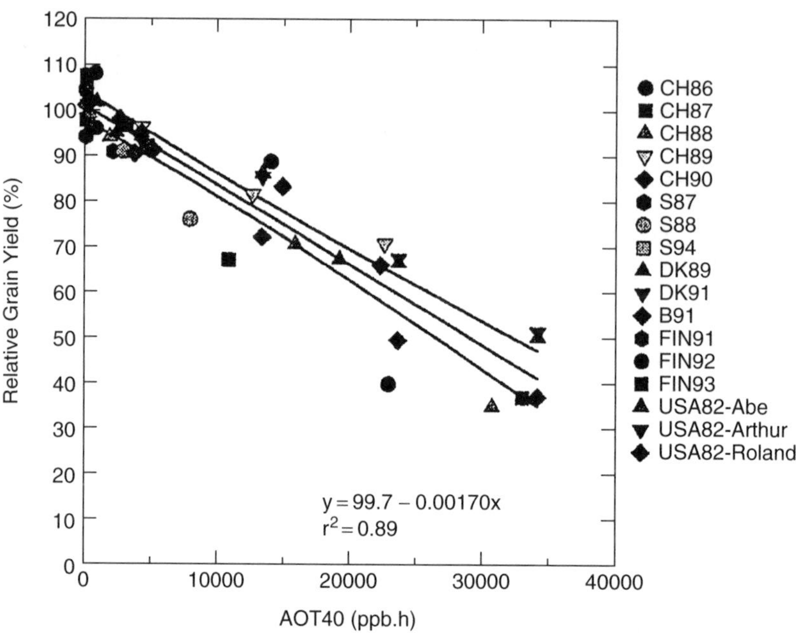

Figure 22.2 Relationship between wheat grain yield, expressed relative to a control treatment, and seasonal ozone exposure, derived from 17 experiments in six countries; the codes refer to the country and year of the experiment. Source: Fuhrer (1996)

averaged over several years has been adopted leading to a critical level of 3 ppm.h (Kärenlampi and Skärby, 1996). This value of 5 percent was adopted as the minimum value that is statistically detectable in the pooled experimental data.

It is important to emphasise that the exposure–response relationship in Figure 22.2, obtained under well-watered conditions in open-top chambers, cannot be used to quantify yield loss under field conditions across locations with different climates. For example, recent analyses of wheat data in Italy suggest that the slope of the exposure–response relationship is significantly smaller than that in Figure 22.2, reflecting the different climatic conditions (Fumigalli *et al.*, 1999). Similarly, Pleijel *et al.* (2000) found different slopes in different years for experiments with wheat in the same location, which reflected the influence of climate on ozone flux into the plants. Ozone flux may be more closely related to ozone effects than the external exposure, expressed using the AOT40 index; however, modelling of ozone flux is complex and highly data-intensive, and so less appropriate as a basis for air quality guidelines in regional-scale policy assessment (Emberson *et al.*, 2000). Wheat was selected for this analysis, because of its agronomic importance across Europe, its known sensitivity to ozone, and because of the extensive data set available. Although the critical level is applied to all crops on a precautionary basis, other crop species may be more sensitive to ozone than wheat, and thus a lower critical level value might be appropriate once a better database of exposure–response data has been developed. Indeed, the relationship used to derive this critical level is from experiments with spring wheat, and may not apply to winter wheat (Pleijel *et al.*, 2000). In summary, the approach adopted, based on experimental exposure–response relationships, has many limitations and uncertainties, because of the problems of extrapolating from experimental conditions, and the critical level value adopted is unlikely to prevent some adverse effects on the most sensitive crop species and cultivars.

For forests, a similar approach has been adopted, using short-term studies with young trees in open-top chambers. Data for beech have been employed to estimate a critical level of 10 ppm.h over a growing season of six months, based on an annual change in above-ground growth of 10 percent, the minimum which can normally be demonstrated to be statistically significant in this type of experiment. However, this empirical approach cannot be used to assess longer-term effects of ozone, as the responses of mature trees may be quite different from these young experimental trees, and an annual growth reduction of 10 percent would lead to massive cumulative decreases in growth. There are limited field data on ozone impacts on mature trees, although a recent epidemiological study of 57 forest stands in Switzerland (Braun *et al.*, 1999) showed a significant linear negative correlation between the stem increment of beech over four years and ozone exposure, expressed as AOT40, at each site. Alternatively, the use of computer models may provide an approach to simulating exposure–response relationships (Ollinger *et al.*, 1997), and hence defining critical levels.

Further methodological problems arise when attempting to apply such an experiment-based approach to effects on semi-natural vegetation, for which the most important impacts of ozone on plant communities may not be on productivity, but through shifts in species composition, loss of biodiversity, and changes in genetic composition. Several studies of mixtures of herbaceous species have demonstrated a shift in the relative proportions of the species in response to ozone, without any effects on the total growth of the mixture (e.g. Ashmore and Ainsworth, 1995), but, since these are relatively short-term experiments on artificial mixtures, it is not possible to derive critical level values for long-term protection of community composition with any certainty. Experimental evidence suggests that certain herbaceous species of semi-natural vegetation are as sensitive as the most sensitive crop species (Ashmore and Davison, 1996; Davison and Barnes, 1998), and thus a similar critical level might be appropriate. However, it is clear that the current critical levels of ozone for forests and semi-natural vegetation are very provisional and not based on an analysis of the long-term cumulative impacts on ecosystems, which characterise the derivation of critical loads (cf. Table 22.1).

Sulphur dioxide

Unlike those for ozone, air quality guidelines for sulphur dioxide have primarily been based on field observation. Values were first proposed by IUFRO (the International Union of Forest Research Organisations) in 1978 (Wentzel, 1983), when standards, expressed as annual mean concentrations, of 50 $\mu g\ m^{-3}$ to protect trees in good growing conditions and 25 $\mu g\ m^{-3}$ to protect trees growing on poor sites were established. The latter value was based on field evidence from Finland and mountain regions of central Europe that conifer species are much more sensitive to SO_2 under cold temperatures.

More recent analysis of the available data has led to the development of a series of critical levels for different types of vegetation (cf. Table 22.2). In each case, the approach has been to identify the field or experimental data which indicate the lowest concentration at which adverse effects have been observed. Values are expressed as both annual and winter means, because of the evidence that effects of SO_2 are greater at low temperatures; in practice, since winter mean concentrations are typically higher than annual means, this is the more stringent criterion. The lowest value in Table 22.2 is set to protect the most sensitive group of lichen species, many of which have been lost from urban areas of Europe due to the effects of air pollution (see Chapter 17). The value of 10 $\mu g\ m^{-3}$ was based on a field study around a newly established rural point source of pollution which demonstrated community change at this concentration (Will-Wolf, 1981). The values for trees and natural vegetation are based on an analysis of the rate of breakdown of Norway spruce stands in Czechoslovakia as a function of SO_2 concentration and altitude (Makela *et al.*, 1987). The lower value of 15 $\mu g\ m^{-3}$ in colder areas reflects greater response observed at higher altitudes. For crops, the critical level of 30 $\mu g\ m^{-3}$ is based on an analysis of experimental data to identify the lowest concentration producing an adverse effect – a study of perennial ryegrass (Bell, 1985).

It should be noted that responses to SO_2 may take the form of curve (b) in Figure 22.1, i.e. there may a positive growth response at low concentrations, especially in regions of sulphur deficiency for crops. This needs to be considered in applying air quality guidelines for different types of vegetation; thus, reducing concentrations to protect the most sensitive lichen species may cause reductions in the yields of crops with a high sulphur demand, if fertiliser applications are not adjusted.

Nitrogen deposition

Nitrogen is a major nutrient limiting plant growth in many ecosystems of high biodiversity, and the initial response to increased nitrogen deposition is often a stimulation of growth. However, for a variety of reasons, continued deposition of nitrogen can cause long-term chemical and biological changes which have adverse effects on ecosystem function or community composition (see Chapter 12). In setting values of critical loads for nitrogen, it is important to recognise that several different adverse effects on soil chemistry need to be considered, including the contribution to soil acidification, nutrient imbalances in the soil and vegetation, and increased nitrate leaching from the soil into drainage waters or groundwater. However, in the context of this chapter, it is impacts on the species composition of different communities which are of primary concern, either because of direct toxic effects on individual sensitive species, or because of differential stimulatory effects on plant species, which may alter the competitive balance and hence community composition. Thus, while the generic response to increased N deposition is best characterised by curve (b) in Figure 22.1, the fact that different species and response variables exhibit different types of response to nitrogen, and the fact that the response to a given deposition rate may change over time, make the formulation of critical loads for N deposition particularly challenging.

Critical loads for nitrogen deposition do not currently account for variation in the physical form or chemical form of the deposition, although these are likely to be important for impacts on vegetation, especially close to pollutant sources, where gas concentrations may be high. In addition to critical loads for total nitrogen deposition, critical levels have been set for concentrations of the two major gases which contribute to dry deposition – ammonia and nitrogen oxides (WHO, 2000). These critical levels apply to all types of vegetation, because of a lack of sufficient data to distinguish clearly different values for different vegetation types, rather than because of a belief that all vegetation types are equally sensitive. In the case of ammonia, the critical level of 8 $\mu g\ m^{-3}$, as an annual mean, is based on an ecotoxicological analysis of experimental data. In the case of nitrogen oxides, it is assumed that both nitric oxide and nitrogen dioxide are phytotoxic, and a value of 30 $\mu g\ m^{-3}$ for the sum of the two pollutants has been set as an annual mean concentration. These values will be of relevance for local air quality management – for example, around intensive livestock units for ammonia or around major roads for nitrogen oxides – rather than regional scale assessments.

Instead of setting a single value for broad vegetation types, as has been done for sulphur dioxide and ozone, different critical loads for nitrogen deposition have been proposed for different ecosystems. These critical loads have been developed on an empirical basis, using evidence from field observations and field manipulation experiments. In contrast to the approach for ozone and sulphur dioxide, it has been explicitly recognised that the evidence available is variable in quality and limited in extent. Table 22.3 provides a selective synopsis of proposed critical load values for different ecosystems in western Europe which are thought to be relatively sensitive. Three important points should be made about Table 22.3.

Firstly, critical load values are defined as a range rather than as a single value. This indicates two factors: the real variation in response between ecosystems within the tabulated categories, and true uncertainty, e.g. in quantifying the deposition of N in field observations or the rate of N addition in experimental studies. In some cases, the causes of the real variation in responses between ecosystems can be identified. For example, in the case of calcareous grasslands, the wide range of values reflects the effect of phosphorus status in modifying the responses to N deposition; thus lower values are applied to N-limited systems and higher values to P-limited systems. In other cases, generalised modifiers have been proposed; for example, critical loads are thought to be lower at low temperatures, because of the slower rates of nutrient cycling.

Secondly, Table 22.3 provides an indication of the reliability of the proposed values. Where values are reliable there is reasonable consistency in published studies, while where values are quite reliable there is some consistency. In contrast, where values are labelled as ‘uncertain’, little or no relevant empirical data are available, and expert knowledge of responses to nitrogen, in comparison with others for which more reliable values can be set, has been used.

Thirdly, Table 22.3 provides an indication of the biological criteria used to define the critical load. It is important to note that these are not consistent – while some refer to specific ecological changes, such as a switch from heather to grass species, others refer to general change. Similarly, while some refer specifically to declines in particular species groups, others simply refer to changes in diversity. Likewise, the same biological effect may not be considered for every ecosystem; for example, it is likely that epiphytic lichen species are among the most sensitive elements of any forest ecosystem; however, they are only included for forests in humid climates, for which the diverse lichen flora is a major reason for their high conservation value.

It may be argued that this approach provides a more honest statement of the ability of scientists, based on the available data, to define appropriate air quality guidelines than the more traditional approach of providing a single value. However, the approach of providing critical loads as a range, rather than a single value, still leaves major problems to be addressed in the process of policy evaluation, especially where the implications of adopting different values within the suggested range have major implications for the costs of emission control.

Conclusions

In summary, there are a number of conceptual approaches to providing air quality guidelines for terrestrial ecosystems. These range from estimating a single air quality guideline for a pollutant, based on the response of the most sensitive known species or community, to estimating a range of guideline values, differentiated by factors such as species or community, nutrient status, climate, and management techniques. In theory, the latter, more sophisticated, approach allows pollution control policy to be better targeted; however, it requires much more information to assign the correct value to each area under consideration. More importantly, all conceptual approaches are severely constrained by the availability of appropriate data on the biological impacts of air pollutants, especially where these are long-term and involve community or ecosystem responses.

References

Ashmore, M.R. and Ainsworth, N. (1995). Effects of ozone and cutting on the species composition of artificial grassland communities. *Functional Ecology*, **9**, 708–712.

Ashmore, M.R. and Davison, A.W. (1996). Towards a critical level of ozone for natural vegetation. In: *Critical Levels for Ozone in Europe. Testing and Finalising the Concepts*, (L. Kärenlampi and L. Skärby, L. eds.), University of Kuopio, Finland. pp. 58–71.

Bell, J.N.B. (1985). SO_2 effects on the productivity of grass species. In: *The Effects of SO_2 on Plant Productivity* (W.E. Winner, H.A. Mooney and R.A. Goldstein, eds.), pp. 209–266. Stanford University Press, Stanford.

Bobbink, R. and Roelofs, J.G.M. (1995). Nitrogen critical loads for natural and semi-natural ecosystems: the empirical approach. *Water Air and Soil Pollution*, **85**, 2413–2418.

Braun, S., Rihm, B. and Flückiger, W. (1999). Growth of mature beech in relation to ozone and nitrogen deposition: an epidemiological approach. In: *Critical Levels for Ozone – Level II* (J. Führer and B. Acherman, eds.). Swiss Agency for the Environment, Forests and Landscapes, Berne, pp. 111–114.

Bull, K.R. (1991). The critical loads/levels approach to gaseous pollution control. *Environmental Pollution*, **69**, 105–123.

Bull, K.R. (1995). Critical loads – possibilities and constraints. *Water Air and Soil Pollution*, **85**, 201–212.
Cresser, M.S. (2000). The critical loads concept: milestone or millstone for the new millennium? *Water Air Soil Pollution*, **249**, 51–62.
Davison, A.W. and Barnes, J.D. (1998). The effects of ozone on wild plants. *New Phytologist*, **139**, 135–151.
Edwards, L. (1998). Limit values. In: *Urban Air Pollution – European Aspects* (J. Fenger, O. Hertel and F. Palmgren, eds.), pp. 419–432. Kluwer Academic, Dordrecht.
Emberson, L.D., Ashmore, M.R., Cambridge, H.M., Simpson, D. and Tuovinen, J.-P. (2000). Modelling stomatal ozone flux across Europe. *Environmental Pollution*, **109**, 403–413.
Fuhrer, J. (1996). The critical level for effects of ozone on crops, and the transfer to mapping. In: *Critical Levels for Ozone in Europe. Testing and Finalising the Concepts* (Kärenlampi, L. and Skärby, L. eds.). University of Kuopio, Finland, pp. 27–43.
Fuhrer, J., Skärby, L. and Ashmore, M. (1997). Critical levels for ozone effects on vegetation in Europe. *Environmental Pollution*, **97**, 91–106.
Fumagalli, I., Ambrogi, R. and Mignanego, L. (1999) Ozone in southern Europe: UN/ECE Experiments in Italy suggest a new approach to critical levels. In: *Critical Levels for Ozone, Level II* (J. Fuhrer and B. Achermann, eds.). Swiss Agency for Environment, Forest and Landscape, Bern, pp. 239–242.
Jäger, H.J., Unsworth, M., de Temmerman, L. and Mathy, P. (eds.) (1993). *Effects of Air Pollution on Agricultural Crops in Europe*. Air Pollution Research Report 45, Commission of the European Communities, Brussels.
Kärenlampi, L. and Skärby, L. (eds.) (1996). *Critical Levels for Ozone in Europe: Testing and Finalising the Concepts*. University of Kuopio, Finland.
Makela, A., Materna, J. and Schopp, W. (1987). *Direct effects of sulfur on forests in Europe: a regional model of risk*. Working paper 87-57, International Institute for Applied Systems Research, Laxenburg, Austria.
Ollinger, S.V., Aber, J.D. and Reich, P.B. (1997). Simulating ozone effects on forest productivity: interactions among leaf- and stand-level processes. *Ecological Applications*, **7**, 1237–1251.
Pleijel, H., Danielsson, H., Karlsson, G.P., Gelang, J., Karlsson, P.E. and Sellden, G. (2000). An ozone flux-response relationship for wheat. *Environmental Pollution*, **109**, 453–462.
Sanders, G.E., Skärby, L., Ashmore, M.R. and Fuhrer, J. (1995). Establishing critical levels for the effects of air pollution on vegetation. *Water Air and Soil Pollution*, **85**, 189–200.
Wentzel, K.F. (1983) IUFRO studies on maximal SO_2 emissions standards to protect forests. In: *Effects of Accumulation of Air Pollutants in Forest Ecosystems* (B. Ulrich and J. Pankrath, eds.), pp. 295–302. D Reidel, Dordrecht.
WHO (1987). *Air Quality Guidelines for Europe*. World Health Organisation, Copenhagen.
WHO (2000). *Air Quality Guidelines for Europe*. Second edition. World Health Organisation, Bilthoven.
Will-Wolf, S. (1981). Structure of corticolous lichen communities before and after exposure to emissions from a 'clean' coal-fired power station. *Bryologist*, **83**, 281–295.

23 Air pollution and climate change

V.C. RUNECKLES

Introduction

Climate change has been the norm throughout the earth's history. Ice ages have been interspersed with interglacial periods with warmer temperatures. Although the causes of these previous natural changes are not completely understood, there is a growing consensus that human activities are contributing to the current period of climate change. In 1996, the World Meteorological Office/United Nations Environment Program's Intergovernmental Panel on Climate Change (IPCC) cautiously stated: 'The balance of evidence suggests a discernible human influence on global climate' (IPCC, 1996). In spite of the nay-sayers, by 2001, new evidence and improved understanding led the IPCC to state unequivocally in its Third Assessment Report that, in its judgment, 'most of the warming observed over the last 50 years is attributable to human activities' (IPCC, 2001). Most recently, further evidence has come from studies of the rise in global ocean temperatures (Barnett *et al.*, 2001; Levitus *et al.*, 2001).

Four features of the present period of change are:

1. increasing concentrations of 'greenhouse gases' (GHGs) in the atmosphere, especially carbon dioxide (CO_2), methane (CH_4), nitrous oxide (N_2O), chlorofluorocarbons (CFCs), and tropospheric ozone (O_3), leading to
2. increasing global mean temperature through radiative forcing and
3. an increased frequency of extreme weather events, and
4. the depletion of stratospheric ozone caused by the release of man-made ozone-depleting substances (ODSs, including CFCs), which has resulted in measurable increases in the surface levels of harmful ultraviolet radiation (UV-B; 280–320 nm) particularly in the southern hemisphere at high latitudes.

It is ironic that although atmospheric CO_2 is essential to most present-day plant and animal life on earth because it provides the carbon input to photosynthesis, it is the significant releases of CO_2 into the atmosphere resulting from activities such as the combustion of fossil fuels and changes in land-use (especially deforestation) that constitute the primary 'discernible human influence' referred to by the IPCC.

Increasing concern over the consequences of the contributory human activities has led to international agreements such as the Montreal Protocol on reducing the use of ODSs and the Kyoto Protocol on the reduction of GHGs.

Air Pollution and Plant Life, second edition. Edited by J.N.B. Bell and M. Treshow. ISBN 0 471 49090 3 (HB), 0 471 49091 1 (PB).

Interest in climate change has generated a vast literature in the last 20 or so years, including the appearance of new scientific journals such as *Climate Research* and *Global Change Biology*. Comprehensive reviews have been published by the German *Bundestag* (1991) and the IPCC (1990, 1992, 1996). The United Nations Environment Program (UNEP) has itself published several reports which include global climate data (e.g. UNEP, 1993), most recently, its 'Global Environment Outlook 2000' (UNEP, 1999), and the draft of the IPCC's Third Assessment Report (IPCC, 2001). The biological impacts have been extensively reviewed by Krupa and Kickert (1989) and in more recent volumes such as those by Graves and Reavey (1996), Luo and Mooney (1999), and Groth and Krupa (2000). Reviews by Wellburn (1994) and those edited by De Kok and Stulen (1998) focus on air pollutant effects. Kickert *et al.* (1999) have provided a comprehensive review of the application of computer simulation modeling to predict the ecological, environmental and societal effects of climate change.

The focus of this chapter is on the interactions of the components of current climate change on the effects of the major air pollutants (ozone, O_3; sulphur dioxide, SO_2; nitrogen oxides, NO_x and hydrogen fluoride, HF) which are discussed in other chapters. Figure 23.1 presents a diagrammatic summary of these interactions. Climatic factors are on the left-hand side of the figure and air pollutants on the right. Ozone occupies a position near the center because, on one hand it is a phytotoxic pollutant and, on the other it is a GHG whose increasing concentrations in the troposphere since

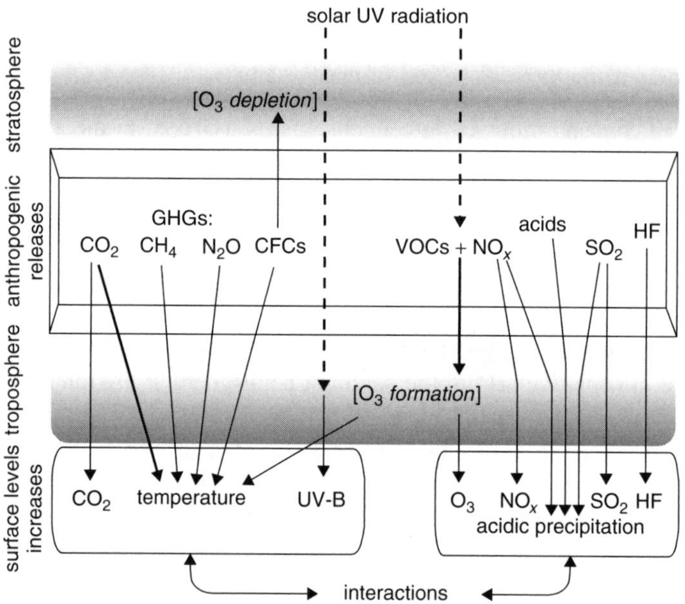

Figure 23.1 The major components of present climate change and their interactions with major air pollutants. Thickness of the broken lines reflects attenuation of surface UV-B radiation by stratospheric and tropospheric O_3. Thickness of the arrows to surface temperature indicate the relative magnitudes of radiative forcing by GHGs. VOCs = volatile organic compound precursors of O_3; $NO_x = NO + NO_2$

pre-industrial times now contribute close to 10% of the total positive radiative forcing due to GHGs (German *Bundestag*, 1991)

Knowledge of the interactions involved is critically important to our understanding of the impact of air pollution *and* climate change (Krupa and Kickert, 1989; Runeckles and Krupa, 1994; Luo and Mooney, 1999; Krupa and Groth, 2000). In order to provide the context in which to discuss these interactions, there follows a brief discussion of increasing CO_2, temperature and UV-B as components of climate change *per se* and the impact of these changes on plants and ecosystems. Although much of the world's land mass is covered by native vegetation, most of the information available has been obtained from studies with agricultural crops and a few temperate forest tree species.

Anthropogenic greenhouse gases and climate change

The spectrum of solar radiation reaching the earth's surface peaks in the visible region at wavelengths between 400 and 600 nm, while the spectrum of the earth's radiation into space is maximal in the infra-red region about 1600 nm. However, both spectra show pronounced absorption bands, especially those due to water vapour and CO_2 in the atmosphere, and it is this absorption of energy that plays an important role in determining surface temperatures and leads to these and other constituents of the atmosphere being referred to as 'greenhouse gases' (GHGs).

Although water vapour is the most important GHG, its effect on global climate is made complex by the fact of its condensation to liquid water or ice crystals, in both of which forms it can reflect or scatter both incoming and outgoing radiation. However, with CO_2, the situation is more straightforward. Evidence from the past clearly illustrates the interrelationship between temperature and atmospheric CO_2 concentration within the earth's complex land–ocean–atmosphere system, for example, over the past 160 000 years as shown in Figure 23.2 and the past 150 years in Figure 23.3 (UNEP, 1993). Half of the 30% rise to more than 360 parts per million (ppm) over the last 250 years has occurred since 1960 (UNEP, 1989).

The *Third Assessment Report* provides the most recent estimate of the global warming caused by these increases in GHGs: 1.4–5.8 °C for the period 1990–2100 (IPCC, 2001), an increase from the previously estimated range: 1.0–3.5 °C (IPCC, 1996). This rate of warming is much greater than that which occurred over the whole of the twentieth century, and is very likely to be the greatest warming in the past 10 000 years. Stabilizing CO_2 at 450 ppm had previously been estimated to result in an increase of 1.5–4.0 °C; stabilization at 550 ppm, an increase of 2.0–5.5 °C; and a doubling of present CO_2 levels, an increase of more than 3.0 °C (IPCC, 1996).

Plant responses to increased carbon dioxide and increased temperature

Carbon dioxide

Long (1991), Crawford and Wolfe (1999), Luo *et al.* (1999) and others have stressed that changes in CO_2 and surface temperature operate interactively with each other and

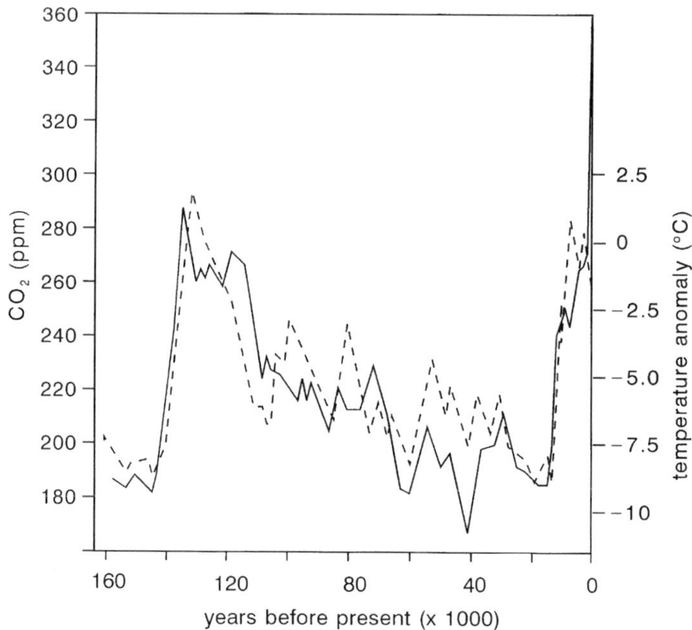

Figure 23.2 Changes in atmospheric CO_2 concentration (solid line) and global temperature (broken line) over the past 160 000 years, based on deuterium and CO_2 analyses of the Vostok (Antarctica) ice-cores. Temperatures are expressed as divergences from a reference mean. Based on data from Barnola *et al.* (1987)

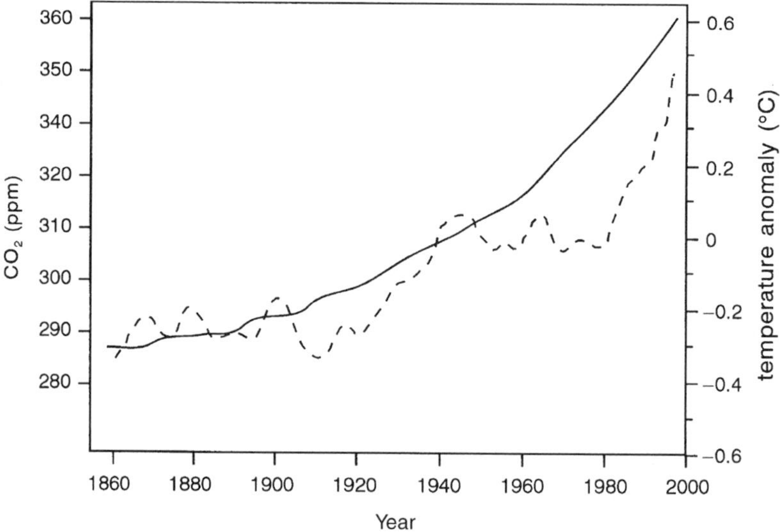

Figure 23.3 Changes in atmospheric CO_2 concentration (solid line) and mean global surface temperature (land and sea; broken line) over the past 150 years. Based on data from UNEP (1989) (CO_2) and UNEP (1993) (temperature)

also with other environmental factors. However, almost all investigations of the effects of increased CO_2 levels (with or without additional factors) have been conducted at close to current ambient temperatures in field exposure systems of various types, or at arbitrary temperatures in controlled environment chambers (CECs).

Numerous growth effects and physiological responses to increased CO_2 have been summarized in many reviews, including those by Allen (1990) and Rogers and Dahlman (1990) for crop species, Bazzaz (1990) with respect to natural ecosystems, and Eamus (1996) and Saxe *et al.* (1998) for forest trees. Several also describe the different experimental methodologies that have been used, and note that marked interspecific and intervarietal differences occur in many responses.

Table 23.1 provides a summary of the main effects of increased CO_2 concentrations on plants. Primary productivity is the outcome of photosynthetic CO_2-assimilation and it has long been known that the rate of assimilation increases with CO_2 concentration. Most plant species (including the major crops, wheat, rice, potatoes, soybean and all tree species) have the C3 pathway of CO_2-assimilation and exhibit significant photorespiration (loss of CO_2 during illumination); these plants show large increases in assimilation rate with increasing CO_2 concentration (Figure 23.4(a)). In contrast, species with the C4 assimilation pathway possess foliar CO_2-concentrating mechanisms that result in photosynthetic carboxylation functioning close to CO_2 saturation, and exhibit little photorespiration. C4 plants contribute approximately 20% of global primary productivity and include the crops, maize, sugarcane and sorghum. In these species the rise in CO_2-assimilation rate is more rapid but tends to reach a lower maximum than in C3 species (Hsaio and Jackson, 1999) (Figure 23.4(a)).

The effect of CO_2 in decreasing stomatal conductance (Table 23.1) is readily observed experimentally and is borne out by the long-term records from fossil and herbarium samples (Beerling and Woodward, 1993). However, any reduction in CO_2 availability for photosynthesis because of decreased conductance (Table 23.1) is more than offset by the increased CO_2 flux rate resulting from the steeper

Table 23.1 Effects of increased atmospheric CO_2 on terrestrial species[1]

Plant response	Usual effect
Stomatal conductance	↓
Transpiration	↓
Photosynthesis	↑ (C3 > C4 species)
Photosynthetic acclimation	~ (frequently ↑)
Growth rates (biomass and elongation)	↑ (C3 > C4 species)
Biomass and crop yield	↑ (C3 > C4 species)
Water-use efficiency (WUE)	↑ (C3 > C4 species)
Leaf area index (LAI, leaf area/unit ground area)	↑
Specific leaf area (SLA, leaf area/unit weight)	↓
Root/shoot ratio	~ (frequently ↑)
Branching and tillering	↑
Fruit size, number and seed/plant	↑

[1] Adapted from Allen (1990). ↓: decrease; ↑: increase; ~: variable effect.

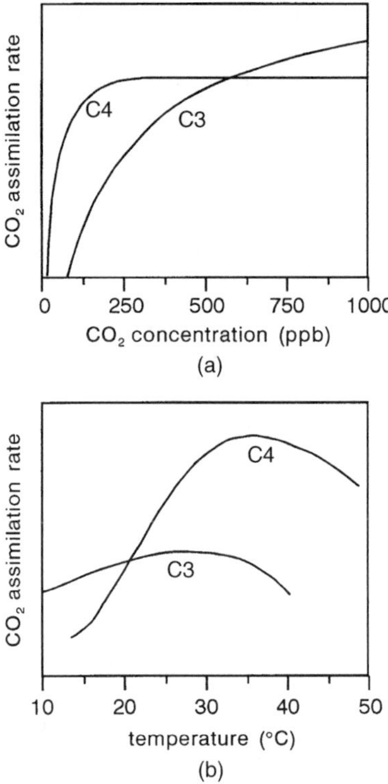

Figure 23.4 Comparison of the effects of (a) CO_2 concentration and (b) temperature on photosynthetic CO_2 assimilation, typical of C3 and C4 species

concentration gradient from the ambient air to the leaf interior (Amthor, 1999). Increased CO_2-assimilation coupled with reduced transpiration leads to improved water-use efficiency (WUE). This improvement is more marked in C3 than in C4 species which already possess high WUEs because of minimal photorespiration. Increased CO_2-assimilation in turn leads to increased overall growth, a response put to commercial use for many years by greenhouse crop growers using CO_2-enrichment to achieve concentrations of 1000–1200 ppm. The increased growth resulting from increased CO_2 concentrations may not be uniformly distributed throughout the plant, and different species show varying degrees of photosynthetic acclimation to higher CO_2 concentrations. Yield increases of C4 plants are usually less impressive than those of C3 species: 22 versus 41% for enrichment to 700 ppm CO_2 (Poorter, 1993). Since C4 species may have evolved in response to low atmospheric CO_2 concentrations (Ehleringer *et al.*, 1991), the current rapid rate of increase in CO_2 levels has led to discussion of its consequences to the adaptability and possibly impaired competitive ability of C4 species, but Henderson *et al.* (1994), and Wand *et al.* (1999) have questioned this.

In general, both coniferous and other tree species show the same responses as crops (Saxe, Ellsworth and Heath, 1998) although there is some inconsistency as to whether species grown for prolonged periods at higher CO_2 levels show any downward acclimation of assimilation (Eamus, 1996).

Much recent crop work has used partial field enclosures (e.g. open-top chambers, OTCs) or Free Air CO_2 Exposure (FACE) systems (Dugas and Pinter, 1994) to determine the effects of long-term exposure to elevated CO_2 levels. OTCs were used in the large, three-year 'ESPACE-wheat' program of the European Community involving 25 experiments at nine sites in eight countries (Jäger *et al.*, 1999). Most of the recent crop studies have been on C3 species such as soybean, wheat and rice, with fewer studies on C4 species such as maize. With regard to tree species, many early studies were conducted using seedlings or young trees in CECs or OTCs or using various types of branch chamber on mature trees (Eamus, 1996), but many authors have questioned the relevance of the findings to mature trees. A FACE system experiment located in a 12-year-old, 10 m tall loblolly pine forest at Duke Forest, NC, USA, commenced in 1994 and has already shown that prolonged exposure to 550 ppm CO_2 can result in up to 43% enhancement of photosynthesis (Ellsworth *et al.*, 1995, Ellsworth *et al.*, 1998).

Scaling up the effects of increased CO_2 from the plant or simple ecosystem level to large ecosystem, regional and global scales is fraught with difficulties (Amthor, 1999) and recourse has to be made to computer simulation models. However, as Kickert *et al.* (1999) point out, all models dealing with aspects of climate change are subject to considerable uncertainties in their output, due to the uncertainties in input data and the validity of the assumptions made. Furthermore, almost all models provide *consequence assessments* addressing the magnitudes of expected changes, rather than *risk assessments* addressing their probability or likelihood.

Temperature

Temperature as one of the environmental factors affecting the responses of plants to air pollutants has been discussed in Chapter 18. After germination, plant growth is dependent on CO_2 assimilation exceeding respiratory and photorespiratory losses. While the purely photochemical reactions of photosynthetic light capture have Q_{10} values close to 1 (Q_{10} is the magnitude of change for a 10 °C rise), for the enzymic reactions of carbon fixation (photosynthesis) and carbon release (respiration and photorespiration), $Q_{10} \approx 2$, and for protein (enzyme) breakdown due to protease action, $Q_{10} \approx 6$. As temperature rises so do CO_2-assimilation and respiration, but photorespiration is also enhanced because the ratio of dissolved O_2 to CO_2 also rises, favoring the oxygenase activity of ribulose-1,5-*bis*-phosphate carboxylase-oxygenase (Rubisco) (however, see the discussion of CO_2-temperature interactions below). But with these and the myriad other enzymic reactions, enzyme denaturation is ever-present and, because of its high Q_{10}, eventually becomes dominant. Hence each enzyme reaction and each sequence of enzyme reactions has a temperature optimum. Thus, with regard to CO_2 assimilation, C3 species typically have temperature optima in the 20 ± 5 °C range while the optima for C4 species are typically 10 °C higher (Figure 23.4(b)). With regard to growth, the optimum temperature range is one in which *all* the individual reactions and vital processes are collectively functioning *optimally* but not necessarily *maximally*.

Different species have evolved with different ways of integrating these cellular level responses, and the optimum growth temperatures for terrestrial species range up to more than 45 °C. C3–C4 differences exist and the temperature–growth curves for these two types resemble those for CO_2 assimilation shown in Figure 23.4(b).

'Temperature determines the start and finish and rate and duration of organ growth and development' (Lawlor, 1998), and individual features of plant development (e.g. shoot and root growth, flowering, pollen tube growth, fruit set and seed development) have different specific temperature optima, so that increases in overall mean temperature lead to changes in the growth and development pattern of a species. Global temperature rise will therefore affect the interspecific relationships within natural communities (Crawford and Wolfe, 1999) and the geography of crop production.

Henderson *et al.* (1994) suggest that projected increases in temperature in Australia will lead to southward expansion in range and greater dominance of the grass flora by C4 species. Evidence from the fossil pollen record suggests past migration rates of temperate forest trees of between 25 and 40 km per century (200 maximum), whereas in temperate North America, a 3 °C rise would lead to a 300 to 375 km northward displacement of isotherms. Many tree species may therefore be unable to migrate as fast as the changing climate (IPCC, 1996). However, any temperature-driven migration of species to higher altitudes and latitudes (with the possibilities of local extinctions) may be opposed by the countervailing resistance to invasion offered by currently extant soil-vegetation systems (IPCC, 1990; Crawford and Wolfe, 1999).

CO_2–temperature interactions

A frequently observed CO_2–temperature interaction is that increased temperature enhances the CO_2-driven stimulation of photosynthetic assimilation in C3 species. This is probably because increasing CO_2 inhibits photorespiration more at warm than cool temperatures (Jordan and Ogren, 1984), and because negative feedback from the products of photosynthesis is reduced by the greater metabolic demands for photosynthates for growth at warmer temperatures (Farrar and Williams, 1994). This effect was demonstrated by the work of Long (1991), whose models of CO_2-assimilation also incorporated the decreased affinity of Rubisco for CO_2 at higher temperatures. In particular, he stressed the essential need to incorporate both CO_2 and temperature rises in predictive modeling of the effects of climate change on photosynthesis at both plant and canopy levels.

However, growth and development processes are more variable in response to temperature than processes such as photosynthesis and respiration (Lawlor, 1998). Hence the photosynthetic interrelationships with CO_2 and temperature do not necessarily translate into increased growth and yield. Thus there are reports of increased temperature either counteracting or enhancing the benefits of growth in increased CO_2. For example, Wheeler *et al.* (1996) used plastic-covered tunnels to establish temperature gradients in field plots of wheat, and reported that a temperature rise of ~1.5 °C was sufficient to cancel the growth and yield increases resulting from enrichment to 700 ppm. Similar findings were reported from the study by Van Oijen *et al.* (1999) in the ESPACE-wheat program. Half of the OTCs used in their study were cooled to negate the 'chamber effect' on temperature (usually a 1 to 3 °C increase above ambient temperature; Heagle *et al.*, 1988). They found that although temperature had no effect on CO_2-enhanced assimilation rates, the enhanced growth and grain yield observed in the cooled chambers was largely cancelled in the ambient chambers – the temperature difference was 2.4 °C. This difference is within the range of the estimated temperature rise in a 700 ppm world. The authors attributed this observation to accelerated phenology and grain filling and lower LAI (leaf area index = area of foliage per unit area of soil) in the warmer chambers.

In contrast, in an OTC study using the perennial grass *Festuca pratensis* in which a temperature increase of 3 °C above ambient was combined with CO_2-enrichment to 700 ppm, both CO_2 and temperature caused increases in total above-ground biomass (Hakala and Mela, 1996). CEC studies with potato (Cao *et al.*, 1994) and soybean (Ziska and Bunce, 1995) similarly showed temperature-enhanced increases in growth in enriched CO_2 atmospheres. Read and Morgan (1996) compared the effects of enriched CO_2 and temperature on two grasses: cool-season *Pascopyrum smithii* (C3) and warm-season *Bouteloua gracilis* (C4). In the C4 species, 750 ppm CO_2 resulted in increased dry matter production at daytime temperatures as high as 35 °C. In the C3 species, CO_2-stimulated growth was greatest at 20 °C but the stimulation was progressively attenuated by increased temperature; at 35 °C growth was only one-third of that in 350 ppm CO_2 at 20 °C.

These results point out that, because of the complexity of the interacting outcomes of the many temperature-sensitive processes involved in growth, including photosynthesis, respiration, photorespiration, diffusive conductance, photosynthate translocation, and various regulatory feedback mechanisms, increased photosynthesis alone is no guarantee of increased growth.

Although the OTC methodology has been widely used for CO_2-enrichment and air pollutant studies in the field, its shortcomings, particularly the increased temperature component of the 'chamber effect', have been widely discussed with regard to the relevance of the results obtained with their use (e.g. Krupa and Groth, 2000). However, even though chamber effects concern factors in addition to temperature, I suggest that, *for studies on climate change*, any measured increase in chamber temperature should be welcomed since it could increase the relevance of the findings, in contrast to the chamberless FACE approach which can only provide data on CO_2 enrichment that may or may not be relevant to the complexities of climate change. Indeed, temperature was one of the covariables measured in the 25 OTC experiments in the ESPACE-wheat program which resulted in a wide range of growth and yield responses of the single cultivar used over three years (e.g. a range of 11–121% increased grain yield). The multiple regression analyses by Bender *et al.* (1999) revealed that temperature was a significant covariable, with a negative coefficient, although a large part of the variability could not be explained.

Plant responses to increased UV-B radiation (280–320 nm)

The effects of UV-B radiation on plants have been the subject of numerous specific reviews such as Allen *et al.* (1998), Caldwell *et al.* (1998), Jordan (1996), Lumsden

Table 23.2 Effects of increased UV-B radiation on terrestrial species[1]

Plant response	Usual effect
Photosynthesis	0 or ↓ (C3 and C4 species, especially at low light intensities)
Stomatal conductance	↓ (especially at low light intensities)
Biomass and crop yield	↓
WUE	↓
LAI	↓
SLA	↓
Crop maturation rate	0
Sensitivity to drought stress	↑
Response to drought stress	↓
Sensitivity to mineral nutrient stress	∼

[1]Modified from Krupa and Kickert (1989) and Runeckles and Krupa (1994). ↓: decrease; ↑: increase; ∼: variable effect; 0: no effect.

(1997) and Tevini (2000). The topic has also been included in reviews of the interactions of factors involved in climate change and tropospheric ozone (O_3) by Krupa and Kickert (1989), Runeckles and Krupa (1994) and Saxe *et al.* (1998). Tevini (2000) provides a succinct description of the experimental difficulties in achieving accurate spectral simulation of the increased UV-B levels attributable to atmospheric O_3 depletion through the use of filtered UV-lamps and the necessary recourse to using action spectra.

Species and cultivars show great variability in sensitivity to UV-B (Caldwell *et al.*, 1998). Thus, while there are numerous reports of adverse effects on photosynthesis, dry matter production, yield, stem elongation, flowering, WUE, SLA and many other aspects of growth and development, there are many if not more reports of no effects of prolonged exposures to UV-B levels predicted to result from 16 to 25% depletions of stratospheric O_3, and occasional stimulations. Table 23.2 provides a summary of effects observed.

At the ecosystem level, relatively little information exists on the effects of UV-B on ecosystems or dominant forest tree species; this is particularly true for tropical forests (Caldwell *et al.*, 1998; Tevini, 2000).

Interactions of CO_2, temperature and UV-B

CO_2–temperature interactions have been discussed above and there have been a few studies of the effects of simultaneous increases in UV-B. Using the O_3 filter technique to reduce natural UV-B levels in CECs (Tevini *et al.*, 1990), Mark and Tevini (1997) found that, in sunflower, a C3 species, a doubling of CO_2 more than compensated for the negative effects of UV-B on photosynthesis, plant biomass and leaf area, but in maize (C4), although it reversed the effect of UV-B on photosynthesis, it merely attenuated the negative effects on growth. This result is in keeping with the generally observed responses of C3 and C4 species discussed previously. On the other hand, an average increase of 4 °C in the presence of enhanced UV-B with or without enriched CO_2 tended to reduce the rates of photosynthesis but more than compensated for the biomass and leaf area losses induced by UV-B at the lower temperature. In a greenhouse study, Teramura *et al.* (1990) studied the combination of CO_2 and UV-B on wheat, rice and soybean and observed no significant effect of UV-B on seed yield in any species, or on plant biomass in the two cereals, and a slight decrease in soybean biomass. CO_2 on the other hand, increased yield and biomass in all species. UV-B reduced the CO_2 enhancement of seed yield of wheat but not of the other species, and had no effect on CO_2-enhanced biomass. Conversely, enriched CO_2 more than compensated for the adverse effects of increased UV-B.

Air pollutants and the components of climate change

The CO_2 and temperature components of climate change may be regarded as *modifiers* of plant metabolism and physiology in contrast to UV-B and air pollutants, increases of which may lead to toxicity. Since increases in surface CO_2 concentration, temperature and UV-B irradiation are concurrent features of the present period of climate change, ideally we need to have an understanding of how the effects of air pollutants are likely to be affected by these concurrent changes Unfortunately, no factorial investigations of the combined effects of all of these factors appear to have been undertaken, probably because of their complexity and expense. Our present knowledge is therefore confined to the results of studies of the effects of individual pollutants with elevated CO_2 or UV-B and the output of a few computer simulation models. Most information relates to effects of elevated CO_2 on plant responses to pollutants (especially O_3), possibly because these can be readily studied in CECs, greenhouse chambers or OTCs without major additional investment. With regard to computer simulation models, the comprehensive review by Kickert *et al.* (1999) describes the current 'state-of-the-art' and the limitations of the current climate change–pollutant models. Earlier, Amthor and Loomis (1996) called for a shift away from photosynthesis-based modeling to the direct use of growth and development parameters including carbon partitioning.

Ozone

The effects of increased CO_2 on the impacts of O_3 on vegetation have recently been reviewed by Polle and Pell (1999) who point out that, although both are gases, CO_2 levels are reasonably constant over lengthy periods of growth whereas O_3 presents a fluctuating stress with dramatic hour-to-hour and day-to-day variability.

Adverse effects of O_3 range from visible foliar injury (necrosis) in sensitive species and cultivars to more subtle biochemical and metabolic changes (Chapter 5) that ultimately manifest themselves as various deleterious effects on plant growth and development (Chapter 6). Table 23.3 lists the elevated CO_2-induced changes to some of the major effects of O_3 reported in the recent literature. Numerous reports have suggested that elevated CO_2 may counteract various adverse effects of O_3, but from Table 23.1

Table 23.3 Recent observations of the effect of increased CO_2 on O_3-induced effects on plants at the metabolic, physiological and whole plant levels. –: decrease; +: increase; 0: no significant effect

Plant response	O_3 effect	CO_2 effect	Species	Reference
Biochemical/metabolic				
Ascorbate peroxidase	–	+	wheat	17
	–	0	sugar maple	16
Glutathione reductase	–	+	wheat	17
	0	0	sugar maple	16
Catalase	–	0	wheat	11
	+	0	sugar maple	16
Superoxide dismutase	0	0	wheat	11
	0	0	sugar maple	16
Rubisco	–	+	soybean	18
Rubisco activity	–	+	soybean	18
Physiological				
Stomatal conductance	–	–	radish	2
	–	+	soybean	13
	–	0	wheat	15
	–	–	wheat	1
	0	0	black cherry	10
	0	0	green ash	10
	0	–	yellow poplar	10
	–	0	white clover (O_3-sensitive)	7
	0	–	white clover (O_3-resistant)	7
	–	–	trembling aspen	22
	0	–	red oak	22
	–	0	*Agropyron smithii*	22
	–	0	*Koeleria cristata*	22
	0	–	*Bouteloua curtipendula*	22
	0	0	*Schizachyrium scoparium*	22
Photosynthesis	–	+	radish	2
	–	+	soybean	4
	–	+	soybean	13
	–	+	wheat	15
	–	+	Scots pine	9
	0	+	black cherry	10
	0	+	green ash	10
	0	+	yellow poplar	10
	0	+	trembling aspen	22
	0	+	red oak	22
	0	+	*Agropyron smithii*	22
	0	+	*Koeleria cristata*	22
	0	+	*Bouteloua curtipendula*	22
	–	+	*Schizachyrium scoparium*	22

Table 23.3 *(continued)*

Plant response	O_3 effect	CO_2 effect	Species	Reference
Photorespiration	–	–	wheat	11
	–	0	soybean	4
Growth, yield				
Total biomass	–	+	soybean	13, 19
	–	+	wheat	5, 12, 17, 21
	0	+	wheat	3
	–	+	maize	21
	0	+	alfalfa	8
	0	+	timothy	8
	–	+	tomato	20
	0	+	black cherry	10
	+	+	green ash	10
	0	+	yellow poplar	10
	–	+	white clover (O_3-sensitive)	7
	0	+	white clover (O_3-resistant)	7
	–	+	trembling aspen	22
	0	0	red oak	22
	–	+	*Agropyron smithii*	22
	–	+	*Koeleria cristata*	22
	–	0	*Bouteloua curtipendula*	22
	0	0	*Schizachyrium scoparium*	22
Seed/grain/fruit Yield	–	+	soybean	6, 13, 14
	–	+	wheat	3, 5, 12 14
	–	+	tomato	20
	–	+	maize	14, 21
Relative growth rate	–	+	trembling aspen	22
	0	+	red oak	22
	–	+	*Agropyron smithii*	22
	–	+	*Koeleria cristata*	22
	0	0	*Bouteloua curtipendula*	22
	0	0	*Schizachyrium scoparium*	22
Specific leaf area (SLA)	–	0	radish	2
	–	+	soybean	18
	–	0	soybean	13
	–	–	white clover (O_3-sensitive)	7
	+	–	white clover (O_3-resistant)	7
	+	–	trembling aspen	22
	+	–	red oak	22
	+	–	*Agropyron smithii*	22
	+	–	*Koeleria cristata*	22
	0	0	*Bouteloua curtipendula*	22
	+	0	*Schizachyrium scoparium*	22

continued overleaf

Table 23.3 (*continued*)

Plant response	O_3 effect	CO_2 effect	Species	Reference
Root/shoot ratio	−	+	radish	2
	+	−	wheat	12
	−	0	white clover (O_3-sensitive)	7
	0	+	white clover (O_3-resistant)	7
	−	−	alfalfa	8
	−	−	timothy	8

Note: changes are based on O_3 effects resulting from <0.15 ppm, and CO_2 effects from approximate doubling of present levels. All are C3 species except corn, *Bouteloua* and *Schizachyrium* which are C4. References: 1, Balaguer *et al.* (1995); 2, Barnes and Pfirrmann (1992); 3, Bender *et al.* (1999); 4, Booker *et al.* (1997); 5, Fangmeier *et al.* (1996); 6, Fiscus *et al.* (1997); 7, Heagle *et al.* (1993); 8, Johnson *et al.* (1996); 9, Kellomäki and Wang (1997); 10, Loats and Rebbeck (1999); 11, McKee *et al.* (1997a); 12, McKee *et al.* (1997b); 13, Mulchi *et al.* (1992); 14, Mulchi *et al.* (1995); 15, Mullholland *et al.* (1997); 16, Niewiadomska *et al.* (1999); 17, Rao *et al.* (1995); 18, Reid *et al.* (1998); 19, Reinert *et al.* (1997); 20, Reinert and Ho (1995); 21, Rudorff *et al.* (1996); 22, Volin *et al.* (1998).

it can be seen that the evidence is far from consistent. However, it is reasonable to say that for *most* species, elevated CO_2 reduces *many* adverse effects of O_3. Certainly, it appears that no synergistic or additive adverse effects on growth have been reported although some physiological processes and features of growth may show additive responses. As Polle and Pell (1999) point out, protection against one adverse effect is not necessarily reflected in protection against others.

Several authors have suggested that protection may be the consequence of reduced O_3 flux resulting from reduced stomatal conductance (Barnes and Pfirrmann, 1992; Balaguer *et al.*, 1995; McKee *et al.*, 1997b), but this may be somewhat species-dependent, since there are numerous examples of adverse effects in spite of significant decreases in stomatal closure.

At the mechanistic level, Rubisco plays a key role in CO_2-assimilation and both O_3 and elevated CO_2 can induce reductions in amount or activity (Table 23.3). Polle and Pell (1999) have reviewed our knowledge of these effects of O_3 and CO_2 and the possible mechanisms involved, but caution that, in spite of the importance of Rubisco, it should not be regarded 'as a unique target for the interaction of the two gases'.

Some of the increased photosynthate produced in elevated CO_2 can be used for maintenance, repair and detoxification and Table 23.3 shows that, although there are reports of CO_2-enhanced ascorbate peroxidase and glutathione reductase which reversed the deleterious effects of O_3, CO_2 has little or no effect on levels of superoxide dismutase (SOD) and catalase. However, Polle and Pell (1999) report that in most tree species CO_2 has been shown to reduce significantly the levels of the last two enzymes which raises the issue of whether trees in high CO_2 atmospheres with reduced antioxidant defences are more susceptible to O_3. Based on their review, they conclude that increased CO_2 appears to confer enhanced *metabolic flexibility* to trees such that the levels of SOD rise with the onset of oxidative stress.

But what if we add temperature and UV-B to the interactions with CO_2? Dealing with temperature first, if the effects of increased CO_2 and increased temperature on growth and yield are largely compensatory in some species (see previous discussion; Van Oijen *et al.*, 1999), does this mean that the protection afforded by CO_2 against O_3 could disappear? Unfortunately the question cannot be answered because surprisingly little information exists on the effects of temperature on plant responses to O_3, other than foliar injury (Guderian *et al.*, 1985). If we consider a species growing at close to its optimum temperature, it will be progressively disadvantaged by increased temperature, because of the increasing catabolic losses. Increased CO_2 can offset this effect by increasing the rate of photosynthesis, but the accelerated utilization of photosynthate for growth processes leads ultimately to decreased accumulated biomass and yield (Van Oijen *et al.*, 1999). Since O_3 adversely affects the photosynthetic output, any temperature-driven acceleration in the utilization of photosynthate would be a disadvantage to normal development. Hence it may be that increased temperature could negate the protection offered by CO_2 against O_3. It is of interest to note that, in attempting to account for the inter-season and inter-country differences observed in the ESPACE-wheat program, Bender *et al.* (1999) developed a multiple linear regression model for yield in which temperature was a highly significant variable with a negative coefficient although, surprisingly, the effect of O_3 was not significant. However, to provide a definitive answer to the question will require investigations specifically addressing the CO_2–temperature–O_3 interaction.

Attempts to simulate wheat growth responses to CO_2, temperature and O_3 in the ESPACE-wheat program using mechanistic crop models were partly successful, but the 'predictions of the variations between sites and years of growth and development parameters and of their responses to CO_2 and O_3 were poor' (Ewart *et al.*, 1999).

With regard to interactions of UV-B and O_3, Krupa and Kickert (1989) and Runeckles and Krupa (1994) noted that the attenuation of UV-B by tropospheric O_3 would tend to counter increased surface level UV-B intensities caused by stratospheric O_3 depletion. They therefore suggested that, in the field, increased exposures to either might well be sequential rather than concurrent. In a study of the salt-marsh grass, *Elymus athericus*, subjected to reciprocal exposures to O_3 and UV-B, van der Staaij *et al.* (1997) observed no interactive effects and no adverse effects of UV-B following 14-day exposures, although an earlier report showed that longer term exposures to UV-B (65 days) caused a 35% loss of biomass (van der Staaij *et al.*, 1993). Recently, Schnitzler *et al.* (1999) reported a CEC study of Norway spruce and Scots pine in which ambient, high altitude levels were compared with near zero UV-B radiation levels, at ambient or twice-ambient exposures to O_3. They found that O_3-induced injury and adverse effects on photosynthesis were more pronounced with near zero UV-B levels, and concluded that UV-B ameliorates the O_3 response. In contrast, studies with increased UV-B levels such as the OTC investigation of two soybean cultivars by Miller *et al.* (1994) found no evidence of interactive effects and no effect of UV-B on yield, in contrast to the earlier work of Teramura *et al.* (1990) using one of the cultivars, Essex. At the moment, therefore, the sparseness of the available data and the often contradictory findings make it impossible to make any definitive statement as to how changing UV-B levels may affect plant response to O_3 and vice versa.

At various organizational levels, Runeckles and Krupa (1994) identified several similarities between plant responses to UV-B and O_3, and at the level of gene expression, Willekens *et al.* (1994) noted that the effects of UV-B, O_3 and SO_2 on several antioxidant genes were very similar.

From the foregoing, it is clear that there are still vast gaps in our knowledge of the various interactions of CO_2, temperature, UV-B and O_3. The gaps will only be filled by extensive and costly systematic investigation, for without more basic information as input, even modeling can be of little help. This is particularly true if we ask: What effect will O_3 have on ecological changes, including species distribution and migration, that will result from increased CO_2, temperature and UV-B? At the moment there is only speculation. Since poleward movements of native and agricultural species appear to be inevitable (IPCC, 1996) and the problems posed by O_3 to plants are to a large degree geographically related to centers of human population with high fossil fuel consumption, the answer may lie in whether or not plant species and people migrate in step!

Although various computer simulation models have been developed for assessing the interactive effects of O_3, CO_2 and temperature, as reviewed by Kickert *et al.* (1999), none has apparently yet been developed to assess the combined impacts of these factors with UV-B.

Sulphur dioxide

Although still locally and, occasionally, regionally important as a gaseous pollutant (see Chapter 8), SO_2 has ceased to be as important in the developed world as in earlier post-Industrial Revolution times. Our knowledge of its interactive effects with climate change factors is even more limited than for O_3. Carlson and Bazzaz (1982) investigated the effects of increased CO_2 on the responses to SO_2 of three C3 (*Chenopodium album*, *Datura stramonium* and *Polygonum pennsylvanicum*) and three C4 species (*Amaranthus retroflexus*, *Setaria faberii* and *Setaria lutescens*). They found that sensitivity to SO_2 (as measured by dry matter production) was decreased by CO_2 for the C3 species but increased for the C4 species. The lack of protective effect of CO_2 on the C4 species was attributed to their stomatal behavior in response to increased CO_2. Protection by CO_2 was also observed in soybean (Carlson, 1983; Lee *et al.*, 1997) and wheat (Deepak and Agrawal, 1999), but no interactions were observed in Norway spruce by Tausz *et al.* (1996).

These data point to a likely protective effect of CO_2 against the adverse effects of SO_2 in other C3 species, whereas C4 species may be at greater risk from SO_2 at elevated CO_2 concentrations. However, it must also be borne in mind that low levels of SO_2 may confer nutritional benefit to plants; low level SO_2 with increased CO_2 might therefore be expected to confer a double benefit.

Oxides of nitrogen

Nitric oxide, NO, and nitrogen dioxide, NO_2, collectively referred to as NO_x, are typical products of combustion and play key roles in photochemical oxidant production. However, their occurrence at phytotoxic levels in ambient air is rare, although they may act interactively with O_3 or SO_2 as discussed in Chapter 14. Much of our knowledge of their effects comes from studies of their phytotoxicity in commercial greenhouses in which they are by-products of the combustion of fuels to provide CO_2 enrichment (see Chapter 7) and hence cover interactions with CO_2. However, there appears to be no information available with regard to NO_x interactions with the other climate change factors, temperature and UV-B.

Mortensen (1985), Saxe and Christensen (1985) and Caporn (1989) reported that NO in CO_2-enriched air (1000–1200 ppm) reduced the growth of several horticultural species but did not determine how CO_2 enrichment affected the responses to NO. The study by Hufton *et al.* (1996) on lettuce grown in 1000 ppm CO_2 revealed that, after an initial inhibitory phase, NO + NO_2 (~2.5:1) eventually stimulated growth and yield, partly because of increased nitrate reductase activity. However, they, too, provide no information about the effect of elevated CO_2 on response to NO. Bruggink *et al.* (1988) found that, with tomato, 1 ppm NO reduced the photosynthesis rate by 38% in 350 ppm CO_2, but only 24% in 1000 ppm CO_2. Conversely, the 5% enhancement observed in 1000 versus 350 ppm CO_2 was increased to 28% in the presence of NO.

As with SO_2, nutritional benefit may occur from low levels of NO_x (Wellburn, 1990).

Other air pollutants

There appear to have been no studies of the interactions of climate change factors with other pollutant gases such as hydrogen fluoride (HF), hydrogen sulphide (H_2S) or ammonia (NH_3), nor with the components of acidic precipitation.

The outlook for the twenty-first century

Although some progress has been made in slowing down the rates of discharge of CO_2 and other GHGs into the atmosphere as a result of international agreements, the refusal of some countries to reduce emissions will inevitably mean that any deceleration in the rate of global warming will be gradual at best. For CO_2 concentrations to stabilize at 450, 650 or 1000 ppm 'would require global anthropogenic CO_2 emissions to drop below 1990 levels within a few decades, about a century, or about two centuries, respectively' (IPCC, 2001).

With regard to air pollutants, as world population increases, centers of population become larger, and fossil fuels continue to provide the major source of energy, the concentration of tropospheric O_3 will continue to rise. On the other hand, reductions in the releases of ODSs will lead to a gradual restoration of the stratospheric O_3 shield. This, together with the increase in tropospheric O_3 will reduce the surface levels of UV-B.

The poleward shift of global isotherms will have significant effects on plant geography and the continuing abilities of present-day agricultural and forest production to be maintained *in situ*. With regard to native vegetation in particular, it seems likely that C4 species will increase their range of dominance in some parts of the world.

If these predictions are correct and in spite of the extent of the investigations of the past decades, we are still poorly equipped to be able to predict how important the relative changes in CO_2, temperature and O_3 will be to the sustained production of food, fiber and forest products. Uncertainties abound, among which, Krupa and Groth (2000) have identified the inadequate capture of the spatial and temporal variabilities of growth-regulating atmospheric parameters, the lack of realism of many experimental exposure systems, and the lack of multivariate investigations. To this I would add that our current knowledge of effects on C4 species is woefully inadequate.

Because the contributory factors to climate change are clearly interactive, future experimental studies need also to be interactive. It is wrong to presume that computer simulation modeling will be able 'to take care of the interactions', although their use may lead to the revelation of unexpected interrelationships. But they do not provide a panacea, because their outputs are still largely dependent upon univariate input data.

In summary, our present understanding suggests that, as far as tropospheric O_3 and SO_2 pollution are concerned, continuing increases in the CO_2 component of climate change are likely to be ameliorative for many (especially C3) species. However, concomitant increases in mean global temperature will probably decrease this protective effect. Reductions in the release of ODSs are expected to lead to a restoration of the stratospheric O_3 layer over the next half-century which will reduce surface UV-B levels in the temperate and polar regions. This will be further reduced by any increase in tropospheric ozone. However, the global impacts of pollutants such as O_3 and SO_2 may well be dramatically changed by the indirect effects of increasing mean global temperatures leading to poleward shifts in both natural and agricultural vegetation with consequent changes in ecosystem structure. Since adverse tropospheric O_3 levels in particular are clearly related to urbanization and 'development' of an increasing global population, the consequences of poleward migration of vegetation on the impact of O_3 on food production and natural ecosystems will depend upon how the geographic distributions of sensitive species change in relation to the anthropogenic sources of pollution.

References

Allen, D.J., Nogués, S. and Baker, N.R. (1998). Ozone depletion and increased UV-B radiation: is there a real threat to photosynthesis? *J. Exper. Bot.* **49**, 1775–1788.

Allen, L.H., Jr. (1990). Plant responses to rising carbon dioxide and potential interactions with air pollutants. *J. Environ. Qual.* **19**, 15–34.

Amthor, J.S. (1999). Increasing atmospheric CO_2 concentration, water use, and water stress: scaling up from the plant to the landscape. In Luo, Y., Mooney, H.A., eds. *Carbon Dioxide and Environmental Stress*, Academic Press, San Diego, CA. pp. 33–59.

Amthor, J.S. and Loomis, R.S. (1996). Integrating knowledge of crop responses to elevated CO_2 and temperature with mechanistic simulation models: model components and research needs. In: Koch, G.W. and Mooney, H.A., eds. *Carbon Dioxide and Terrestrial Ecosystems*, Academic Press, San Diego, CA. pp. 317–345.

Balaguer, L., Barnes, J.D., Panicucci, A. and Borland, A.M. (1995). Production and utilization of assimilates in wheat (*Triticum aestivum* L.) leaves exposed to elevated O_3 and/or CO_2. *New Phytol.* **129**, 557–568.

Barnes, J.D. and Pfirrmann, T. (1992). The influence of CO_2 and O_3, singly and in combination, on gas exchange, growth and nutrient status of radish (*Raphanus sativus* L.). *New Phytol.* **121**, 403–412.

Barnett, T.P., Pierce, D.W. and Schnur, R. (2001). Detection of anthropogenic climate change in the world's oceans. *Science* **292**, 270–274.

Barnola, J.H., Raynaud, D., Korotkevitch, Y.S. and Lorius, C. (1987). Vostok ice core: a 160,000-year record of atmospheric CO_2. *Nature* **329**, 408–414.

Bazzaz, F.A. (1990). The response of natural ecosystems to the rising global CO_2 levels. *Annu. Rev. Ecol. Syst.* **21**, 167–196.

Beerling, D.J. and Woodward, F.I. (1993). Ecophysiological responses of plants to global environmental change since the last glacial maximum. *New Phytol.* **125**, 641.

Bender, J., Hertstein, U. and Black, C.R. (1999). Growth and yield responses of spring wheat to increasing carbon dioxide, ozone and physiological stresses: a statistical analysis of 'ESPACE–wheat' results. *Eur. J. Agron.* **10**, 185–195.

Booker, F.L., Reid, C.D., Brunschon-Harti, S., Fiscus, E.L. and Miller, J.E. (1997). Photosynthesis and photorespiration in soybean [*Glycine max* (L.) Merr.] chronically exposed to elevated carbon dioxide and ozone. *J. Exper. Bot.* **48**, 1843–1852.

Bruggink, G.T., Wolting, H.G., Dassen, J.H.A. and Bus, V.G.M. (1988). The effect of nitric oxide fumigation at two CO_2 concentrations on net photosynthesis and stomatal resistance of tomato (*Lycopersicon lycopersicum* L. cv. Abunda). *New Phytol.* **110**, 185–191.

Caldwell, M.M., Björn, L.O., Bornman, J.F., Flint, S.D., Kulandaivelu, G., Teramura, A.H. and Tevini, M. (1998). Effects of increased solar ultraviolet radiation on terrestrial ecosystems. *J. Photochem. Photobiol.* **46**, 40–52.

Cao, W., Tibbitts, T.W. and Wheeler, R.M. (1994). Carbon dioxide interactions with irradiance and temperature in potatoes. *Adv. Space Res.* **14**, 243–250.

Caporn, S.J.M. (1989). The effects of oxides of nitrogen and carbon dioxide enrichment on photosynthesis and growth of lettuce (*Lactuca sativa* L.). *New Phytol.* **111**, 473–481.

Carlson, R.W. (1983). The effect of SO_2 on photosynthesis and leaf resistance at varying concentrations of CO_2. *Environ. Pollut. (Ser. A)* **30**, 309–321.

Carlson, R.W. and Bazzaz, F.A. (1982). Photosynthetic and growth response to fumigation with SO_2 at elevated CO_2 for C3 and C4 plants. *Oecologia* **54**, 50–54.

Crawford, R.M.M. and Wolfe, D.W. (1999). Temperature: cellular to whole-plant and population responses. In: Luo, Y. and Mooney, H.A., eds. *Carbon Dioxide and Environmental Stress*, Academic Press, San Diego, CA. pp. 61–106.

De Kok, L.J. and Stulen, I., eds. (1998). *Responses of Plant Metabolism to Air Pollution and Global Change*. Backhuys Publishers, Leiden, Netherlands.

Deepak, S.S. and Agrawal, M. (1999). Growth and yield responses of wheat plants to elevated levels of CO_2 and SO_2, singly and in combination. *Environ. Pollut.* **104**, 411–419.

Dugas, W.A. and Pinter, P.J., eds. (1994). The free-air carbon dioxide (FACE) cotton project: a new field approach to assess the biological consequences of climate change. *Agric. For. Meteorol.* **70**, 1–342.

Eamus, D. (1996). Responses of field grown trees to CO_2 enrichment. *Commonwealth For. Rev.* **75**, 39–47.

Ehleringer, J.R., Sage, R.F., Flanagan, L.B. and Pearcy, R.W. (1991). Climate change and the evolution of C4 photosynthesis. *Trends Ecol. Evol.* **6**, 95–99.

Ellsworth, D.S., Oren, R., Huang, C., Phillips, N. and Hendrey, G.R. (1995). Leaf and canopy responses to elevated CO_2 in a pine forest under free-air CO_2 enrichment. *Oecologia* **104**, 139–146.

Ellsworth, D.S., LaRoche, J. and Hendrey, G.R. (1998). Elevated CO_2 in a prototype free-air CO_2 enrichment facility affects photosynthetic nitrogen use in a maturing pine forest. *Brookhaven National Laboratory Report vol. BNL 52545*. Brookhaven National Laboratory, Upton, NY.

Ewart, F., van Oijen, M. and Porter, J.R. (1999). Simulation of growth and development processes of spring wheat in response to CO_2 and ozone for different sites and years in Europe using mechanistic crop simulation models. *Eur. J. Agron.* **10**, 231–247.

Fangmeier, A., Grüters, U., Hertstein, U., Sandhage-Hofmann, A., Vermehren, B. and Jäger, H.-J. (1996). Effects of elevated CO_2, nitrogen supply and tropospheric ozone on spring wheat. I. Growth and yield. *Environ. Pollut.* **91**, 381–390.

Farrar, J.F. and Williams, M.L. (1991). The effects of increased atmospheric carbon dioxide and temperature on carbon partitioning, source-sink relations and respiration. *Plant Cell Environ.* **14**, 819–830.

Fiscus, E.L., Reid, C.D., Miller, J.E. and Heagle, A.S. (1997). Elevated CO_2 reduces O_3 flux and O_3-induced yield losses in soybeans: possible implications for elevated CO_2 studies. *J. Exper. Bot.* **48**, 307–313.

German Bundestag, ed. (1991). *Protecting the Earth. A Status Report with Recommendations for a New Energy Policy. Volume 1*. Deutscher Bundestag, Referat Offentlichkeirsarbeit, Bonn.

Graves, J. and Reavey, D. (1996). *Global Environmental Change*. Longmans, Harlow, Essex, UK.

Groth, J.V. and Krupa, S.V. (2000). Crop ecosystem responses to climatic change: interactive effects of ozone, ultraviolet-B radiation, sulphur dioxide and carbon dioxide on crops. In Reddy, K.R. and Hodges, H.F., eds., *Climate Change and Global Crop Productivity*, CABI Publishing, New York. pp. 387–405.

Guderian, R., Tingey, D.T. and Rabe, R. (1985). Effects of photochemical oxidants on plants. In: Guderian, R., ed. *Air Pollution by Photochemical Oxidants*, Springer-Verlag, Berlin. pp. 127–333.

Hakala, K. and Mela, T. (1996). The effects of prolonged exposure to elevated temperatures and elevated CO_2 levels on the growth, yield and dry matter partitioning of field-sown meadow fescue. *Agric. Food Sci. in Finland* **5**, 285–298.

Heagle, A.S., Kress, L.W., Temple, P.J., Kohut, R.J., Miller, J.E. and Heggestad, H.E. (1988). Factors influencing ozone dose-yield response relationships in open-top field chamber studies. In: Heck, W.W., Taylor, O.C. and Tingey, D.T., eds. *Assessment of Crop Loss from Air Pollutants*, Elsevier Applied Science, London. pp. 141–179.

Heagle, A.S., Miller, J.E., Sherrill, D. and Rawlings, J.O. (1993). Effects of ozone and carbon dioxide mixtures on two clones of white clover. *New Phytol.* **123**, 751–762.

Henderson, S., Hattersley, P., Caemmerer, S. and von Osmond, C.B. (1994). Are C4 pathway plants threatened by global climatic change? In Schulze, E.-D. and Caldwell, M.M., eds. *Ecophysiology of Photosynthesis*, Springer-Verlag, Berlin. pp. 531–549.

Hendrey, G.F., Ellsworth, D.S., Lewin, K.F. and Nagy, J. (1999). A free-air enrichment system for exposing tall forest vegetation to elevated atmospheric CO_2. *Global Change Biol.* **5**, 293–309.

Hsaio, T.C. and Jackson, R.B. (1999). Interactive effects of water stress and elevated CO_2 on growth, photosynthesis, and water use efficiency. In: Luo, Y. and Mooney, H.A., eds. *Carbon Dioxide and Environmental Stress*, Academic Press, San Diego, CA. pp. 3–31.

Hufton, C.A., Besford, R.T. and Wellburn, A.R. (1996). Effects of NO ($+NO_2$) pollution on growth, nitrate reductase activities and associated protein contents in glasshouse lettuce grown hydroponically in winter with CO_2 enrichment. *New Phytol.* **133**, 495–501.

IPCC (Intergovernmental Panel on Climate Change) (1990). *Climate Change: The IPCC Scientific Assessment*. Houghton, J.T., Jenkins, G.J. and Ephraums, J.J., eds. Cambridge University Press, Cambridge, UK.

IPCC (Intergovernmental Panel on Climate Change) (1992). *Climate Change 1992: The Supplementary Report to the IPCC Scientific Assessment*. Houghton, J.T., Callander, B.A. and Varney, S.K., eds. Cambridge University Press, Cambridge, UK.

IPCC (Intergovernmental Panel on Climate Change) (1996). *Climate Change 1995: The Science of Climate Change*. Houghton, J.T., Meira Filho, L.G., Callander, B.A., Harris, N., Kattenberg, A. and Maskell, K., eds. Cambridge University Press, Cambridge, UK.

IPCC (Intergovernmental Panel on Climate Change) (2001). *Climate Change 2001: The Scientific Basis*. Third Assessment Report, Shanghai Draft 21-01-2001, United Nations Environment Programme (UNEP), GRID, Arendal, Norway.

Jäger, H.J., Hertstein, U. and Fangmeier, A. (1999). The European Stress Physiology and Climate Experiment – Project 1: wheat (ESPACE-wheat); introduction, aims and methodology. *Eur. J. Agron.* **10**, 155–162.

Johnson, B.G., Hale, B.A. and Ormrod, D.P. (1996). Carbon dioxide and ozone effects on growth of a legume-grass mixture. *J. Environ. Qual.* **25**, 908–916.

Jordan, B.R. (1996). The effects of ultraviolet-B radiation on plants: A molecular perspective. *Adv. Botan. Res.* **22**, 97–162.

Jordan, D.B. and Ogren, W.L. (1984). The CO_2/O_2 specificity of ribulose 1,5-bisphosphate carboxylase/oxygenase. Dependence on ribulose bisphosphate concentration, pH and temperature. *Planta* **161**, 308–313.

Kellomäki, S. and Wang, K.-Y. (1997). Effects of elevated O_3 and CO_2 on chlorophyll fluorescence and gas exchange in Scots pine during the third growing season. *Environ. Pollut.* **97**, 17–27.

Kickert, R.N., Tonella, G., Simonov, A. and Krupa, S.V. (1999). Predictive modeling of effects under global change. *Environ. Pollut.* **100**, 87–132.

Krupa, S.V. and Groth, J.V. (2000). Global climate change and crop responses: uncertainties associated with the current methodologies. In: Agrawal, S.B. and Agrawal, M., eds. *Environmental Pollution and Plant Responses*, Lewis, Boca Raton, FL. pp. 1–18.

Krupa, S.V. and Kickert, R.N. (1989). The greenhouse effect: impacts of ultraviolet-B (UV-B), carbon dioxide (CO_2), and ozone (O_3) on vegetation. *Environ. Pollut.* **61**, 263–393.

Lawlor, D.W. (1998). Plant responses to global change: temperature and drought stress. In: De Kok, L.J. and Stulen, I., eds. *Responses of Plant Metabolism to Air Pollution and Global Change*, Backhuys Publishers, Leiden. pp. 193–207.

Lee, E.H., Pausch, R.C., Rowland, R.A., Mulchi, C.L. and Rudorff, B.F.T. (1997). Responses of field-grown soybean (cv. Essex) to elevated SO_2 under two atmospheric CO_2 concentrations. *Environ. Exper. Bot.* **37**, 85–93.

Levitus, S., Antonov, J.I., Wang, J., Delworth, T.L., Dixon, K.W. and Broccoli, A.J. (2001). Anthropogenic warming of the earth's climate system. *Science* **292**, 267–270.

Loats, K.V. and Rebbeck, J. (1999). Interactive effects of ozone and elevated carbon dioxide on the growth and physiology of black cherry, green ash, and yellow-poplar seedlings. *Environ. Pollut.* **106**, 237–248.

Long, S.P. (1991). Modification of the response of photosynthetic productivity to rising temperature by atmospheric CO_2 concentrations: has its importance been underestimated?. *Plant Cell Environ.* **14**, 729–739.

Lumsden, P.J., ed. (1997). *Plants and UV-B: Responses to Environmental Change*. Cambridge University Press, Cambridge, UK.

Luo, Y. and Mooney, H.A., eds., (1999). *Carbon Dioxide and Environmental Stress*. Academic Press, San Diego, CA.

Luo, Y., Canadell, J. and Mooney, H.A. (1999). Interactive effects of carbon dioxide and environmental stress on plants and ecosystems: a synthesis. In: Luo, Y. and Mooney, H.A., eds. *Carbon Dioxide and Environmental Stress*, Academic Press, San Diego, CA. pp. 393–408.

Luo, Y., Medlyn, B., Hui, D., Ellsworth, D., Reynolds, J. and Katul, G. (2001). Gross primary productivity in Duke Forest: Modeling synthesis of CO_2 experiment and eddy-flux data. *Ecol. Applic.* **11**(1), 239–252.

Mark, U. and Tevini, M. (1997). Effects of elevated ultraviolet-B-radiation, temperature and CO_2 on growth and function of sunflower and maize seedlings. *Plant Ecol.* **128**, 224.

McKee, I.F., Bullimore, J.F. and Long, S.P. (1997a). Will elevated CO_2 concentrations protect the yield of wheat from O_3 damage? *Plant Cell Environ.* **20**, 77–84.

McKee, I.F., Eiblmeier, M. and Polle, A. (1997b). Enhanced ozone-tolerance in wheat grown at an elevated CO_2 concentration: ozone exclusion and detoxification. *New Phytol.* **137**, 275–284.

Miller, J.E., Booker, F.L., Fiscus, E.L., Heagle, A.S., Pursley, W.A., Vozzo, S.F. and Heck, W.W. (1994). Ultraviolet-B radiation and ozone effects on growth, yield, and photosynthesis of soybean. *J. Environ. Qual.* **23**, 83–91.

Mortensen, L.M. (1985). Nitrogen oxides produced during CO_2 enrichment. I. Effects on different greenhouse plants. *New Phytol.* **101**, 103–108.

Mulchi, C.L., Slaughter, L., Saleem, M., Lee, E.H., Pausch, R. and Rowland, R. (1992). Growth and physiological characteristics of soybean in open-top chambers in response to ozone and increased atmospheric CO_2. *Agric. Ecosys. Environ.* **38**, 107–118.

Mulchi, C.L., Rudorff, B., Lee, E.H., Rowland, R. and Pausch, R. (1995). Morphological responses among crop species to full-season exposures to enhanced concentrations of atmospheric CO_2 and O_3. *Water Air Soil Pollut.* **85**, 1379–1386.

Mulholland, B.J., Craigon, J., Black, C.R., Colls, J.J., Atherton, J. and Landon, G. (1997). Impact of elevated atmospheric CO_2 and O_3 on gas exchange and chlorophyll content in spring wheat (*Triticum aestivum* L.). *J. Exper. Bot.* **48**, 1853–1863.

Niewiadomska, E., Gaucher-Veilleux, C., Chevrier, N., Mauffette, Y. and Dizengremel, P. (1999). Elevated CO_2 does not provide protection against ozone considering the activity of several antioxidant enzymes in the leaves of sugar maple. *J. Plant Physiol.* **155**, 70–77.

Polle, A. and Pell, E.J. (1999). Role of carbon dioxide in modifying the plant response to ozone. In: Luo, Y. and Mooney, H.A., eds. *Carbon Dioxide and Environmental Stress*, Academic Press, San Diego, CA. pp. 193–213.

Poorter, H. (1993). Interspecific variation in the growth response to an elevated ambient CO_2 concentration. *Vegetatio* **105/105**, 77–97.

Rao, M.V., Hale, B.A. and Ormrod, D.P. (1995). Amelioration of ozone-induced oxidative damage in wheat plants grown under high carbon dioxide. *Plant Physiol.* **109**, 421–432.

Read, J.J. and Morgan, J.A. (1996). Growth and partitioning in *Pascopyrum smithii* (C3) and *Bouteloua gracilis* (C4) as influenced by carbon dioxide and temperature. *Ann. Bot.* **77**, 487–496.

Reid, C.D., Fiscus, E.L. and Burkey, K.O. (1998). Combined effects of chronic ozone and elevated CO_2 on Rubisco activity and leaf components in soybean (*Glycine max*). *J. Exper. Bot.* **49**, 1999–2011.

Reinert, R.A. and Ho, M.C. (1995). Vegetative growth of soybean as affected by elevated carbon dioxide and ozone. *Environ. Pollut.* **89**, 89–96.

Reinert, R.A., Eason, G. and Barton, J. (1997). Growth and fruiting of tomato as influenced by elevated carbon dioxide and ozone. *New Phytol.* **137**, 411–420.

Rogers, H.H. and Dahlman, R.C. (1990). Influence of more CO_2 on crops. *Proc. 83rd Ann. Mtg. Air, Waste Management Assoc.*, Pittsburgh, PA, June 1990. Paper 90–151.1.

Rudorff, B.F.T., Mulchi, C.L., Lee, E.H., Rowland, R. and Pausch, R. (1996). Effects of enhanced O_3 and CO_2 enrichment on plant characteristics in wheat and corn. *Environ. Pollut.* **94**, 53–60.

Runeckles, V.C. and Krupa, S.V. (1994). The impact of UV-B radiation and ozone on terrestrial vegetation. *Environ. Pollut.* **83**, 191–213.

Saxe, H. and Christensen, O.V. (1985). Effects of carbon dioxide with and without nitric oxide pollution on growth, morphogenesis and production time of pot plants. *Environ. Pollut. (Ser. A)* **38**, 159–169.

Saxe, H., Ellsworth, D.S. and Heath, J. (1998). Tree and forest functioning in an enriched CO_2 atmosphere. *New Phytol.* **139**, 395–436.

Schnitzler, J.-P., Langebartels, C., Heller, W., Liu, J., Lippert, M., Dòhring, T., Bahnweg, G. and Sandermann, H. (1999). Ameliorating effect of UV-B radiation on the response of Norway spruce and Scots pine to ambient ozone concentrations. *Global Clim. Change* **5**, 83–94.

Tausz, M., De Kok, L.J., Stulen, I. and Grill, D. (1996). Physiological responses of Norway spruce trees to elevated CO_2 and SO_2. *J. Plant Physiol.* **148**, 362–367.

Teramura, A.H., Sullivan, J.H. and Ziska, L.H. (1990). Interaction of elevated ultraviolet-B radiation and CO_2 on productivity and photosynthetic characteristics in wheat, rice, and soybean. *Plant Physiol.* **94**, 470–475.

Tevini, M. (2000). UV-B effects on plants. In: Agrawal, S.B. and Agrawal, M., eds. *Environmental Pollution and Plant Responses*, Lewis Publ., Boca Raton, FL. pp. 83–97.

Tevini, M., Mark, U. and Saile, M. (1990). Plant experiments in growth chambers illuminated with natural sunlight. In: Payer, H.D., Pfirrmann, T and Mathy, P, eds. *Environmental Research with Plants in Closed Chambers*, Air Pollution Research Report No. 26. Commission of the European Communities, Brussels.

UNEP (United Nations Environment Program) (1989). *Environmental Data Report*, Second edition, 1989–1991. Basil Blackwell, Oxford, UK.

UNEP (United Nations Environment Program) (1993). *Environmental Data Report*, Fourth edition, 1993–94. Blackwell, Oxford, UK.

UNEP (United Nations Environment Program) (1999). *Geo-2000: Global Environmental Outlook 2000*. Earthscan, London.

van der Staaij, J.W.M., Lenssen, G.M., Stroetenga, M. and Rozema, J. (1993). The combined effects of elevated CO_2 levels and UV-B radiation on growth characteristics of *Elymus athericus* (= *E. pycnanthus*). *Vegetatio* **104/105**, 433–439.

van der Staaij, J.W.M., Tonneijck, A.E.G. and Rozema, J. (1997). The effect of reciprocal treatments with ozone and ultraviolet-B radiation on photosynthesis and growth of perennial grass *Elymus athericus*. *Environ. Pollut.* **97**, 281–286.

Van Oijen, M., Schapendonk, A.H.C.M., Jansen, M.J.H., Pot, C.S. and Maciorowski, R. (1999). Do open-top chambers overestimate the effects of rising CO_2 on plants? An analysis using spring wheat. *Glob. Change Biol.* **5**, 411–421.

Volin, J.C., Reich, P.B., Givnish, T.J. (1998). Elevated carbon dioxide ameliorates the effects of ozone on photosynthesis and growth: species respond similarly regardless of photosynthetic pathway or plant functional group. *New Phytol.* **138**, 315–325.

Wand, S.J.E., Midgley, G.F., Jones, M.H. and Curtis, P.S. (1999). Responses of wild C4 and C3 grass (Poaceae) species to elevated atmospheric CO_2 concentrations: a meta-analytic test of current theories and perceptions. *Glob. Change Biol.* **5**(6), 723–741.

Wellburn, A.R. (1990). Why are atmospheric oxides of nitrogen usually phytotoxic and not alternative fertilizers? *New Phytol.* **115**, 395–429.

Wellburn, A. (1994). *Air Pollution and Climate Change: the Biological Impact*. Longman Scientific, Technical, Harlow, Essex, UK.

Wheeler, T.R., Batts, G.R., Ellis, R.H., Hadley, P. and Morrison, J.I.L. (1996). Growth and yield of winter wheat (*Triticum aestivum*) in response to CO_2 and temperature. *J. Agric. Sci.* **127**, 37–48.

Willekens, H., Van Camp, M., Van Montagu, M., Inzé, D., Langebartels, C. and Sandermann, H., Jr. (1994). Ozone, sulfur dioxide and ultraviolet B have similar effects on mRNA accumulation of antioxidant genes in *Nicotiana plumbaginifolia* L. *Plant Physiol.* **106**, 1007–1014.

Ziska, L.H. and Bunce, J.A. (1995). Growth and photosynthetic response of three soybean cultivars to simultaneous increases in growth temperature and CO_2. *Physiol. Plant.* **94**, 575–584.

24 Conclusions and future directions

J.N.B. BELL

In Chapter 2 we looked at the history of the subject of this book, covering around 150 years, starting in Europe, moving to North America and subsequently elsewhere in the world. In this final chapter I shall look back across the earlier parts of this book and reflect on the conclusions of each contribution and present my own views as to future directions. This will be the only chapter in which the first person is employed – because it is my personal view.

Where shall I start? In the simplest terms it is worth thinking about whether there are major phytotoxic pollutants which have not been considered so far. In my personal journey over the last 30 years, I have seen the shift from SO_2 and coal-smoke to NO_x and O_3 becoming the pollutants of most concern. Early experiments in which filtration of ambient air improved plant performance led to the realisation that more than one pollutant was operating in the real world. We still have difficulties in disentangling the pollutants concerned, which is a serious weakness for formulation of air pollution control policy. There are two types of pollutants which I suggest should be viewed as candidates for future research and which have been largely overlooked in the past. One of these is NO, for which the case has been argued eloquently by Terry Mansfield in Chapter 7. The other is the general category of volatile organic compounds, of which only ethylene has really attracted any interest so far as direct impacts on vegetation are concerned, but where there are suspicions of other, as yet untested, compounds having deleterious effects on plant life, as outlined in Chapter 10. It is noteworthy that some of the pollutants are not removed by charcoal filtration and thus studies utilising this technique have effectively precluded any information on their impacts. In the case of NO, as Terry Mansfield's chapter suggests, there may be even more subtle effects in terms of a signal for protection against pathogen attack.

Concerns over transboundary and urban air pollution have led to massive improvements in our understanding of sources, atmospheric chemistry and dispersion processes, as outlined in Chapter 3. There is little doubt that much further progress will be made in the future, not the least in developing models which integrate emissions inventories, dispersion in complex rural and urban terrain and impacts on receptors, including vegetation. Much still remains to be determined in the field of atmospheric chemistry, with the need to understand the non-linear processes involved in production of secondary pollutants. In Chapter 4 the role of vegetation in deposition in processes is considered in depth. This highlights the naivety of our approaches to air pollution

Air Pollution and Plant Life, second edition. Edited by J.N.B. Bell and M. Treshow. ISBN 0 471 49090 3 (HB), 0 471 49091 1 (PB).

impacts on plants at the time when I started research in this field some 30 years ago. It was widely assumed that the only parameter of interest was atmospheric concentration. Nowadays it is very apparent that the parameter of importance is the flux into and onto the plant, and that this can be utilised to refine air quality standards and guidelines according to species and local environmental conditions, as outlined by Mike Ashmore in Chapter 22. Furthermore, we now understand that fluxes are much more complex than realised hitherto, with 2-way processes and co-deposition in some cases. An important topic highlighted by Dave Fowler in Chapter 4 is the chemical interaction of O_3 with the external surface of vegetation – certainly an important topic for future research.

Reminiscing yet again, I look at the wealth of detail on O_3 and its effects at the biochemical, cell, physiological, whole plant and community levels described in Chapters 5 and 6. Thirty years ago this was a topic of almost no interest outside North America. Indeed, in the United Kingdom the popular belief still held that O_3 levels were very high at the seaside and that breathing in this gas 'was good for you'. We have come a very long way since then, with O_3 recognised as the most serious gaseous air pollutant in terms of impacts on plant life. Steve Long and Shawna Naidu point to the massive recent strides in knowledge of the impacts of O_3, not the least the signalling associated with stress induced by this pollutant. They point towards future developments in understanding the genetic nature of O_3 responses. A major area for research could well be genetic engineering of O_3 resistance in crops grown in polluted locations. Chapter 6 demonstrates the requirement for much more research into the impacts of O_3 on native vegetation, where widespread effects on community composition could be predicted for this ubiquitous phytotoxin. It is also clear that even for agricultural crops much deeper understanding is required for elucidation of dose-response relationships, not the least incorporating the concept of fluxes rather than concentrations, as mentioned earlier.

In recent years there has been a major loss of interest in the effects of SO_2 on vegetation in the developed world, accompanying emission controls and switches in fuel type. In many ways this is unfortunate because SO_2 is a growing problem in a vast amount of the developing world and Allan Legge and Sagar Krupa also highlight issues remaining in a developed country. This refers to Alberta in Canada, where major emissions of sulphur gases occur associated with the oil industry. This was brought home to me vividly when I recently visited northern Alberta with Allan and witnessed the massive operations to extract oil from the sands where a substantial proportion of the world's proven reserves lie and where major expansion is likely to take place in the near future. The significance of SO_2 for the performance of lichens and bryophytes is also a major ongoing issue, as outlined by Jeff Bates in Chapter 17. This is an under-researched area and there are suspicions that the Critical Level of only 10 $\mu g\ m^{-3}$ for cyanobacterial lichens may not be sufficient to protect these species. Unlike for higher plants, short-term peak exposures may be of importance and future research should address this matter. Another field in which I feel future research should be directed is towards improvements in the use of lichens and bryophytes as biomonitors of changing air quality appropriate for different circumstances and different parts of the world.

Chapter 11 highlights the lack of research into the impacts of particulates on plant life, as opposed to human health. This is a difficult area, because of the very wide range of chemical and physical characteristics of particulates, including their size and shape which all have the potential to affect vegetation in different ways. Many developing countries are extremely dusty places, at some times of the year at least, and a fruitful area of research would appear to be the combined impacts of particulates and gaseous pollutants – a very poorly researched topic. The effects of gaseous pollutant mixtures are described in Chapter 14. Here it is noted that there is a real paucity of data on the effects of two or more pollutants at realistic ambient concentrations and this is certainly a subject worthy of much greater attention.

The effects of wet deposited acidity continue to attract interest, despite the sharp downward trends in nitrogen and sulphur oxide emissions that are continuing to take place in the developed world. John Innes and John Skelly have debunked the idea that acid deposition is causing massive forest declines in Europe and North America. Yet I anticipate that research will continue on soil acidification, which to my mind is as clearly demonstrated for sensitive forest soils in many parts of Europe, as is lake acidification. Are we really sitting on a 'time-bomb' of falling pH and rising aluminium levels in European forests? As pointed out in Chapter 16 research on effects of acid deposition on aquatic life will primarily be concerned with rates of ecosystem recovery as emissions of acid gases continue to fall. Thirty years ago the occurrence of widespread lake acidification was only just becoming detected. But it is very much more recently that widespread eutrophication resulting from both reduced and oxidised forms of nitrogen has become detected. Now that there is growing evidence of major shifts in the species composition of low nitrogen ecosystems as described in Chapter 12, which are being impacted by increased nitrogen deposition, research needs to be expanded in this area with a holistic approach to impacts at all trophic levels.

Gina Mills in Chapter 18 has again highlighted the importance of understanding fluxes of pollutants to vegetation in terms of modification by climatic factors. There is little doubt that such effects and those of various soil parameters in modifying plant response will remain a priority area of research for many years to come. However, I feel that of even greater priority is the elucidation of the effects of air pollutants on abiotic and biotic stresses, detailed in Chapters 19 and 20, respectively. Massive sums of money are spent annually in both the developed and the developing world in protecting crops against familiar abiotic stresses, such as drought, frost and salinity, and biotic stresses, such as herbivorous insect attack and infection by viral, fungal and bacterial pathogens. Thus any impacts of air pollution on these stresses could have enormous economic consequences. Chapters 19 and 20 demonstrate clearly that ambient levels of the common phytotoxic pollutants can markedly change the effects of such stresses, with the potential to impact on vegetation health. This is an area of fundamental importance, which has been grossly neglected so far and must be a topic for substantive future research programmes.

The entire history of air pollution has been one of increasing scale, from the purely local to regional to international and, finally, to the global dimension. Chapters 21 and 23 emphasise the significance of the latter. Rising background levels of NO_x and O_3 world-wide have massive potential for impacts on crops, trees and native vegetation in areas hitherto unstudied. In addition SO_2 is a growing problem in many parts of the developing world. All this bodes ill for the populations of developing countries whose food security is threatened both in terms of yield and quality. In addition there must be threats to the biodiversity of the ecosystems of such places, exacerbating the severe pressures imposed by land use changes. All this must be placed in the perspective of the air pollution induced changes resulting from rising levels of greenhouse gases and stratospheric ozone depleting substances. It is these global and developing country problems that must be the top priority for future research into the topic of this book. All the air quality guidelines that have been developed from the many years of research in North America and Europe, which are discussed in Chapter 22, are only strictly applicable to the developed world. There is the urgent requirement for major research programmes which will enable the development of guidelines which are appropriate for the crops, cultivars, pollution climates, agronomic practices and environmental conditions of the developing world.

Appendix Conversion factors between different units of gaseous pollutant concentrations

There are a number of different units employed when reporting concentrations of gaseous pollutants, some of which are synonymous. In this book no attempt has been made to standardise these units, which instead have been reported on the basis of their use in the references cited. Thus in order to facilitate comparison between concentrations quoted in different units, the appropriate conversion factors are quoted below. It should be noted that there is no meaningful conversion for particulates and that the conversion factors for gases are given for standardised conditions of 25 °C (298.15 K) and one atmosphere of pressure.

On a volume/volume basis, concentrations are usually reported as ppb = nl l^{-1} = nmol mol^{-1}, while on a mass to air volume basis the most common unit is μgm^{-3}.

	μgm^{-3} to ppb	ppb to μgm^{-3}
NH_3	×1.44	×0.69
HF	×1.22	×0.82
NO	×0.81	×1.23
NO_2	×0.53	×1.88
O_3	×0.51	×1.96
SO_2	×0.38	×2.60

Air Pollution and Plant Life, second edition. Edited by J.N.B. Bell and M. Treshow. ISBN 0 471 49090 3 (HB), 0 471 49091 1 (PB).

Index